PRODUCTION CHEMICALS FOR THE OIL AND GAS INDUSTRY

PRODUCTION CHEMICALS FOR THE OIL AND GAS INDUSTRY

Malcolm A. Kelland

CRC Press
Taylor & Francis Group
Boca Raton   London   New York

CRC Press is an imprint of the
Taylor & Francis Group, an **informa** business

CRC Press
Taylor & Francis Group
6000 Broken Sound Parkway NW, Suite 300
Boca Raton, FL 33487-2742

© 2009 by Taylor and Francis Group, LLC
CRC Press is an imprint of Taylor & Francis Group, an Informa business

No claim to original U.S. Government works

Printed in the United States of America on acid-free paper
10 9 8 7 6 5 4 3 2 1

International Standard Book Number: 978-1-4200-9290-5 (Hardback)

This book contains information obtained from authentic and highly regarded sources. Reasonable efforts have been made to publish reliable data and information, but the author and publisher cannot assume responsibility for the validity of all materials or the consequences of their use. The authors and publishers have attempted to trace the copyright holders of all material reproduced in this publication and apologize to copyright holders if permission to publish in this form has not been obtained. If any copyright material has not been acknowledged please write and let us know so we may rectify in any future reprint.

Except as permitted under U.S. Copyright Law, no part of this book may be reprinted, reproduced, transmitted, or utilized in any form by any electronic, mechanical, or other means, now known or hereafter invented, including photocopying, microfilming, and recording, or in any information storage or retrieval system, without written permission from the publishers.

For permission to photocopy or use material electronically from this work, please access www.copyright.com (http://www.copyright.com/) or contact the Copyright Clearance Center, Inc. (CCC), 222 Rosewood Drive, Danvers, MA 01923, 978-750-8400. CCC is a not-for-profit organization that provides licenses and registration for a variety of users. For organizations that have been granted a photocopy license by the CCC, a separate system of payment has been arranged.

Trademark Notice: Product or corporate names may be trademarks or registered trademarks, and are used only for identification and explanation without intent to infringe.

Visit the Taylor & Francis Web site at
http://www.taylorandfrancis.com

and the CRC Press Web site at
http://www.crcpress.com

Contents

Preface xv
Author Biography xvii

Chapter 1 Introduction and Environmental Issues 1
    1.1 Production Chemistry Overview 1
    1.2 Factors That Affect the Choice of Production Chemicals 5
    1.3 Environmental and Ecotoxicological Regulations 8
        1.3.1 OSPAR Environmental Regulations for Oilfield Chemicals 9
        1.3.2 European REACH Regulations 10
        1.3.3 U.S. Environmental Regulations 10
        1.3.4 Environmental Regulations Elsewhere 11
    1.4 Designing Greener Chemicals 11
        1.4.1 Bioaccumulation 11
        1.4.2 Reducing Toxicity 12
        1.4.3 Increasing Biodegradability 12
    1.5 Mercury and Arsenic Production 15
    References 16

Chapter 2 Water and Gas Control 21
    2.1 Introduction 21
    2.2 Resins and Elastomers 22
    2.3 Inorganic Gels 22
    2.4 Cross-Linked Organic Polymer Gels for Permanent Shut-Off 24
        2.4.1 Polymer Injection 25
            2.4.1.1 Metal Ion Cross-Linking of Carboxylate-Containing Acrylamides and Biopolymers 26
            2.4.1.2 Gels Using Natural Polymers 27
            2.4.1.3 Organic Cross-Linking 27
            2.4.1.4 Polyvinyl Alcohol or Polyvinylamine Gels 30
            2.4.1.5 Problems Associated with Polymer Gel Water Shut-Off Treatments 30
            2.4.1.6 Other Improvements for Cross-Linked Polymer Gels 31
        2.4.2 In Situ Monomer Polymerization 31
    2.5 Viscoelastic Surfactant Gels 32

	2.6	Disproportionate Permeability Reducer or Relative Permeability Modifier 32
		2.6.1 Emulsified Gels as DPRs.. 33
		2.6.2 Hydrophilic Polymers as RPMs 33
		2.6.2.1 Types of Polymer RPM 34
		2.6.2.2 Hydrophobically Modified Synthetic Polymers as RPMs 36
		2.6.2.3 Cross-Linked Polymer RPMs 38
	2.7	Water Control Using Microparticles 39
	2.8	Gas Shut-Off..40
		2.8.1 Gas Well Foamers for Liquid Unloading 41
	References 41

Chapter 3 Scale Control 53

	3.1	Introduction 53
	3.2	Types of Scale.. 53
		3.2.1 Calcium Carbonate Scale 54
		3.2.2 Sulfate Scales 55
		3.2.3 Sulfide Scales 56
		3.2.4 Sodium Chloride (Halite) Scale 57
		3.2.5 Mixed Scales 57
	3.3	Nonchemical Scale Control... 57
	3.4	Scale Inhibition of Group II Carbonates and Sulfates 59
		3.4.1 Polyphosphates 61
		3.4.2 Phosphate Esters.. 61
		3.4.3 Nonpolymeric Phosphonates and Aminophosphonates.. 62
		3.4.4 Polyphosphonates.. 65
		3.4.5 Phosphino Polymers and Polyphosphinates 66
		3.4.6 Polycarboxylates... 67
		3.4.6.1 Biodegradable Polycarboxylates 70
		3.4.7 Polysulfonates.. 71
	3.5	Sulfide Scale Inhibition 73
	3.6	Halite Scale Inhibition... 74
	3.7	Methods of Deploying Scale Inhibitors................................... 75
		3.7.1 Continuous Injection 76
		3.7.2 Scale Inhibitor Squeeze Treatments.......................... 76
		3.7.2.1 Scale Inhibitor Squeeze Treatments Combined with Other Well Treatments....... 79
		3.7.3 Nonaqueous or Solid Scale Inhibitors for Squeeze Treatments...80
		3.7.3.1 Oil-Miscible Scale Inhibitors..................... 81
		3.7.3.2 Totally Water-Free Scale Inhibitors in Organic Solvent Bends 81

Contents vii

		3.7.3.3	Emulsified Scale Inhibitors	81
		3.7.3.4	Solid Scale Inhibitors (for Squeezing and Otherwise)	82
	3.7.4	Placement of Scale Inhibitor in a Squeeze Treatment		83

- 3.8 Performance Testing of Scale Inhibitors 84
- 3.9 Chemical Scale Removal 85
  - 3.9.1 Carbonate Scale Removal 85
  - 3.9.2 Sulfate Scale Removal 87
  - 3.9.3 Sulfide Scale Removal 89
  - 3.9.4 Lead Scale Removal 91
- References 91

Chapter 4 Asphaltene Control 111

- 4.1 Introduction 111
- 4.2 Asphaltene Dispersants and Inhibitors 114
- 4.3 Low Molecular Weight, Nonpolymeric Asphaltene Dispersants 116
  - 4.3.1 Low-Polarity Nonpolymeric Aromatic Amphiphiles 117
  - 4.3.2 Sulfonic Acid-Based Nonpolymeric Surfactant ADs 118
  - 4.3.3 Other Nonpolymeric Surfactant ADs with Acidic Head Groups 120
  - 4.3.4 Amide and Imide Nonpolymeric Surfactant ADs ... 123
  - 4.3.5 Alkylphenols and Related ADs 125
  - 4.3.6 Ion-Pair Surfactant ADs 126
  - 4.3.7 Miscellaneous Nonpolymeric ADs 127
- 4.4 Oligomeric (Resinous) and Polymeric AIs 128
  - 4.4.1 Alkylphenol-Aldehyde Resin Oligomers 129
  - 4.4.2 Polyester and Polyamide/Imide AIs 131
  - 4.4.3 Other Polymeric Asphaltene Inhibitors 135
- 4.5 Summary of ADs and AIs 137
- 4.6 Asphaltene Dissolvers 137
- References 140

Chapter 5 Acid Stimulation 149

- 5.1 Introduction 149
- 5.2 Fracture Acidizing of Carbonate Formations 149
- 5.3 Matrix Acidizing 150
- 5.4 Acids Used in Acidizing 150
  - 5.4.1 Acids for Carbonate Formations 150
  - 5.4.2 Acids for Sandstone Formations 151
- 5.5 Potential Formation Damage from Acidizing 152

	5.6	Acidizing Additives		153
		5.6.1	Corrosion Inhibitors for Acidizing	153
			5.6.1.1 General Discussion	153
			5.6.1.2 Nitrogen-Based Corrosion Inhibitors	154
			5.6.1.3 Oxygen-Containing Corrosion Inhibitors Including Those with Unsaturated Linkages	155
			5.6.1.4 Corrosion Inhibitors Containing Sulfur	158
		5.6.2	Iron Control Agents	159
		5.6.3	Water-Wetting Agents	161
		5.6.4	Other Optional Chemicals in Acidizing Treatments	161
	5.7	Axial Placement of Acid Treatments		162
		5.7.1	Solid Particle Diverters	163
		5.7.2	Polymer Gel Diverters	164
		5.7.3	Foam Diverters	166
		5.7.4	Viscoelastic Surfactants	167
	5.8	Radial Placement of Acidizing Treatments		171
		5.8.1	Oil-Wetting Surfactants	172
		5.8.2	Weak Organic Acids	172
		5.8.3	Weak Sandstone-Acidizing Fluorinated Agents	172
		5.8.4	Buffered Acids	172
		5.8.5	Gelled or Viscous Acids	173
		5.8.6	Foamed Acids	173
		5.8.7	Temperature-Sensitive Acid-Generating Chemicals and Enzymes	173
		5.8.8	Emulsified Acids	174
	References			175
Chapter 6	Sand Control			185
	6.1	Introduction		185
	6.2	Chemical Sand Control		185
	References			187
Chapter 7	Control of Naphthenate and Other Carboxylate Fouling			189
	7.1	Introduction		189
	7.2	Naphthenate Deposition Control Using Acids		190
	7.3	Low-Dosage Naphthenate Inhibitors		191
	References			192
Chapter 8	Corrosion Control during Production			195
	8.1	Introduction		195
	8.2	Methods of Corrosion Control		197
	8.3	Corrosion Inhibitors		198

8.4	Film-Forming Corrosion Inhibitors		200
	8.4.1	How Film-Forming Corrosion Inhibitors Work	200
	8.4.2	Testing Corrosion Inhibitors	201
	8.4.3	Efforts to Develop More Environment-Friendly Film-Forming Corrosion Inhibitors	202
	8.4.4	Classes of Film-Forming Corrosion Inhibitors	203
		8.4.4.1 Phosphate Esters	204
		8.4.4.2 Amine Salts of (Poly)Carboxylic Acids	205
		8.4.4.3 Quaternary Ammonium and Iminium Salts and Zwitterionics	206
		8.4.4.4 Amidoamines and Imidazolines	207
		8.4.4.5 Amides	211
		8.4.4.6 Polyhydroxy and Ethoxylated Amines/Amides	212
		8.4.4.7 Other Nitrogen Heterocyclics	213
		8.4.4.8 Sulfur Compounds	213
		8.4.4.9 Polyaminoacids and Other Polymeric Water-Soluble Corrosion Inhibitors	216
References			217

Chapter 9 Gas Hydrate Control 225

9.1	Introduction		225
9.2	Chemical Prevention of Hydrate Plugging		227
	9.2.1	Thermodynamic Hydrate Inhibitors	227
		9.2.1.1 Operational Issues with THIs	230
	9.2.2	Kinetic Hydrate Inhibitors	231
		9.2.2.1 Introduction to KHIs	231
		9.2.2.2 Vinyl Lactam KHI Polymers	233
		9.2.2.3 Hyperbranched Polyesteramide KHIs	235
		9.2.2.4 Compatibility of KHIs	236
		9.2.2.5 Pyroglutamate KHI Polymers	236
		9.2.2.6 Poly(di)alkyl(meth)acrylamide KHIs	237
		9.2.2.7 Other Classes of KHIs	238
		9.2.2.8 Performance Testing of KHIs	240
	9.2.3	Anti-Agglomerants	242
		9.2.3.1 Emulsion Pipeline AAs	243
		9.2.3.2 Hydrate-Philic Pipeline AAs	243
		9.2.3.3 Performance Testing of Pipeline AAs	247
		9.2.3.4 Natural Surfactants and Nonplugging Oils	247
		9.2.3.5 Gas-Well AAs	248
9.3	Gas Hydrate Plug Removal		248
	9.3.1	Use of Thermodynamic Hydrate Inhibitors	249
	9.3.2	Heat-Generating Chemicals	249
References			250

Chapter 10 Wax (Paraffin) Control 261

- 10.1 Introduction 261
  - 10.1.1 Wax Deposition 262
  - 10.1.2 Increased Viscosity and Wax Gelling 263
- 10.2 Wax Control Strategies 264
- 10.3 Chemical Wax Removal 266
  - 10.3.1 Hot-Oiling and Related Techniques 266
  - 10.3.2 Wax Solvents 266
  - 10.3.3 Thermochemical Packages 267
- 10.4 Chemical Wax Prevention 269
  - 10.4.1 Test Methods 269
  - 10.4.2 Wax Inhibitors and Pour-Point Depressants 270
  - 10.4.3 Ethylene Polymers and Copolymers 273
  - 10.4.4 Comb Polymers 274
    - 10.4.4.1 (Meth)acrylate Ester Polymers 274
    - 10.4.4.2 Maleic Copolymers 276
  - 10.4.5 Miscellaneous Polymers 278
  - 10.4.6 Wax Dispersants 279
  - 10.4.7 Polar Crude Fractions as Flow Improvers 281
  - 10.4.8 Deployment Techniques for Wax Inhibitors and PPDs 281
- References 282

Chapter 11 Demulsifiers 291

- 11.1 Introduction 291
- 11.2 Methods of Demulsification 293
- 11.3 Water-in-Oil Demulsifiers 293
  - 11.3.1 Theory and Practice 293
  - 11.3.2 Test Methods and Parameters for Demulsifier Selection 295
  - 11.3.3 Classes of Water-in-Oil Demulsifier 296
    - 11.3.3.1 Polyalkoxylate Block Copolymers and Ester Derivatives 298
    - 11.3.3.2 Alkylphenol-Aldehyde Resin Alkoxylates 298
    - 11.3.3.3 Polyalkoxylates of Polyols or Glycidyl Ethers 300
    - 11.3.3.4 Polyamine Polyalkoxylates and Related Cationic Polymers 301
    - 11.3.3.5 Polyurethanes (Carbamates) and Polyalkoxylate Derivatives 302
    - 11.3.3.6 Hyperbranched Polymers 302
    - 11.3.3.7 Vinyl Polymers 303
    - 11.3.3.8 Polysilicones 304

Contents   xi

                           11.3.3.9 Demulsifiers with Improved
                                     Biodegradability 304
                           11.3.3.10 Dual-Purpose Demulsifiers 306
         References 307

Chapter 12 Foam Control 313

         12.1 Introduction 313
         12.2 Defoamers and Antifoams 313
                12.2.1 Silicones and Fluorosilicones 314
                12.2.2 Polyglycols 315
         References 316

Chapter 13 Flocculants 319

         13.1 Introduction 319
         13.2 Theory of Flocculation 320
         13.3 Flocculants 321
                13.3.1 Performance Testing of Flocculants 322
                13.3.2 Cationic Polymers 323
                         13.3.2.1 Diallyldimethylammonium Chloride
                                    Polymers 323
                         13.3.2.2 Acrylamide or Acrylate-Based
                                    Cationic Polymers 324
                         13.3.2.3 Other Cationic Polymers 325
                         13.3.2.4 Environment-Friendly Cationic
                                    Polymeric Flocculants 327
                         13.3.2.5 Dithiocarbamates: Pseudocationic
                                    Polymeric Flocculants with Good
                                    Environmental Properties 328
                13.3.3 Anionic Polymers 330
         References 330

Chapter 14 Biocides 335

         14.1 Introduction 335
         14.2 Chemicals for Control of Bacteria 337
         14.3 Biocides 338
                14.3.1 Oxidizing Biocides 339
                14.3.2 Nonoxidizing Organic Biocides 341
                         14.3.2.1 Aldehydes 342
                         14.3.2.2 Quaternary Phosphonium Compounds 343
                         14.3.2.3 Quaternary Ammonium Compounds 345
                         14.3.2.4 Cationic Polymers 345
                         14.3.2.5 Organic Bromides 346
                         14.3.2.6 Metronidazole 347

		14.3.2.7	Isothiazolones (or Isothiazolinones) and Thiones	347
		14.3.2.8	Organic Thiocyanates	348
		14.3.2.9	Phenolics	349
		14.3.2.10	Alkylamines, Diamines, and Tramines	349
		14.3.2.11	Dithiocarbamates	349
		14.3.2.12	2-(Decylthio)ethanamine and Its Hydrochloride	350
		14.3.2.13	Triazine Derivatives	350
		14.3.2.14	Oxazolidines	350
		14.3.2.15	Specific Surfactant Classes	351
	14.4	Biostats (Control "Biocides" or Metabolic Inhibitors)		351
		14.4.1	Anthraquinone as Control Biocide	351
		14.4.2	Nitrate and Nitrite Treatment	352
		14.4.3	Other Biostats	353
	14.5	Summary		354
References				354

Chapter 15 Hydrogen Sulfide Scavengers 363

15.1 Introduction 363
15.2 Nonregenerative H_2S Scavengers 365
    15.2.1 Solid Scavengers 366
    15.2.2 Oxidizing Chemicals 366
    15.2.3 Aldehydes 367
    15.2.4 Reaction Products of Aldehydes and Amines, Especially Triazines 369
    15.2.5 Metal Carboxylates and Chelates 371
    15.2.6 Other Amine-Based Products 372
15.3 Summary 373
References 373

Chapter 16 Oxygen Scavengers 377

16.1 Introduction 377
16.2 Classes of Oxygen Scavengers 377
    16.2.1 Dithionite Salts 378
    16.2.2 Hydrazine and Guanidine Salts 378
    16.2.4 Hydroxylamines and Oximes 378
    16.2.5 Activated Aldehydes and Polyhydroxyl Compounds 379
    16.2.6 Catalytic Hydrogenation 379
    16.2.7 Enzymes 379
    16.2.8 Bisulfite, Metabisulfite, and Sulfite Salts 380
References 381

Contents   xiii

Chapter 17 Drag-Reducing Agents 383
           17.1   Introduction 383
           17.2   Drag-Reducing Agent Mechanisms 384
           17.3   Oil-Soluble DRAs 385
                  17.3.1   Background 385
                  17.3.2   Oil-Soluble Polymeric DRAs 386
                          17.3.2.1   Polyalkene (Polyolefin) DRAs 386
                          17.3.2.2   Poly(meth)acrylate Ester DRAs 388
                          17.3.2.3   Other Oil-Soluble DRA Polymers 389
                          17.3.2.4   Overcoming Handling, Pumping, and
                                         Injection Difficulties with UHMW
                                         DRA Polymers 389
                          17.3.2.5   Oil-Soluble Polymeric DRAs in
                                         Multiphase Flow 390
                  17.3.3   Oil-Soluble Surfactant DRAs 390
           17.4   Water-Soluble DRAs 391
                  17.4.1   Water-Soluble Polymer DRAs 391
                          17.4.1.1   Polysaccharides and Derivatives 392
                          17.4.1.2   Polyethyleneoxide Drag-Reducing
                                         Agents 392
                          17.4.1.3   Acrylamide-Based DRAs 393
                  17.4.2   Water-Soluble Surfactant DRAs 395
                  17.4.3   Drag Reduction and Corrosion Inhibition ... 398
           References 398

APPENDIX 1: OSPAR Environmental Regulations for Oilfield Chemicals 405
           A.1   United Kingdom and the Netherlands North Sea
                  ECOTOX Regulations 408
           A.2   Norwegian Offshore ECOTOX Regulations 409
           References 410

Index 413

Preface

It struck me a few years ago that there was a lack of general literature providing an overview of all the various issues of oilfield production chemistry. Certainly, there was not a book that focuses on the structures of production chemicals and their environmental properties that could be helpful to service companies and chemical suppliers in designing better or greener products. Although I was sure that there were others who could do a better job at writing such a book, I decided to have a go.

This book is primarily a handbook of production chemicals and as such should be useful to oil and gas companies, oilfield chemical service companies, and chemical suppliers. The introduction and main points in each chapter should also be useful for university students wishing to study oilfield production chemistry. If you are working for a chemical supply company and are unfamiliar with the oil and gas industry, I would recommend reading up on the basics of upstream oil and gas production before delving into this book.

I have limited the book to sixteen chapters on production chemicals and an introductory chapter, which also includes environmental issues. Some of the production chemicals are specifically for use downhole such as acid stimulation and water and gas shut-off chemicals. I have not included all stimulation chemicals such as those used in proppant fracturing, as these are not usually considered production chemicals. I have included chemicals used in water injection wells for enhanced oil recovery, such as oxygen scavengers and biocides, but I have not discussed polymers and surfactants, which are used to further enhance oil recovery (EOR), or tracers. Polymers and surfactants are not very widely used for EOR today. However, if the oil price continues to remain very high, their use may become a more economically rewarding and prevalent EOR technique.

In each chapter, I have begun by introducing the problem for which there is a production chemical (e.g., scale, corrosion). Then, I have briefly discussed all methods to treat the problem, both chemical and nonchemical. This is followed by a thorough discussion of the structural classes of production chemicals for that particular chapter usually with a brief discussion on how they can be performance tested. I have also mentioned the environmental properties of known chemicals where such data are available. I have also endeavored to mention whether a chemical or technique has been successfully used in the field, whenever a report is available.

I have included many references at the end of each chapter so the reader can look up the details on the synthesis, testing, theory, or application of each type of production chemical. I have endeavored to be as thorough and up-to-date as possible in the literature, not wishing to leave out any structural class of production chemical, whether they have been used in the field or not. The references are from patents or patent applications, books, journals, and conference proceedings that are readily obtainable. Many patents and patent applications claim a wide spectrum of production chemical structures. This is often standard practice from service companies and chemical suppliers not wishing to divulge the chemistry of their best products to their competition. This lack of specificity may not be so helpful to the reader, so I have therefore chosen to mention only preferred chemical structures in the patent, particularly those synthesized and/or tested.

I have deliberately not mentioned in the text of each chapter the names of authors or companies and institutes behind the articles and patents so as to be as impartial as possible. However, all this information can be gleaned by looking up the references. Nearly all the patent data can be obtained on line for free from the Internet except a few older patents. There are a number of production chemical articles, which do not disclose any chemical names or structures, rather they only give laboratory and/or field test data and sometimes environmental data on chemical A, chemical B, and so forth. These articles I have deemed as less useful to the reader and cannot be correlated with any patents where chemical structures are given, and, therefore, I have omitted many of them from the references. Those that are mentioned are usually included because they contain useful information on test procedures and/or environmental data.

This book is an overview of production chemicals and does not discuss the actual handling and application of the chemicals in the laboratory or field. Thus, the author cannot be held responsible for the consequences of handling or using any of the chemicals discussed in this book.

I hope you will find this book useful in your studies or work. That, at least, was my intention in deciding to write it. I welcome any feedback you may have on its contents.

I would like to thank Barbara Glunn at CRC Press for all her encouragement and help in managing this book project. I would also like to thank Jim McGovern for all his help through the editing stages. I particularly wish to thank Alan Hunton of Humber Technical Services for his input and comments on many of the chapters. They have been most valuable. I would also like to thank Nick Wolf at ConocoPhillips for comments on the introductory chapter, Abel De Oliveira at Dow Brazil S.A. for comments on demulsifiers, Roald Kommedal of the University of Stavanger, Norway, for his comments on biodegradation testing, and Jean-Louis Peytavy and Renaud Cadours of Total for information on H_2S capture with amines. Finally, but by no means least, I thank my wife Evy for her patience while I spent many long evenings at the computer.

My email address for comments: malcolm.kelland@uis.no.

Malcolm Kelland
University of Stavanger, Norway

Author Biography

Malcolm A. Kelland grew up in South Croydon, in the south of Greater London, UK, attending Whitgift School. He obtained a first class honors degree in chemistry and a D.Phil. in organometallic chemistry from Oxford University, UK. He worked at RF-Rogaland Research (now the International Research of Stavanger, IRIS), Norway, from 1991 to 2000 mostly on production chemistry projects, leading the production chemistry group for three of those years. He was involved in a variety of research projects on issues such as gas hydrates, scale, corrosion, wax, and drag reducers. He has particularly been involved in designing and testing low-dosage hydrate inhibitors (LDHIs) and is author and co-author on a number of patents on this subject. After a brief spell at a minor production chemical service company, he moved to the University of Stavanger, Norway, in 2001, where he is currently an associate professor of inorganic chemistry. He teaches chemistry, environmental science, and inorganic chemistry. His current research is in designing and testing more environmentally friendly LDHIs as well as smaller projects on new scale, asphaltene, and corrosion inhibitors. He is married with three children and enjoys gardening and playing badminton, squash, and rugby fives.

1 Introduction and Environmental Issues

1.1 PRODUCTION CHEMISTRY OVERVIEW

Production chemistry issues occur as a result of chemical and physical changes to the well stream fluids, as it is transported from the reservoir through the processing system. The well stream fluids may consist of a mixture of liquid hydrocarbon (oil or condensate), gaseous hydrocarbon (raw natural gas), and associated water. This mixture passes from the reservoir, through the tubular string and wellhead, and then along flowlines to the processing plant where the various phases are separated. As the fluids will experience a significant drop in pressure, a change in temperature, and considerable agitation, there will be predictable and sometimes unpredictable changes in state that impact on the efficiency of the overall operation. Downstream of the processing plant, the oil will be exported to the refinery, the gas will be processed, and the water will be treated to remove impurities: these processes can lead to further complications.

In general, production chemistry problems are one of four types:

- Problems caused by fouling. This is defined as the deposition of any unwanted matter in a system and includes scales, corrosion products, wax (paraffin), asphaltenes, biofouling, and gas hydrates.
- Problems caused by the physical properties of the fluid. Foams, emulsions, and viscous flow are examples.
- Problems that affect the structural integrity of the facilities and the safety of the workforce. These are mainly corrosion-related issues.
- Problems that are environmental or economic. Oily water discharge can damage the environment and the presence of sulfur compounds such as hydrogen sulfide (H_2S) has environmental and economic consequences.

The resolution of these problems can be made by the application of nonchemical techniques and through the use of properly selected chemical additives. Among the very many nonchemical techniques that are available, the following examples are commonly used:

- Insulation (retaining heat to delay the onset of waxes or gas hydrate formation)
- Heating a flowline (prevention and remediation of waxes and gas hydrates)
- Heating in a separator (resolution of emulsions)

- Lowering of pressure (remediation of gas hydrates)
- Maintenance of high pressure (delaying asphaltene flocculation, carbonate scale, naphthenates)
- Use of corrosion-resistant materials and coatings (minimizing corrosion)
- Increase flowrate/turbulence (minimize asphaltenes, waxes, biofilm)
- Decrease flowrate (minimize foaming, emulsion formation)
- Increase separator size (improve oil/water separation)
- Use electric fields (increase coalescence of water droplets in o/w emulsion)
- Centrifugation (separation of o/w emulsions)
- Membranes and fine filters (removal of fines, colloidal particles, and specific ions)
- Pigging of flowlines (prevent build-up of solids in pipes)
- Scraping tools (deposit removal, especially downhole)
- Milling or drilling/reaming (removal of scale deposits downhole)
- Application of vacuum (removal of gases from water)
- Screens and plugs (for water shut-off downhole)
- Screens and gravel packs (for sand control)

A good facilities design and correct choice of materials can significantly reduce production chemistry issues later in field life. Unfortunately, crude oil production is characterized by variable production rates and unpredictable changes to the nature of the produced fluids. It is therefore essential that the production chemist can have a range of production chemical additives available that may be used to rectify issues that would not otherwise be fully resolved. Modern production methods, the need to upgrade crude oils of variable quality, and environmental constraints demand chemical solutions.

Oilfield production chemicals are therefore required to overcome or minimize the effects of the production chemistry problems listed above. In summary, they may be classified as:

- Inhibitors to minimize fouling and solvents to remove preexisting deposits
- Process aids to improve the separation of gas from liquids and water from oil
- Corrosion inhibitors to improve integrity management
- Chemicals added for some other benefit, including environmental compliance.

Many production chemistry fouling problems relate to so-called flow assurance, a term coined in the nineties to describe the issues involved in maintaining produced fluid flow from the well to the processing facilities. Flow assurance chemical issues usually relate to solids deposition problems (fouling) such as wax (paraffins), asphaltenes, scale, naphthenate, and gas hydrates in flowlines. There are two general chemical strategies for prevention of these deposits: either to use a dispersant, which allows solid particles to form but disperses them in the production stream without deposition, or to prevent solids forming by using an inhibitor. Most new large fields are being found offshore in ever-increasing water depths and/or colder environments.

In addition, smaller offshore fields are often "tied back" to existing platforms or other infrastructure requiring long, subsea multiphase flow pipelines. The extremes of high pressure and low subsea temperatures and long fluid residence times place greater challenges on flow assurance, particularly mitigating gas hydrate and wax (paraffin) deposition. As with other production chemistry issues, a strategy for prevention of gas hydrate deposition must be worked out at the field-planning stage. Chemically, one can use thermodynamic hydrate inhibitors or the more recently developed low-dosage hydrate inhibitors (LDHIs). Wax deposition may not be fully prevented by using wax (paraffin) inhibitors, but regular mechanical pigging may possibly help keep the pipeline wax-free.

Upstream of the wellhead, production chemistry deposition problems that can occur include scale and asphaltene deposition, and even wax deposition if the temperature in the upper part of the well is low. In offshore and/or cold onshore environments, gas hydrates can also form upstream of a subsea wellhead or in the flowline if it is shut in. Removal of a hydrate plug upstream of a subsea well is usually done by melting with thermodynamic hydrate inhibitors and/or by heating. Most subsea pipeline hydrate plugs are removed by depressurization, although other techniques are available. To reduce the amount of water that needs to be handled at the processing facilities, the production of water can be reduced either mechanically or chemically, using so-called water shut-off treatments that block water flow. This may also alleviate scale-formation problems. Normally, treatment with a scale inhibitor, downhole and/or topside, is required to prevent scaling. Asphaltene, wax, and inorganic scales can all be removed using various chemical dissolver treatments. Naphthenate problems and related emulsion problems can be reduced by careful acidification or with the use of more recently discovered naphthenate low-dosage inhibitors. Acid stimulation treatments, either by fracture or matrix acidizing, are designed to enhance hydrocarbon production. They are generally used to remove part of the natural rock formation (sandstone or carbonate), but they can also remove deposited carbonate and sulfide scales. Corrosion during acid stimulation is a major concern and requires special corrosion inhibitors that tolerate and perform well under very acidic conditions. Other downhole chemical treatments include water and gas shut-off and sand consolidation.

The separated oil, gas, and water streams must meet certain minimum specifications for impurities. Process aids, including demulsifiers and antifoams are used to enhance the performance of the processing plant and ensure that the specifications can be met. However, because of the presence of asphaltenes, resins, naphthenates, and other natural surfactants in the oil and the high shear at the wellhead and mixing during transportation, some or all of the produced water will be emulsified with the liquid hydrocarbon phase. These emulsions require resolving at the surface processing facilities in the separators. Efficient operation of these facilities will provide the operator with oil of "export" quality. Demulsifier chemicals are always used for this process, usually together with other nonchemical techniques such as heat or electric treatment. The separated water normally has too much dissolved and dispersed hydrocarbons (and/or oil-in-water emulsion) as well as suspended particles for discharge into the environment to be allowed. Therefore, the separated water is treated with flocculants (also called deoilers, water clarifiers, or reverse emulsion breakers).

The flocculated impurities (or "flocs") are separated out to leave purer produced water, which can then be discharged. Several other technologies are now available for separating both dispersed and dissolved (water-soluble) hydrocarbons, and even some production chemicals, from water. Onshore disposal of produced water may have higher environmental demands, sometimes including a limit on the concentration of certain salts. In such cases, reinjection of produced water is carried out. This has also been done offshore, for example, in the North Sea.[1] Foam can also be a problem in the gas–oil separators of the processing facilities for which defoamers and antifoam chemicals have been designed.

Corrosion occurs wherever metals are in contact with water, and this problem affects both the internal surfaces and the exposed external surfaces of facilities and pipelines. The rate of corrosion varies in proportion to the concentrations of water-soluble acid gases such as carbon dioxide, CO_2, and hydrogen sulfide (H_2S) and in proportion to aqueous salinity. It is potentially a serious issue in high-temperature wells, and in this situation, special corrosion-resistant alloys (CRAs) may be economically advantageous. Reducing the concentration of H_2S can alleviate corrosion problems and methods to do this such as the use of H_2S scavengers, biocides/biostats, and nitrate/nitrite injection are dealt with in Chapters 14 and 15. Batch or continuous treatment with corrosion inhibitors is normally needed to control corrosion to within an acceptable limit for the predicted lifetime of the field.

A number of other miscellaneous production chemicals are used in the upstream oil and gas industry. Although H_2S scavenger chemicals reduce corrosion, they may be deployed specifically to avoid refinery problems or for environmental reasons, i.e., to reduce toxicity. Similarly, it can be argued that flocculants also improve environmental compliance by reducing toxic contaminants in separated water. Drag-reducing agents (DRAs) do not fit into the normal classifications for production chemicals, as they do not impact on solids formation, affect corrosion, or change emulsions or foams: their function is simply to provide additional flow in a pipeline or injection well.

Production chemicals can be injected downhole, at the wellhead, or between the wellhead and the processing facilities (separator system). Some production chemicals, such as corrosion inhibitors, wax inhibitors, and sometimes scale inhibitors and biocides, are dosed to oil export lines. Corrosion inhibitors may also be used in gas lines.

Injection downhole can be either via a capillary string or gas-lift system if available. Batch treatment is commonly practiced for downhole locations, and this includes the technique of squeezing to place a chemical within the reservoir. Downhole chemical treatments can be bullheaded, that is, pumped from the platform, boat, or truck into the well via wellhead or production flowlines. Examples are acid stimulation, water shut-off, and scale-inhibitor squeeze or scale-dissolver treatments. Squeeze treatments are discussed in Section 3.7.2. Various diversion methods can be used to gain better placement of the squeeze treatment in the relevant zones. If squeezing is considered risky (i.e., if precise placement is required or the treatment fluids are expected to cause damage and, thus, loss of production), well-intervention techniques such as coiled tubing can be used to place the chemical in the desired zones. However, this is costly especially for subsea interventions. Production chemicals can also be placed in the near-well area during fracturing operations.[2]

Introduction and Environmental Issues

Besides squeeze treatments, other techniques have been developed for controlled release of a production chemical in the well. For example, solid particles that slowly release the desired chemical as the produced fluids flow over them can be placed in the bottom of the well (rathole) or farther up, or if the particles are very small they can be squeezed into the near-well area.[3–5] Development of these techniques is well-known for scale control. Further details and references can be found in Chapter 3 on scale control.

Production chemical service companies source chemical components from bulk chemical suppliers, specialty chemical suppliers, or from their own dedicated manufacturing facility. Products can be simple (e.g., methanol) or complex formulations with several active ingredients in a solvent.

Most production chemical service companies also sell chemicals used in water injection wells, hydrotesting, and other maintenance and utility systems. Chemicals for water injection systems include oxygen scavengers to reduce corrosion, biocides to reduce microbially enhanced corrosion and hydrogen sulfide production, water-based drag reducers to increase the injection rate, scale and corrosion inhibitors, and antifoams. Polymers and/or surfactants designed to enhance oil recovery further may also be injected. Oil-based DRAs are usually used to increase the transportation capacity of crude oil pipelines or reduce the need for boosting stations.

Production chemistry issues for a field are managed by the field operator. In many producing regions, it is a production chemistry service company that is charged with the responsibility for optimizing and carrying out chemical treatments. The strategy adopted should be to develop a comprehensive chemical treating program for the whole of the production process, which would include selection of appropriate chemicals and their dose rate, consideration of compatibility issues (see Section 1.2), injection point placement, field life cycle needs, etc. The production chemist will be mindful of the contribution of nonchemical techniques that are in use and the impact that they have on the demands for chemical addition.

Production chemistry problems ought to be determined during the field-development stage by the operator, often in collaboration with service companies to find the best solutions. This is particularly relevant for offshore fields where the cost of work-overs or remediation treatments will be high.

1.2 FACTORS THAT AFFECT THE CHOICE OF PRODUCTION CHEMICALS

A number of factors affect the choice of production chemicals. These include:

- Performance
- Price
- Stability
- Health and safety in handling and storage
- Environmental restrictions
- Compatibility issues

Generally, an operator wants a product that performs satisfactorily at an affordable price. The overall performance may be based on more than one test. For example, a scale inhibitor for squeeze treatments may be an excellent inhibitor, but because of poor adsorption onto rock, it may give a poor squeeze lifetime. Thus, an inhibitor with lower inhibition performance may be preferred if it adsorbs better to the rock. For some production chemicals, such as scale, wax, asphaltene, corrosion, and LDHIs, an operator may ask several service companies to submit a chosen product that either they or an independent company will test to rank the performance of the products. The operator may not necessarily choose the highest-performing product for field application but one that they consider performs satisfactorily and suits them best economically. However, a cheap product may appear to be economical in the short term, but if its performance is significantly worse than a more expensive product, it may turn out to be more expensive in the long run to the operator if it causes more production upsets, more frequent well or pipeline workovers, and lost production.

Production chemical formulations must remain stable for the intended lifetime during transportation and storage before being injected. In cold environments, the product must not get too viscous or freeze to avoid injection problems. Conversely, it may be very hot at the field location, so the product must not degrade too rapidly or undergo phase changes, which may affect its performance. One can imagine a possible dilemma for regions where good biodegradation of the production chemical is an environmental requirement. Thus, the operator wants a product that degrades fast in seawater but does not degrade during storage at the field location.

All developed countries have regulations regarding the classification of chemicals according to the hazards and risks they may pose to the safety and health of users. Essential information is found on the material safety data sheet for a chemical, which must accompany shipment of all potentially hazardous products according to national laws. An example is volatile and toxic solvents, which, if breathed in, can cause health problems depending on the dosage and exposure time. These include some nonpolar aromatic solvents used, for example, in wax and asphaltene inhibitor formulations or demulsifiers. Many chemical suppliers and service companies have already made efforts to replace such solvents with safer, less toxic solvents with lower volatility and/or higher flash point.

There are a number of other operational issues with the use of production chemicals, which can be lumped under the heading "compatibility." They include the following:

- Will the use of a production chemical cause or worsen other production chemistry issues? Conversely, could it work synergistically with other production chemicals?
- Is it compatible with all materials found along the production line?
- Will it cause downstream problems?
- Are there any injection problems—viscosity, cloud point, foaming?
- Is it compatible with other production chemicals used simultaneously?
  - Does one production chemical affect the performance of another and vice versa?
  - Can it be coinjected with other production chemicals?

Regarding the first subcategory, there are a few well-known issues. For example, some film-forming corrosion inhibitors can make emulsion and foam problems worse in the separators. The use of thermodynamic hydrate inhibitors such as methanol and glycols can make scale deposition worse. Triazine-based hydrogen sulfide scavengers will increase the pH in the produced water, which can worsen the potential for carbonate scale formation. Acids used in downhole acid stimulation can cause asphaltene precipitation (sludging), but there are additional chemicals that can be used in the formulations to reduce this. Conversely, quaternary gas hydrate anti-agglomerants can contribute to corrosion inhibition, sometimes to such an extent that a separate corrosion inhibitor is not needed. Some suppliers have available combined-scale and corrosion-inhibitor formulations, which have the advantage of using only one storage tank, one pump, and one injection line. Many of these are simply mixtures of compatible, individual products, although some multifunctional, single-component products are available. Drag reducers can also improve corrosion inhibition under some conditions. All these examples are discussed further in the relevant chapters in this book.

New production chemicals may have to be checked to determine if they are compatible with all materials along the production line, including elastomers used in seals. This may be as supplied, neat/concentrated chemicals, or after dilution in the produced fluids following injection.

Some production chemicals that follow the oil or gas phase can cause downstream problems, such as polluting catalysts used in the refinery. An example is the thermodynamic hydrate inhibitor methanol. Too much methanol pollution can lower the value of the hydrocarbons. The operator may also need to check that oil-soluble production chemicals, such as wax and asphaltene inhibitors and some gas hydrate anti-agglomerants, do not cause downstream problems such as fouling in the crackers.

Subsea wellhead injection of production chemicals is carried out by injection from the platform along one or more small umbilical flowlines to the wellhead. The length of umbilicals can be up to 50 km (30 miles) and their diameters can range from 1/4 to 1 in. (0.6–2.5 cm). The viscosity of the injected chemical formulation must be low enough to be pumped along this flowline at seabed temperatures. In cold waters, this sets limits on the concentration of some production chemicals in the product and the solvents in which they are dissolved. Incompatibilities can occur between different production chemicals when they mix in the same umbilical flowline, usually accidentally. For example, neat corrosion inhibitor injected into a line previously used to dose wax inhibitor or scale inhibitor can cause a blockage unless adequate flushing with a solvent or water is carried out. During long-term shutdowns, production chemicals in the umbilical tubes are usually replaced with solvent packages to ensure their integrity. As a rule, subsea injection points are placed at least three pipe diameters away from each other to avoid immediate mixing of different production chemicals and possible incompatibilities. Improved test methods for qualifying subsea injection chemicals at injection pressures, which could be up to 700 bar, and seafloor temperatures have been reported.[6] Topside injection of DRAs has, in the past, caused problems since these chemicals are ultrahigh–molecular weight polymers, which can be extremely viscous in solution. However, new ways of formulating and injecting these materials have been developed that have overcome these difficulties (see Chapter 17 for details).

Some production chemicals affect the performance of other chemicals. Thus, performance tests of any production chemical must be carried out in the presence of any other chemicals planned to be used simultaneously. For example, many film-forming corrosion inhibitors are known to reduce the performance of many scale inhibitors and demulsifiers and some can almost totally kill the performance of kinetic hydrate inhibitors. The reverse problems are also known.

Some production chemicals are water-soluble, while others are oil-soluble, which may mean that both products cannot be injected through the same line. For example, an operator may need coinjection at the wellhead of an oil-soluble wax or asphaltene inhibitor and a water-soluble scale inhibitor. However, there have been efforts to find mutual solvents or develop emulsion-based products, which can overcome this problem.[7–8]

1.3 ENVIRONMENTAL AND ECOTOXICOLOGICAL REGULATIONS

There is increasing focus on making the handling and deployment of oilfield chemicals less hazardous, as well as reducing the discharge of environmentally unacceptable oilfield chemicals, usually in produced water, to the environment. Sometimes, it has proven difficult to find "greener" chemicals with the same high performance. At the same time, many of the new fields are being found in harsher conditions such as colder or deeper water or deeper, hotter reservoirs, which may require special high-performing chemicals to cope with the harsh conditions.

Most regions of the world have environmental regulations concerning the use and discharge of oilfield chemicals. Production chemicals are added to well streams and can therefore end up in the water, liquid hydrocarbon, or gas phase. Some low-boiling solvents, such as methanol, will partly follow the gas phase and can pollute the natural gas, lowering its economic value. However, most production chemicals will follow either the produced water or liquid hydrocarbon phase or partition between them. If they follow the liquid hydrocarbon phase, they will not normally be discharged to the environment unless a leakage or other accident occurs. It is the discharge of produced water containing residual oil as well as production chemicals that constitute the greatest risk to the environment during normal production. Residual oil in separated produced water can contain toxic carcinogens such as polyaromatic hydrocarbons (PAHs); therefore, there are regulations on its discharge level. For example, in the North Sea, this level is currently 30 ppm, but for some regions it can be as low as 5–10 ppm, which may be difficult to achieve. For onshore or coastal oilfields, there can be very strict requirements on the quality of the produced water. Often, reinjection of the produced water is carried out to avoid any environmental issues with discharging the water. Reinjection of produced water is also a good option offshore and is also used when reservoir pressure support is required for enhanced oil recovery. Produced water treatment technologies aimed at reducing the levels of aliphatic hydrocarbons, benzene, heavy aromatic compounds (PAHs), alkylated phenols, and some production chemicals have been reviewed.[9] There may also be restrictions on chemicals in well treatments onshore, for example, in shallow gas reservoirs, if there is a potential for fluids to contaminate freshwater aquifers and water wells.

1.3.1 OSPAR Environmental Regulations for Oilfield Chemicals

Regulations for the discharge of produced water containing production chemicals vary from location to location. The required ecotoxicological tests on all components of production chemicals proposed for use in the North Sea offshore region are laid out in the OSPAR guidelines for the North-East Atlantic, implemented in 2001 under a Harmonised Mandatory Control Scheme.[10–12] This testing is done well in advance of deployment, and irrespective of the fate of the chemical, i.e., whether it ends up in the produced water, gas, or oil phase. The chemical, physical, and ecotoxicological test data are submitted to the relevant national pollution authorities, which assess the data. The result of the assessment is that the chemical will be ranked as to its suitability for deployment offshore. The end-user (the field operator) must then seek a permit from the authorities for use of the chemical over a time-limited period. In addition, there is also a downstream requirement for each platform from the authorities that discharges of oil and chemicals must be controlled within limits. In Norway, the operator is required to have the produced water regularly analyzed for toxic components found in the oil, such as benzene and phenols, as well as the amount of production chemicals. The operator is obliged to keep the discharges of these chemicals within the stipulated levels given by the pollution authorities.

Three categories of ecotoxicological tests on production chemicals are required by OSPAR:

- Acute toxicity[13]
- Bioaccumulation
- Biodegradation in seawater

A brief discussion of these tests is given in Appendix 1 at the end of the book.

It is the environmental properties of each individual production chemical (and not the finished product) in a proposed formulation that has to be determined. Some service companies own ecotoxicological test laboratories and report ecotoxicological data for oilfield chemicals from their own laboratories to the authorities. A few chemicals are banned from use offshore North Sea on other environmental grounds, such as endocrine disrupters that cause chronic problems in marine organisms.[14–16] Examples are the alkylphenols. Chemicals listed on OSPAR's PLONOR list can be used in the North Sea without special approval (PLONOR means "pose little or no risk" to the environment).[17]

The required OSPAR ecotoxicological testing is the same for all North Sea countries but there are small differences in the way the data are interpreted and implemented. The regulations in the North Sea countries are under constant revision, so the reader should check with the relevant national pollution authorities for any updates. Currently, the assessment system used in Norway differs somewhat from that used in the other North Sea oil-producing countries, the United Kingdom, Denmark, and the Netherlands (see Appendix 1). Particularly, since the late nineties, there has been more focus on the biodegradation of oilfield chemicals, especially those that also have a tendency to bioaccumulate. Thus, new chemicals in this category with < 20% 28-day biodegradation (by OECD306) are unlikely to be allowed offshore and those already in use should be prioritized for substitution.

1.3.2 EUROPEAN REACH REGULATIONS

A new set of regulations called REACH may eventually supersede the OSPAR regulations. REACH is a new European Community regulation on chemicals and their safe use (EC1907/2006). It deals with the registration, evaluation, authorization, and restriction of chemical substances.[18–20] Three useful papers on these new regulations regarding chemicals for use in the upstream oil industry were published in 2007.[21–23]

The new law entered into force on 1 June 2007, but all REACH provisions will be phased in over 11 years. Essentially, there is a control on manufacture, import, and export of chemicals within the EU. The big issue will be around large tonnages of chemicals and the consequent costs of getting environmental tests run. Polymers are, for the moment, exempt, but not the monomers used. Consequently, chemical manufacturers need to consider these costs when deciding which existing chemicals to continue to support and which new ones to introduce. Certainly, any new chemistry for use in Europe needs checking against these new rules. The REACH regulations have already had an effect on the choice of oilfield production chemicals. For example, the manufacture in the Netherlands of a quaternary surfactant anti-agglomerant LDHI, used offshore in the same country, has been discontinued because of the cost of the new REACH regulations (see Section 9.2.3.2 in the chapter on Gas Hydrate Control for details of the chemistry).

1.3.3 U.S. ENVIRONMENTAL REGULATIONS

U.S. environmental regulations are different from the North Sea and also vary for different locations within the United States. U.S. environmental regulations are administered by the U.S. Environmental Protection Agency. For example, discharges from offshore oil and gas platform/production facilities operating in the U.S. Gulf Coast Outer Continental Shelf (and part of the Gulf of Mexico, or GoM) are covered under general National Pollution Discharge Elimination System (NPDES) permits. These NPDES discharge permits are water-quality-based, that is, the criteria for compliance with these permits are based on the toxicity of the effluent rather than hazard assessments on each specific chemical.[24] Operators in the GoM must submit the following toxicity data to the agency on their whole effluent:

- Produced water: 7-day no observed effect concentration (NOEC)
  - mysid shrimp (*Mysidopsis bahia*)
  - silverside minnow (*Menidia beryllina*)
- Miscellaneous discharges of seawater or freshwater to which chemicals have been added: 48-hour NOEC
  - mysid shrimp
  - silverside minnow

Introduction and Environmental Issues 11

A critical dilution factor (CDF) is assigned for produced water discharges for each facility. Effluent samples must be obtained according to the permit, and the produced water must not be toxic to marine organisms (silverside minnow and mysid shrimp) at or below the CDF over the specified period. However, whole effluent toxicity testing requirements in the GoM for produced water discharges are proposed to be changed to include compliance with sublethal effects.

1.3.4 Environmental Regulations Elsewhere

The North Sea region is generally considered to have the most complex environmental regulations for oilfield chemicals based on toxicity, biodegradation, and bioaccumulation of individual substances. However, in regions where the environmental regulations are less well-defined, regulatory agencies or oil companies may ask the service company to provide products that meet the classification requirements of the North Sea, with the general thinking that, if it is green enough for the North Sea, it has a good chance of being permitted for use elsewhere. In fact, where there are no clear local regulations governing produced water discharges, some regulatory agencies (or oil companies) are adopting the North Sea criteria as their global standard (in part or whole). On occasions, local authorities will modify the ecotoxicological tests to include one or more indigenous species.

1.4 DESIGNING GREENER CHEMICALS

Many articles have been written on this subject, although not specifically for the oilfield chemical industry. If we take the OSPAR regulations as our standard, there are three factors one can try to improve to design a greener chemical: the rate of biodegradation, a lower bioaccumulation potential (lower $\log P_{ow}$ value), and a lower toxicity. These are discussed briefly in the next three subsections.

1.4.1 Bioaccumulation

For many classes of production chemicals, there is little you can do about the bioaccumulation potential. For example, scale inhibitors must function in the water phase and therefore must be water-soluble and will consequently have low bioaccumulation. However, if the product is designed to work in the oil phase (such as wax and asphaltene inhibitors), it will have a high $\log P_{ow}$ value. Some production chemicals such as corrosion inhibitors for production lines and hydrate anti-agglomerants can be both water- or oil-soluble. Traditionally, many film-forming corrosion inhibitors for production flowlines, such as fatty acid imidazoline surfactants, have partitioned significantly into the oil phase. However, as discussed in Chapter 8, much work has been done to develop more water-soluble (and less toxic) corrosion inhibitors, for example, by increasing the hydrophilicity of the head group in a surfactant or reducing the length of the hydrophobic tail. Some classes of corrosion inhibitor have been developed that contain no hydrophobic tail and are water-soluble (see Chapter 8). Consideration of bioaccumulation potential is restricted to organic molecules with molecular weight below 700 (although this value varies in some countries).

1.4.2 Reducing Toxicity

Polymers generally have low toxicity, however, some cationic polymers are known to be biocides and could therefore be toxic to other marine organisms (see Chapter 14). For surfactants, anionic and nonionic surfactants are generally less toxic than cationic surfactants. The most common class in the latter category is quaternary ammonium surfactants, which includes pyridinium and quinolinium surfactants. Long-chain tertiary amines and imidazolines can also be included, as they will be partially quaternized (cationic) in acidic produced fluids.[13] The toxicity of cationic surfactants can be reduced in a number of ways:

- Placing anionic groups near the cationic center in the surfactant so as to neutralize the charge on the head, that is, make the surfactant amphoteric
  - This has been done with imidazoline-based corrosion inhibitors, for example, by adding carboxylate groups derived from acrylic acid (see Chapter 8)
- Blending a cationic surfactant with an anionic molecule (as an ion pair)
  - This has been claimed for gas hydrate anti-agglomerants (see Chapter 9)
- Reducing the length of the hydrophobic tail of a surfactant to about eight carbons or less
  - This has been done with film-forming corrosion inhibitors, although this may compromise the performance of the chemical (see Chapter 8).

Biosurfactants can be highly biodegradable and low in toxicity and could be of interest as certain oilfield chemicals. However, most of the current products are too expensive for oilfield applications.

1.4.3 Increasing Biodegradability

A useful review entitled "Designing Small Molecules for Biodegradability" (i.e., not including polymers) has been published.[25] The authors give a brief list of structural factors that generally increase resistance to aerobic biodegradation, although they state there will always be exceptions. These are given here:

1. Halogens (presumably bonded to carbon)—especially fluorine or chlorine and especially if more than three in the molecule
2. Carbon chain branching if extensive—quaternary carbon is particularly problematic
3. Tertiary amines (and presumably also quaternary ammonium)
4. Nitro, nitroso, azo, and arylamine groups
5. Polycyclic residues such as polycyclic aromatic hydrocarbons
6. (Some) heterocyclic residues, for example imidazoles
7. Aliphatic ether bonds, except in ethoxylates

Factor 2 is illustrated with alkylaryl surfactants. If the alkyl group is a highly branched "tetrapropylene," it will be poorly biodegraded, but a linear alkyl group

Introduction and Environmental Issues 13

will degrade much faster. An example of factor 2 is illustrated with polypropoxylates. The extra methyl branch reduces the biodegradability compared with polyethoxylates. Thus, capping a polyethoxylated molecule with polypropoxylate, as is sometimes done for some surfactant and demulsifier classes, actually reduces the biodegradability. A few small molecules with quaternary carbon are quite biodegradable such as the natural compounds vitamin A, cholesterol, and puntothenic acid and the synthetic pentaerythritol ($C(CH_2OH)_4$).

Some heterocyclic residues are fairly biodegradable such as the solvent N-ethyl pyrrolidone (NEP). In contrast, polyvinylpyrrolidone and polyvinylcaprolactam (used as kinetic hydrate inhibitors) are very poorly biodegraded even though they contain the same ring structure as NEP. This is probably because enzyme attack at the polymer rings is very limited due to steric crowding.

The same review lists a few features that generally increase biodegradation rates for small molecules. These are:

- Groups susceptible to enzymatic hydrolysis—chiefly esters (including phosphate esters) and also amides, but this is more equivocal
- Oxygen atoms in the form of hydroxyl, aldehyde, or carboxylic acid groups, and probably also ketone, but not ether, except in ethoxylate groups
- Unsubstituted linear alkyl chains (especially those with more than four carbons) and phenyl rings—these are good places for attack by oxygenase (oxygen-inserting) enzymes. It is the next best structural factor to include if the molecule does not already have an oxygen "handle."

A class of production chemical that nicely illustrates some of these rules of thumb is the twin-tail quaternary surfactant hydrate anti-agglomerants (AAs) discussed in Chapter 9. The first generation of these twin-tail AAs had a structure similar to that in Figure 1.1, with a quaternary ammonium group bonded to two butyl groups (a key structural feature for performance) and two long hydrocarbon tails. The molecule is toxic (as with many quaternary ammonium surfactants) and shows poor biodegradation. To increase the biodegradation, the second-generation twin-tail AA had an ester group placed in the long tails (Figure 1.2). This molecule does degrade at the ester group, leaving two long-chain fatty carboxylic acids and a small, less

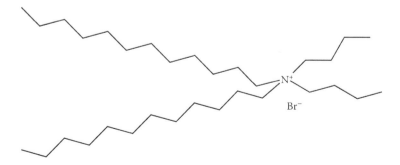

FIGURE 1.1 First-generation twin-tail gas hydrate anti-agglomerant.

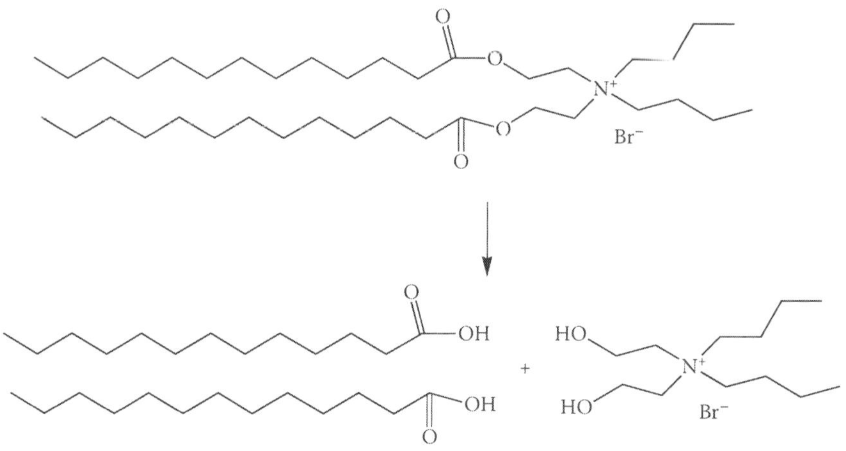

FIGURE 1.2 Second-generation twin-tail gas hydrate anti-agglomerant and first biodegradation products.

surface-active and less-toxic quaternary ammonium compound with two hydroxyl groups. The linear fatty carboxylic acids that are formed biodegrade quite rapidly, but the small quaternary ammonium compound is very persistent even in a test longer than the standard 28 days by OECD306. These data were at least part of the reason why the chemical was not allowed for use offshore in Norway, although it has since been used offshore in the Netherlands. If the fatty carboxylic acid tails had been considerably branched, the rate of their biodegradation would be expected to be less. The third-generation twin-tail AA of this type had an extra methyl group placed near the ester groups to enhance performance (Figure 1.3). This branching might be expected to affect the rate of enzymatic hydrolysis of the ester groups by causing steric hindrance.

For polymers, it is well known that many classes with a polyvinyl backbone are poorly biodegraded, although this does depend on the type and size of the side groups.[26-27] Examples are polyacrylates and esters thereof (used in scale, wax, and asphaltene inhibitors), polyvinylcaprolactam (a kinetic hydrate inhibitor), polyacrylamides (used in water shut-off treatments and as drag reducers), and polydimethyldiallylammonium chloride (a flocculant). The use of groups susceptible to enzymatic hydrolysis (primarily esters and amides) and ethoxylate groups is useful to enhance biodegradation of polymers. Thus, many natural polymers such as polysaccharides

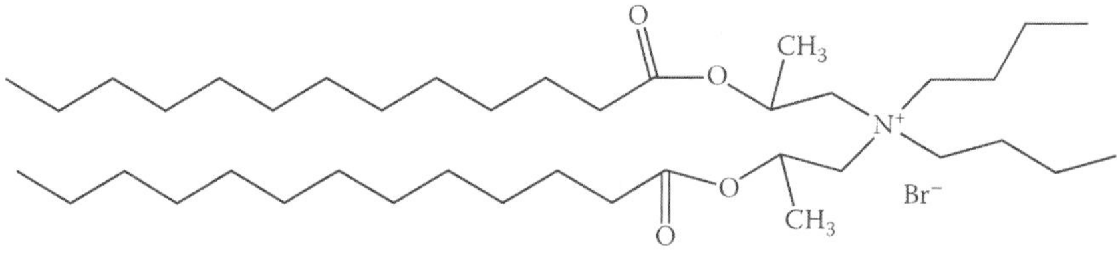

FIGURE 1.3 Third-generation twin-tail gas hydrate anti-agglomerant.

and polyaminoacids show good biodegradation, although lignin is an exception. To illustrate this, biodegradable scale inhibitors based on polyaminoacids (such as polyaspartates) and polysaccharides (such as derivatives of inulin) have been developed. However, there have been unofficial reports that a few polyvinyl-based scale inhibitors (e.g., based on maleic acid chemistry) also biodegrade fairly rapidly (see Chapter 3 for more information). In addition, a range of more biodegradable demulsifiers has been claimed using various readily biodegradable polymeric core structures, which are then ethoxylated and propoxylated (see Section 11.3.3.9).

Biodegradable polymers for use as plastics is a well-researched subject.[28-30] Most of them have polyester backbones, such as polylactide acid and poly-3-hydroxybutyrate. Derivatives of these polymers do not appear to have been explored as potentially biodegradable oilfield chemicals, either in water- or oil-based applications. However, other graft polymers with polyester backbones have been explored for use as demulsifiers and kinetic hydrate inhibitors (see Chapters 9 and 11, respectively).

1.5 MERCURY AND ARSENIC PRODUCTION

A production chemistry issue, which is not discussed elsewhere in this book, is the production of toxic metals and metalloids, such as mercury and arsenic.

Mercury originates from volcanic rocks that underlie hydrocarbon reservoirs and is naturally associated with the production of these hydrocarbons. It is soluble in hydrocarbons as organometallic compounds such as dimethylmercury ($HgMe_2$) and other dialkylmercury species and even as mercury metal and inorganic mercury salts. Mercury associated with asphaltene was also found to be a significant fraction of the total mercury concentration in a number of crude oils examined.[31] The organometallic mercury compounds are particularly toxic, more so than inorganic mercury or the metal. These species are volatile and follow both the liquid and particularly gas hydrocarbon phases in parts-per-billion levels. Mercury and its organometallic compounds can react with sulfur compounds during production, leading to mercury sulfide deposits usually in the oil-processing equipment. Metallic mercury can accumulate in aluminum alloy heat exchangers, causing corrosion and catastrophic failure of the equipment.[32-33] Mercury in processed crude oil or condensate has a detrimental impact on refining by poisoning catalysts, degrading materials, and reducing product quality, especially refined products used as feedstocks for petrochemical manufacture.

Measuring mercury concentrations in crude oil, condensate, and natural gas is now becoming more important as mercury's impact on production and processing systems becomes better understood.[34] Many technologies have been developed to remove mercury compounds from hydrocarbon streams. The most common method today is the use of adsorbents.[35-36] Examples are adsorption on silver or copper sulfide on an alumina support,[37] adsorption on cuprous or stannous metal salts,[38] activated carbon adsorbent,[39] metal oxides or sulfide adsorbents,[33,40] an ion-exchange resin containing active thiol-groups,[41] polymers with thiol groups.[42] An oil-soluble sulfur compound can first be added to the hydrocarbon stream followed by trapping of the mercury–sulfur compounds on an adsorbent.[43] However, suspended (colloidal) forms of mercury, such as mercuric sulfide, evade capture by some sorbent beds. Alkali polysulfides and oil-soluble organic dithiocarbamate or sulfurized isobutylene have been proposed to remove mercury from crude oils, natural gas condensates, and other liquid hydrocarbons.[44–45]

Some produced waters can also contain low levels of inorganic mercury cations, which can be a danger to the environment if discharged, as the mercury (especially when converted to alkylmercury species) can bioaccumulate through the food chain, ending up in fish and, eventually, humans if they eat the fish. Chemical precipitation, coagulation, and activated carbon adsorption methods can be used, but reinjection of the water in deep dry wells or depleted production wells can also be carried out.[46–47] The best chemical precipitation technique for removal of water-soluble mercury species (and arsenic) is to add sulfur compounds. Mercury (I) and (II) ions are soft acids, which bond strongly to sulfur-containing soft bases such as sulfides, thiols (mercaptans), thiocarbamates, and thiocarbonates.[48] A sequential desander, oxidant (NaOCl), Fe^{3+}, thiol, flocculant water treatment process has been operating relatively successfully at two South-East Asian fields since 2003.[49–50] Branched polymeric dithiocarbamates can be used to bind and precipitate mercury ions with simultaneous flocculation and water clarification.[51] The dithiocarbamate polymers are made from reaction of ethylene dichloride with ammonia, and then derivatizing them to contain at least 5 mol% of dithiocarbamate salt groups. In one field application, an additional anionic flocculant was used to help settle the precipitate.[52] Sulfur compounds such as thiols and dithiocarbamates will also form insoluble complexes with other metal ions such as Ag^+, Pb^{2+}, Cu^{2+}, and Cd^{2+}.

Arsenic may also present in natural gas streams as arsenic hydrides, such as arsine (AsH_3), which is a very toxic gas.[53] Arsenic compounds may also be found in accompanying produced waters and are also toxic. There are several methods of removing arsenic compounds from natural gas and water to meet environmental requirements. These include pyrolysis, adsorption, and absorption. Many of the adsorbent methods referenced for mercury removal also work to remove arsenic from hydrocarbon streams. Regarding chemical methods, reaction with an oxidizing agent has been used to remove arsenic compounds from natural gas. If arsenic is present in water, it is predominantly as As(III) (arsenite) with only a minor amount of As(V) (arsenate). As(III) is the most harmful of the two species. As with mercury, As(III) can be removed by precipitation with sulfur compounds such as mercaptans, followed by flocculation.[49,54] Arsenic-containing corrosion inhibitors used to be included in acid stimulation packages, but they have been replaced by more environmentally friendly organic inhibitors and intensifiers (see Chapter 5 for details).

REFERENCES

1. T. A. Hjelmas, S. Bakke, T. Hilde, E. A. Vik, and H. Grüner, "Produced Water Reinjection: Experiences from Performance Measurements on Ula in the North Sea," SPE 35874 (paper presented at the Third International Conference on Health, Safety & Environment in Oil & Gas Exploration & Production, New Orleans, La., 9–12 June 1996).
2. S. Szymczak, G. Brock, J. M. Brown, D. Daulton, and B. Ward, "Beyond the Frac: Using the Fracture Process as the Delivery System for Production Chemicals Designed to Perform for Prolonged Periods of Time," SPE 107707 (paper presented at the Rocky Mountain Oil & Gas Technology Symposium, Denver, Colo., 16–18 April 2007).

3. C. Berkland, M. Cordova, J.-T. Liang, G. P. Willhite, International Patent Application WO/2008/030758.
4. S. E. Campbell, U.S. Patent 7135440, 2006.
5. M. W. Kendig, M. M. Hon, and L. F. Warren, U.S. Patent Application 20050201890.
6. R. Alapati, L. Sanford, T. Williams, and S. Gao, "New Test Method to Qualify Production Chemicals for Subsea Injection," SPE 114924 (paper presented at the SPE Annual Technical Conference and Exhibition, Denver, Colo., 21–24 September 2008).
7. J. Yang, V. Jovancicevic, U.S. Patent Application 20060166835.
8. R. J. Dyer, U.S. Patent Application 20080047712.
9. B. L. Knudsen, M. Hjelsvold, T. K. Frost, M. B. E. Svarstad, P. G. Grini, C. F. Willumsen, and H. Torvik, "Meeting the Zero-Discharge Challenge for Produced Water," SPE 86671 (paper presented at the SPE International Conference on Health, Safety, and Environment in Oil and Gas Exploration and Production, Calgary, Alberta, Canada, 29–31 March 2004).
10. OSPAR Guidelines for Completing the Harmonised Offshore Chemical Notification Format (2005-13) (http://www.ospar.org).
11. S. Glover and I. Still, "HMCS (Harmonised Mandatory Control Scheme) and the Issue of Substitution," Ninth Annual International Petroleum Environmental Conference, October 22–25, 2002.
12. M. Thatcher and G. Payne, "Impact of the OSPAR Decision on the Harmonised Mandatory Control System on the Offshore Chemical Supply Industry," Proceedings of the Chemistry in the Oil Industry VII Symposium, RSC, Manchester, UK, 13–14 November 2001.
13. G. M. Rand, *Fundamentals of Aquatic Toxicology: Effects, Environmental Fate and Risk Assessment*, 2nd ed., Philadelphia, PA: Taylor & Francis, 1995.
14. G.-G. Ying, Fate, "Behavior and Effects of Surfactants and Their Degradation Products in the Environment," *Environment International* 32 (2006): 417.
15. J. Beyer, A. Skadsheim, M. A. Kelland, K. Alfsnes, and S. Sanni, "Ecotoxicology of Oilfield Chemicals: The Relevance of Evaluating Low-Dose and Long-Term Impact on Fish and Invertebrates in Marine Recipients," SPE 65039 (paper presented at the SPE International Symposium on Oilfield Chemistry, Houston, TX, 13–16 February 2001).
16. J. M. Getliff and S. G. James, "The Replacement of Alkylphenol Ethoxylates to Improve the Environmental Acceptability of Drilling Fluid Additives," SPE 35982 (paper presented at the International Conference on Health, Safety & Environment, New Orleans, La., June 1996).
17. OSPAR List of Substances/Preparations Used and Discharged Offshore Which Are Considered to Pose Little or No Risk to the Environment (PLONOR). Reference number: 2004-10 (http://www.ospar.org).
18. http://ec.europa.eu/environment/chemicals/reach/reach_intro.htm.
19. http://europa.eu.int/comm/enterprise/reach/index.htm or http://ecb.jrc.it/REACH/.
20. Council Directive 67/548/EEC in order to Adapt it to Regulation (Ec) of the European Parliament and of the Council Concerning the Registration, Evaluation, Authorisation and Restriction of Chemicals (http://www.reach-compliance.eu/).
21. G. Dougherty, "REACH—The Big Picture" (paper presented at the RSC/EOSCA Chemistry in the Oil Industry X Symposium, November 2007).
22. S. W. Longworth, "REACH—From a Supplier's Perspective" (paper presented at the RSC/EOSCA Chemistry in the Oil Industry X Symposium, November 2007).
23. J. Hislop and D. Knight, "REACH—Intelligent Safety Evaluation and Prediction of Properties" (paper presented at the RSC/EOSCA Chemistry in the Oil Industry X Symposium, November 2007).
24. Final NPDES Permit for the Western Portion of the GoM OCS (GMG290000) (http://www.epa.gov/region6/water/npdes/genpermit/index.htm).

25. R. S. Boethling, E. Sommer, and D. DiFiore, "Designing Small Molecules for Biodegradability," *Chemical Reviews* 107 (2007): 2207.
26. D. K. Platt, *Biodegradable Polymers, Rapra Market Report*, Cambridge, UK: Woodhead Publishing, 2006.
27. G. T. Howard, "Biodegradation of Polyurethane: A Review," *International Biodeterioration & Biodegradation* 49 (2002): 245.
28. NIIR Board, *The Complete Book on Biodegradable Plastics and Polymers (Recent Developments, Properties, Analysis, Materials & Processes)*, Delhi, India: Asia Pacific Business Press, 2006.
29. E. S. Stevens, *Green Plastics*, Princeton, NJ: Princeton University Press, 2001.
30. C. E. Cooke, U.S. Patent Application 20080015120.
31. S. M. Wilhelm, L. Liang, and D. Kirchgessner, "Identification and Properties of Mercury Species in Crude Oil," *Energy & Fuels* 20 (2006): 180–186.
32. S. M. Wilhelm and N. Bloom, "Mercury in Petroleum," *Fuel Processing Technology* 63 (2000): 1–27.
33. M. Abu El Ela, I. S. Mahgoub, M. H. Nabawi, and M. Abdel Azim, "Mercury Monitoring and Removal at Gas-Processing Facilities: Case Study of Salam Gas Plant," SPE 106900, *SPE Projects, Facilities & Construction* 3(1) (2008): 1.
34. K. J. Grice, L. D. Van Orman, L. A. Young, and C. R. Manning, "Evaluation of Portable Mercury Vapor Monitors and Their Response to a Range of Simulated Oil Processing Environments," SPE 111837 (paper presented at the SPE International Conference on Health, Safety, and Environment in Oil and Gas Exploration and Production, Nice, France, 15–17 April 2008).
35. M. Abu El Ela, M. H. Nabawi, and M. Abdel Azim, "Behavior of the Mercury Removal Absorbents at Egyptian Gas Plant," SPE 114521 (paper presented at the CIPC/SPE Gas Technology Symposium 2008 Joint Conference, Calgary, AB, 16–19 June 2008).
36. M. R. Sainal, T. Mat, A. Shafawi, and A. J. Mohamed, "Mercury Removal Project: Issues and Challenges in Managing and Executing a Technology Project," SPE 110118 (paper presented at the SPE Annual Technical Conference and Exhibition, Anaheim, CA., 11–14 November 2007).
37. T. Y. Yan, U.S. Patent 4909926, 1990.
38. T. Torihata, E. Kawashima, U.S. Patent 4946582, 1990.
39. J. D. McNamara, U.S. Patent 5336835, 1994.
40. M. Roussel, P. Courty, J.-P. Boitiaux, and J. Cosyns, U.S. Patent 4911825, 1990.
41. H. A. M. Duisters, and P. C. Van Geem, U.S. Patent 4950408, 1990.
42. F. R. Van Buren, L. Deij, G. Merz, H. P. Schneider, U.S. Patent 553844, 1994.
43. T. F. Degnan and S. M. LeCours, U.S. Patent 6350372, 2002.
44. T. Y. Yan, U.S. Patent 4915818, 1990.
45. T. C. Frankiewicz and J. Gerlach, U.S. Patent 6685824, 2004.
46. P. Dhanasin, W. Utoomprurkporn, M. Hungspreugs, and T. Soponkanabhorn, "Monitoring of Mercury Content of Fishes Near the Bongkot Platform, Gulf of Thailand," SPE 96492 (paper presented at the SPE Asia Pacific Health, Safety and Environment Conference and Exhibition, Kuala Lumpur, Malaysia, 19–21 September 2005).
47. M. Rangponsumrit, S. Athichanagorn, and T. Chaianansutcharit, "Optimal Strategy of Disposing Mercury Contaminated Waste," SPE 103843 (paper presented at the International Oil & Gas Conference and Exhibition, China, Beijing, China, 5–7 December 2006).
48. G. S. Elfline, U.S. Patent 4612125, 1986.
49. D. L. Gallup and J. B. Strong, "Removal of Mercury and Arsenic from Produced Water" (paper presented at the 13th Annual International Petroleum Environmental Conference, San Antonio, TX, October 17–20, 2006).
50. T. C. Frankiewicz and J. Gerlach, U.S. Patent 6117333, 2000.

Introduction and Environmental Issues 19

51. K. S. Siefert, P. L. Choo, J. W. Sparapany, J. H. Collins, J. H., U.S. Patent 5346627, 1994.
52. M. L. Braden, "Mercury: Real Problems...Not Roman Mythology" (paper presented at 13th Annual International Petroleum Environmental Conference, San Antonio, TX, October 17–20, 2006).
53. A. Block-Bolten and L. Glowacki, "Natural Gas Separation from Arsenic Compounds," SPE 19078 (paper presented at the SPE Gas Technology Symposium, Dallas, TX, 7–9 June 1989).
54. T. J. Pinnavaia, J. I. Dulebohn, and E. J. McKimmy, U.S. Patent 7404901, 2008.

2 Water and Gas Control

2.1 INTRODUCTION

This chapter discusses chemical methods to control water and gas production. Some of these techniques can be used to shut off either water or gas.

As oilfields mature, they produce increasing quantities of water. Handling such large volumes of water is an expensive business and can even cause bottlenecking at the surface-processing facilities. One report published in 2007 illustrated the cost using a typical North Sea field of 50 wells, with each well producing 5,000 bbl of water per day. If the cost of treating each barrel is $0.50, the daily water-handling cost for the oilfield would equate to $125,000, ($45.6 million per annum).[1] Water production costs are not just associated with handling the water. Water production can cause scale formation, fines migration, sandface failure, corrosion of tubulars, and even kill wells by hydrostatic loading. Thus, it pays to defer water production for as long as possible.[2-3]

The most common techniques for shutting off water-producing zones are mechanical isolation, cement squeezing, and polymer gel treatments, which require zonal isolation. Disproportionate permeability reducer (DPR) or relative permeability modifier (RPM) treatments can be bullheaded, and so they are an alternative to zonal isolation treatments.

Production wells are initially perforated near the bottom of the pay zone. When bottom water begins to dominate the fluid production, these perforations are sealed off with a cement squeeze, a packer, or a plug. The well is reperforated above the sealed zone, and oil production is resumed. This process continues until the entire pay zone has been watered out. Squeezing with cement is often not sufficient by itself: a success rate of about 30% is achieved. The reason for this is that the size of standard cement particles restricts penetration of material into smaller channels, fractures, and highly permeable zones. The use of organic polymer gel can give good penetration and a treatment can last for several years. However, a combination treatment (cement/polymer gel) could be even more effective. Gel is first placed into the formation to the required lateral depth. Then, the near-wellbore channels are sealed with a tail-in of cement. The cement helps lock the gel or polymer in the formation and helps prevent residual polymer production.

Another water shut-off method is to place a membrane in the well made of a material that will swell on contact with water but not with oil.[4] When water production takes place, the swelled material will block off the water-producing zone. However, this method does not give good penetration depth away from the wellbore.

There is a range of chemical techniques for shutting off or reducing water production from a well, which will be discussed in this chapter. An overview of commercial products available up to 2001 has been published.[5] Water shut-off treatments have also been used very successfully in injector wells to obtain a better directional profile. Water shut-off chemical techniques include:

- resins and elastomers
- inorganic gels
- cross-linked organic polymer gels
    - monomer-based
    - polymer-based
- viscoelastic surfactants (VESs)
- DPR or RPM
    - includes emulsified gels
- microparticles and colloids

Gel treatments are also useful for gas shut-off. A technique that works for gas shut-off but not water shut-off is the use of stabilized foams. This is reviewed in Section 2.8 (Gas Shut-Off).

2.2 RESINS AND ELASTOMERS

Resins and elastomers such as those based on phenolics, epoxies, or furfuryl alcohol can be used to shut off water-producing zones. They have sufficient physical strength to seal fractures, channels, and perforations. They are relatively expensive. Therefore, their use normally is confined to within the first radial foot of a wellbore. Lower concentrations of these resins or elastomers can be used for sand consolidation.[5] Phenolics are resins based on phenols and formaldehydes. A composition containing phenolformaldehyde and at least one of sodium bisulfite, sodium metabisulfite, or mixtures thereof together with various deployment procedures has been claimed as a method of reducing the deleterious effects of water production.[6] The sulfite salt has been found to be critical for delay of viscosity set up for an adequate time to enable flow of the aqueous polymer to the desired site. A styrene-butadiene latex, activated by brine, is also available. Epoxy is a product of the reaction between epichlorohydrin and bisphenol A and a common hardener is diethylenetriamine. A controlled catalyst system using trichlorotoluene and pyridine with furfuryl alcohol has also been used.[7] A consolidation fluid for reducing particle and water production comprises a hardenable resin component, a liquid-hardening agent, a silane-coupling agent, and a surfactant has been claimed.[8] Typical resins are bisphenol A–epichlorohydrin resin, polyepoxide resin, novolak resin, polyester resin, phenol-aldehyde resin, urea-aldehyde resin, furan resin, urethane resin, and glycidyl ethers.

2.3 INORGANIC GELS

Inorganic gels based on silicates have been widely used for water shut-off. The general method is to use a water-soluble sodium silicate solution and react it with a gelling agent to generate a gel. The silicate solution is typically not compatible with formation waters, since sodium silicate reacts with divalent calcium ions instantly to generate gel. In this approach, two solutions are injected in any order and are separated by a slug of an inert aqueous spacer liquid. A common gelling agent is ammonium sulfate.[9] However, this technology cannot generate uniform gels to plug the porous medium and cannot place the gel deep into the formation. Several staged treatments are also required in pumping the fluids using these techniques.

A common inorganic gel sealant is based on internally activated sodium silicates (IAS).[10-16] An IAS system is generally placed as a water-thin, freshwater-based solution that consists of a silicate source and an activator that is designed to trigger gelation of the silicate at a designated time. The gel times of IAS systems are controlled by the pH and temperature. When sodium silicate is acidified to a pH value of less than 10.6, a three-dimensional network of silica polymers forms that creates a gel. A slight decrease in the pH greatly reduces gel time. As a result, gel times are difficult to control by the addition of acid at the surface. However, the target pH can be reached in the formation when materials are added that slowly release acids. The addition of an alcohol will also cause high-pH sodium silicate solutions to gel. To regulate the gelling time, an organic material (activator) that reacts with water to produce an alcohol and/or acid is often mixed with the silicate solution.

Examples of activators that release acid or alcohol at elevated temperatures in the aqueous formation fluids include halogenated organic compounds such as trichloroacetic acid and salts and esters thereof: these release hydrochloric acid (HCl). Another method of generating an acid is to use ester, lactone, or amide in situ hydrolysis, such as diesters of dicarboxylic acids (e.g., dialkyl succinate) solubilized as microemulsions.[17-19] Activators that generate alcohols include polyhydroxy compounds such as glycerol, mannitol, or sugar, or reducing sugars such as lactose, xylose, and so forth.[20-21]

Another silicate gel activator system uses urea/formaldehyde resins, which decompose to release urea in the formation thereby gelling the silicate.[22-23] Addition of a chelate such as EDTA greatly increases the divalent ion tolerance of the silicate gelling system since silicates can be made to gel by the addition of cations.[24] Silicate gels are stable at up to 200°C (392°F) and hence are applicable for high-temperature wells. They have also been used in gas shut-off treatments (see Section 2.8 on gas shut-off).

Organosilicate (or siloxane) blocking gels can be generated in situ via hydrolysis of oil-soluble organosilanes on contact with water. At low concentrations, they are useful for sand consolidation but at higher concentrations, they have been claimed for water shut-off treatments.[25] Typical organosilanes are 3-aminopropyltriethoxysilane and bis-(triethoxy silylpropyl)amine, or mixtures thereof.

Another inorganic gel system that has been used in the field is based on in situ hydrolysis of trivalent cationic solutions.[26-27] Iron(III) is the easiest to use. Iron(III) ions are only stable in solution in acid solution. They hydrolyze easily when the pH drops by spontaneous aging in the formation, causing precipitation of iron(III) hydroxide species. These particles can flocculate and form a water block. The new blocking material has excellent stability under field conditions, and yet simple remediation is possible in case of placement failures. Further, the novel method is characterized by a self-controlling chemical mechanism. Injectivity problems did not arise using this technique in the field, even in low-permeability and tough formations.

A related system comprises injecting sequentially an aqueous caustic solution and an aqueous solution containing a polyvalent cation, such as Mg^{2+}, into the near-wellbore environment of a subterranean formation interposed by a hydrocarbon spacer.[28] The spacer causes the two aqueous solutions to mix in the near-wellbore environment and an insoluble precipitate, such as $Mg(OH)_2$, is formed, which preferentially reduces the permeability of the relatively high permeable zones in the near wellbore environment thereby improving conformance and flow profiles of fluids subsequently injected into or produced from the formation. A chemically related technique, which has been used for water shut-off via coiled tubing in the North Sea, uses magnesium oxychloride slurried in water.[29–30] The composition undergoes a rapid phase transition to form a substantially solid mass with a near-linear relationship between the time required for the phase transition to occur and the composition temperature at which the phase transition occurs, thereby permitting accurate determination of the set time.

Partially hydrolyzed aluminum chloride has been used as an inorganic gel precursor.[5] This system precipitates an aluminum hydroxide gel when an activator responds to temperature and raises the system pH above a certain value. Activators are selected from sodium cyanate with urea or hexamethylenetetramine.[31–32] A gel forms since aluminum and hydroxyl ions link with each other in such a way as to form an amorphous, irregular 3D impermeable network. High concentrations of sulfate and carbonate ions can destabilize the system. Sodium aluminate $NaAl(OH)_4$ has been used as a water-control treatment in gas wells. On contact with formation water, the pH drops and an aluminum hydroxide gel forms.

2.4 CROSS-LINKED ORGANIC POLYMER GELS FOR PERMANENT SHUT-OFF

Strong cross-linked polymer gels were the first chemical water shut-off treatments to be used in the field. Basically, a 100% water-producing zone is isolated and an aqueous polymer solution containing a cross-linker is pumped into the near-wellbore. As the temperature of the solution increases, cross-linking takes place and a gel is formed, blocking the pores. These gels block both water and oil production, so careful planning and zone isolation are needed to avoid plugging oil-producing zones. Even so, the industry now has a good grasp of the technology, and the success rate for water shut-off gel treatments is about 80%. Polymer gel water shut-off treatments can be used in oil or gas wells.

The polymer solution (plus cross-linker) can be made in advance and pumped into the formation, or monomer solutions can be pumped with a polymerization initiator and the polymer formed in the near-wellbore during shut-in at elevated temperature where it also cross-links. Polymer gel treatments can be combined with sand consolidation treatments[33] or huff-puff surfactant stimulation techniques.[34] A huff-puff is a stimulation process in which a surfactant solution (normally anionic or nonionic) is first injected into a single production well (huff) and then the well is shut off for a few days to allow surfactant imbibition and lastly the well is put back into production (puff). The injected surfactant can scour the residual oil in contact with surfactant near the wellbore. In addition, the surfactants that are injected into a carbonate reservoir can be imbibed into the carbonate matrix to alter the matrix from oil-wet to water-wet, which allows the aqueous phase to penetrate the oil-rich matrix and push out formerly by-passed oil. Polymer gel treatments have also been combined with cementlike rigid-setting materials or noncement particulates for better conformance control of high water-cut wells.[35–36] The system can be bullheaded into the well and the temperature range of the particle gel system is 21–177°C (70–350°F). The system can be easily washed out of the wellbore, as compared with cement, which must be drilled out.

2.4.1 Polymer Injection

A variety of water-soluble polymers can form solid gels when they are cross-linked.[37,38] Most cross-linked polymer gel water shut-off treatments practiced today use ready-made polymers that become cross-linked and gel in the formation at elevated temperatures. The most common gel systems that have been investigated are as follows:

- Partially hydrolyzed polyacrylamide (PHPA, or acrylate/acrylamide copolymers)
    - Cross-linked with metal ions
    - Cross-linked with organic compounds
- Cationic or sulfonated polyacrylamides (PAMs)
    - Cross-linked with metal ions
    - Cross-linked with organic compounds
- PAM
    - Cross-linked with organic compounds
- Acrylamide/*t*-butyl acrylate copolymer
    - Cross-linked with polyethyleneimine (PEI)
- Biopolymers
    - Cross-linked with metal ions
- Polyvinyl alcohol or polyvinylamine
    - Cross-linked with organic compounds

Polyacrylamides or biopolymers cross-liked at high pH with borate ions and/or metal oxide cross-linkers have been claimed as gelling diverting agents for water shut-off treatments.[39] The gelation time can be controlled by varying the cross-linker concentration or by injecting a preflush to cool down the formation. Generally, the molecular weights of the polymers are very high (millions of daltons). PAM has neutral amide side groups and is difficult to cross-link. PHPA contains carboxylate groups in the form of acrylate monomers. The percentage of carboxylate groups may vary from 0 to 60%. These carboxylic acid groups can be made to cross-link using various organic compounds or trivalent or tetravalent metal ions at elevated temperature.

2.4.1.1 Metal Ion Cross-Linking of Carboxylate-Containing Acrylamides and Biopolymers

Polyvalent metal ions will cross-link with carboxylate groups found in PHPA, acrylate copolymers, or certain natural polymers. Chromium(VI), such as in dichromate ions, can be used as a cross-linker precursor that is reduced to chromium(III) cross-linker by a reducing agent.[40] This system was much used in the early days of water shut-off treatments. However, this system does not give sufficient control over the gelation time. H_2S in the reservoir may also give premature reduction to chromium(III) and subsequent cross-linking.

The most common metal ion cross-linkers to have been investigated are Cr(III), Ti(IV), Al(III), and Zr(IV) ions. Al(III) ions as cross-linker is rarely used because the cross-linking reaction cannot be controlled or delayed as well as other systems.[41] For example, a field trial with aluminum citrate that needed precooling of the near-wellbore to avoid premature gelling has been reported.[5] Cr(III) has been much used in the field to cross-link PHPA.[42–43] However, it is banned from use in the North Sea, the reason being that Cr(VI) compounds are carcinogenic, even though one can argue that there is little chance that Cr(III) will get oxidized to Cr(VI) in the environment.

To gain some control over the gelation times and obtain deeper penetration of the gel, carboxylate salts of Cr(III) are used.[44] An early and today much-used system is PHPA and chromium(III) acetate (Figure 2.1).[45–46] This system is relatively insensitive to pH and salinity and can be used to form water-blocking gels at up to 124°C (255°F). High molecular weight PHPA (ca. 5×10^6 Da) is used for blocking fractures whereas lower molecular weights are used for blocking matrix pores.

Another commercial system uses PHPA and Cr(III) propionate as cross-linker and is claimed to be more temperature resistant than the PHPA/Cr(III) acetate system.[47] PHPA/Cr(IIII) lactate or malonate have also been shown to have useful gelling times over the range 90–135°C (275°F).[48–49] Since the metal ion release for cross-linking with a PHPA is dictated by the strength of interaction between the metal ion and the protective anion, the release of the metal ion can be controlled by pH adjustments

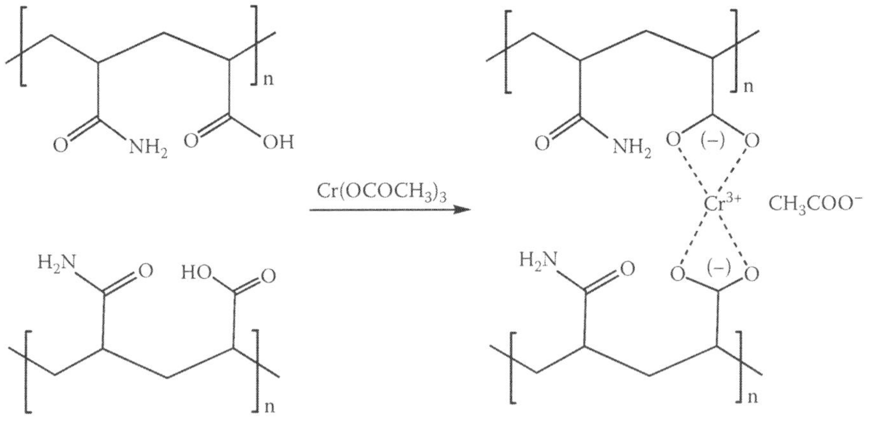

FIGURE 2.1 Cross-linking of PHPA with Cr(III) acetate.

or by adding chelants to the metal/polymer solution.[50] NaF can also be added as a stability additive to PHPA/Cr(III) gelation solutions.[51] A laboratory study has shown improved performance for fracture-problem water shut-off polymer gels that are formulated with a combination of high and low molecular weight polymers. These gels are intended for application to fractures or other high-permeability anomalies that are in direct contact with petroleum production wells.[52]

Besides chromium(III), zirconium(IV) is the most used metal cation cross-linker, for example, zirconium(IV) lactate has been deployed.[53] The polymer can be PHPA or weakly hydrolyzed and sulfonated PAM.[54] Metal ion cross-linked gels for high-temperature applications can be made using sulfonated polymers with additional amounts of open-chain and especially cyclic vinylamides such as N-vinyl pyrrolidone or N-vinyl caprolactam. These polymers also have good compatibility with divalent cations. Addition of small amounts of phosphonic acid groups can improve their adsorption characteristics.[55] Improved polymer gel thermal stability can also be obtained by using organic cross-linkers (see below).

2.4.1.2 Gels Using Natural Polymers

Gels based on metal cross-linked natural polymers have also been studied and applied in the field. However, results have generally been poorer than with synthetic polymers as the gels are not very robust, degrading too quickly.[56] Thus, they are best used for low-temperature applications. For example, results at 90°C (194°F) with cellulose-based and xanthan gels showed short half-lives ranging from less than a day (complete syneresis) up to 45 days depending on concentration and polymer type. Intermediate stabilities were observed with scleroglucan, while the most stable gels with half-lives of more than a year were observed at 90°C (194°F) with lignosulfonate and phenoformaldehyde materials.[57] The triple helix structure of scleroglucan may be responsible for its superior thermal stability over xanthan.[58] Xanthan, cross-linked with Cr(III) ions has been used successfully in many low-temperature fields to plug water-thief zones and divert the water to underswept oil-bearing zones.[59] An extracellular polysaccharide produced by *Alcaligenes* bacteria has been cross-linked with Cr(IIII) salts to form firm gels that are stable for long periods at elevated temperatures. Field results have been reported.[60-61] Aldehydes, such as glutaraldehyde, have been proposed as cross-linker for polysaccharide such as scleroglucan for reservoir temperatures up to 130°C (266°F).[62] Hexamethylenetetraamine can be used as an aldehyde precursor.[63,64]*

2.4.1.3 Organic Cross-Linking

Metallic cross-linking of carboxylate polymers such as PHPA is not suitable for very high temperature applications. In these reservoirs, excessive polymer gel hydrolysis will occur.[65-66] Also, syneresis will take place via additional unwanted cross-linking with divalent cations such as Mg^{2+} and Ca^{2+}. In soft waters, PAM gels are more stable.[67]

---

* Gels formed by cross-linking of polymers with vicinal hydroxyl groups, such as galactomannan and polyvinylalcohol, with boron-containing compounds gave a loss of viscosity at elevated pressures, but this was not observed with Ti(IV) and Zr(IV) cross-linkers.

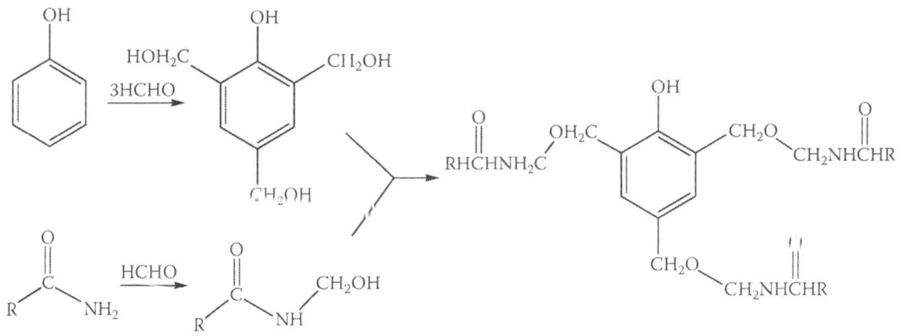

FIGURE 2.2 Possible gelation mechanism for acrylamides with phenol and formaldehyde.

PHPA can be cross-linked with mixtures of phenolic compounds and an aldehyde for higher-temperature applications. The simplest cross-linker system would be phenol and formaldehyde (Figure 2.2).[68–73] To avoid the use of toxic phenol or formaldehyde, precursors can be used that generate phenol or formaldehyde in the reservoir. Hexamethylenetetraamine is used to generate formaldehyde and resorcinol, salicyl alcohol, phenyl esters such as phenyl acetate, furfuryl alcohol, aminobenzoic acids, and salicylic acid are used as more environmentally friendly phenol precursors.[74–76] Unhydrolyzed PAM can also be cross-linked with hexamethylenetetraamine, aldehydes such as teraphthalaldehyde and hexanal (with solubilizing surfactants), and glutaric acid.[77] The gels can be made more stable by adding a secondary cross-linker such as hydroquinone, dihydroxynaphthalene, or gallic acid. In fact, phenolic compounds can form thermosetting gels with formaldehyde alone, without the need for an acrylamide polymer. A system thought to be based on resorcinol and formaldehyde has been tried in the field.[78] Cationic PAM cross-linked with glyoxal has also been used in the field.

An alternative approach to delayed gel systems is to use an acrylamide/acrylic ester copolymer. The most studied and applied copolymer is acrylamide/t-butyl acrylate ester copolymer (PAtBA). Both the acrylamide and t-butyl ester groups will hydrolyze in the reservoir at elevated temperatures producing acrylate groups.[79] The t-butyl ester hydrolyzes more readily. The resulting acrylate copolymer could be cross-linked with metal ions but these gels are not stable at high temperatures for long periods. A better cross-linker is a polyamine such as PEI forming amide cross-links (Figure 2.3).[80–82] Gels formed with PEI and PAtBA were stable in the laboratory at 156°C (313°F) for many months. Another study obtained good stability at 177°C.[83] These gelants were shown to propagate eight times farther than chromium-based gelants under equivalent conditions. The reason for using the t-butyl acrylate ester is that it reacts more slowly with PEI than small less-steric esters such as methyl acrylate ester thereby avoiding premature gelling and deeper penetration. Many field applications with this system have been carried out.[84]

Regarding the cross-linking reaction of the PAtBA/PEI system, the ester groups on the PAtBA polymer provide masked cross-linking sites. These groups either hydrolyze or thermolyze according to pH and temperature. At conditions of low pH and temperature, the copolymer hydrolyzes and forms PHPA and t-butyl alcohol.[85]

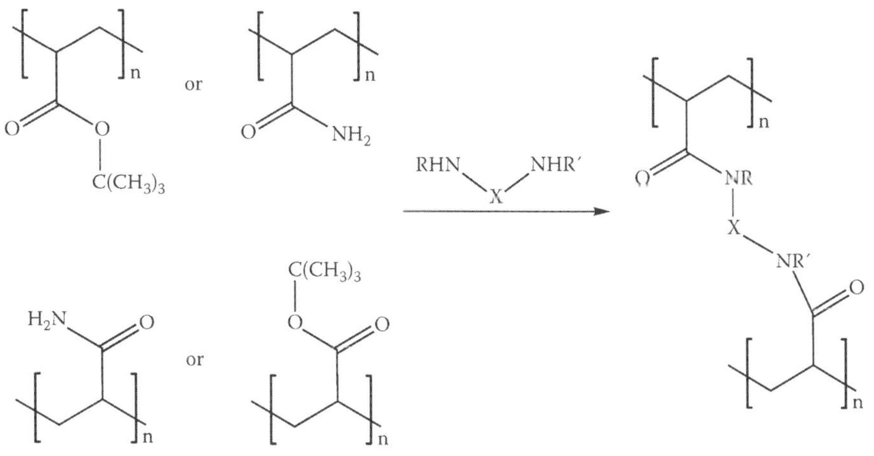

FIGURE 2.3 Hydrolysis and cross-linking of acrylamide/*t*-butyl acrylate ester copolymer with PEI. X is the branched backbone of the PEI. The nitrogen atoms are either primary or secondary amines.

At high temperatures, the copolymer thermolyzes producing PHPA and isobutene. After the breakage of the ester groups, carboxylate groups are formed. At < 100°C (212°F) and < 20 h, nucleophilic attack by an imine nitrogen from the PEI on the carbonyl carbon attached to the ester occurs resulting in a covalent bond. At high temperatures, another mechanism may be operating possibly by attack of amine groups on the carbonyl moiety in acrylamide.

In the Norwegian sector of the North Sea, PEI is targeted for phase-out as a cross-linker for acrylamide or acrylate ester polymers for environmental reasons. A study has shown that chitosan can be used as an alternative more environmentally acceptable cross-linker to form stable blocking gels at up to 120°C (248°F).[86] Chitosan is formed from hydrolysis of the natural polymer chitin and contains primary amine groups as well as hydroxyl groups. It is primarily the amine groups that cross-link to form amide groups (Figure 2.4).

Since the chitosan cross-linker is only a minor component of the gel composition, the total system is still predominantly nonbiodegradable due to poor biodegradability of the synthetic base polymer. To avoid this problem, it has been claimed to use a polysac-

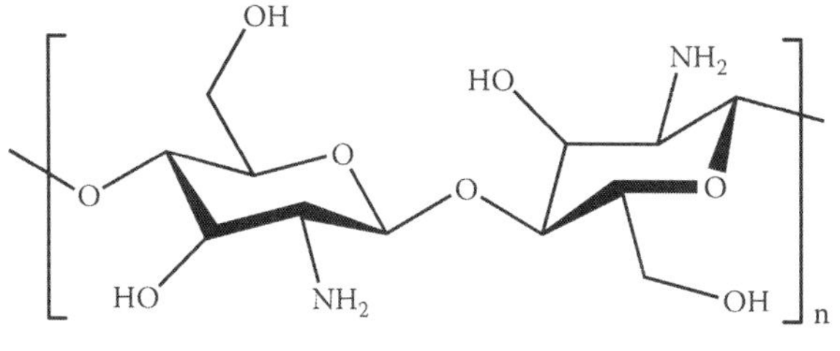

FIGURE 2.4 Structure of chitosan.

charide-based polymer, such as starch, with an oxidizing agent, which is capable of at least partially oxidizing the polysaccharide-based polymer making it cross-linkable.[87]

2.4.1.4 Polyvinyl Alcohol or Polyvinylamine Gels

Polyvinyl alcohol (PVA) or polyvinylamine (PVAm) can be cross-linked with phenols/naphthols and aldehydes to form very stable gels.[88–89] However, there are few reports of their use in the field.[90–91] A typical aldehyde is formaldehyde. PVA is made industrially by hydrolysis of polyvinyl acetate and PVAm is made industrially by hydrolysis of polyvinylformamide.[92–93] PVA plus partially methylated melamine-formaldehyde resin has also been claimed.[94] PVA can also be cross-linked using only a dialdehyde such as glyoxal or glutaraldehyde. The reaction proceeds via a hemiacetal.[95–96] The only metal ion that is reported to be able to cross-link PVA is titanium(IV).[97]

2.4.1.5 Problems Associated with Polymer Gel Water Shut-Off Treatments

There are a number of potential problems with the use of polymer gels.[5] These include:

- syneresis
- precipitation
- chemical degradation
- mechanical degradation (shear degradation)

Syneresis refers to a collapse of the polymer gel structure and has been observed in many polymer systems. This is characterized by a loss of adhesion, reduction in volume, and expulsion of water. Possible causes are excess cross-linker, polymer hydrolysis, and divalent ions such as Ca^{2+}. Where too much cross-linker is present, cross-linking continues well past the point of gelation. This causes the polymer gel to contract in volume (synerese) expelling water as it does so. Depending on the composition, a syneresed gel may occupy as little as 5% of the initial solution volume. Acrylamide polymers can hydrolyze to acrylate groups in the hot reservoir, which gives more sites for cross-linking and possible syneresis. High concentrations of divalent cations in the formation can provide more cross-linking than that expected from the injected metal cross-linker. The permeability reduction obtained in rock pores remains technologically useful even with very high percentages of syneresis. In contrast, in a fracture system, the effects of syneresis are expected to be much more important than in matrix applications.

Polymer precipitation is the result of interaction between divalent cations and the carboxylate groups present within the hydrolyzed polymer. As the percentage of carboxylate groups increases, the solubility decreases in the presence of these cations. Eventually, precipitation occurs and the polymer is then unavailable to form a gel.

A polymer may gel correctly in the formation and later thin out due to chemical bond breakage. This results in short field life of the treatment. Probable causes are high reservoir temperatures, oxygen contamination, or peroxide contamination followed by free-radical generation. The effect is a loss of polymer molecular weight and structure. Gel stability can be improved by adding sulfonated monomers such as AMPS.

Water and Gas Control

Many water shut-off polymers have very high molecular weights resulting in long, linear polymers. Some shear degradation may occur as polymer passes through a pump, in turbulent flow through pipes and tubing, through perforation tunnels and through formation pores. A shear-degraded polymer will have a lower molecular weight and will be less effective in forming a gel.

2.4.1.6 Other Improvements for Cross-Linked Polymer Gels

Cationic PAMs have been used to give better polymer adsorption to carbonate rock and prevent premature back production.[98] Improved gel strength can be obtained by adding an amount of an inert colloidal particulate material such as colloidal silica.[76] Deeper placement of the gelling formulation can be obtained by using preflush of a chemical that lets the polymer flow over it into deeper areas. The chemical is selected from the group consisting of tetramethylammonium salts, polyvinylpyrrolidone, polyethyleneoxide, PEI, and nonionic, amphoteric, anionic, and cationic surfactants.[99] The chemical attaches to adsorption sites on surfaces within the porosity of the formation but slowly washes off the sites as another fluid flows through the treated formation. Thereafter, a water-flow–resisting polymer is introduced into the formation so that it flows deeply into the porosity of the formation before the previously introduced chemical washes off and attaches to the adsorption sites as the chemical washes off.

An example of a self–cross-linking water-soluble polymer consists of the monomer units acrylamide, acrolein, and 4-vinylphenol (Figure 2.5).[100] Foamed gels have also been claimed as a water shut-off technique.[101] The foamed gel is formed from a cross-linkable carboxylate-containing polymer, a cross-linking agent containing a reactive transition metal cation, a polyvinyl alcohol, an aqueous solvent, and an added gas.

2.4.2 IN SITU MONOMER POLYMERIZATION

Instead of using a preformed polymer, monomer solutions plus cross-linker can be pumped with a polymerization initiator and the polymer formed in the near-wellbore during shut-in at elevated temperature.[102–104] The second method has been used in the field significantly less than the former method.[105–107] One of the reasons for this is that acrylamide monomer, a major component in these polymers, is classified as

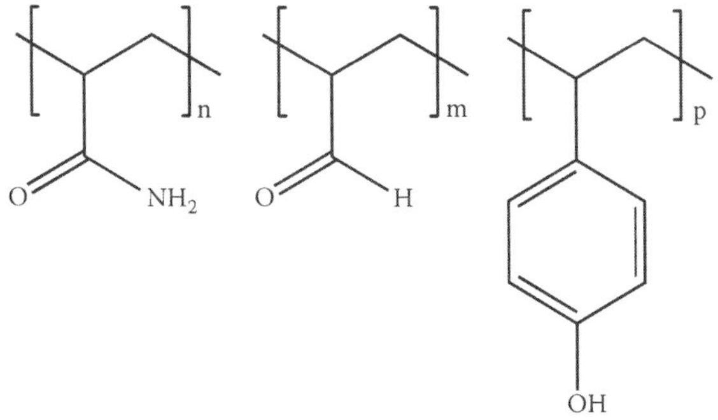

FIGURE 2.5 Acrylamide/acrolein/4-vinylphenol terpolymer.

carcinogenic. Another reason is that the polymerization initiators, azo or peroxy compounds, are expensive. In addition, it is difficult to obtain long-enough polymerization delay times with these systems. This latter problem has been overcome by using an emulsion.[108] A typical emulsion system uses N-methylolacrylamide or N-methylolmethacrylamide monomer, a diperoxide polymerization initiator such as 2,2-bis(*tert*-butylperoxy)butane, an emulsifier such as a polyoxyethylenated sorbitan ester, a polymerization retarder such as *para*-*t*-butylcatechol and an aliphatic hydrocarbon.

2.5 VISCOELASTIC SURFACTANT GELS

Viscoelastic surfactant–based fluid systems that form viscous plugs have been claimed for use in water and/or gas shut-off applications.[109] The fluid systems may include brine, a viscosity enhancer, as well as the VES, and, optionally, a stabilizer for high-temperature applications. Suitable VESs include nonionic, cationic, amphoteric, and zwitterionic surfactants (see Section 5.7.4 for examples). Amine oxide VESs are a preferred class as they have the potential to offer more gelling power per unit weight, making them more cost-effective than other fluids of this type. Preferred stabilizers are magnesium oxide, titanium(IV) oxide, and aluminum oxide. The viscosity enhancer may include pyroelectric particles, piezoelectric particles, and mixtures thereof.

A chemical method of plugging water-producing zones during a well-fracturing operation is the use of specific viscoelastic anionic surfactants (VASs), which are a class of selective permeability blockers.[110–111] Viscoelastic anionic surfactants produce shear-thinning gels in the presence of cations, which are easily pumped and can permeate into porous, permeable rocks. An example of a VAS is the family of alkyl sarcosinates (Figure 2.6).[110–111] Once in the formation pores, the viscosity could increase by as much as 100 times, thereby restricting fluid movement. Contact with hydrocarbons breaks the gel, greatly reducing the viscosity. This frees up only the pores with residual hydrocarbon saturation, leaving them clear and strongly water-wet. Conversely, pores with high water saturation remain plugged with the gel.

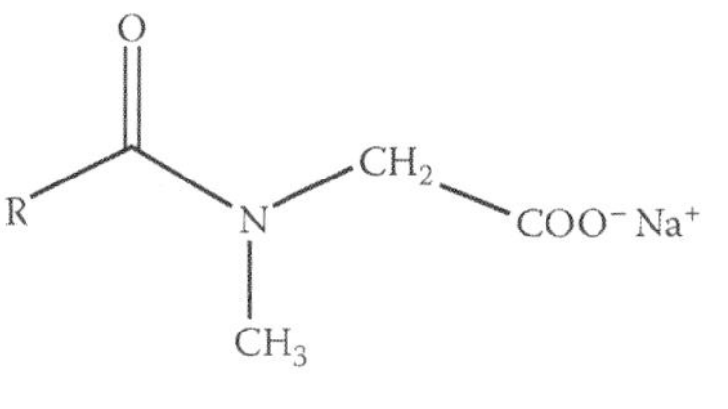

FIGURE 2.6 Alkyl sarcosinates.

2.6 DISPROPORTIONATE PERMEABILITY REDUCER OR RELATIVE PERMEABILITY MODIFIER

DPRs are distinguished by some authors from RPMs although the effect is the same, that is, to reduce water permeability without significantly affecting oil permeability. DPRs are sometimes described as chemicals that plug pores but they do not precipitate, swell, or viscosify as much in the presence of hydrocarbons as they would do in a water environment.[90] The net effect is a reduction of water permeability by a larger factor than that to oil. DPR and RPM treatments can be bullheaded and are an alternative to polymer gel treatments, which require zonal isolation. Examples of DPRs are rosin wood derivatives, which gel in the presence of water. One can also use oil-soluble silicon compounds, which react with water in the formation to form water-blocking silicate gels. In this way, the oil-bearing zones are not appreciably affected.[112] Examples of silicon compounds include alkyl silicates such as methylorthosilicate or ethylsilicate, and halosilanes. These silicon compounds can be used in combination with polymeric RPMs (see below) for improved water blocking.[113] The DPR can also be combined with scale inhibitor squeeze treatments.[114]

2.6.1 Emulsified Gels as DPRs

To optimize the DPR effect, it is important to place the gel at oil saturation higher than the residual. Based on this fact, a new DPR technique has been developed using a thermally sensitive water-in-oil emulsion containing a gelling formulation.[115] Successful field applications have been reported in which the gel system used was acrylamide/*t*-butyl acrylate ester copolymer with PEI as the cross-linker.[116] For a significant reduction in oil permeability, the gel should occupy only a fraction of the treated volume. An important feature of the emulsion is that it breaks spontaneously before a gel is formed. Under static conditions after shut-in, the water phase separates and gels up, leaving the oil phase mobile. Emulsified systems are easier to handle and predict than the previously evaluated coinjection of oil and gelant.[117] In water-wet cores, the permeability reduction was much stronger when using a gelant with the saturation of gelant in the oil (25%), since an aqueous gelant in a water-wet media blocks narrow passages like pore throats. Using emulsified gelants, it is possible to obtain a measurable permeability reduction instead of a complete blocking. The reason is that the oil (in the emulsion) helps to keep some channels open so that it is possible for oil to flow through the core without first having to break the gel mechanically.

2.6.2 Hydrophilic Polymers as RPMs

RPMs are water-soluble, hydrophilic polymers that adsorb to rock surfaces in the formation. Useful reviews have been published.[90,118]

The required effects of an RPM are to water-wet the pores and use the adsorbed hydrated polymer layer to reduce the pore throat channel size significantly. Thus, maximizing the hydrated polymer volume is important. The net effect is a reduction of water permeability by a larger factor than that of oil. If the loss of productivity index can be compensated by higher drawdown, the treated well can also produce more oil or more gas, hence a stimulation technique.[119] If the RPM enters mainly fractures, there will be little effect on the relative water permeability as the fracture channels are much wider than pore throats. RPM R&D grew rapidly after the mid-nineties since treatments can be bullheaded and do not require coiled tubing and zonal isolation. Thus, they are cheaper than polymer gel treatments, which require zonal isolation. However, polymer gel water shut-off treatments can last for several years if carried out correctly whereas RPM treatments usually have a shorter lifetime of a year or less.[120] The treatment success rate in the field with RPMs is still not as high as polymer gel treatments, and may not be due to product failure but to poor candidate selection. A methodology to evaluate well candidates for RPM and other matrix applications has been published,[121–122] as well as the feasibility of using RPMs to control water production in gas wells. RPM injection followed by a polymer gelling treatment has been claimed for blocking water production in gas wells.[123]

Although there is still some contention over the mechanism of RPMs, a general mechanism has been suggested, where RPM is governed by segregated or preferred flow of oil and water at the pore level.[124] The water-based RPM fluid will be located in the parts of the pores preferred for water flow with little impact on the oil flow. The swollen adsorbed polymer layer, being hydrophilic but oleophobic, would dehydrate and shrink when hydrocarbons flowed through the pore, and so would not impair oil/gas production. The polymer layer can also have a drag effect on the flowing water but minimal effect on flowing oil. The authors of the mechanism also suggest that RRF (defined as the ratio between permeability before and after treatment) could be determined by the gel saturation in the porous media.

To be effective, RPMs must enter the complete zone and strongly adhere to the rock.[1] Therefore, removing oil deposits in the near-wellbore is essential, usually by the use of a surfactant preflush. Better still, the surfactant can be added to the polymer RPM formulation.[125] If the RPM material works correctly, then the impact on the oil zone is minimal.[126] However, it should also be possible to clean up the treatment if the oil-producing zones are damaged, usually by adding a polymer breaker such as a persalt.

Besides the well-known use of hydrophilic polymers, there are very few reports or claims on the use of non-VESs to reduce water production. One patent covers the use of imidazoline compounds, which are fatty acid, oxyethyl derivatives of imidazole to control water production in sandstone reservoirs.[127]

2.6.2.1 Types of Polymer RPM

A wide range of hydrophilic polymers has been studied as RPMs. They can be divided into two categories, synthetic vinyl polymers and natural polymers. Examples of RPM polymers that have been deployed without the need for cross-linking include

- PAM[128–129]
- PHPA[130]
- Xanthan[131]
- Scleroglucan[132]
- Cationic acrylamide polymers with methacrylamidopropyltrimethylammonium chloride, dimethyl diallyl ammonium chloride, or dimethylaminoethylmethacrylate (DMAEMA) monomers[133–135]
- Cationic modified natural polymers, such as cationic starches[136]
- Amphoteric terpolymers of acrylic acid, acrylamide, and diallyl dimethyl ammonium chloride monomers[137–139]
- Acrylamide/AMPS copolymers[140]
- Vinyl sulfonate/acrylamide copolymers and vinyl sulfonate/vinyl formamide/acrylamide terpolymers[141]
- Acrylamide/N-vinyl pyrrolidone copolymers[142]

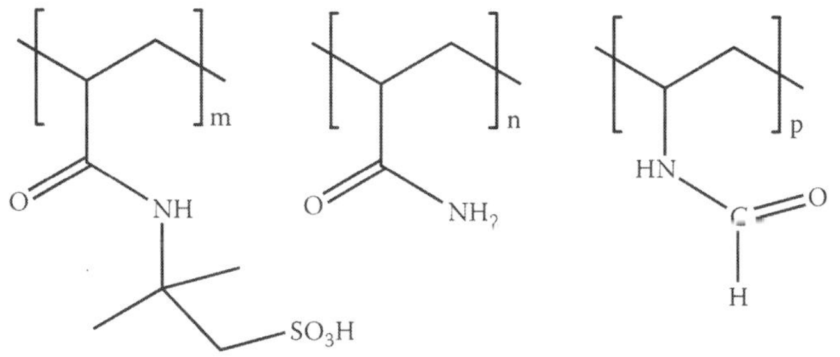

Water and Gas Control

FIGURE 2.7 Acrylamide/AMPS/vinyl formamide terpolymer RPMs.

- Acrylamide/2-acrylamido-2-methyl propane sulfonic acid (AMPS)/vinyl formamide terpolymers (Figure 2.7)[143]
- Brush polymers consisting of a polymeric back-bone grafted with polyethylene oxide, such as poly[dialkylaminoacrylate-co-acrylate-g-poly(ethyleneoxide)].[99,144]

Early RPMs used simple PAM or PHPA or natural polymers such as xanthan or scleroglucan. However, treatment lifetimes were fairly short with these polymers as they were back-produced or degraded too quickly. Addition of formaldehyde as biocide gave better thermal stability to scleroglucan. However, biopolymers have injectivity problems as a result of the high viscosity of the polymeric solution and its tendency to flocculate. Cationic acrylamide or amphoteric acrylamide polymers have been used for better rock adsorption, shear sensitivity, and temperature and salt tolerance.

PAMs have a poor efficacy and reduced duration of the treatment due to the limited thermal stability; PAMs modified by the introduction of cationic groups have a good efficacy and show good rock adsorption and salt tolerance.[133–134] Their temperature stability can be improved by using methacrylamidopropyltrimethylammonium chloride monomer instead of acrylamidopropyltrimethylammonium chloride to make the copolymer.[135] Several hundred jobs have been carried out with the anionic acrylamide/AMPS/vinyl formamide terpolymer of moderate molecular weight, mostly in low-oil-producing wells. This has largely replaced the use of amphoteric copolymers.[145] The vinyl formamide functions as an anchoring monomer. Initial fracturing treatment field trials successfully used this RPM as an additive to organoborate cross-linked pad stages. Vinyl polymers containing sulfonated monomers have even more stability than nonsulfonated polymers such as PAM and PHPA. They also have better compatibility with divalent metal ions than PAM and PHPA, which, at reservoir temperatures, are prone to further hydrolyze to acrylate groups. It has been mentioned that polymers can selectively reduce the relative permeability to water with respect to oil, and this property does not depend on the type of polymer (as long as it is hydrophilic) or the rock type.[146]

In one RPM treatment improvement, the volume occupied by the hydrated polymer can be tuned by varying the salinity of the aqueous solution.[147] Thus, when a PHPA (for example) dissolved in a high-salinity brine (which causes it to shrink because of electrostatic shielding) is pumped into a well, it will swell when lower salinity formation water contacts the adsorbed polymer. It has also been claimed to inject two polymers sequentially and with opposite charge into sandstone formations to increase polymer retention.[148] Thus, the negatively charged formation is first contacted with an anionic polymer, then a cationic polymer. As a result of the contact of the polymers, coacervation (phase separation) occurs between the anionic and cationic polymers, which reduces the amount of the anionic polymer removed from the formation by fluids produced there from. Sequential injection of a cationic polymer, then an anionic polymer would presumably be useful in positively charged carbonate reservoirs.

As mentioned earlier, maximizing the volume occupied by the hydrated polymer is important for water control. To this effect, brush (or comb) polymers consisting of a polymeric backbone grafted with polyethylene oxide have been designed, which appear to give a greater success rate in the field. The brush side chains will collapse on contact with hydrocarbons but become fully hydrated, occupying a larger volume, on contact with water.[99,149]

RPMs can be used to control water production in several different combined treatments such as hydraulic fracturing. The similarity between some polymeric scale inhibitors and polymer RPMs has not gone unnoticed. Combined RPM and scale inhibitor squeeze treatments have been carried out.[150–151] A combined scale inhibitor treatment and water control treatment has been described that requires fewer steps than the sum of each treatment procedure practiced separately.[152] RPM water control treatments can also be combined with sand consolidation treatments or acidizing treatments. For example, one sand consolidation/RPM combined treatment comprises the steps of applying to a subterranean formation a preflush fluid, applying aqueous surfactant fluid (containing RPM), applying a low-viscosity consolidating fluid (containing surfactant), and applying an postflush fluid.[153]

2.6.2.2 Hydrophobically Modified Synthetic Polymers as RPMs

Laboratory results on sandstone cores showed that hydrophobic modification of PAM increases significantly mineral coverage and attenuate surface roughness thereby giving better RPM effect than a totally hydrophilic polymer.[154] Rather than reaching a plateau adsorption, as is common for hydrophilic polymers, hydrophobic modification appears to produce a continued growth in adsorption with increased polymer concentration. This behavior is attributed to associative adsorption of polymer chains on previously adsorbed layers of polymers.[155] The effect is enhanced oil recovery by improving oil mobility and reducing water mobility. Examples of claimed hydrophobically modified RPMs include a polymer of DMAEMA partially quaternized with an alkyl halide, wherein the alkyl halide has an alkyl chain length of six to 22 carbons (Figure 2.8).[156] Other examples are an acrylamide/octadecyldimethylammoniumethyl methacrylate bromide copolymer, a dimethylaminoethyl methacrylate/vinyl pyrrolidone/hexadecyldimethylammoniumethyl methacrylate bromide terpolymer and an acrylamide/2-acrylamido-2-methyl propane sulfonic acid/2-ethylhexyl methacrylate terpolymer.[157–158]

The use of cross-linkable, hydrophobically modified water-soluble polymers with surfactants as RPM treatments has been claimed.[100,159] The polymer can be bullheaded with a VES to form a physically stabilized structure in the oil- and water-producing

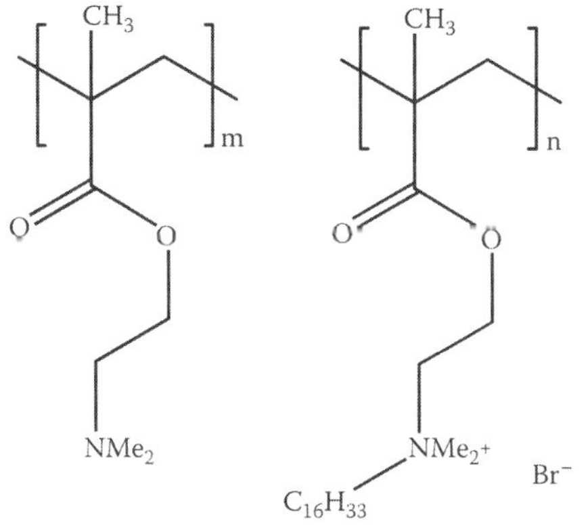

Water and Gas Control

FIGURE 2.8 Example of a hydrophobically modified RPM/DMAEMA copolymer partially quaternized with $C_{16}H_{33}Br$, $m > 10n$ for water solubility.

rock layers. When hydrocarbons pass over these structures, they break down, allowing the flow of produced hydrocarbons. The physically stabilized structures are not broken down in the water-producing zones. The polymer can be formulated with a cross-linker, which will gel over time in the formation, forming a more permanent barrier to water production. An example consists of a solution of the VES N-erucyl-N,N-bis(2-hydroxyethyl)-N-methylammonium chloride (Figure 2.9) with PAM hydrophobically modified with 3 mol% of the hydrophobe n-nonyl acrylate in a solution of 0.5 M sodium chloride. The development of the viscoelasticity of the surfactant is delayed by the addition of urea phosphate (a hydrogen-bonding modifier) and the hydrophobically modified PAM is cross-linked by the addition of acetaldehyde. Hydrophobically modified acrylamide/acrylate copolymers could be cross-linked with metal ions such as Cr(III) or Zr(IV). A second example uses the VES solution formed by potassium oleate in a potassium chloride electrolyte solution. Coinjection with a vinyl polymerization initiator such as 2,2′-azo(bisamidinopropane)dihydrochloride forms a polymerized surfactant gel. The resulting solution of polymerized surfactants is slightly less viscoelastic than the original monomeric solution but the observed viscoelasticity is insensitive to contact with hydrocarbons. The gel formed by the polymerized surfactant retains its viscoelasticity after prolonged contact with water.

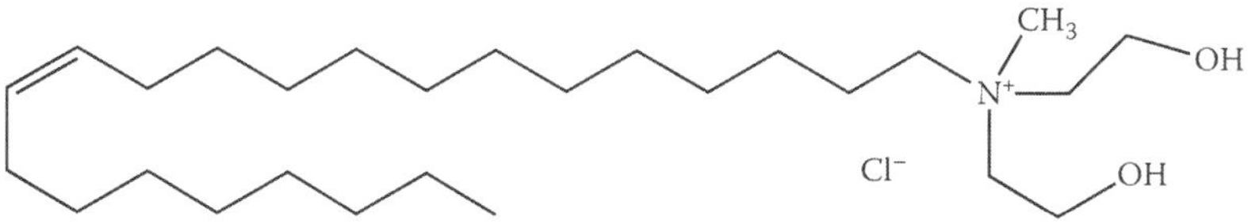

FIGURE 2.9 N-erucyl-N,N-bis(2-hydroxyethyl)-N-methylammonium chloride.

2.6.2.3 Cross-Linked Polymer RPMs

Cross-linking hydrophilic polymers is useful not only for total water shut-off gel treatments as discussed earlier but also for RPM treatments in high-permeability formations. A high degree of cross-linking produces a total blocking gel. Partial cross-linking produces a three-dimensional polymer network, which bridges the pore spaces. This gives a stronger effect than normal RPM polymers, which only alter the surface properties of the formation.[160]

An early example that has been used in the field is a cationic PAM cross-linked with glyoxal.[161] Polysaccharides cross-linked with aldehydes have also been claimed.[162] A more recent system that has been developed uses a sulfonated polymer cross-linked with metal ions. The cross-linker is Zr(IV) ions, chelated for control purposes. Typical polymers are acrylamide/AMPS terpolymers with a little vinyl phosphonic acid monomer to increase rock adsorption (Figure 2.10).[163] The polymer molecular weight must be not be too high (> 10,000,000 Da), or else it is difficult to handle, nor must the molecular weight be too low (< 300,000 Da) or a gel would form blocking both the water- and oil-producing layers.[164] Instead, the polymer is of medium molecular weight (1–3 million) and cross-links to form an open, hydrated, flexible network. When produced water comes into contact with the polymer network, it inflates and stiffens, providing resistance to water flow and greater water permeability reduction. If hydrocarbons pass through the polymer network, it collapses due to water displacement giving minimal resistance to the flow of hydrocarbons. This system has been bullheaded very successfully in a high–barium sulfate–scaling field to reduce water production. This RPM treatment is useful for high-permeability formations at high temperatures. For lower permeability formations, it may not be necessary to cross-link the polymer into a 3D network.[165] However, field trials where iron(II) ions, and not iron(III) ions, were present gave poor RPM effect.

Colloidal particles of acrylamide-based gels with neutral organic covalent cross-linkers have been reported as RPMs for controlling water production.[166–169] This technology differs from the other cross-linked RPMs in that cross-linking is done before the formulation is injected. The microgels can be varied in size (0.3–2.0 μm), consistency, and chemistry. Due to their large size, they do not easily penetrate low-permeability thereby fulfilling one of the challenges of water shut-off technology. The microgels have excellent shear, thermal, and chemical stability. The first field

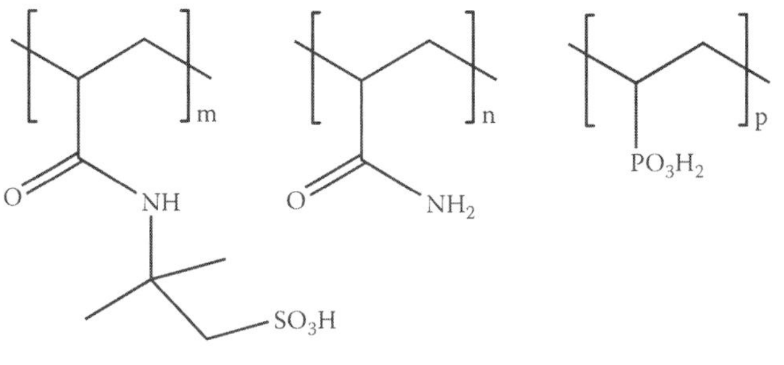

FIGURE 2.10 Metal ion cross-linkable polymer RPM for high permeability formations: acrylamide/AMPS/vinyl phosphonic acid terpolymer.

application was on a gas storage reservoir.[170] Other successful applications in the North Sea have been carried out.

2.7 WATER CONTROL USING MICROPARTICLES

The idea of using solid particles to shut off water production has been around a long time but has not been used much in the field. For example, one patent claims the use of solid spheres of water-insoluble polymer formed in situ from an oil-in-water emulsion of monomers, such as divinylbenzene and styrene, and polymerization initiator.[171] The plasticlike solid spheres will penetrate into the more permeable formation zone and stop fluid flow. The emulsion can also be utilized before commencing a carbon dioxide flood during profile control. A later and related patent claims that the emulsion should contain a nonpolymerizable liquid carrier and a minor proportion of polymerizable monomers such as acrylates or styrene.[172] In this way, when the discontinuous phase of the emulsion is miscible with the ambient fluid in the matrix, undesirable particle formation is minimized since the emulsion droplets will become diluted. The particles produced by the polymerization in the discontinuous phase of the emulsion may also serve as reservoirs for well-treatment chemicals, for example, scale inhibitors.

Related to the use of RPMs and DPRs are water-swellable cross-linked polymer particles. For example, a dispersion of polymer particles made from vinyl amides (such as acrylamide or N-vinyl pyrrolidone) with cationic monomers has been claimed for water shut-off treatments.[173] The particles are made by invert polymer emulsion processes, resulting in particles that are much smaller in size than that of the formation pores. Upon injection, the particles become trapped in the formation, and upon flowback of water, the particles swell and adsorb onto the formation, forming a film and restricting further fluid flow.

Another patent claim relates to coated particles.[174] The method comprises the steps of coating a particulate solid material with an organic polymer, which reacts with water and swells when contacted therewith. This reduces the flow of water through a pack of the resulting polymer-coated particulate solid. The organic polymer can be a copolymer of a vinyl silane such as vinyltrimethoxysilane or methacrylatetrimethoxysilane and one or more water-soluble organic monomers such as 2-hydroxyethyl acrylate, (meth)acrylamide, and N-vinylpyrrolidone (Figure 2.11). Alternatively, the

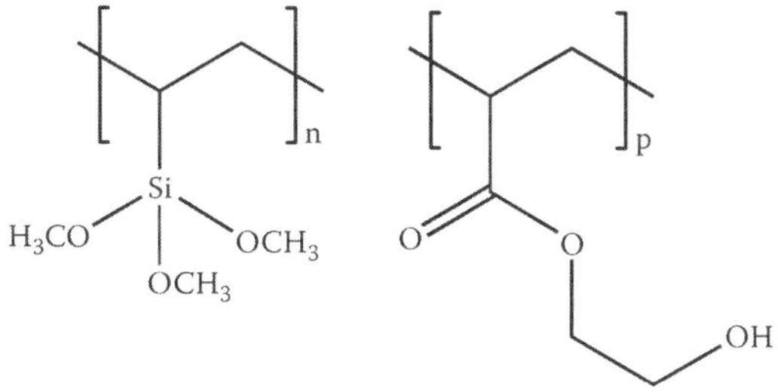

FIGURE 2.11 Example of a swelling polymer for coated particles using vinyltrimethoxysilane and 2-hydroxyethyl acrylate.

organic polymer is a cationic water-soluble polymer such as polydialkyldiallyl polymers or the quaternized ammonium salt of polyDMAEMA copolymers. The particle size could be small enough to penetrate pore throats, or it could be the size of proppants and be used in fracturing operations.

Another method that could be used for water shut-off is the use of kaolinite clay particles.[175] Testing has also been carried out to assess the potential of injecting relatively high concentrations of kaolinite with a fixation agent as a water shut-off technique. The concept is based on the ability of kaolinite to cause formation damage coupled to the ability of the kaolinite fixation agent to maintain clay in place. So far, it has been demonstrated that the use of this technology can withstand a differential pressure of 150 bar without any impairment to sealant performance.

2.8 GAS SHUT-OFF

Some of the techniques used to control water production can also be used to control gas production.[5] For example, organic polymer gels have been used successfully, as well as inorganic gels such as sodium silicates.[176–180] Foamed cement or cement particle gels have also been used.[36,181]

Another method that has been used with good success to control gas in high-GOR oil producers is the use of foams.[172–184] Field tests show that foam can be effective in controlling gas coning, injector-to-producer breakthrough of gas in high-permeability zones, and, at least in some situations, gas cusping. A foam confined inside the pore network of the rock consists of thin liquid films spanning the pores that make the gas phase discontinuous. This drastically reduces gas mobility essentially without influencing liquid relative permeabilities. Foam is generated by gas (usually nitrogen) displacing a suitable surfactant foaming-agent solution. A foam whose purpose is to block gas from entering into a production well should ideally be formed in situ wherever gas breakthrough may occur and then remain stagnant, maintaining the strongest possible reduction of gas mobility for the longest period. After one series of field applications, it was concluded that high permeability reservoirs require gas/foamer coinjection to improve placement of a foam block in the reservoir. In less permeable reservoirs, in situ foam generation/placement by the surfactant alternating gas technique (SAG) is sufficient to achieve a good result.[185]

One essential property of the foam is that it is stable in the presence of crude oil. In one field application, foams generated in nonaqueous solutions (such as kerosene) were washed out by crude oil flowing beneath the foam barrier.[186] Aqueous foams based on α-olefin sulfonate (AOS) surfactants were shown to be poor foam blockers in the presence of crude oil.[187] Better foam stability was obtained with AOS combined with fluorosurfactants or AOS with polymers such as PAM. Gas shut-off treatments with these mixed foaming agents have met with good success.[188–189] Gelling or cross-linking the foam has also been claimed to create a more rigid foam.[190] However, one laboratory study indicated that addition of polymer did not give any benefit over the use of AOS surfactant alone and could be detrimental if used in too high a concentration.[191] A surfactant system with a source of calcium has been evaluated for gas shut-off purposes in oil carbonate reservoirs.[192]

2.8.1 Gas Well Foamers for Liquid Unloading

Liquid buildup in gas wells causes an additional back-pressure that can reduce gas productivity and, in worse cases, stop production completely. Besides mechanical techniques, a well-known chemical method to remove the liquid (usually water) is to inject a foaming surfactant.[193–195] The surfactant is injected at the bottom of the well, where it mixes with the liquid and gas, lowering the surface tension and forming a foam of lower density than the bulk liquid, which can then be produced from the well.[196] The surfactant can be applied as a liquid concentrate, either continuously or batchwise or it may be applied as "foam sticks." The latter is more common for low-volume gas-producing wells: the surfactant is compounded within a wax "candle," which is simply lowered down the well.

A variety of surfactants, anionic, cationic, zwitterionic, and nonionic, can be used but they should not damage the reservoir or cause emulsion problems if there are liquid hydrocarbons present.[197] Some surfactants also have the benefit of corrosion inhibition properties.[198] Typical surfactants are AOSs, alcohol ether sulfates, and betaines, the latter two categories being more biodegradable.[199] Traditional foaming surfactants tend to be ineffective as the condensate-to-water ratio increases.[200] A novel foamer specifically designed to unload condensate from wells has been reported.[201] The use of foaming surfactants to unload liquid in gas wells is very common. About 40% of gas wells worldwide suffer from this problem and are therefore producing at below optimum rate.

REFERENCES

1. O. Vazquez, M. Singleton, K. S. Sorbie, and R. Weare, "Sensitivity Study on the Main Factors Affecting a Polymeric RPM Treatment in the Near-Wellbore Region of a Mature Oil-Producing Well," SPE 106012 (paper presented at the SPE International Symposium on Oilfield Chemistry, Houston, TX, 28 February–2 March 2007).
2. R. S. Seright, R. H. Lane, and R. D. Sydansk, "A Strategy for Attacking Excess Water Production," SPE 84966, *SPE Production & Facilities* 18(3) (2003): 158.
3. J. Hibbeler and P. Rae, "The Environmental Benefits of Reducing Unwanted Water Production," SPE 96582 (paper presented at the SPE Asia Pacific Health, Safety and Environment Conference and Exhibition, Kuala Lumpur, Malaysia, 19–21 September 2005).
4. D. W. Ross, U.S. Patent Application, 20060175065, 2006.
5. A. H. Kabir, "Chemical Water & Gas Shutoff Technology—An Overview," SPE 72119 (paper presented at the SPE Asia Pacific Improved Oil Recovery Conference, Kuala Lumpur, Malaysia, 6–9 October 2001).
6. G. E. Anderson and W. J. Heaven, International Patent Application WO/2008/083468.
7. D. D. Sparlin and R. W. Hagen Jr., "Controlling Water in Production Operations," *World Oil* (1984).
8. P. D. Nguyen and D. L. Brown, U.S. Patent 7028774, 2006.
9. B. B. Sandiford, U.S. Patent 4004639, 1977.
10. G. D. Herring, J. T. Milloway, and W. N. Wilson, "Selective Gas Shut-Off Using Sodium Silicate in the Prudhoe Bay Field, AK," SPE 12473 (paper presented at the SPE Formation Damage Control Symposium, Bakersfield, CA, 13–14 February 1984).

11. T. A. T. Lund, H. I. Berge, S. Espedal, R. Kristensen, T. A. Rolfsvaag, and G. Stromsvik, "The Technical Performance and Interpretation of Results from a Large Scale Na-Silicate Gel Treatment of a Production Well on the Gullfaks Field," Proceedings of the 8th European Symposium on Increased Oil Recovery, Vol. 2, Vienna, 1995.
12. R. S. Seright and J. Liang, "A Survey of Field Applications of Gel Treatments for Water Shut-Off," SPE 26991 (paper presented at the SPE Permian Basin Oil and Gas Recovery Conference, Midland, TX, 16–18 March 1994).
13. J. J. Jurinak and L. E. Summers, "Oilfield Applications of Colloidal Silica Gel," *SPE Production Engineering* 6(4) (1991): 406.
14. M. A. Hardy, D. W. van Batenburg, and C. W. Botermans, "Use of Temperature Simulations in Water-Control Design," SPE 60896, *SPE Production & Facilities* 15(1) (2000): 14.
15. R. Boreng and O. B. Svendsen, "A Successful Water Shut Off: A Case Study from the Statfjord Field," SPE 37466 (paper presented at the SPE Production Operations Symposium, Oklahoma City, OK, 9–11 March 1997).
16. M. R. Islam and S. M. Farouq Ali, "Use of Silica Gel for Improving Waterflooding Performance of Bottom-Water Reservoirs," *Journal of Petroleum Science and Engineering* 8 (1993): 303.
17. B. Vinot, R. S. Schechter, and L. W. Lake, "Formation of Water-Soluble Silicate Gels by the Hydrolysis of a Diester of Dicarboxylic Acid Solubilized as Microemulsions," SPE 14236, *SPE Reservoir Engineering* 4(3) (1989): 291.
18. B. Vinot, G. Berrod, and J.-L. Brun, U.S. Patent 4799549, 1989.
19. T. Huang and P. M. McElfresh, U.S. Patent Application 20040031611, 2004.
20. W. H. Smith and E. F. Vinson, U.S. Patent 4640361, 1987. T. M. Vickers and L. J. Powers, U.S. Patent 4384894, 1983.
21. E. A. Elphingstone, H. C. McLaughlin, and C. W. Smith, U.S. Patent 4293440, 1981.
22. M. A. H. Laramay, U.S. Patent 5320171, 1994.
23. H. A. Nasr-El-Din and K. C. Taylor, *Journal of Petroleum Science and Engineering* 48 (2005): 141–160.
24. K. C. Taylor and H. A. Nasr-El-Din, U.S. Patent 6660694, 2003.
25. H. K. Kotlar, International Patent Application WO/2005/124099.
26. I. Lakatos, J. Lakatos-Szabo, B. Kosztin, Gy. Palasthy, P. Kristof, "Application of Iron-Hydroxide–Based Well Treatment Techniques at the Hungarian Oil Fields," SPE 59321 (paper presented at the SPE/DOE Improved Oil Recovery Symposium, Tulsa, OK, 3–5 April 2000).
27. B. Kosztin, Gy. Palasthy, F. Udvari, L. Benedek, I. Lakatos, and J. Lakatos-Szabo, "Field Evaluation of Iron Hydroxide Gel Treatments," SPE 78351 (paper presented at the SPE European Petroleum Conference, Aberdeen, UK, 29–31 October 2002).
28. R. D. Sydansk, U.S. Patent 4304301, 1981.
29. B. H. Tomlinson, U.S. Patent 7044222, 2006.
30. D. Barclay, K. Lawson, B. Mullins, and B. Cardno, "Stand Alone Coiled Tubing Water Shutoff Operations Reinstate Well on a Normally Unattended Installation," SPE 100132 (paper presented at the SPE/ICoTA Coiled Tubing and Well Intervention Conference and Exhibition, Woodlands, TX, 4–5 April 2006).
31. A. Parker and C. Davidson, U.S. Patent 4889563, 1989.
32. A. Parker, European Patent EP0266808, 1988.
33. S. M. Lahiliah, *Journal of Applied Polymer Science* 106 (2007): 2076.
34. Y. Wang, B. Bai, and H. Gao, "Enhanced Oil Production Through a Combined Application of Gel Treatment and Surfactant Huff'n'Puff Technology," SPE 112495 (paper presented at the SPE International Symposium and Exhibition on Formation Damage Control, Lafayette, LA, 13–15 February 2008).

35. C. Deolarte, J. Vasquez, E. Soriano, and A. Santillan, "Successful Combination of an Organically Crosslinked Polymer System and a Rigid-Setting Material for Conformance Control in Mexico," SPE 112411 (paper presented at the SPE International Symposium and Exhibition on Formation Damage Control, Lafayette, LA, 13–15 February 2008).
36. J. Vasquez, L. Eoff, D. Dalrymple, and J. van Eijden, "Shallow Penetration Particle-Gel System for Water and Gas Shutoff Applications," SPE 114885 (paper presented at the SPE Annual Technical Conference and Exhibition, Denver, CO, 21–24 September 2008).
37. A. Moradi-Araghi, "A Review of Thermally Stable Gels for Fluid Diversion in Petroleum Production," *Journal of Petroleum Science and Engineering* 26 (2000): 1.
38. (*a*) G. P. Karmakar and C. Chakraborty, "Improved Oil Recovery Using Polymeric Gelants: A Review," *Indian Journal of Chemical Technology* 13 (2006): 162. (*b*) J. K. Fink, *Oilfield Chemicals*, Gulf Professional, Elsevier, 2003.
39. C. Harrison, M. Luyster, L. Moore, B. B. Prasek, and R. D. Ravitz, International Patent Application WO/2009/026021.
40. S. M. Vargas-Vasquez and L. B. Romero-Zerón, "A Review of the Partly Hydrolyzed Polyacrylamide Cr(III) Acetate Polymer Gels," *Petroleum Science and Technology* 26 (2008): 481.
41. A. Stavland and H. C. Jonsbråten, "New Insight into Aluminium Citrate/Polyacrylamide Gels for Fluid Control," SPE 35381 (paper presented at the SPE/DOE 10th Symposium on Improved Oil Recovery, Tulsa, OK, 21–24 April 1996).
42. R. D. Sydansk and G. P. Southwell, "More Than 12 Years of Experience with a Successful Conformance-Control Polymer Gel Technology," SPE 49315 (paper presented at the SPE Annual Technical Conference and Exhibition, New Orleans, LA, 27–30 September 1998).
43. P. Shu, "Gelation Mechanism of Chromium(III), Oilfield Chemistry-Enhanced Recovery and Production Stimulation," in *ACS Symposium Series*, eds. J. K. Borchardt and T. F. Yen, Washington, DC: American Chemical Society, 1989, 137.
44. A. Sabhapondit and A. Borthakur, "A Comparative Study of Gelation Behaviour of Some Acrylamide Copolymers with Cr(III) Crosslinker," *Journal of Polymer Materials* 20(3) (2003): 309.
45. R. D. Sydansk, U.S. Patent 5421411, 1995.
46. R. D. Sydansk, "Acrylamide-Polymer/Chromium (III)-Carboxylate Gels for Near Wellbore Matrix Treatments," SPE 20214, *SPE Advanced Technology Series* 1(1) (1993): 146.
47. P. D. Moffitt, "Long-Term Production Results of Polymer Treatments in Producing Wells in Western Kansas," *Journal of Petroleum Technology* (1993): 356.
48. T. P. Lockhart and P. Albonico, "New Chemistry for the Placement of Chromium(III)/Polymer Gels in High-Temperature Reservoirs," SPE 24194, *SPE Production & Facilities* 9(4) (1994): 273.
49. M. Bartosek, A. Mennella, T. P. Lockhart, C. Emilio, R. Elio, and C. Passucci, "Polymer Gels for Conformance Treatments: Propagation of Cr(III) Crosslinking Complexes in Porous Media," SPE 27828 (paper presented at the SPE/DOE Improved Oil Recovery Symposium, Tulsa, OK, 17–20 April 1994).
50. J. P. Feraud, S. Karlstad, and L. Aasberg, Water Control through Placement of Polyacrylamide Gel in Propped Fracture in the Gullfaks Field," Paper 23 (paper presented at the 8th International Oil Field Chemical Symposium, Geilo, Norway, 2–5 March 1997).
51. R. D. Sydansk, U.S. Patent 6189615, 2001.
52. R. D. Sydansk, A. M. Al-Dhafeeri, Y. Xiong, and R. S. Seright, "Polymer Gels Formulated with a Combination of High- and Low-Molecular-Weight Polymers Provide Improved Performance for Water-Shutoff Treatments of Fractured Production Wells," SPE 89402, *SPE Production & Facilities* 19(4) (2004): 229.

53. P. D. Moffitt, A. Moradi-Araghi, I. Ahmed, and V. R. Janway, "Development and Field Testing of a New Low Toxicity Polymer Crosslinking System," SPE 35173 (paper presented at the SPE Permian Basin Oil and Gas Recovery Conference, Midland, TX, 27–29 March 1996).
54. G. Chauveteau, R. Tabary, M. Renard, and A. Omari, "Controlling In-Situ Gelation of Polyacrylamides by Zirconium for Water Shutoff," SPE 50752 (paper presented at the SPE International Symposium on Oilfield Chemistry, Houston, TX, 16–19 February 1999).
55. C. Kayser, G. Botthof, K. H. Heier, A. Tardi, M. Krull, and M. Schaefer, U.S. Patent Application 20060019835, 2006.
56. M. Zettlitzer, W. Schuhbauer, and N. Kohler, "Laboratory Evaluation of a New, Selective Water Control Treatment and Its Implementation in a North Sea Well," *Petroleum Geosciences* 2(4) (1996): 325.
57. S. S. Nagra, J. P. Batycky, R. E. Nieman, and J. B. Bodeux, "Stability of Waterflood Diverting Agents at Elevated Temperatures in Reservoir Brines," SPE 15548 (paper presented at the SPE Annual Technical Conference and Exhibition, New Orleans, LA, 5–8 October 1998).
58. B. Kalpakci, Y. T. Jeans, N. F. Magri, and J. P. Padolewski, "Thermal Stability of Scleroglucan at Realistic Reservoir Conditions," SPE 20237 (paper presented at the SPE/DOE Enhanced Oil Recovery Symposium, Tulsa, OK, 22–25 April 1990).
59. M. R. Avery, L. A. Burkholder, and M. A. Gruenenfelder, SPE 14114 (paper presented at the SPE International Meeting on Petroleum Engineering, Beijing, China, 17–20 March 1986).
60. E. T. Strom, J. M. Paul, C. H. Phelps, and K. Sampath, "A New Biopolymer for High-Temperature Profile Control: Part 1—Laboratory Testing," SPE 1963, *SPE Reservoir Engineering* 6(3) (1991): 360.
61. K. Sampath, L. G. Jones, E. T. Strom, C. H. Phelps, and C. S. Chiou, "A New Biopolymer for High-Temperature Profile Control: Part II-Field Results," SPE 19867 (paper presented at the SPE Annual Technical Conference and Exhibition, San Antonio, TX, 8–11 October 1989).
62. N. Kohler, A. Zaitoun, U.S. Patent 5082577, 1992.
63. B. B. Sandiford, H. T. Dovan, and R. D. Hutchins, U.S. Patent 5486312, 1996.
64. R. D. Hutchins, H. T. Dovan, and B. B. Sandiford, "Field Applications of High Temperature Organic Gels for Water Control," SPE 35444 (paper presented at the SPE/DOE Improved Oil Recovery Symposium, Tulsa, OK, 21–24 April 1996).
65. R. G. Ryles, "Chemical Stability Limits of Water-Soluble Polymers Used in Oil Recovery Processes," SPE 13585, *SPE Reservoir Engineering*, February (1988): 23.
66. A. Moradi-Aragahi and P. H. Doe, "Hydrolysis and Precipitation of Polyacrylamides in Hard Brines at Elevated Temperatures," *SPE Reservoir Engineering*, May (1987): 189.
67. P. M. DiGiacomo and C. M. Schramm, "Mechanism of Polyacrylamide Gel Syneresis Determined by C-13 NMR," SPE 11787 (paper presented at the SPE Oilfield and Geothermal Chemistry Symposium, Denver, CO, 1–3 June 1983).
68. D. O. Falk, U.S. Patent 4485875, 1985.
69. B. L. Swanson, 4440228, 1984.
70. R. Banerjee, B. Ghosh, and K. C. Khilar, *Oil Gas-European Magazine* 32(4) (2006): 184.
71. A. Moradi-Araghi, G. Bjornson, and P. H. Doe, "Thermally Stable Gels for Near-Wellbore Permeability Contrast Modifications," SPE 18500, *Advanced Technical Series* 1(1) (1993): 140.
72. R. Hutchins and M. Parris, "New Crosslinked Gel Technology," (paper presented at the PNEC 4th International Conference on Reservoir Conformance, Profile Control, Water and Gas Shut-Off, Houston, TX, 10–12 August 1998).

73. R. L. Clampitt, H. M. Al-Rikabi, and M. K. Dabbous, "A Hostile Environment Gelled Polymer System for Well Treatment and Profile Control," SPE 25629 (paper presented at the SPE Middle East Oil Show, Bahrain, 3–6 April 1993).
74. A. Moradi-Araghi, U.S. Patent 4994194, 1991.
75. A. Moradi-Araghi, U.S. Patent 5179136, 1993.
76. M. D. Parris and R. D. Hutchins, U.S. Patent 6011075, 2000.
77. H. T. Dovan and R. D. Hutchins, "Delaying Gelation of Aqueous Polymers at Elevated Temperatures Using Novel Organic Crosslinkers," SPE 37246 (paper presented at the SPE International Symposium on Oilfield Chemistry, Houston, TX, 18–21 February 1997).
78. P. W. Chang, I. M. Goldman, and K. J. Stingley, "Laboratory Studies and Field Evaluation of a New Gelant for High-Temperature Profile Modification," SPE 14235 (paper presented at the SPE Annual Technical Conference and Exhibition, Las Vegas, NV, 22–26 September 1985).
79. G. A. Al-Muntasheri, I. A. Hussein, H. A. Nasr-El-Din, *Journal of Petroleum Science and Engineering* 55 (2007): 56.
80. J. C. Morgan, P. L. Smith, and D. G. Stevens, "Chemical Adaptation and Deployment Strategies for Water and Gas Shut-off Gel Systems, RSC 6th International Symposium on Chemistry in the Oil Industry, Ambleside, UK, 14–17 April 1997).
81. G. A. Al-Muntasheri, H. A. Nasr-El-Din, K. R. Al-Noaimi, and P. L. J. Zitha, "A Study of Polyacrylamide-Based Gels Crosslinked with Polyethyleneimine," SPE 105925 (paper presented at the SPE International Symposium on Oilfield Chemistry, Houston, TX, 28 February–2 March 2007).
82. G. A. Al-Muntasheri, P. L. J. Zitha, and H. A. Nasr-El-Din, "Evaluation of a New Cost-Effective Organic Gel System for High Temperature Water Control," IPTC-11080 (paper presented at the International Petroleum Technology Conference, Dubai, UAE, 4–6 December 2007).
83. J. Vasquez, E. D. Dalrymple, L. Eoff, B. R. Reddy, and F. Civan, "Development and Evaluation of High Temperature Conformance Polymer Systems," SPE 93156 (paper presented at the SPE International Symposium on Oilfield Chemistry, Houston, TX, 2–4 February 2005).
84. M. Hardy, W. Botermans, A. Hamouda, J. Valdal, and J. Warren, "The First Carbonate Field Application of a New Organically Crosslinked Water Shutoff Polymer System," SPER 50738 (paper presented at the SPE International Symposium on Oilfield Chemistry, Houston, TX, 16–19 February 1999).
85. M. B. Hardy, C. W. Botermans, and P. Smith, "New Organically Cross-Linked Polymer System Provides Competent Propagation at High Temperatures in Conformance Treatments," SPE 39690 (paper presented at the SPE/DOE Symposium on Improved Oil Recovery, Tulsa, OK, 1998).
86. B. R. Reddy, L. Eoff, E. D. Dalrymple, K. Black, D. Brown, and M. Rietjens, "A Natural Polymer-Based Cross-Linker System for Conformance Gel Systems," SPE 84937, *SPE Journal*, June (2003): 99.
87. B. R. Reddy, L. S. Eoff, and E. D. Dalrymple, U.S. Patent 7007752, 2006.
88. S. M. Lahalih and E. F. Ghloum, "Rheological Properties of New Polymer Compositions for Sand Consolidation and Water Shutoff in Oil Wells," *Journal of Applied Polymer Science* 104 (2007): 2076.
89. S. M. Skjæveland, A. Skauge, L. Hinderaker, and C. D. Sisk eds., *RUTH Program Summary*, Stavanger, Norway: NDP (Norwegian Petroleum Directorate), 1996 (http://www.npd.no/engelsk/projects/ruth/ruth.htm).
90. G. Di Lullo and P. Rae, "New Insights into Water Control—A Review of the State of the Art," SPE 77963 (paper presented at the SPE Asia Pacific Oil and Gas Conference and Exhibition, Melbourne, Australia, 8–10 October 2002).

91. M. Mahajan, N. Rauf, T. Gilmore, and A. Maylana, "Water Control and Fracturing: A Reality," SPE 101019 (paper presented at the SPE Asia Pacific Oil & Gas Conference and Exhibition, Adelaide, Australia, 11–13 September 2006).
92. D. H. Hoskin and P. Shu, U.S. Patent 4896723, 1990.
93. P. Shu, U.S. Patent 4964463, 1990.
94. C. Victorius, U.S. Patent 5061387, 1991.
95. M. L. Marrocco, U.S. Patent 4664194, 1987.
96. M. L. Marrocco, U.S. Patent 4498540, 1985.
97. P. Shu, U.S. Patent 4678032, 1987.
98. M. R. Avery, T. A. Wells, P. W. Chang, and J. P. Millican, "Field Evaluation of a New Gelant for Water Control in Production Wells," SPE 18201 (paper presented at the SPE Annual Technical Conference and Exhibition, Houston, TX, 2–5 October 1988).
99. E. D. Dalrymple, L. S. Eoff, B. R. Reddy, D. L. Brown, and P. S. Brown, U.S. Patent 6364016, 2002.
100. S. N. Davies, T. G. J. Jones, S. Olthoff, and G. J. Tustin, U.S. Patent 6920928, 2005.
101. R. D. Sydansk, U.S. Patent 5834406, 1998.
102. P. D. Nguyen; L. Sierra, E. D. Dalrymple, and L. S. Eoff, U.S. Patent Application 20070012445, 2007.
103. K. A. Rodrigues, U.S. Patent 5335726, 1994.
104. K. A. Rodrigues, U.S. Patent 5358051, 1994.
105. H. Lane, "Design, Placement and Field Quality Control of Polymer Gel Water & Gas Shut-Off Treatments," PNEC 97#3 (paper presented at the 3rd International Conference on Reservoir Conformance, Profile Control, Water and Gas Shut-Off, Houston, TX, 6–8 August 1997).
106. D. Dalrymple, J. T. Tarkington, and H. James, "A Gelation System for Conformance Technology," SPE 28503 (paper presented at the SPE Annual Technical Conference and Exhibition, New Orleans, LA, 25–28 September 1994).
107. P. Woods, K. Schramko, D. Turner, D. Dalrymple, and E. Vinson, "In-Situ Polymerization Controls CO_2/Water Channelling at Lick Creek," SPE 14958 (paper presented at the SPE/DOE Fifth Symposium on Enhanced Oil Recovery, Tulsa, OK, 20–23 April 1986).
108. M.-C. P. Leblanc, J. A. Durrieu, J.-P. P. Binon, G. G. Provin, and J.-J. Fery, U.S. Patent 4975483, 1990.
109. T. Huang, J. B. Crews, and J. R. Willingham, U.S. Patent Application 20080248978.
110. G. Di Lullo, A. Ahmad, P. Rae, L. Anaya, and R. Ariel Meli, "Toward Zero Damage: New Fluid Points the Way," SPE 69453 (paper presented at the SPE Latin American and Caribbean Petroleum Engineering Conference, Buenos Aires, Argentina, 25–28 March 2001).
111. G. F. DiLullo, P. J. Rae, and A. J. K. Ahmad, U.S. Patent 6767869, 2004.
112. G. P. Karmakar, C. A. Grattoni, and R. W. Zimmerman, "Relative Permeability Modification Using an Oil-Soluble Gelant to Control Water Production," SPE 77414 (paper presented at the SPE Annual Technical Conference and Exhibition, San Antonio, TX, 29 September–2 October 2002).
113. L. J. Kalfayan and J. C. Dawson, British Patent GB 2399364.
114. J. C. Dawson, L. J. Kalfayan, P. H. Javora, M. Vorderbruggen, J. W. Kirk, and Q. Qu, U.S. Patent Application 20060065396, 2006.
115. A. Stavland and S. Nilsson, International Patent Application WO/2001/021726.
116. A. Stavland, K. I. Andersen, B. Sandøy, T. Tjomsland, and A. A. Mebratu, "How To Apply a Blocking Gel System for Bullhead Selective Water Shutoff: From Laboratory to Field," SPE 99729 (paper presented at the SPE/DOE Symposium on Improved Oil Recovery, Tulsa, OK, 22–26 April 2006).

117. S. Nilsson, A. Stavland, and H. C. Jonsbraten, "Mechanistic Study of Disproportionate Permeability Reduction," SPE 39635 (paper presented at the SPE/DOE Improved Oil Recovery Symposium, Tulsa, OK, 19–22 April 1998).
118. D. Dalrymple, L. Eoff, B. R. Reddy, and C. W. Botermans, "Relative Permeability Modifiers for Improved Oil Recovery: A Literature Review" (paper presented at the PNEC 5th International Conference on Reservoir Conformance, Profile Control, Water and Gas Shut-Off, Houston, TX, 8–10 November 1999).
119. J. T. Liang, R. L. Lee, and R. S. Seright, "Gel Placement in Production Wells," SPE 20211, *SPE Production & Facilities* 8 (1993): 276.
120. A. Zaitoun, N. Kohler, D. Bossie-Codreanu, and K. Denys, "Water Shutoff by Relative Permeability Modifiers: Lessons from Several Field Applications," SPE 56740 (paper presented at the SPE Annual Technical Conference and Exhibition, 3–6 October, Houston, TX, 1999).
121. J. Novotny, "Matrix Flow Evaluation Technique for Water Control Applications," SPE 30094 (paper presented at the SPE European Formation Damage Conference, The Hague, Netherlands, 15–16 May 1995).
122. D. J. Ligthelm, "Water Shut Off in Gas Wells: Is There Scope for a Chemical Treatment?," SPE 68978 (paper presented at the SPE European Formation Damage Conference, The Hague, Netherlands, 21–22 May 2001).
123. K. Munday, U.S. Patent 6516885, 2003.
124. A. Stavland and S. Nilsson, "Segregated Flow Is the Governing Mechanism of Disproportionate Permeability Reduction in Water and Gas Shutoff," SPE 71510 (paper presented at the SPE Annual Technical Conference and Exhibition, New Orleans, LA, 30 September–3 October 2001).
125. C. W. Botermans, D. W. van Batenburg, J. Bruining, "Relative Permeability Modifiers: Myth or Reality?," SPE 68973 (paper presented at the SPE European Formation Damage Conference, The Hague, Netherlands, 21–22 May 2001).
126. R. D. Sydansk and R. S. Seright, "When and Where Relative Permeability Modification Water-Shutoff Treatments Can Be Successfully Applied," SPE 99371 (paper presented at the SPE/DOE Symposium on Improved Oil Recovery, Tulsa, OK, 22–26 April 2006).
127. D. Dalrymple, U.S. Patent 5146986, 1992.
128. A. Zaitoun, N. Kohler, U.S. Patent 4842071, 1989.
129. C. Tielong, Z. Yong, P. Kezong, and P. Wanfeng, "A Relative Permeability Modifier for Water Control of Gas Wells in a Low-Permeability Reservoir," *SPE Reservoir Engineering* 11(3) (1996): 168.
130. N. Kholer, R. Tabary, A. Zaitoun U.S. Patent 4718491, 1988.
131. A. Zaitoun, N. Kohler, B. K. Maitin, and M. Zettlitzer, "Preparation of a Water Control Polymer Treatment at Conditions of High Temperature and Salinity," *Journal of Petroleum Science and Engineering* 7 (1992): 67.
132. K. Denys, C. Fichen, A. Zaitoun, "Bridging Adsorption Of Cationic Polyacrylamides In Porous Media," SPE 64984 (paper presented at the SPE International Symposium on Oilfield Chemistry, Houston, TX, 13–16 February 2001).
133. L. Chiappa, M. Andrei, T. P. Lockhart, G. Burrafato, and G. Maddinelli, International Patent Application WO/2002/097236.
134. P. D. Nguyen, S. Ingram, L. Sierra, E. D. Dalrymple, and L. S. Eoff, U.S. Patent Application 20070029087, 2007.
135. L. Chiappa, M. Andrei, T. P. Lockhart, G. Burrafato, and G. Maddinelli, U.S. Patent 7188673, 2007.
136. J. L. Boles and G. Mancillas, U.S. Patent 4476931, 1984.

137. F. O. Stanley, M. E. Hardianto, and P. S. Tanggu, "Improving Hydrocarbon/Water Ratios in Producing Wells—An Indonesian Case History Study," SPE 36615 (paper presented at the SPE Annual Technical Conference and Exhibition, Denver, CO, 6–9 October 1996).
138. D. D. Dunlap, J. L. Boles, and R. J. Novotny, "Method for Improving Hydrocarbon/Water Ratios in Producing Wells," SPE 14822 (paper presented at the SPE Formation Damage Control Symposium, Lafayette, LA, 26–27 February 1986).
139. G. Push, "Practical Experience with Water Control in Gas Wells by Polymer Treatments," *EAPG Improved Oil Recovery European Symposium, Vienna, Austria* 2 (1995): 48–56.
140. M. Ranjbar, P. Czolbe, N. Kohler, "Comparative Laboratory Selection and Field Testing of Polymers for Selective Control of Water Production in Gas Wells," SPE 28984 (paper presented at the SPE International Symposium on Oilfield Chemistry, San Antonio, TX, 14–17 February 1995).
141. D. Coehlo, "Development and Application of Selective Polymer Injection to Control Water," (paper presented at the 5th Latin American and Caribbean Petroleum Engineering Conference and Exhibition, Rio de Janeiro, Brazil, 30 August–3 September 1997).
142. J. C. Dawson, H. V. Le, and S. Kesavan, U.S. Patent 6228812, 2001.
143. L. Eoff, E. D. Dalrymple, B. R. Reddy, and D. Everett, "Structure and Process Optimization for the Use of a Polymeric Relative-Permeability Modifier in Conformance Control," SPE 64985 (paper presented at the SPE International Symposium on Oilfield Chemistry, Houston, TX, 13–16 February 2001).
144. E. D. Dalrymple, L. S. Eoff, B. R. Reddy, D. L. Brown, P. S. Brown, U.S. Patent 6364016, 2002.
145. S. G. Nelson, L. J. Kalfayan, and W. M. Rittenberry, "The Application of a New and Unique Relative Permeability Modifier in Selectively Reducing Water Production," SPE 84511 (paper presented at the SPE Annual Technical Conference and Exhibition, Denver, CO, 5–8 October 2003).
146. P. Barriau, H. Bertin, A. Zaitoun, "Water Control in Producing Wells: Influence of Adsorbed Polymer Layer on Relative Permeabilities and Capillary Pressure," SPE 35447 (paper presented at the SPE/DOE Symposium on Improved Oil Recovery, Tulsa, OK, 21–24 April 1996).
147. A. Zaitoun and N. Kohler, "Improved Polyacrylamide Treatments for Water Control in Producing Wells," SPE 18501 (paper presented at the SPE International Symposium on Oilfield Chemistry, Houston, TX, 8–10 February 1989).
148. D. Dalrymple and E. Vinson, U.S. Patent 4617132, 1986.
149. L. Eoff, E. D. Dalrymple, B. R. Reddy, and D. Everett, "Structure and Process Optimization for the Use of a Polymeric Relative-Permeability Modifier in Conformance Control," SPE 64985 (paper presented at the SPE International Symposium on Oilfield Chemistry, Houston, TX, 13–16 February 2001).
150. J. C. Dawson, L. J. Kalfayan, P. H. Javora, M. Vorderbruggen, J. W. Kirk, and Q. Qu, U.S. Patent Application 20060065396, 2006.
151. R. Castano, J. Villamizar, O. Diaz, S. A. Hocol, M. Avila, S. Gonzalez, E. D. Dalrymple, S. Milson, and D. Everett, "Relative Permeability Modifier and Scale Inhibitor Combination in Fracturing Process at San Francisco Field in Colombia, South America," SPE 77412 (paper presented at the SPE Annual Technical Conference and Exhibition, San Antonio, TX, 29 September–2 October 2002).
152. P. Powell, M. A. Singleton, and K. S. Sorbie, US 6913081, 2005.
153. (*a*) E. D. Dalrymple, L. Eoff, B. R. Reddy, and J. Venditto, U.S. Patent 7182136, 2007. (*b*) P. D. Nguyen, L. Sierra, E. D. Dalrymple, and L. S. Eoff, U.S. Patent Application 20070012445, 2007.

154. M. Cordova, J. L. Mogollon, H. Molero, and M. Navas, "Sorbed Polyacrylamides: Selective Permeability Parameters Using Surface Techniques," SPE 75210 (paper presented at the SPE/DOE Improved Oil Recovery Symposium, Tulsa, OK, 13–17 April 2002).
155. E. Volpert, J. Selb, F. Candau, N. Green, J. F. Argillier, and A. Audibert, *Langmuir* 14 (1998): 1870.
156. E. D. Dalrymple, L. Eoff, B. R. Reddy, and J. Venditto, U.S. Patent 7182136, 2007.
157. L. S. Eoff, B. R. Reddy, and E. D. Dalrymple, U.S. Patent 6476169, 2002.
158. L. S. Eoff, E. D. Dalrympel, B.R Reddy, and F. Zamora, U.S. Patent 7114568.
159. T. G. J. Jones and G. J. Tustin, U.S. Patent 6194356, 2001.
160. J. C. Dawson, L. J. Kalfayan, and G. Brock, U.S. Patent Application 20040177957.
161. M. J. Faber, G. J. P. Joosten, K. A. Hashmi, and M. Gruenenfelder, "Water Shut-off Field Experience with a Relative Permeability Modification System in the Marmul Field (Oman)," SPE 39633 (paper presented at the SPE/DOE Improved Oil Recovery Symposium, Tulsa, OK, 19–22 April 1998).
162. N. Kohler and A. Zaitoun, U.S. Patent 5082577, 1992.
163. K. H. Heier, C. Kayser, A. Tardi, M. Schaefer, R. Morschhaeuser, J. C. Morgan, and A. M. Gunn, U.S. Patent 7150319, 2006.
164. J. J. Wylde, G. D. Williams, and C. Shields, "Field Experiences in Application of a Novel Relative Permeability Modifier Gel in North Sea Operations," SPE 97643 (paper presented at the SPE International Improved Oil Recovery Conference in Asia Pacific, Kuala Lumpur, Malaysia, 5–6 December 2005).
165. J. Morgan, A. Gunn, G. Fitch, H. Frampton, R. Harvey, D. Thrasher, R. Lane, R. McClure, K. H. Heier, and C. Kayser, "Development and Deployment of a 'Bullheadable' Chemical System for Selective Water Shut Off Leaving Oil/Gas Production Unharmed," SPE 78540 (paper presented at the Abu Dhabi International Petroleum Exhibition and Conference, Abu Dhabi, UAE, 13–16 October 2002).
166. Y. Feng, R. Tabary, M. Renard, C. Le Bon, A. Omari, G. Chauveteau, "Characteristics of Microgels Designed for Water Shutoff and Profile Control," SPE 80203 (paper presented at the SPE International Symposium on Oilfield Chemistry, Houston, TX, 5–7 February 2003).
167. G. Chauveteau, R. Tabary, N. Blin, M. Renard, D. Rousseau, R. Faber, "Disproportionate Permeability Reduction by Soft Preformed Microgels," SPE 89390 (paper presented at the SPE/DOE Symposium on Improved Oil Recovery, Tulsa, OK, 17–21 April 2004).
168. D. Rousseau, G. Chauveteau, M. Renard, R. Tabary, A. Zaitoun, P. Mallo, O. Braun, and A. Omari, "Rheology and Transport in Porous Media of New Water Shutoff/Conformance Control Microgels," SPE 93254 (paper presented at the SPE International Symposium on Oilfield Chemistry, The Woodlands, TX, 2–4 February 2005).
169. C. Cozic, D. Rousseau, and R. Tabary, "Broadening the Application Range of Water Shutoff/Conformance Control Microgels: An Investigation of Their Chemical Robustness," SPE 115974 (paper presented at the SPE Annual Technical Conference and Exhibition, Denver, CO, 21–24 September 2008).
170. A. Zaitoun, R. Tabary, D. Rousseau, T. Pichery, S. Nouyoux, P. Mallo, and O. Braun, "Using Microgels to Shut Off Water in a Gas Storage Well," SPE 106042 (paper presented at the SPE International Symposium on Oilfield Chemistry, Houston, TX, 28 February–2 March 2007).
171. C. H. Phelps, E. T. Strom, and M. L. Hoefner, U.S. Patent 5048607, 1991.
172. H. K. Kotlar, B. Schilling, and J. Sjoblom, U.S. Patent 7270184, 2007.
173. J. C. Dawson, H. V. Le, and S. Kesavan, U.S. Patent 5735349, 1998.
174. P. D. Nguyen and B. T. Dewprashad, U.S. Patent 6109350, 2000.

175. N. Fleming, K. Ramstad, A.-M. Mathisen, Alex Nelson, and S. Kidd, "Innovative Use of Kaolinite in Downhole Scale Management: Squeeze Life Enhancement and Water Shutoff," SPE 113656 (paper presented at the SPE 9th International Conference on Oilfield Scale, Aberdeen, UK, 28–29 May 2008).
176. J. F. Hurtado, A. Milne, and E. Olivares, "Shallow Gas Shut-Off Using Rigid Polymer Gel in Lake Maracaibo, Venezuela," SPE 94532 (paper presented at the SPE Latin American and Caribbean Petroleum Engineering Conference, Rio de Janeiro, Brazil, 20–23 June 2005).
177. M. A. Llamedo, F. V. Mejias, E. R. González, J. Espinoza, E. M. Valero, and N. Calis, SPE 96696 (paper presented at the Offshore Europe, Aberdeen, UK, 6–9 September 2005).
178. L. Perdomo, H. Rodríguez, M. Llamedo, L. Oliveros, E. González, O. Molina, and C. Giovingo, "Successful Experiences for Water and Gas Shutoff Treatments in North Monagas, Venezuela," SPE 106564 (paper presented at the Latin American & Caribbean Petroleum Engineering Conference, Buenos Aires, Argentina, 15–18 April 2007).
179. G. D. Herring, J. T. Milloway, and W. N. Wilson, "Selective Gas Shut-Off Using Sodium Silicate in the Prudhoe Bay Field, AK," SPE 12473 (paper presented at the SPE Formation Damage Control Symposium, Bakersfield, CA, 13–14 February 1984).
180. G. Burrafato, E. Pitoni, G. Vietina, L. Mauri, and L. Chiappa, "Rigless WSO Treatments in Gas Fields. Bullheading Gels and Polymers in Shaly Sand: Italian Case Histories," SPE 54747 (paper presented at the SPE European Formation Damage Conference, The Hague, Netherlands, 31 May–1 June 1999).
181. E. Ali, F. E. Bergren, P. DeMestre, E. Biezen, and J. van Eijden, "Effective Gas Shutoff Treatments in a Fractured Carbonate Field in Oman," 102244 (paper presented at the SPE Annual Technical Conference and Exhibition, San Antonio, TX, 24–27 September 2006).
182. J. E. Hanssen, S. Ekrann, U.S. Patent 4903771, 1990.
183. J. E. Hanssen and M. Dalland, "Foams for Effective Gas Blockage in the Presence of Crude Oil," SPE 20193 (paper presented at the SPE/DOE Enhanced Oil Recovery Symposium, Tulsa, OK, 22–25 April 1990).
184. M. N. Bouts and M. Dalland, "Foam Treatments Against Unwanted Gas Production in Oil Producers," *Chemistry in the Oil Industry VI*, Manchester, UK: Royal Society of Chemistry, 1998, 132.
185. V. O. Chukwueke, M. N. Bouts, and C. E. van Dijkum, "Gas Shut-Off Foam Treatments," SPE 39650 (paper presented at the SPE/DOE Improved Oil Recovery Symposium, Tulsa, OK, 19–22 April 1998).
186. M. N. Bouts, A. S. de Vries, M. Dalland, and J. E. Hanssen, "Design of Near Well Bore Foam Treatments for High GOR Producers," SPE 35399 (paper presented at the SPE/DOE Improved Oil Recovery Symposium, Tulsa, OK, 21–24 April 1996).
187. M. Dalland, J. E. Hanssen, and T. S. Kristiansen, "Oil Interaction with Foams under Static and Flowing Conditions in Porous Media," *Colloids and Surfaces A: Physicochemical and Engineering Aspects* 82 (1994): 129.
188. S. Thach, K. C. Miller, Q. J. Lai, G. S. Sanders, J. W. Styler, and R. H. Lane, "Matrix Gas Shut-Off in Hydraulically Fractured Wells Using Polymer Foams," SPE 36616 (paper presented at the SPE Eastern Regional Meeting, Columbus, OH, 23–25 October 1996).
189. J. Van Houwelingen, "Chemical Gas Shut-Off Treatments in Brunei," SPR 57268 (paper presented at the SPE Asia Pacific Improved Oil Recovery Conference, Kuala Lumpur, Malaysia, 25–26 October 1999).
190. R. D. Sydansk, "Polymer-Enhanced Foams: Laboratory Development and Evaluation," SPE 25168 (paper presented at the SPE International Symposium on Oilfield Chemistry, New Orleans, LA, 1993).

191. J. E. Hanssen and M. Dalland, "Increased Oil Tolerance of Polymer-Enhanced Foams: Deep Chemistry or Just 'Simple' Displacement Effects?" SPE 59282 (paper presented at the SPE/DOE Improved Oil Recovery Symposium, Tulsa, OK, 3–5 April 2000).
192. A. M. Al-Dhafeeri, H. A. Nasr-El-Din, H. K. Al-Mubarak, and J. Al-Ghamdi, "Gas Shut-Off Treatment in Oil Carbonate Reservoirs in Saudi Arabia," SPE 114323 (paper presented at the SPE Annual Technical Conference and Exhibition held in Denver, CO, 21–24 September 2008).
193. B. P. Price and B. Gothard, "Foam Assisted Lift—Importance of Selection and Application," SPE 106465 (paper presented at the Production and Operations Symposium, Oklahoma City, OK, 31 March–3 April 2007).
194. S. Ramachandran, J. Bigler, and D. Orta, "Surfactant Dewatering of Production and Gas Storage Wells," SPE 84823 (paper presented at the SPE Eastern Regional Meeting, Pittsburgh, PA, 6–10 September 2003).
195. M. J. Willis, D. Horsup, and D. Nguyen, "Chemical Foamers for Gas Well Deliquification," SPE 115633 (paper presented at the SPE Asia Pacific Oil and Gas Conference and Exhibition, Perth, Australia, 20–22 October 2008).
196. S. Ramachandran, J. Collins, C. Gamble, P. Schorling, J. G. R. Eylander, and C. Wittfield, Proceedings of the Chemistry in the Oil Industry IX, Royal Society of Chemistry, UK, 31 October–2 November 2005, 205.
197. W. Jelinek and L. L. Schramm, "Improved Production From Mature Gas Wells by Introducing Surfactants into Wells," SPE 11028 (paper presented at the International Petroleum Technology Conference, Doha, Qatar, 21–23 November 2005).
198. S. Campbell, S. Ramachandran, and K. Bartrip, "Corrosion Inhibition/Foamer Combination Treatment to Enhance Gas Production," SPE 67325 (paper presented at the SPE Production and Operations Symposium, Oklahoma City, OK, 24–27 March 2001).
199. S. T. Davis and V. Panchalingam, U.S. Patent Application 20060128990.
200. J. Yang, V. Jovancicevic, and S. Ramachandran, *Colloids and Surfaces A: Physicochemical and Engineering Aspects* 309 (2007): 177.
201. D. Orta, S. Ramanchandran, J. Yang, M. Fosdick, T. Salma, J. Long, J. Blanchard, A. Allcorn, C. Atkins, and O. Salinas, "A Novel Foamer for Deliquification of Condensate-Loaded Wells," SPE 107980 (paper presented at the Rocky Mountain Oil & Gas Technology Symposium, Denver, CO, 16–18 April 2007).

3 Scale Control

3.1 INTRODUCTION

Scale formation is the deposition of sparingly soluble inorganic salts from aqueous solutions.[1] There is another type of scale containing metal ions in which the anions are organic carboxylates or naphthenates. This is discussed in Chapter 7. Scale can deposit on almost any surface so that once a scale layer is first formed it will continue to get thicker unless treated (Figure 3.1).[2] Scales can block pore throats in the near–well bore region or in the well itself causing formation damage and loss of well productivity. They can deposit on equipment in the well, such as electric submersible pumps or sliding sleeves, causing them to malfunction. Scale can occur anywhere along the production conduit narrowing the internal diameter and blocking flow and, finally, scale can form in the processing facilities. Next to corrosion and gas hydrates, scale is probably one of the three biggest water-related production problems and needs to be anticipated in advance to determine the best treatment strategy. For some fields, scale control can be the single biggest operational cost.[3]

3.2 TYPES OF SCALE

The most common scales encountered in the oil industry in order of prevalence are:

- calcium carbonate (calcite and aragonite)
- sulfate salts of calcium (gypsum), strontium (celestite), and barium (barite)—radium may also be found in the lattice, especially that of barium sulfate
- sulfide scales—iron (II), zinc and lead (II) salts are the most common
- sodium chloride (halite)

Most minerals are less soluble as the temperature decreases (although calcium carbonate is an exception as discussed later). Thus, long-distance, cold water tie-backs can give enhanced pipeline scale problems as the produced fluids cool to seabed temperatures. Further, the deposition of all these inorganic scales is exacerbated by the presence of organic thermodynamic hydrate inhibitors (THIs) such as methanol or small ethylene glycols, methanol being worst.[4] As more and more fields are being developed in deeper and/or colder water, the use of THIs will increase to combat hydrate formation, and with this, the number and severity of the scale problems will also increase. The challenges facing scale control in deep water fields has been reviewed.[5]

Other more exotic scales are sometimes encountered in the oilfield. They include iron carbonate (siderite, mainly from corrosion), calcium fluoride (fluorspar, as a by-product of HF acidizing), silicate salts[1,6] and trona ($Na_3H(CO_3)_2 \times 2H_2O$).[7]

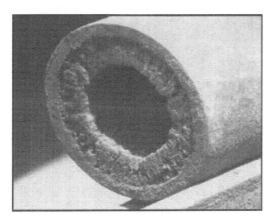

FIGURE 3.1   Scale deposits in a conduit.

Silica/silicate scaling is sometimes encountered in steam flood operations.[8] Methods that may be used to control silica/silicate scaling in steam flood operations include dilution with fresh water, reducing the pH of the water, treating the water with reducing, complexing, and sequestering agents, removing silica from water by lime softening, precipitation of silica in water with metals or cationic surfactants, and, lastly, treating the water with geothermal silica scale inhibitors/dispersants.[1] Radioactive lead and polonium metals and salt scales have been known to form in some instances, mostly in gas wells.

3.2.1   Calcium Carbonate Scale

Calcium bicarbonate ($CaHCO_3$) is very water-soluble, but calcium carbonate ($CaCO_3$) is not. Formation water usually contains bicarbonate ions as well as calcium ions. Calcium carbonate can deposit as a consequence the following equilibrium moving to the right:

$$2HCO_3^- \leftrightarrow CO_3^{2-} + H_2O + CO_2 (g)$$

Thus, if the pressure drops, by Le Chatelier's Principle, the above equilibrium will move to the right to try to increase the pressure by forming more CO_2 gas. As a result, more carbonate ions are formed and the pH rises.[9] At some point, the concentration of carbonate ions may be high enough that calcium carbonate precipitates:

$$Ca^{2+} + CO_3^{2-} \rightarrow CaCO_3 (s)$$

The critical drop in pressure may occur anywhere in the production system, for example, across the perforations, at a choke or anywhere in the production tubing, downhole, or topside. Often, the CO_2 content of the aqueous fluids in the well is high enough and, thus, the pH low enough, that calcium carbonate will not deposit in the well but rather further downstream, beyond the choke at the wellhead, where the pressure has dropped sufficiently. Further, calcium carbonate scaling may occur only after several years in the life of a field, as it is only then that the pressure drops in the production line to a level where carbonate scales will form. As the pressure drops further scale formation will move further upstream (i.e., further into the producing well). The thermodynamics of calcium carbonate scaling can be predicted through various commercial computer programs. Kinetic models have also been developed to predict where and when calcium carbonate scaling occurs.[10]

Formation damage due to the reprecipitation of carbonate scales has been known to occur from spent acid solutions after matrix stimulation treatments. Chemicals that dissolve and chelate calcium carbonate can break this reprecipitation cycle.[11]

In brines with high iron(II) ions, it is possible to obtain iron carbonate deposition in addition to calcium carbonate scale.[12] A laboratory study showed that common calcium carbonate scale inhibitors, such as phosphinopolycarboxylic acid (PPCA) and bis-hexamethylene triamine-penta(methylene phosphonic) acid, were not effective at preventing ferrous carbonate precipitation, but citrate ions did perform well.[12] However, other calcium carbonate scale inhibitors are effective on iron(II) carbonate.[12] Corrosion inhibition will also help reduce iron(II) carbonate scaling.

3.2.2 Sulfate Scales

Group II metal ions, except magnesium, can all form sparingly soluble sulfate scales by mixing of sulfate ions and metal ions as given below:

$$M^{2+} + SO_4^{2-} \rightarrow MSO_4 \text{ (s)}$$

The solubility of the sulfates decreases as you go down the group such that barium sulfate is most insoluble and the hardest to control. Sulfate scales are usually formed when formation water and injected seawater mix. When this occurs in the near–well bore region of the producing wells, it causes precipitation of sulfate scales as formation damage. Also, two different non-scaling well fluids may mix in topside flowlines and cause a topside sulfate scale problem. It is the high concentration of sulfate ions in seawater (roughly 2,800 ppm) mixing with group II metal ions in the formation water, which causes sulfate scale to form. Sulfate scaling is usually a problem in seawater-flooded reservoirs, but it has also been known to occur due to use of seawater in workovers of production wells. Freshwater workovers can be used to counteract this problem. There is increasing evidence that, as a result of a waterflood, reactions in the reservoir modify produced water compositions so that they differ from those expected for simple mixtures of injection and formation water. These reactions affect both the injection water and the mixtures of injection water and formation water, and include dissolution and precipitation of sulfates and carbonates, and ion exchange. Scale management plans and decisions are often based on the assumption that produced water will have the composition of simple mixtures of injection and formation water. Where reactions occur, this assumption will be incorrect, so that scale predictions, inhibitor testing, and assessment of squeeze-treatment performance (discussed later) can be compromised.

Calcium sulfate (anhydrite or gypsum) is the easiest sulfate scale to deal with being slightly soluble in water and soluble in many chelate dissolvers, while barium sulfate is the worst to deal with, being very hard and dissolved at a reasonable rate in only the very best dissolvers. Due to its high insolubility (very low solubility product), it does not take a very high concentration of barium ions in the formation water for barium sulfate scale (barite) to deposit. Produced water can also be saturated in barium sulfate and if the temperature drops during production, even more barium sulfate scale can precipitate out.[13] Strontium and barium ions can coprecipitate in the presence of sulfate ions to form a mixed sulfate scale. Formation water also contains low concentrations of radioactive radium, which coprecipitate in the lattice of barium and strontium sulfate scales. Hence, removal of naturally occurring radioactive material (NORM scale) is an environmental problem that must be dealt with. As with carbonate scales, computer programs are available to determine the sulfate scale potential. In addition, prediction and monitoring of seawater breakthrough from seawater injection wells is very important so as to know when to treat a production well for sulfate scaling. Understanding the geochemistry in the reservoir has been shown to be helpful in determining the scale control strategy.[14] A process for enhanced hydrocarbon recovery from a reservoir by actively forming scale (mostly easily sulfate scale) within the reservoir has been claimed.[14] The severity of sulfate scale depends on the ratio of formation water to seawater breakthrough and upon the degree of supersaturation. Thus, late in a field life there may be little or no sulfate scaling as the produced water is predominantly seawater. But when seawater first breaks through the rate, the severity of sulfate scaling can be dramatic. In one well in a UK North Sea field, production dropped from 30,000 bpd to zero in a period of 24 h.[15]

3.2.3 Sulfide Scales

Sulfide scales are less common than carbonate and sulfate scales but can still cause serious problems if not controlled. Some hydrogen sulfide is naturally present in many formation waters, but for oil wells, the bulk of this comes from the activity of sulfate-reducing bacteria (SRBs) on the sulfate ions in injected seawater. The SRB reduce sulfate ions to hydrogen sulfide (H_2S) which is in equilibrium with hydrosulfide and sulfide ions:

$$H_2S + H_2O \leftrightarrow H_3O^+ + HS^-$$

$$HS^- + H_2O \leftrightarrow H_3O^+ + S^{2-}$$

Iron(II) ions, formed mainly by corrosion of steel either in the injector or producing wells, can react with the sulfide ions, forming iron sulfide scale.

$$Fe^{2+} + S^{2-} \rightarrow FeS\ (s)$$

Similarly, other even more sparingly soluble sulfide scales, such as zinc sulfide (ZnS, zinc blend) and lead sulfide (PbS, galena), can be formed if the formation water contains these metals ions and mixes with sulfide ions.[16] Mixed zinc/lead sulfide scale appears to be quite common in high-pressure, high-temperature (HPHT) wells in the North Sea.[17] Zinc sulfide scale can also be formed as an effect of using zinc-based completion brines. In sour gas wells, the deposition of galena can be explained by local supersaturation (caused by a sudden temperature or pressure drop) where sulfide anions react with lead cations in the produced water. Concentrations of up to 150 mg lead cations per liter brine have been detected in some formation waters.[18]

3.2.4 Sodium Chloride (Halite) Scale

Sodium chloride is much more soluble than the scales described above and increases somewhat in solubility with increasing temperature.[19–20] Some formation waters contain very high concentrations of this salt, particularly in HPHT reservoirs. Thus, the formation water may sometimes be saturated in sodium chloride. As the temperature of the produced water decreases, sodium chloride may precipitate out. The kinetics of this process is very fast such that a conduit can block very quickly if not treated. Even wells with very low water-cut (< 0.5%) may rapidly salt-up overnight. Water flash-off into the gas phase as pressure decreases during production will concentrate solutions of sodium chloride, which may eventually also lead to halite precipitation.

3.2.5 Mixed Scales

Scales can often be layered and of mixed composition, for example, containing both carbonate and sulfate scales, at the appropriate field conditions. They can be oily and may even contain other deposits such as asphaltenes making remedial chemical treatment more complicated. If the asphaltenes contain overly inorganic scale deposits, they can render aqueous scale dissolvers ineffective.[1]

3.3 NONCHEMICAL SCALE CONTROL

It has been recognized that effective scale management means making decisions early in the field development phase and continuously reviewing these throughout the field lifetime. Scale control is needed primarily in the production facilities, but seawater injection and produced water reinjection may also need scale control. Iron ions are sometimes removed before water injection.

There are three basic approaches to mitigating scale formation:

1. desulfation of injected seawater
2. scale control/inhibition
3. let scale form and remove the scale physically or chemically

Method 1 will only prevent sulfate and sulfide scales, although hydrogen sulfide may be naturally present in the formation water and cause some sulfide scale (softening of injected water, by removing calcium ions to avoid calcium carbonate scale, is not practiced in the oil production industry). Desulfating the injected seawater is done using membrane nanofiltration. Alternately, aquifer water, if available and low in sulfate ions, can be injected. The use of desulfation facilities requires considerable capital investment but can be the best option for large fields with severe predicted sulfate scale formation.[21] Not all the sulfate ions can be removed by desulfation (ca. 2,700 ppm in seawater is reduced to 40–100 ppm) but enough that sulfate and sulfide (from SRB activity) scale problems are considerably reduced. In addition, desulfation drastically reduces reservoir souring (H_2S production) and will probably affect the degree of microbial corrosion. It should also be mentioned that sulfate stripping occurs naturally in calcium carbonate reservoirs due to deposition of calcium sulfate in the reservoir. Up to 95% sulfate has been removed this way leading to almost no barium sulfate scaling in the producing wells.[22–23] A novel method has been claimed whereby scale particles are injected into the injector wells. These act as seeding/nucleation sites for scale formation before the scaling ions reach the production wells, thus, avoiding scale deposits in these wells.[24] Scale control is particularly effective when the mean particle size of the seed crystals is less than 2.5 µm.[25] Another method that has been laboratory and successfully field-tested is to induce a randomly pulsed high-frequency electrical signal into the piping system, which causes the scale crystals to form in the produced fluid rather than on the walls or surfaces of downhole and topside equipment.[26] The electrical field does not prevent precipitation, it only changes the physical location where precipitation occurs. By inducing an electric field across the pipe diameter, ions are drawn together in clusters in solution.

Lowering the ion solubility below that of saturation also causes existing scale deposits to redissolve, cleaning pipelines from scale, particularly calcium carbonate. Hard scale may be slow to breakdown where there is low water volume and little variation of temperature, flow, hardness, and pressure. In such cases, the equipment is best fitted from new or after chemical cleaning.

The use of acoustic waves has also been proposed as a method of scale control.[27] One method comprises arranging a liquid whistle for producing acoustic waves in the fluid in the tubular, and allowing at least part of the fluid that flows through the tubular to pass through the liquid whistle to generate acoustic waves. Because the fluid flowing through the conduit drives the acoustic whistle, no external power source is needed. Various coatings, that prevent or delay scale deposition, and sometimes also inhibit corrosion, have been investigated.[28–29] Examples are epoxy resins, fluoropolymers, silicones, and polysilazanes. These may be useful to protect key components in the production system (such as downhole safety valves and key parts of intelligent well systems).

However, if scale begins to form on the coated surface, the surface will be altered and more scale can more easily deposit. The smooth, coated surface will also be eroded during turbulent fluid flow carrying sand and other particles, which will probably make scale deposition more likely.

Scale control using magnetic fields has had some success, particularly outside the oil industry such as in heat exchangers.[30–36] However, there are many reports of failure also. Permanent magnets or electronically generated magnetic fields can be used. Such equipment has been used in the oil industry particularly for calcium carbonate scale control, for example, offshore in the Netherlands, but is generally looked upon with skepticism due to the lack of a robust mechanism for its effect. The results obtained from research to date suggest that there are at least three possible

Scale Control

magnetic treatment mechanisms. The first is magnetohydrodynamic effects and the second, agglomeration of ferromagnetic and super paramagnetic microparticles. The occurrence of small pH changes at the wall of a pipe has also been claimed as a possible third mechanism as well as changes in crystal morphology. Magnetic scale prevention may involve a combination of these mechanisms, and probably additional phenomena. Thus, magnets or electronic devices can be effective on calcium carbonate scale at low supersaturation values.

In this chapter, chemical scale inhibition and chemical scale removal will be discussed. The sections on stimulation with acids in Chapter 5 on acid stimulation should also be read alongside chemical scale removal. Controlling the volume of produced water can also alleviate the severity of scaling (see Chapter 2 on water and gas control).

3.4 SCALE INHIBITION OF GROUP II CARBONATES AND SULFATES

Scale inhibitors are water-soluble chemicals that prevent or retard the nucleation and/or crystal growth of inorganic scales. A brief review of the early years of chemical scale inhibition has been published.[37] As a broad generalization, polymers are good nucleation inhibitors and dispersants. When scale crystallization occurs, they adsorb onto the crystal surfaces and are consumed in the lattice, thereby slowing growth when tested below their threshold levels.[38–39] Small, nonpolymeric inhibitors, such as the well-known aminophosphonates, tend to be good at preventing crystal growth, by blocking active growth sites, but if tested below their threshold levels, are less likely to prevent nucleation. Thus, increasing the test dose rate of a polymer will make sure it stops growth and similarly increasing the concentration of an aminophosphonate will make sure it stops nucleation. Actually, the key function of a scale inhibitor is to prevent deposition of scales and for this nucleation inhibition, crystal growth inhibition and even scale dispersion can be the functioning mechanism.[40] Studies have shown that only 3–5% of the surface of a carbonate or sulfate scale crystal needs to be covered by some polymeric inhibitors for complete inhibition.[41] For small aminophosphonate inhibitors on barite scale, the necessary coverage for nucleation inhibition was shown to be 16%.[42] Models and mechanisms for determining scale inhibitor inhibition efficiency, for example, on barium sulfate scale, have been developed.[43] For some sulfate scale inhibitors, such as the small aminophosphonates, the inhibition mechanism is more complicated, whereby a calcium-scale inhibitor complex leads first to calcium inclusion in the lattice, which distorts the lattice and inhibits further growth.[44] Thus, aminophosphonates have been shown to be poor sulfate scale inhibitors at very low $[Ca^{2+}]$.[39] On the other hand, polymeric polycarboxylate inhibitors are shown to be effective even at very low $[Ca^{2+}]$, indicating that the formation of multiple bonds between the polymer and the crystal surface allows for stronger adsorption and, thereby, inhibition.

There are many ways to deploy scale inhibitors in the field and these will be discussed later. First, we will look at the various classes of scale inhibitors that have been used as they apply to the various types of scale. The first thing to note is that most oilfield scales such as carbonates and sulfates consist of divalent anions, that is, CO_3^{2-} and SO_4^{2-}, together with divalent metal cations. Regarding either subcritical nuclei inhibition or crystal growth inhibition, to bind to a scale particle, the scale inhibitor must interact either with the produced water anions or cations. Usually several of these interactions are necessary to hold the inhibitor tightly on the surface so molecules with several similar functional groups and proper spacing of these groups are needed so they interact with the lattice ions on the crystal surface.

To bind well to anions, oppositely charged cations are needed. The only easy way to put several cations in a molecule is through quaternary ammonium, phosphonium, or sulfonium groups. However, polyquaternary ammonium salts, such as polydiallyldimethylammonium chloride and polyacrylamidopropyltrimethylammonium chloride, are poor scale inhibitors, probably due to the mismatch of size between the quaternary groups and the cations, such as calcium, in the scale lattice. However, incorporating quaternary groups into anionic scale inhibitor polymers can be beneficial for adsorption onto formation rock in squeeze treatments. This is discussed later in this chapter on scale inhibitor deployment techniques.

There are several anionic groups attached to an organic molecule that can interact well with group II cations on the scale crystal surface. The most important of these are:

- phosphate ions ($-OPO_3H^-$)
- phosphonate ions ($-PO_3H^-$)
- phosphinate ions ($-PO_2H^-$)
- carboxylate ions ($-COO^-$)
- sulfonate ions ($-SO_3^-$)

Thus, molecules preferably with two or more of these ions, or mixtures of these ions, built into the structure can be good inhibitors for many oilfield scales. The molecules can be prepared in the acid form (e.g., carboxylic acid, phosphonic acid), but it is in the anionic dissociated form, usually as sodium salts, that they are most active as scale inhibitors. The anionic groups are all bound via carbon atoms to the main backbone of the molecule except in polyphosphates.

Below is a list of the most common classes of scale inhibitors containing these ions or acids:

- polyphosphates
- phosphate esters
- small, nonpolymeric phosphonates and aminophosphonates
- polyphosphonates
- polycarboxylates
- phosphino polymers and polyphosphinates
- polysulfonates

Various copolymers and terpolymers with carboxylic, phosphonic acid, and/or sulfonic acid groups are also good scale inhibitors and will be discussed also. Because of the similarities between the functional groups, many scale inhibitors are capable of inhibiting several types of scale.

Many classes of traditional scale inhibitors, such as acrylate-based polymers or aminophosphonates, have poor biodegradability. In addition, produced water discharges containing N- or P-containing inhibitors can enrich the environment in nutrients, which causes an environmental disequilibrium (i.e., eutrophication). Since the early nineties, there has been an increased drive to find more environment-friendly inhibitors. Several of these are now commercially available and are discussed below where appropriate.

There is some overlap between the classes as described, for example, polymers containing both carboxylic and phosphonic or sulfonic acid groups could come under two headings. In such cases, priority has been given to phosphonic acid groups over sulfonic acid groups over carboxylic acid groups. In addition, there are numerous patents detailing synergy between two classes of inhibitors, for example, small aminophosphonates and polycarboxylic acids. An overview of the various detection methods for scale inhibitor concentrations has been published.[45]

3.4.1 Polyphosphates

Polyphosphate anions, such as found in sodium tripolyphosphate or sodium hexametaphosphate, have long been known to be calcium carbonate scale inhibitors (Figure 3.2; hexametaphosphate is also a very good inhibitor of $BaSO_4$ scale). However, they are mainly used in boiler water treatment at low calcium concentrations as there are more thermally stable and more compatible products for oilfield scale inhibition. Polyphosphates also show some activity as corrosion inhibitors. Although it is not commercial, citric acid phosphate is a natural crystal growth inhibitor in mammalian soft tissue with low toxicity. It is a combination of the phosphate and carboxylate groups, which make it a good inhibitor.[46]

3.4.2 Phosphate Esters

Phosphate esters ($ROPO_3H_2$) are well known as environment-friendly scale inhibitors, particularly for calcium carbonate and calcium sulfate scales but also for barium sulfate if the conditions are not very acidic. They are generally not the most powerful class of scale inhibitors.[47] They are made by reacting phosphoric acid with alcohols. By varying the length of the alkyl tail in the alcohol, phosphate esters can be made that are either water- or oil-soluble. Phosphate esters are more tolerant of acid conditions than polyphosphates and are generally compatible with high-calcium brines. Phosphate esters have limited thermal stability. Thermal stability up to 95°C (203°F) for topside application for sulfate scale and temperatures of up to 110°C (230°F) for

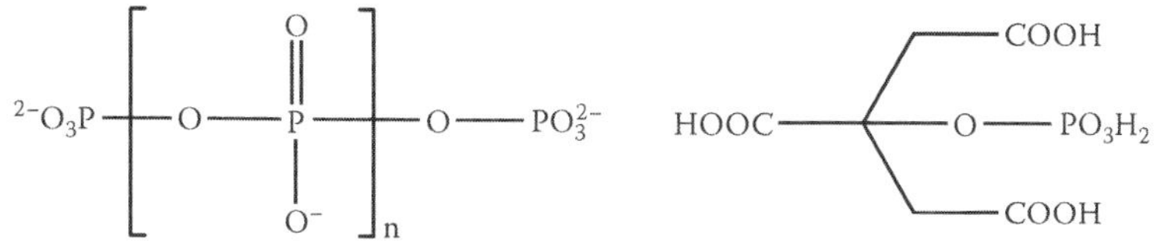

FIGURE 3.2 Polyphosphate anion (left) and citric acid phosphate (right).

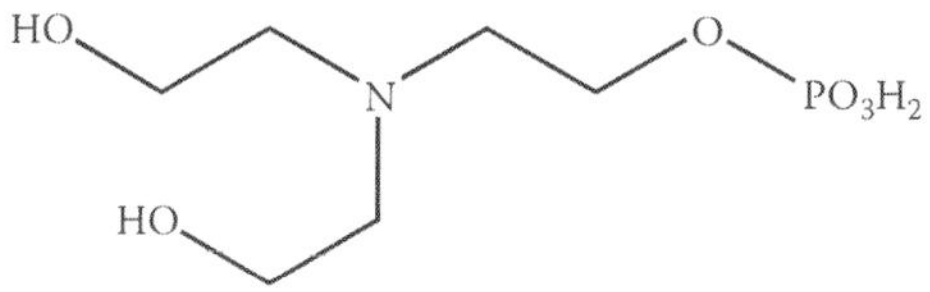

FIGURE 3.3 Triethanolamine phosphate ester.

carbonate scale control have been claimed.[48] Phosphate esters can also be used as a squeeze chemical at temperatures up to about 100°C (212°F). Triethanolamine phosphate monoester, used at least as far back as the eighties, has good biodegradability but is only useful up to temperatures of about 80°C (176°F) due to hydrolysis instability (Figure 3.3). Phosphate esters also show corrosion inhibition activity.[49]

A method of introducing phosphate groups into scale inhibitor polymer is to use the monomer ethylene glycol methacrylate phosphate.[50] This monomer can, for example, be copolymerized with vinyl carboxylic or vinyl sulfonic monomers. The phosphate functionality provides the polymeric inhibitor with good adsorption/desorption characteristics in squeeze treatments, allowing the polymer to be retained in the reservoir and providing extended treatment lifetimes.

3.4.3 Nonpolymeric Phosphonates and Aminophosphonates

Nonpolymeric molecules with only carboxylate and/or sulfonate groups are poor scale inhibitors, but this is not the case with molecules containing phosphonate groups. However, it is known that phosphonates tend to have a lower "cutoff" temperature than many polymeric scale inhibitors, below which they are much less effective. There are a number of scale inhibitors with one phosphonic acid group and several carboxylic acid groups. The most common example is 2-phosphonobutane-1,2,4-tricarboxylic acid (PBTCA), although variations such as phosphonosuccinic acid and 1-phosphonopropane-2,3-dicarboxylic acid are also useful inhibitors (Figure 3.4).[51] PBTCA is mostly used as a calcium carbonate scale inhibitor. Salts of PBTCA have also been proposed as sulfate scale dissolvers, although there are better products available (see section on chemical scale removal).[52]

Improved routes to PBTCA and related molecules such as phosphonosuccinic acid have led to the synthesis of other phosphonate scale inhibitors. Phosphonosuccinic acid is a poor scale inhibitor, but mixtures with oligomers of this molecule show good scale inhibitor performance on carbonate and sulfate scales (Figure 3.5). These molecules can be transformed into phosphonocarboxylic acid esters, which are oil-soluble scale inhibitors, which can also function as asphaltene dispersants.[53]

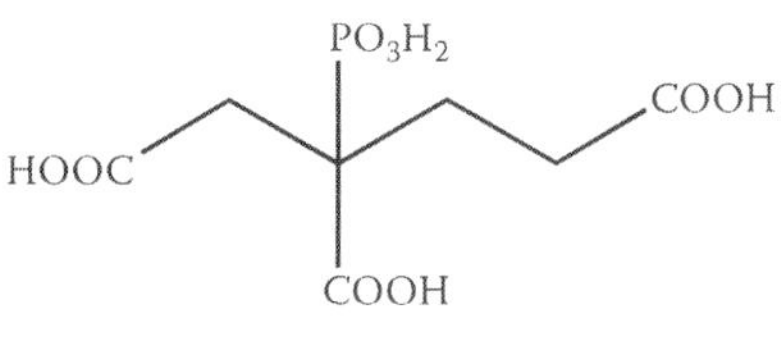

FIGURE 3.4 2-Phosphonobutane-1,2,4-tricarboxylic acid.

Scale Control

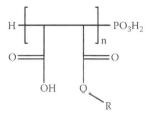

FIGURE 3.5 Phosphonosuccinic acid oligomers (R = H) and esters (R = alkyl, alkylaryl, or alkenyl): $n = 2$–10.

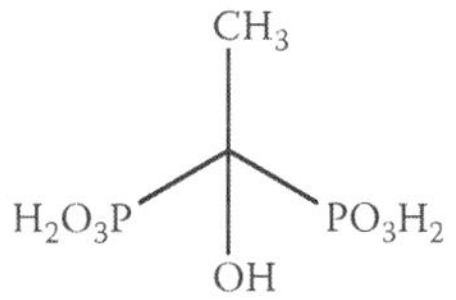

FIGURE 3.6 1-Hydroxyethane-1,1-diphosphonic acid.

Another common phosphonate scale inhibitor is 1-hydroxyethane-1,1-diphosphonic acid (HEDP) also known as etidronic acid (Figure 3.6).

However, the introduction of an amine group into a phosphonate molecule to obtain a $-NH_2$-C-PO(OH)$_2$ group increases the metal binding abilities of the molecule by amine and phosphonate interactions. There are a range of commercial aminophosphonate scale inhibitors used mainly for carbonate and sulfate scale inhibition. The most important molecules are shown in Figure 3.7.

One of the smallest aminophosphonate scale inhibitors is ethanolamine-N,N-bis (methylene phosphonates) (EBMP) formed by the reaction of ethanolamine, formaldehyde, and phosphorous acid. Actually the reaction produces a mixture comprising approximately 50% of EBMP and 50% of a cyclic ester of EBMP. The ester does not contribute to the scale inhibition so it can be hydrolyzed in base to EBMP to effectively double the scale inhibition properties of the original mixture.[54] Both EMBP and its amine oxide are good oilfield scale inhibitors, even for silica scale.[55]

A simple and cheap, but not the most effective inhibitor, is aminotris(methylenephosphonic acid) (ATMP). This can bind to metal ions via all three phosphonic acid groups and the lone pair on the tertiary nitrogen atom. Moving up in molecular weight, we have 1,2-diaminoethanetetrakis(methylenephosphonic acid) (EDTMP) a good all-round phosphonate scale inhibitor. The next molecule up in size is diethylenetriaminepentakis(methylenephosphonic acid) (DTPMP). DTPMP is an excellent carbonate and sulfate scale inhibitor and probably the most used phosphonate scale inhibitor in the oil industry. Another aminophosphonate is dihexamethylenetriaminepentakis(methylenephosphonic acid). This has improved calcium tolerance (i.e., it does not precipitate in the presence of high calcium concentrations) and is useful for high-temperature applications of 120 to > 140°C (248 to > 284°F). Yet another phosphonate inhibitor is hexamethylenediaminetetrakis(methylenephosphonic acid), which has a different distance between the phosphonic acid groups

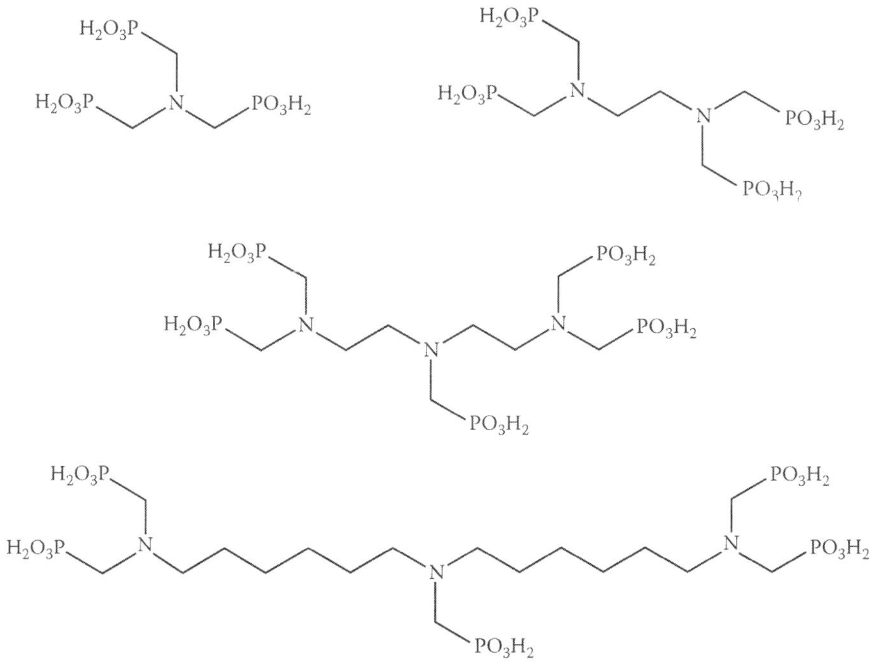

FIGURE 3.7 Aminotris(methylenephosphonic acid), diethylenetriaminepentakis(methylenephosphonic acid), 1,2-diaminoethanetetrakis(methylenephosphonic acid), and dihexamethylenetriaminepentakis(methylenephosphonic acid).

than DTPMP. Phosphonate derivatives of N,N'-bis(3-aminopropyl)ethylene diamine or 1,7-bis(3-aminopropyl)ethylene diamine have been claimed as superior scale inhibitors for high barium brines.[56]

Most aminophosphonates have low biodegradability but are low in toxicity and bioaccumulation. In areas with strict environmental regulations, there has been a move to replace the aminophosphonates with greener alternatives. However, a low molecular weight phosphonate has been found to be environmentally acceptable for North Sea applications.[48] Other biodegradable phosphonate scale inhibitors have been reported.[48b] Discharge of phosphonates into the environment can however lead to unwanted eutrophication in lakes and coastal areas.

Aminomethylenephosphonates derived from small polyglycol diamines such as triethyleneglycol diamine have also been claimed as good carbonate and sulfate scale inhibitors with improved compatibility.[57] In the presence of iron(II) ions, aminophosphonates with hydroxyl or carboxylic acids groups performed significantly better than the traditional aminophosphonates (such as shown in Figure 3.7) used alone or in blends with polycarboxylate scale inhibitors.[58]

Aminoacid alkylphosphonic acids have been claimed as carbonate and sulfate scale inhibitors.[59–60] They are made by reacting aminoacids with formaldehyde and phosphorous acid. Examples are d,l-leucine bis (methylene phosphonic acid) and l-phenyl alanine bis (methylene phosphonic acid).

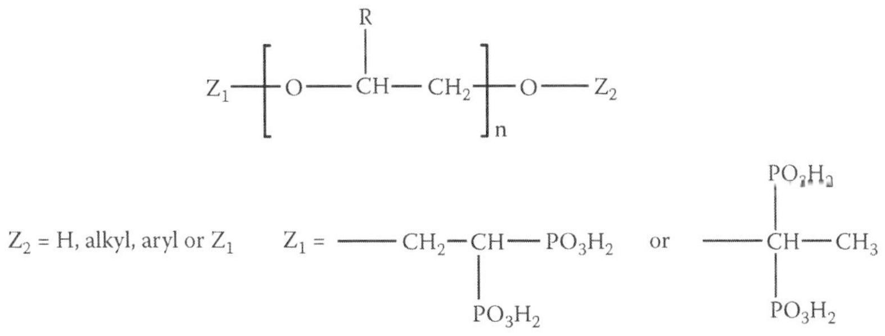

FIGURE 3.8 Preferred polyglycol phosphonates ($R = H$ or CH_3 or a mix of these).

Another class of phosphonate scale inhibitors is based on the reaction of vinyl phosphonic acid (VPA) or vinylidene-1,1-diphosphonic acid (VDPA) with a small polyglycol. The preferred proposed structures are shown in Figure 3.8. An example is the reaction product of a 1:1 mix of triethylene glycol and VDPA. Alternatively, inhibitors in this class can be made by reacting an alkylene oxide (e.g., ethylene oxide or propylene oxide) with a hydroxyphosphonate (or its salt or ester) derived from VPA or VDPA. These inhibitors are further claimed to be partially biodegradable and have corrosion inhibition properties.[61]

3.4.4 POLYPHOSPHONATES

There are two main classes of polyphosphonates, those with a polyamine backbone and those with a polyvinyl backbone. The scale inhibitors with a polyamine backbone are made by reacting a small polyalkyleneamine such as triethylenetetramine with epichlorohydrin and then reacting the amine groups with formaldehyde and phosphorous acid to give the final product, a N-phosphonomethylated amino-2-hydroxypropylene polymer having a molecular weight of between about 300 and 5,000 (Figure 3.9).[62] They are particularly useful for barite scale inhibition and for squeeze applications.[63-64]

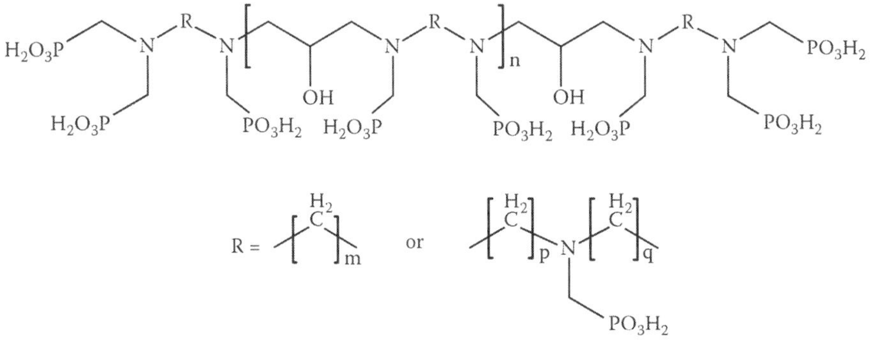

FIGURE 3.9 N-phosphonomethylated amino-2-hydroxypropylene polymers.

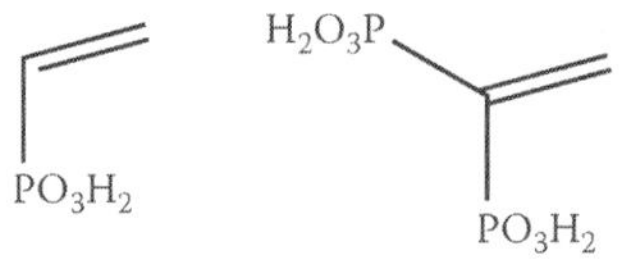

FIGURE 3.10 Vinyl phosphonic acid and vinyl diphosphonic acid monomers.

Polyphosphonates with polyvinyl backbones can be prepared using vinyl phosphonic acid (VPA) or vinyl diphosphonic acid (VDPA) monomers together with any number of other monomers such as acrylic acid, maleic acid, vinyl sulfonic acid, and so forth (Figure 3.10). Copolymers of VPA with unsaturated dicarboxylic anhydrides (ring-opened in water to carboxylic acids) have been claimed as Ba/Sr scale inhibitors. Examples are VPA/isobutylene/maleic anhydride copolymers hydrolyzed to give carboxylic acid groups.[65] VPA and VDPA are fairly expensive so they are mostly used to make phosphonate end-capped phosphino polymers (see Section 3.4.5).

Other phosphonate polymers include the reaction product of a carbonyl compound or imine and hypophosphorus acid, reacted further with vinyl monomers such as acrylic acid to produce polymers. Oxidation gives phosphonate polymers.[66] Polyacrylates with phosphonate end-groups and amidophosphonates have also been claimed as scale inhibitors.[67]

3.4.5 Phosphino Polymers and Polyphosphinates

The most common phosphino polymer used in the oil industry is PPCA, which contains a single phosphino group attached to two polyacrylic or polymaleic chains (Figure 3.11). For example, the reaction of hypophosphite ions with maleic acid produces phosphinicosuccinic acid oligomers together with various smaller molecules. The molar ratio of the products can be varied. These mixtures, and especially the phosphinicosuccinic acid oligomers, are useful carbonate scale squeeze inhibitors.[68] The presence of the phosphorus atom makes PPCA polymers easier to analyze than polycarboxylic acids and gives them better performance (especially for barium sulfate), calcium compatibility and rock adsorption (squeeze lifetime).[69] However, phosphinate groups do not bind as well to rock as phosphonate acid groups.

It is possible to introduce several phosphinate groups into an oligomer using hypophosphorous acid and alkyne chemistry. Using acetylene, one obtains a mixture of ethane-1,2-bisphosphinic acid and diethylenetriphosphinic acid, which is a useful scale inhibitor (Figure 3.12).[70]

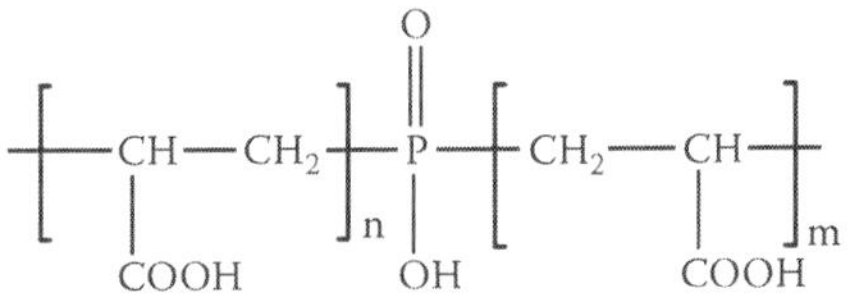

FIGURE 3.11 Polyphosphinocarboxylic acid.

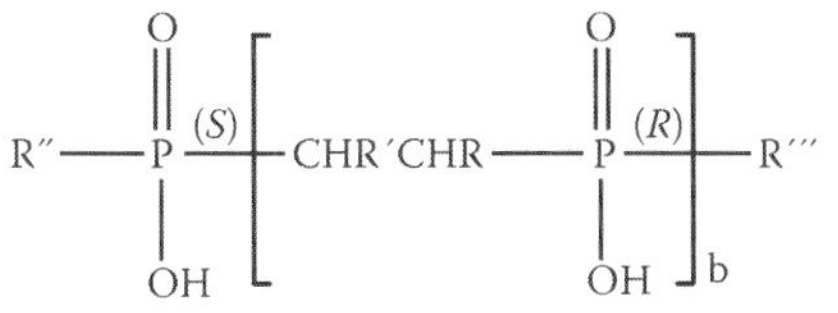

FIGURE 3.12 Phosphino telomers. ($b = 1–2$).

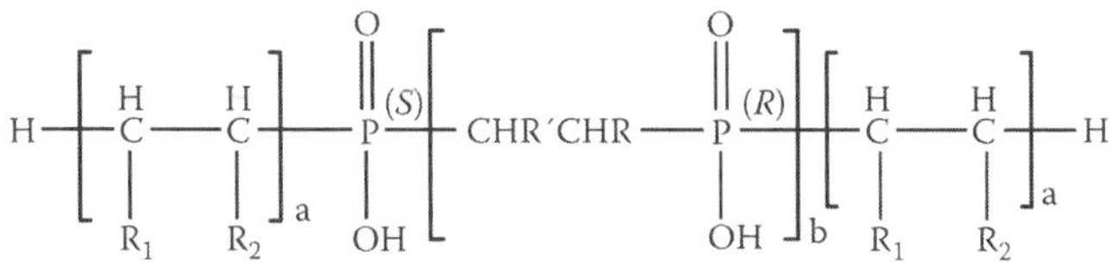

FIGURE 3.13 Polyphosphinates: in most preferred examples, one of R_1 and R_2 are a carboxylic, sulfonic or phosphonic acid group, or R_1 and R_2 can both be carboxylic acid groups.

These oligomers (or telomers) can be reacted with vinyl monomers to form polyphosphino polymers with improved performance (Figure 3.13).[71]

Aminophosphinate polymers can also be made using the phosphino telomers (Figure 3.14).[72–73]

Phosphonate-end–capped phosphino polymers are particularly useful for barite scale (Figure 3.15). The polymers show good absorption to rock and better thermal stability than random copolymers containing equivalent phosphorus. These end-capped polymers have over 20% biodegradability in the OECD 306 28-day biodegradation test. End-capped VDPA polymers are commercially available.[74–75]

3.4.6 POLYCARBOXYLATES

Salts of polycarboxylic acids have been used as scale inhibitors almost as long as polyphosphates. The most common classes of polycarboxylic acids are based on polyacrylic acid, polymethacrylic acid, and polymaleic acid (Figure 3.16). All these linear polymers have a carbon backbone. Polyacrylates are generally poorly biodegradable, but some polymaleates can be fairly biodegradable. In fact, a polycarboxylic acid scale inhibitor with > 60% biodegradation by OECD306 has recently been approved for use in the North Sea.[76] As with most polymeric scale inhibitors, the number of active repeating units in the polymer needs to be at least 15–20 for optimum scale inhibition, otherwise, not enough active groups will bind to the scale crystal surface to hold it in place. For polyacrylates, this means a molecular weight of at least 1,000–1,500. Most polymeric scale inhibitors have molecular weights in the range 1,000–30,000 as the performance drops off at high molecular weight. It is known that the biodegradability of acrylic acid–containing polymers, which are generally poorly biodegraded, is greatly improved by reducing the weight average molecular weight (Mw) below 700 Da; however, this is usually at the expense of the performance. Branched or partially cross-linked polymers are claimed to have improved performance over linear polymers, both for carboxylated and sulfonated

FIGURE 3.14 Aminophosphinate polymers.

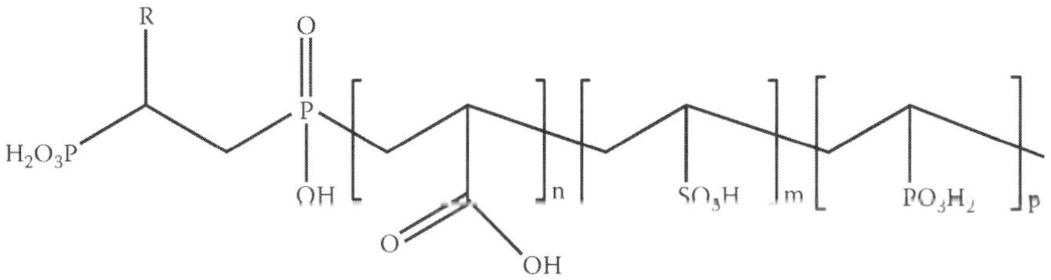

FIGURE 3.15 Phosphonate-end-capped polymers (R = H or preferably PO_3H_2 and n, m, and p can be zero or any number).

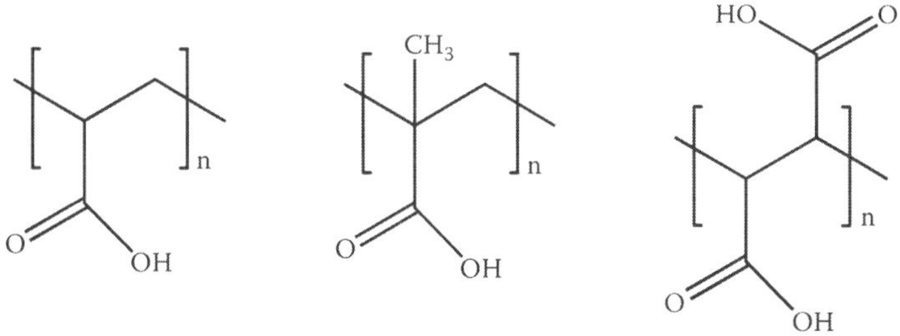

FIGURE 3.16 Polyacrylic, polymethacrylic, and polymaleic acids.

polymers.[77] Interestingly, acrylic acid/isoprene and related copolymers are claimed to have better biodegradability than polyacrylic acids, although addition of the hydrophobic monomer will probably reduce the compatibility of the polymer with produced waters.[78]

The monomers acrylic acid and maleic acid are relatively low cost, raw materials, and also occur in a wide variety of copolymer and terpolymer scale inhibitors with additional sulfonic or phosphonic acid groups. Within the category of polycarboxylic acids, acrylic acid/maleic acid copolymers are very common. Maleic acid does not polymerize easily by itself so, to increase the polymer size and performance, it is usually copolymerized. Copolymers with hydrophobic monomers have been claimed as good barium/strontium sulfate scale inhibitors.[79] The method of manufacture of a polymeric scale inhibitor is extremely important to the end performance. Therefore, manufacturers who have found routes to good scale inhibitors keep to the same recipes. Other unsaturated carboxylic acid monomers include methacrylic acid, crotonic acid, itaconic acid, glutaconic acid, tiglic acid, and angelic acid, but these are generally more expensive (except methacrylic acid) and are less widely used to make scale inhibitors.

The addition of a percentage of amide or hydroxyl groups to polycarboxylates has been claimed to improve their performance as scale inhibitors and also increase their calcium tolerance.[80] Examples are maleic acid copolymers with some maleamide or maleimide groups or acrylic acid polymers with acrylamide, methacrylamide, N,N-dimethyl acrylamide or hydroxypropyl acrylate monomers.[81] Sulfonated groups or an unsaturated polyglycol monomer can also be introduced for higher calcium tolerance.

Acrylic copolymers with cationic monomers such as methacryloxyethyltrimethylammonium chloride have biocidal and anticorrosion properties besides being scale inhibitors.[82] Incorporation of a quaternary amine monomer into a carboxylic acid–based polymer has also been shown to increase rock adsorption giving longer squeeze lifetimes.[83–84]

3.4.6.1 Biodegradable Polycarboxylates

Besides a highly biodegradable polyvinyl-based polycarboxylic acid mentioned on page 67, another class of polycarboxylates are the polyaminoacids or peptides, exemplified by polyaspartate salts (Figure 3.17).[85–88] This polymer can be made from l-aspartic acid or from maleic anhydride and ammonia via polysuccinimide, which hydrolyzes in base to polyaspartate. Polyaspartate is a highly biodegradable scale inhibitor with good performance against carbonate and sulfate scales. It is also a fairly good corrosion inhibitor.[87] The structure is complex due to some degree of branching, irrespective of the manufacturing procedure. It contains both α- and β-groups, the α-groups containing pendant carboxylic acid groups one carbon atom further away from the peptide backbone than the β-groups. Polyaspartate is now used as a scale inhibitor in the oil industry, particularly in regions where the environmental regulations normally require >20% biodegradability such as the North Sea basin. It has also been used in squeeze treatments at up to ca. 85°C (185°F),[89] although an improved polyaspartate for squeezing at up to 120°C (248°F) is about to be commercially available in 2008. Polyaspartate has also been claimed to reduce fines migration after a squeeze treatment.[90] Another patent claims the biosynthesis of polyaspartate, or similar polyaminoacid scale inhibitors, downhole by injecting genetically engineered, thermophilic microorganisms and suitable nutrients.[91]

Derivatives of polyaspartic acid can also make useful scale inhibitors. Polyaspartates with monomer units containing oxygenated hydrocarbonamides with the N-2-hydroxyalkylaspartamide structure, and optionally other amino acids have been claimed as scale inhibitors with good calcium compatibility and biodegradability.[92] Variations on polyaspartates, using other amino acids such as glutamic acid, have been investigated as scale inhibitors, but it does not appear that they have been commercialized probably due to the higher cost of other amino acids. Polyaspartates are also useful low-toxicity corrosion inhibitors and are used in the field as combination

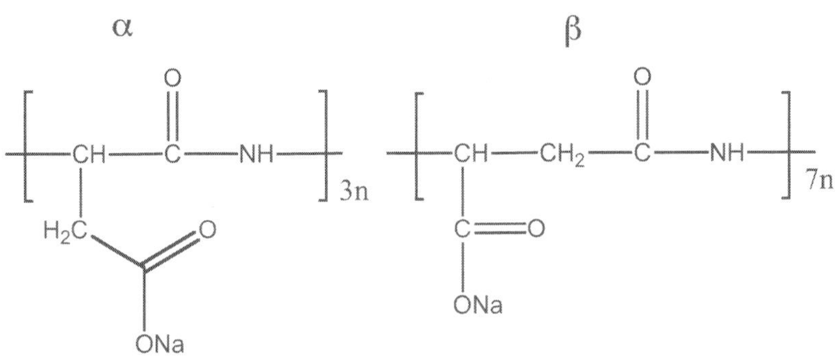

FIGURE 3.17 Sodium polyaspartate. The ratio α/β is approximately 3:7 in this case, but this factor and the degree of branching can be varied depending on the manufacturing process.

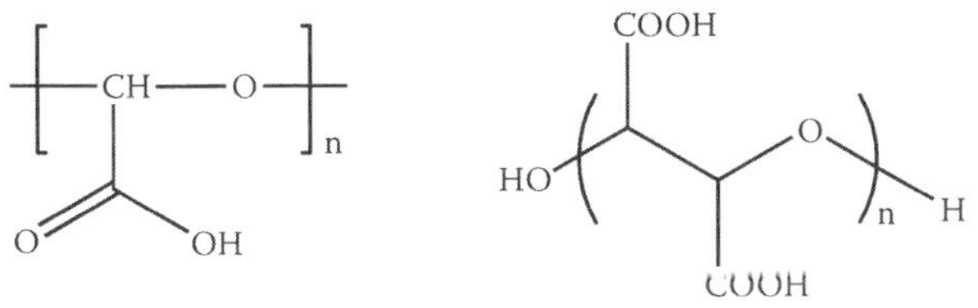

FIGURE 3.18 Polyglyoxylic acid (left) and polytartaric acid (right).

inhibitors against scale and corrosion.[93-94] Polyamino acids with hydroxamic acid groups have been claimed as biodegradable, scale and corrosion inhibitors.[95]

Polytartaric acid is another polymer with better biodegradation than the polyvinyl carboxylic acids such as polyacrylic acid.[96] Polyglyoxylic acid is a carboxylic acid polymer with a heteroatom backbone, which may give it some biodegradability, although this has not been reported (Figure 3.18). Homopolymers or copolymers with pendant polyglycol groups for greater calcium compatibility are used as carbonate scale inhibitors.[97] Polyepoxysuccinic acid has also been claimed as a potentially green scale and corrosion inhibitor.[98]

A carboxylated derivative of an oligosaccharide, carboxymethylinulin, has been commercialized as a biodegradable oilfield scale inhibitor, particularly for use in environmentally sensitive areas.[99-101] It is a fairly good carbonate scale inhibitor but poorer sulfate scale inhibitor. A feasibility study for its use as a squeeze inhibitor showed promise but concluded that the polymer needed optimising.[102] The hydroxyl groups in all oligopolysacharides and polysaccharides can be derivatized to carboxylic acid groups using base and chloroacetic acid. An example is carboxymethylcellulose (CMC). However, solutions of these polycarboxylates can be too viscous for injection purposes. Carboxymethylinulin is an exception, possibly due to being a fairly low molecular weight molecule (Figure 3.19). However, polysaccharides can be selectively oxidized to give carboxylic acid groups and then controllably degraded by cooking the polysaccharide, forming a lower molecular weight version of the modified polysaccharide with carboxyl and aldehyde functional groups. Such polymers can also function as scale inhibitors.[103] An example is peroxide-depolymerized carboxyalkyl polysaccharide (preferably based on polygalactomannans found naturally in guar, locust bean gum, etc.) having from 0.5 to 3.0 degrees of substitution of COOH groups per sugar unit.[103] Polysaccharide derivatives of aloe with carboxylic and alcohol groups have also been claimed as scale inhibitors.[104a] Other biodegradable scale inhibitors have also recently been reported.[104b]

3.4.7 POLYSULFONATES

Most commercial sulfonated polymeric scale inhibitors have a polyvinyl backbone. The commonest vinyl monomers used are salts of vinyl sulfonic acid (VS), acrylamido(methyl)propylsulfonic acid (AMPS), allyloxy-2-hydroxypropyl sulfonic acid, and styrene sulfonic acid (SSA) (Figure 3.20). Allyl sulfonic acid/maleic acid copolymers have also been claimed as sulfate scale inhibitors.[105] Among the most used monomers, polymers with vinyl sulfonic acid appear to give the best

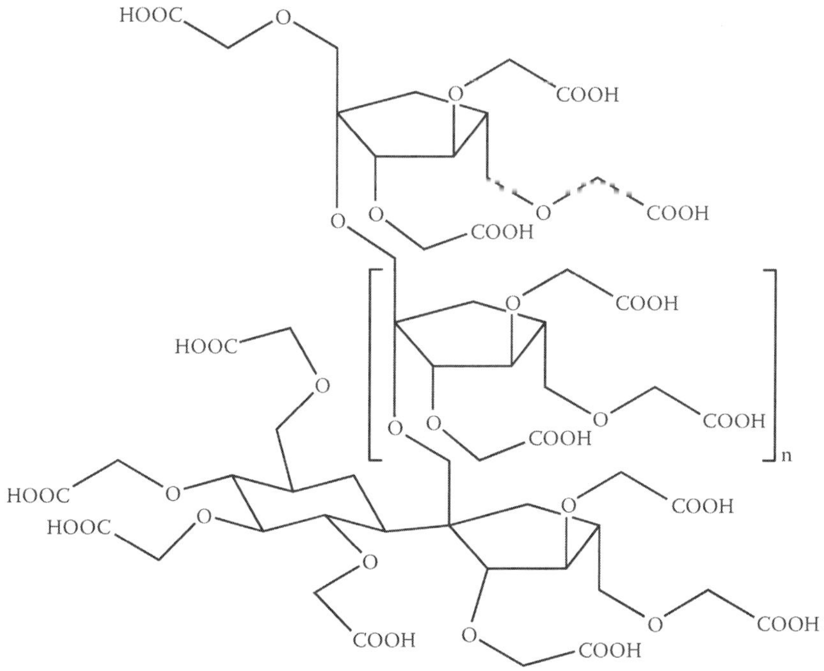

FIGURE 3.19 Carboxymethylinulin.

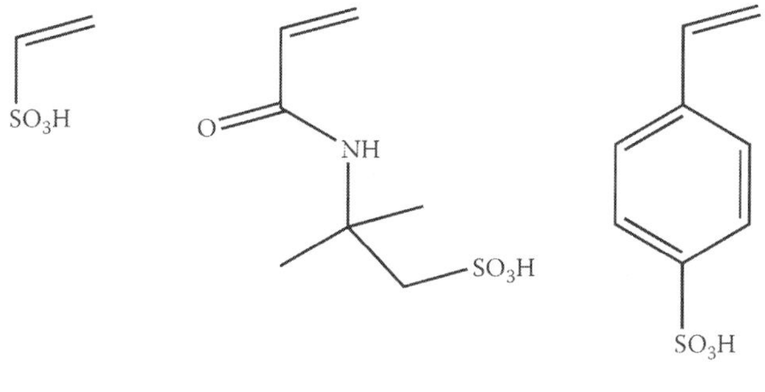

FIGURE 3.20 Vinyl sulfonic acid (left), acrylamido(methyl)propylsulfonic acid (middle) and styrene sulfonic acid (right).

performance against barite scale. Many types of copolymers with one of these monomers are commercially available, for example, copolymers of VS and acrylic acid, or AMPS and acrylic acid.[106] Terpolymers of meth(acrylic) acid or maleic acid, AMPs and a cationic monomer such as diallyldimethylammonium chloride have also been claimed as well as copolymers of vinyl sulfonate and polyalkylene glycol mono- or di-methacrylates.[107–108] Polyvinylsulfonate (PVS) has a lower pK_a value than phosphonic or carboxylic based scale inhibitors and has lower stability constants with Ca or Mg. Therefore, it can work as a scale inhibitor at lower pH values, principally by nucleation inhibition.

Polysulfonates have been used particularly for high-temperature squeeze applications for sulfate scale because of their high thermal stability and calcium tolerance. PVS and VS and VS/SSA copolymers are more thermally stable than AMPS-based polymers.[109] However, several other classes of scale inhibitors once thought to be unstable at high temperature in solution may be applicable for squeeze applications because their stability is increased once they are adsorbed to the formation rock.[110] Polysulfonates do not adsorb as strongly to rock as, for example, phosphonates and therefore squeeze lifetimes will be shorter with these polymers. PVS works very well at low temperatures (4–5°C, 39–41°F), better than many other classes of scale inhibitor.[44]

3.5 SULFIDE SCALE INHIBITION

The most common sulfide scales are salts of iron(II), zinc and lead(II) ions. Compared with carbonate or sulfate scale inhibition, research on sulfide scale inhibition has increased significantly since the turn of the millennium. Sulfide scaling can be limited by avoiding biogenic production of hydrogen sulfide (H_2S) by using biocides or nitrate/nitrite injection (see Chapter 14). However, naturally occurring H_2S in the formation can lead to sulfide scales. Alternatively, H_2S scavengers can be used on the production side to chemically remove H_2S (see Chapter 15). However, the commonly used triazine-based H_2S scavengers will raise the pH and may lead to a worsening of sulfide and carbonate scaling.[111]

Iron sulfide scales can build up in either seawater injector wells or producing wells. Injection of aminocarboxylate chelates such as nitrilotriacetic acid (NTAA) has been used to sequester iron and prevent iron sulfide scaling in moderate temperature sour gas wells. Laboratory studies showed that NTAA was decomposed in 5 h above 149°C (300°F) and was therefore unsuitable for very high temperature applications.[112] Another report claims that an aminocarboxylate (no definitive structure given) was best at inhibiting iron sulfide scale apparently via sequestration.[113] Treatments that use organic acids to chelate the iron are required to be added in stoichiometric amounts (or usually greater) with respect to the level of dissolved iron. At such treatment rates, these acids have the capacity to be very corrosive on the production hardware.

Blends of tris(hydroxymethyl)phosphines (THP) and tetrakis-hydroxymethylphosphonium salts (e.g., THPS) and sufficient of a chelate (aminocarboxylates or aminophosphonates) have also been claimed to inhibit as well as dissolve iron sulfide scale.[114] An improvement on the use of THP or THP salts that avoids polymerization side reactions at low pH is to use these phosphine compounds in a blend with ammonia or a small primary amine, for example, methylamine[115] (see Chapter 14 for more references to THPS).

Rather than sequester iron(II) ions, direct inhibition of iron sulfide scale is possible, although there are few reports of this technique.[116] For example, the common scale inhibitors DETPMP and PPCA have been shown to inhibit FeS scale formation.[117] A clarification phenomenon was observed after sulfide formation using the DETPMP phosphonate. This involves the formation of black FeS followed by clarification after 24 h, which indicates subsequent "inhibition," although probably by direct chelation rather than by threshold (substoichiometric) inhibition.

Another method to prevent iron sulfide deposition is to disperse the scale particles rather than inhibit their formation. Some iron sulfide dispersant products on the market can often move the scale to the hydrocarbon phase where they can have the propensity to form complex emulsions, and create additional operational problems. New dispersants (the structures were not reported) have been developed, which makes the iron sulfide both water-wet and deactivated.[118] Evidence has been provided that shows how the treatment keeps the particle size small so that is unlikely to associate and deposit in production systems. The concept was proven in a limited field trial where the treatment successfully prevented the formation of pads in a primary separator and the deposition of solids in a heater treater.

Several reports of zinc sulfide inhibition have been published. An early report is the use of hydroxyethylacrylate/acrylic acid copolymer as a zinc sulfide threshold inhibitor in oil-well-production processes.[119] In a more recent report, small aminophosphonates (DTPMP, HEDP, ATMP) performed fairly well, needing 50–100 ppm for good zinc sulfide inhibition.[120] The authors state that it is unclear whether the inhibition is due to a lowering of the pH by the inhibitor or whether threshold scale inhibition is taking place. Improved results were obtained with polymeric inhibitors. A phosphonate-maleic copolymer, an AMPS-maleic copolymer and an AMPS-acrylic acid copolymer were better than a polyacrylic acid. In general, sulfide scale inhibition required up to ten times the dosage as sulfate-scale inhibition. Other workers have reported the improved effect of polymeric scale inhibitors over small phosphonates.[121–122] Chelates such as Na_4EDTA and Na_5EDTA also performed well, but, as they only work stoichiometrically, they may be expensive to use at high metal sulfide concentrations. Successful squeeze treatments with PbS/ZnS polymeric scale inhibitors for HPHT wells have been reported.[123]

3.6 HALITE SCALE INHIBITION

Halite (sodium chloride) deposition can be controlled by continuously diluting the produced water supersaturated in halite.[124] For example, dilution downhole with freshwater via a macaroni string can be carried out. However, this may require large volumes of low salinity water, compatible with the produced water and deoxygenated to limit corrosion. Dilution with seawater can lead to sulfate scaling unless treated with a sulfate scale inhibitor.

Halite contains only monovalent anions so the scale inhibitors discussed earlier used for scales with divalent anions (carbonate, sulfate, etc.) do not work on halite scale.[125] There are two classes of chemicals that have been known for some time to inhibit (i.e., change the crystal morphology of) halite scale:

- hexacyanoferrate salts, $M_4Fe(CN)_6$, M = Na or K
- nitrilotrialkanamides and quaternary salts, $N[(CH_2)_nCONH_2]_3$, $n = 1$ or 2

Halite "inhibition" with these chemicals usually requires significantly higher doses than inhibitors for sulfate and carbonate scale inhibition. However, new products have been developed, which, from laboratory studies, are claimed to perform well at

Scale Control

significantly lower concentrations. Details of their chemistry have not been reported, although two products are inorganic and one is organic nitrogen-based. Other inhibition and field applications have also been reported.[126] Among these three, the organic-based product has also shown excellent environmental profile and acceptable adsorption characteristics under reservoir conditions. Results from pH tests suggest that the solution pH has a profound effect on the amount of adsorption, which has been valuable information in developing optimized squeeze packages.

Potassium hexacyanoferrate (HCF) has long been known as a modifier of halite crystals.[127] An octahedron of chloride ions with sodium in the middle can be replaced by a $Fe(CN)_6^{4-}$ ion.[128] In concentrations ranging from 2.48×10^{-4} up to 2.85×10^{-3} M HCF was able to increase the solution critical supersaturation (up to 8%) resulting in a significant crystallization inhibition effect.[129] HCF has been successfully used in the field to greatly reduce the deposition of halite scale in gas compression equipment.[130] It is normally applied at a concentration of about 250 mg/l. The fact that laboratory tests gave even better performance is attributed to iron ions incompatibility in the field. HCF modifies the halite crystals such that any deposits are easily removed by periodic washing with low salinity water. No downhole applications with HCF have yet been reported. HCF reacts with iron(III) salts in the produced water to produce a Prussian blue color. To prevent this from happening, sequestering agents such as the trisodium salt of nitrilotriacetic acid and alkali metal citrates can be added.[131]

In the class of nitrilotrialkanamides, both nitrilotriacetamide and nitrilotripropionamide have been reported to inhibit halite precipitation (Figure 3.21).[132] The hydrochloride or hydrogen sulfate salts can also be used.[133-134] The crystal modification effects are correlated with the fit between the molecular structures of the additives and the lattice of the crystals.[135-136] Salts of nitrilotriacetamide have been used successfully in the field to reduce halite scale formation, both topside and downhole in squeeze treatments.[137-138] The commercial product usually contains ammonium chloride, which is a side-product in the synthesis of nitrilotriacetamide. The chloride ion is a common ion to halite formation and would be best removed from the product, but this would raise the price considerably. Nitrilotriacetamide is stable at high temperatures and is reasonably biodegradable.

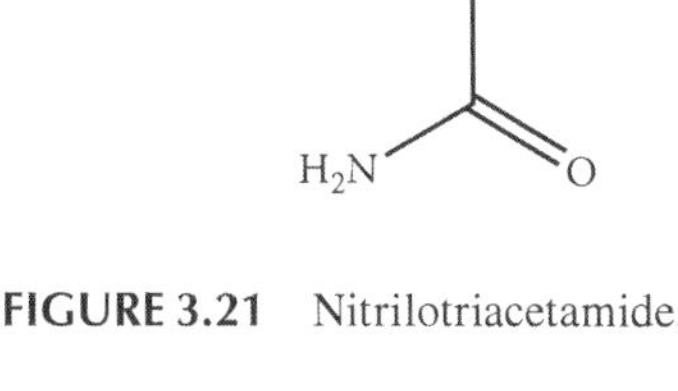

FIGURE 3.21 Nitrilotriacetamide.

3.7 METHODS OF DEPLOYING SCALE INHIBITORS

There has been little advance in designing higher performing scale inhibitors in recent years; where there has been advance has generally been in developing more biodegradable products. However, the scale-control area where most advances can still be made is technology concerning the application of scale inhibitors. There are a variety of methods of applying scale inhibitors in the field, but the main alternatives are as follows:[1]

- continuous injection
- squeeze treatment
- solid, slow-release scale inhibitor compositions

3.7.1 Continuous Injection

Continuous injection of scale inhibitor may be needed in injector wells, especially for produced water reinjection, or in producing well streams. Continuous injection in the injector wells has also been carried out to prevent scaling in the producing wells.[139] Continuous injection into produced waters is usually carried out topside at the wellhead, where other production chemicals, such as corrosion inhibitors, may also be injected. In fact, many scale inhibitors are not compatible with certain corrosion inhibitors. Scale inhibitors can also be injected downhole if a capillary string is available or via the gas lift injection system.[140–141] In gas lift injection, it is important to add a low–vapor-pressure solvent (vapor pressure depressant, VPD) such as a glycol to the aqueous scale inhibitor solution to avoid excessive solvent evaporation and "gunking" of the scale inhibitor. In addition, glycol or some other hydrate inhibitor may be needed to suppress gas hydrate formation.[142] There is one field report where use of a VPD did not prevent gunking. In this case, a solid scale inhibitor was dissolved in a very high boiling solvent and successfully deployed through the gas lift system.[143] A scale dissolver blended with a scale inhibitor has also been deployed in a gas lift system.[144] Downhole injection of some scale inhibitors can lead to increased downhole corrosion rates.[145]

3.7.2 Scale Inhibitor Squeeze Treatments

The basic idea in a scale inhibitor squeeze treatment is to protect the well downhole from scale deposition and formation damage. The inhibitor will, of course, continue to work above the wellhead, protecting the pipeline from scaling, but a further dose of a scale inhibitor may be needed topside. In a squeeze treatment, a solution of the scale inhibitor is injected into the well above the formation pressure whereby the inhibitor solution will be pushed into the near-well formation rock pores.[146] The well is then usually shut-in for a period of hours to allow the inhibitor to be retained, by various mechanisms, in the rock matrix. When the well is put back onstream again, produced water will pass the pores where the chemical has been retained dissolving some of it (Figure 3.22). In this way, the produced water should contain enough scale inhibitor to prevent scale deposition. When the concentration of the inhibitor

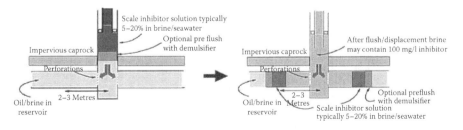

FIGURE 3.22 Scale inhibitor squeeze treatment.

falls below the MIC (minimum inhibitor concentration that prevents scale deposition) the well should be resqueezed. Naturally, long squeeze lifetimes will keep the overall downhole scale treatment costs to a minimum. Squeeze modeling programs are commercially available to assist in the design of scale inhibitor squeeze treatments.[147] Understanding the chemistry of the rock-scale inhibitor interactions can also help in the squeeze design.[148-149] The ability to analyze the residual scale inhibitor accurately, or more importantly, the active inhibiting components of the inhibitors can be difficult in some oilfield-produced waters, but a number of methods are available. Complementary and more direct methods of analysis such as scanning electron microscopy, stressed tests, and in-line monitoring are also available.[1] Different tagged-scale inhibitors have been proposed for use in subsea completions where several individual wells flow into a common seabed pipeline and on to the nearest production platform. Analysis of the tagging allows the operator to know the concentration of scale inhibitor in the produced fluid from each well to know when each well needs resqueezing.[150] Squeeze treatments can also be carried out with other production chemicals such as asphaltene inhibitors.

The traditional "adsorption" scale inhibitor squeeze treatment uses a water-based scale inhibitor solution usually as a 5–20% active solution in KCl or seawater. The scale inhibitor is adsorbed onto the formation rock during the shut-in. A preflush (such as 0.1% inhibitor in KCl or seawater, optionally with a demulsifier) can be used to clean and prepare the near–well bore for the scale inhibitor itself. An afterflush is used to push the scale inhibitor deeper into the formation.[151] This treatment process is accepted for wells at significant water-cut (> 10%). In the region of 0–10 % water-cut, more unconventional treatment methods need to be evaluated. Phosphonate groups generally adsorb better than carboxylate groups, which are better than sulfonate groups. Thus, squeeze lifetimes with sulfonated inhibitors such as PVS may be lower than with inhibitors with many phosphonate groups due to low retention in the rock matrix. However, the adsorption must not be so strong that the inhibitor is back-produced below the MIC. With adsorption squeeze treatments, it is common for a fair percentage (25–35%) of the scale inhibitor to be produced back immediately after the well is put on production. Thus, the squeeze lifetime is reduced to significant early loss of scale inhibitor. The ideal situation would be to have the scale inhibitor produced back at a constant MIC level from day 1.

There are a number of techniques, which have been developed to increase scale inhibitor retention on the rock formation and thus enhance the lifetime of a squeeze treatment. They include:

- precipitation squeeze treatment
- use of some transition metal ions and Zn^{2+} ions
- raising the pH in situ
- mutual solvents to change the rock wettability
- blends with cationic polymers
- incorporating cationic monomers in the scale inhibitor polymer structure
- cross-linked scale inhibitors
- use of kaolinite or other clay that enhances inhibitor adsorption
- scale inhibitor microparticles

A common way to increase the retention of the scale inhibitor in the near–well bore is to carry out a precipitation squeeze treatment.[152-154] Many scale inhibitors are incompatible at elevated temperatures and reservoir pH with high calcium or magnesium concentrations. Thus, by injecting these cations or iron(II) ions (or dissolution of these ions from the reservoir minerals) with the scale inhibitor an inhibitor-cation complex can be precipitated in the near–well bore, giving better retention than using the inhibitor alone.[155-156] In situ dissolution of these ions from the reservoir minerals can also enhance inhibitor retention.

It is important not to form the precipitated complex in the perforations, and cause well damage, by using a suitably sized overflush. A solid source of cations or polycations, such as microparticles of a basic anion exchange resin, can be squeezed first into the formation followed by the scale inhibitor to increase the retention of the inhibitor.[157] Highly sulfonated polymers such as PVS do not easily form precipitates with calcium salts and therefore have to be deployed by some other technique such as an adsorption squeeze, although sulfonated polymers tend to desorb rather quickly giving short squeeze lifetimes.

The use of Zn^{2+} ions in squeeze formulations has been shown in laboratory studies to significantly increase the retention of the inhibitor.[158] Furthermore, Zn^{2+} showed significant synergistic effect on barite scale inhibition by certain scale inhibitors. Together with phosphonate, Zn^2 also demonstrated a synergistic effect on corrosion inhibition. Stronger solution complexes are observed for common scale inhibitors with Zn^{2+} than that with alkaline earth metal ions, such as Ca^{2+}, Mg^{2+}, and Ba^{2+}. This stronger solution complex formation with Zn^{2+} ions may attribute to the observed enhanced scale inhibition efficiency.

Another method to increase scale inhibitor retention in a precipitation squeeze, and thus the squeeze lifetime, is to raise the pH of the scale inhibitor solution in situ in the near–well bore. In this way, acid groups in the inhibitor molecules become anions and can complex more easily with cations, precipitating out as calcium/magnesium complexes. For example, the scale inhibitor solutions can be blended with urea or an alkyl urea derivative.[159-162] At elevated temperatures in the formation, the ureas decompose to release the basic gas ammonia or an alkylamine, which raises the pH of the scale inhibitor solution. Other base-generating amides such as acetamide or dimethylformamide can also be used.[163]

However, urea only breaks down above 85°C (185°F), so it cannot be used in low-temperature wells by itself. For such cases, two approaches have been designed. First, a novel multistage precipitation squeeze treatment can be carried out.[164] A low-pH blend of phosphonate scale inhibitor and calcium ions was pumped (actually as a temperature sensitive invert emulsion) alternately with high-pH solutions of sodium carbonate (containing a low dose of carbonate scale inhibitor). Mixing of the aqueous phases caused precipitation of the Ca-inhibitor complex on the rock. Second, an enzyme can be used to decompose the urea at temperatures as low as 40°C (104°F).[165] Both methods have been successfully used in the field.

Another method to increase the squeeze lifetime is to use a mutual solvent (small nonionic amphiphile, NIA) such as the small alkyl glycols (e.g., butyl triglycol ether).[166-167] Mutual solvents enhance inhibitor retention by making rock more water-wet and help in well clean-up rates.[168-170] They can also remove trapped water

Scale Control

(water blocks) caused by an all-aqueous squeeze treatment, thus providing stimulation.[171] Hence, a preflush with a mutual solvent is often used.[169] The squeeze lifetime enhancement is far superior in a precipitation squeeze to the performance of the same mutual solvent when used in an adsorption scale squeeze method. The mutual solvent enhancement technique has been combined successfully in the field with the pH-modifying technique, using urea.[172]

A method claimed to increase the squeeze lifetime is to precondition the rock surface with a cationic polymer such as polydiallyldimethylammonium chloride (polyDADMAC) originally used as a clay stabilizing/sand control additive.[173–174] The positively charged surface is then better able to adsorb negatively charged scale inhibitor ions. It has also been found that incorporating cationic monomers into a scale inhibitor polymer such as PPCA also gives a product, which is retained on the rock above the MIC for a longer period.[83,175] It is thought that by "fixing" the polymers to the surface of the formation (e.g., by electrostatic attraction between positively charged monomers and negatively charged sandstone), the inhibitor adsorption can be enhanced. A polymer composed of acrylate and a quaternary monomer gave a high biodegradation of over 60% (unusual for a polyvinyl polymer) in the OECD306 28-day seawater test.[176] The amount of the positively charged quaternary monomer had to be optimized to allow the adsorbed inhibitors to be released to the formation brine.

Another method to enhance squeeze lifetimes is to use a cross-linked scale inhibitor.[177–178] This has been shown to double the squeeze lifetime of carboxylic polymeric scale inhibitors in the laboratory, but has not been tried in the field.[179] The technique is a hybrid of a scale squeeze treatment and a cross-linked polymer water shut-off treatment. In fact, a water shut-off treatment that can also provide scale inhibition by incorporating a scale inhibitor into the cross-linked polymer matrix has been claimed.

Another method that enhances scale inhibitor retention, and also can be used for water shut-off, is the use of kaolinite clay.[180] Kaolinite is a commonly occurring clay type within sandstone reservoirs and contributes towards increasing squeeze lifetimes by increasing the available surface for scale inhibitor adsorption. For those fields that produce from high-permeability, low–clay-content sandstone reservoirs, a frequently encountered problem is that of relatively short squeeze lifetimes due to the relatively low-scale inhibitor adsorption achieved upon quartz and feldspar grains that are typically present. It has been demonstrated by coreflooding that alteration of the near–well bore mineralogy by kaolinite injection can increase inhibitor adsorption and, therefore, provide the potential for significantly enhanced squeeze lifetime within clean, high-permeability reservoirs. Injection of low concentrations of kaolinite could be achieved within such reservoirs without causing significant formation damage and thereby affecting well productivity. It is possible that other clay types or different materials could be more efficient.

3.7.2.1 Scale Inhibitor Squeeze Treatments Combined with Other Well Treatments

Scale inhibitors have been added to fracturing fluids to provide protection against scale-related damage during the early stages of water breakthrough.[181–183] However, adding a water-soluble scale inhibitor to a water-based fracturing fluid is not always straightforward as studies have shown.[184] However, permeable proppants impregnated with scale inhibitor have been successfully used in the field to deliver scale protection during and after a fracturing operation.[185-190] Simultaneous fracturing and scale inhibitor squeeze treatments in gas reservoirs have also been carried out.[191]

Combining HCl acid stimulation chemicals and scale inhibitors is not a straightforward process either due to compatibility problems among the acid, the acid additives, the scale inhibitor, and the spent acids containing higher cation salinities. However, certain scale inhibitors are not only compatible with HCl but also that they retain their ability to adsorb onto reservoir rock under highly acidic conditions.[192] Hence, a scale inhibitor could be deployed directly in the acid system, negating the need for a separate scale inhibition treatment.[193] Combined acidizing scale removal and scale inhibition treatments have also been reported.[194] However, one study showed that chelates used as sulfate scale dissolvers seriously impair the performance of a range of scale inhibitors such as phosphonates, polycarboxylates, and phosphate esters. Only highly sulfonated polymers such as PVS appeared to be immune to the chelates.[195]

In addition, a relative permeability modifier (for water shut-off) and scale inhibitor–combined treatment has been successfully carried out.[196] A combined scale-inhibitor treatment and water-control treatment requiring fewer steps than the sum of each treatment procedure practiced separately has been claimed.[197] Rather than carry out two separate well treatments, it has been claimed to carry out a water shut-off and scale-squeeze treatment in a single step. Thus, a carboxylate polymeric scale inhibitor and cross-linker is injected into the formation, which becomes cross-linked in the near–well bore providing a water shut-off gel. When the gel breaks down, the scale inhibitor is then available to control scaling.[198] Other production chemicals besides scale inhibitors can be retained in the formation in the polymer gel.

3.7.3 Nonaqueous or Solid Scale Inhibitors for Squeeze Treatments

One problem with downhole water-based squeeze treatments is that the aqueous solutions of scale inhibitor may change the wettability of the rock. Once water-wet, the water permeability of the rock will have been changed, sometimes permanently, so that a water channel may eventually open up into a water pocket, leading to the so-called water coning effect, wherein a well is irreversibly damaged. Such a well will never again return to full productivity and new perforations need therefore be sunk to economically extract oil from the field. Relative permeability changes may also take place with water-based squeeze treatments. Thus, avoidance of the use of water-based squeezed products in low water cut wells or water-sensitive wells is important to avoid loss of oil production. In addition, wells with high water cuts with lifting problems would benefit from lower density, nonaqueous squeeze treatments.

To overcome this problem, a variety of nonaqueous scale-inhibitor compositions have been developed.[199-201] These inhibitors can be:

- oil-miscible (often referred to as oil-soluble)
- totally water-free materials in organic solvent blends
- invert emulsions, microemulsions, or nanoemulsions
- encapsulated products

Scale Control

3.7.3.1 Oil-Miscible Scale Inhibitors

Scale inhibitor compositions in hydrocarbon solvents have been developed that avoid the practice of injecting aqueous solutions or calcium-sensitive scale inhibitors. They are often referred to as oil-soluble scale inhibitors. In fact, the composition is an alkylamine and the acid form of the scale inhibitor, which form an ion pair, which is oil-soluble. The blend can be diluted in a cheap hydrocarbon solvent such as diesel or kerosene. An example is a blend of a tertiary alkyl primary amine with an aminophosphonic acid such as DTPMP. The acid form of the inhibitor contains water, as sold, so that the final oil-soluble inhibitor composition does actually contain a low percentage of water (5–10%).[202] A more preferred amine is 2-ethylhexylamine as it has better environmental properties.[203] When injected, the oil-soluble scale inhibitor will partition into the formation water and be adsorbed onto the rock. The overflush used for deeper penetration of the inhibitor can be hydrocarbon-based such as diesel. Oil-soluble scale inhibitors are now successfully field-proven in water-sensitive wells.[204–205]

Oil-soluble scale inhibitors can also be used in continuous injection. In such cases, they can be coinjected with THIs such as methanol and glycols products, which are not compatible with some scale inhibitors. Alternatively, other oil-soluble production chemicals such as wax and asphaltene inhibitors can be blended with the oil-soluble scale inhibitors to avoid using more than one umbilical.

3.7.3.2 Totally Water-Free Scale Inhibitors in Organic Solvent Bends

Phosphorous-tagged, polymeric, sulfonated scale inhibitors have been found to dissolve in blends of organic solvents without any water present at all.[206–207] They work in a similar way to the oil-miscible scale inhibitors described above by partitioning back into the formation water on squeezing. From here, they can be precipitated onto the formation rock. In addition, the scale inhibitor precipitation can be further controlled and enhanced through emulsification with a calcium-loaded aqueous medium also containing organic additives to aid and enhance the precipitation process. Although this involves incorporating water into the product package, the invert emulsion provides an oil-continuous matrix and is considered by the authors to be significantly less damaging to water-sensitive formations than standard oil-miscible and emulsified scale inhibitor packages.

3.7.3.3 Emulsified Scale Inhibitors

Invert emulsions are water-in-oil macroemulsions. They can be tuned to break at reservoir temperatures, thereby releasing the aqueous phase containing a water-soluble scale inhibitor or other production chemical.[169,208–211] A demulsifier with a cloud point of about 40–60°C (104–140°F) can be injected after the macroemulsion and cause the macroemulsion to break releasing the scale inhibitor.[212] Microemulsion squeeze treatments have also been reported.[213–214] An emulsified precipitation scale inhibitor squeeze treatment using urea for enhanced precipitation has been successfully carried out in the field.[206] Temperature-sensitive nanoemulsions containing up to 3% scale inhibitor, which are prepared using biodegradable nonionic surfactants at lower dosages than those used for making invert or microemulsions have been qualified in the laboratory, but field trials have not yet been reported. The nanoemulsions show significantly better stability at room temperature compared with other emulsions.[215]

The water-in-oil emulsion does not need to break if it can be retained in the pores over a long period. This normally requires a low-permeability reservoir. Thus liquid-liquid microencapsulation, slowly degrading emulsion products have been developed.[216,217] This droplet entrapment technique is useful for squeezing many other production chemicals such as biocides, H_2S scavengers, corrosion inhibitors, and so forth. Field applications have been reported.[218] For high-permeability reservoirs, where trapping will not take place, an invert emulsion precipitation squeeze can be carried out to circumvent this problem.[219]

3.7.3.4 Solid Scale Inhibitors (for Squeezing and Otherwise)

A technique that avoids the use of water in squeeze treatments is to use very fine, coated particles containing scale inhibitor. A suspension of these microencapsulated scale inhibitor particles in a liquid hydrocarbon can be squeezed into the formation and trapped in the pores, slowly releasing scale inhibitor into the produced water when put back on production.[220] The technology has successfully been used to control sulfate scaling.[221]

There are various ways of producing particles containing scale inhibitor including 100%, solid products, encapsulated products, and highly porous materials that can capture the inhibitor.[222–225] An improved method of generating particles containing scale inhibitor is by thermal radical-initiated polymerization of suitable monomers in an inverse emulsion or suspension containing the scale inhibitor.[226] In this way, the rate of leaching of the inhibitor, precursor, or generator may be controlled or selected by choice of particle properties (e.g., swellability, porosity, degradability, size, molecular weight, degree of cross-linking), which, in conjunction with properties of the downhole environment (e.g., temperature, pH, salinity), will govern the leaching or release rate.

A suspension of scale inhibitor particles can also be injected at water injector wells. From here, the suspension will percolate through the subterranean formation towards a production well and controllably release the scale inhibitor from the particles in the near–well bore region of the production well.[227] Suitably, the particles of the controlled-release scale inhibitor comprise an esterifiable scale inhibitor cross-linked with a polyol via ester cross-links.

Solid particles containing scale inhibitors or other production chemicals can also be used in other situations besides squeeze treatments. Examples for fracturing operations were discussed earlier in Section 3.7.2.1. Scale-inhibitor–impregnated proppants have also been used in gravel-pack completions.[228] A weighted solid-scale inhibitor capsule, suspended in a carrier brine has been pumped to the wellhead and allowed to fall, under gravity, into the sump. On reaching the sump, the diffusion of scale inhibitor from the capsule established a concentration gradient, which delivered a near-constant level of inhibitor over the lifetime of the treatment.[229] Scale inhibitors particles can also be placed behind screens in gravel packs where they will gradually release the inhibitor on contact with produced water or as proppants in fracture operations.[230–231] Scale inhibitor particles can also be placed behind sliding sleeves, which are opened when water is detected.[232]

Calcium-DTPMP submicron-sized particles, prepared directly by chemical precipitation with the assistance of PPCA, have been shown to provide enhanced retention of the phosphonate scale inhibitor in laboratory coreflooding.[233] Adsorption of PPCA to Ca-DTPMP particles increases their negative surface charge and decreases particles deposition in porous media.

3.7.4 Placement of Scale Inhibitor in a Squeeze Treatment

Placement of the scale inhibitor in a squeeze application is important so that all the water-producing zones are treated.[146–147] This is especially true in long, deviated, or horizontal wells where you do not want all the inhibitor to be squeezed into the heel of the well but to penetrate also at the toe of the well.[234] If significant permeability or pressure variations are present in the interval to be treated, treatment fluid will enter the zones with the higher permeability and lower pressure, leaving little fluid to treat the other zones, which can potentially be the water-producing zones. The challenge is even greater in long horizontal wells with significant permeability and pressure contrast. To achieve a more uniform fluid coverage, the original flow distribution across an interval often needs to be altered. The methods used to alter this are called "diversion" methods. The purpose is to divert the flow from one portion of the interval to another. A section on chemical diverter techniques is given in Chapter 5 on acid stimulation. Some of the techniques listed there do not appear to have been investigated for scale-inhibitor placement: these include viscoelastic surfactants, biodegradable. The main two methods of diversion that have been used in scale inhibitor squeeze applications are wax diverters and viscosified aqueous solutions. Foamed scale inhibitor treatments have recently been patented.[329] The foams collapse to liquid shortly after delivery.

For wax diverters, particles of wax are injected, which block the perforations at the heel of the well and divert the fluid towards the toe. The beads will later melt as they reach the temperature of the formation avoiding any formation damage.[235–237]

The second method is to use a temporary, self-diverting viscosified (or gelled) scale inhibitor solution.[238] Use of polymers such as xanthan or succinoglycan in the scale inhibitor solution giving increased viscosities lead to self-diversion, useful for better scale inhibitor placement in horizontal and complex wells.[239] Lightly viscosified shear-thinning fluids based on xanthan have been successfully used in laboratory studies at up to 170°C (338°F).[240] Succinoglycan breaks irreversibly above about 70°C (158°F) in seawater giving no permeability loss, so its use is limited to fairly low-temperature well applications. In laboratory tests, it also gave increased squeeze lifetimes compared with a nonviscosified treatment. Xanthan is a shear-thinning polysaccharide that breaks thermally when heated above 115°C (239°F), depending on the salinity. Salinity increases the breaking temperature. For xanthan, thermal polymer degradation usually with an oxidant breaker is needed to "break" the viscous gel completely and return fluid viscosity approximate to conventional aqueous-based squeeze treatment deployment. Theoretical modeling indicated that to achieve best placement across multiple-fractured zones that the viscosified scale inhibitor solution should be a lightly shear-thinning fluid of about 20 cp and injected at low flow rates

(1–10 bpm).[241] On one North Sea field, a viscous scale inhibitor squeeze treatment was only partly successful for a deviated well. The placement of the scale inhibitor seemed to be good, but the well was damaged from plugging during injection.[242]

Emulsifying the scale inhibitor also gives a higher viscosity (see section on emulsified scale inhibitors).[243] This has been done successfully in the field for better placement of acid stimulation treatments. Guar polymers have been claimed to give benefits over the use of xanthan as a viscosifier.[244] Only about 0.3 wt.% polymer is needed. A polycationic guar showed particularly good results and may also give enhanced inhibitor retention. Chapter 5 on acid stimulation should be consulted for more information on other diverter techniques.

3.8 PERFORMANCE TESTING OF SCALE INHIBITORS

The most common types of laboratory tests carried out for new scale inhibitor applications include the following:[246]

- static bottle tests
- tube blocking test
- compatibility tests
- thermal ageing (in solution or on rock)
- static adsorption tests
- dynamic adsorption tests—core flooding including permeability changes

Static bottle tests are used as a rough screening method to rank the performance of sulfate scale inhibitors. This is most usually done for barium sulfate inhibition (another method of evaluation of scale inhibitor efficiency consists of determining a supersaturation level of any scale forming compound in a given water, at defined conditions, in the presence of a specified amount of scale inhibitor or without inhibitor).[245] For continuous injection field applications, the next step is to carry out dynamic tube-blocking tests with the highest performing candidates from the static tests.[246] Tube-blocking tests will be the first step for carbonate scale inhibitors. In the dynamic tube-blocking test, inhibitors are tested at decreasing concentrations until a given pressure drop occurs across the tube. Prescaling of the tube is often carried out to get more repeatable results. Using this technique, the minimum inhibitor concentration (MIC) necessary for complete inhibition can be ascertained and thus the performance of inhibitors can then be compared. The MIC is a guideline figure for use in the field and its effect needs monitoring once in use.[58,247] It is important to do inhibitor performance testing at the pH of the produced water in the presence of the main produced cations. For example, Fe(II) ions can affect the performance of many scale inhibitors. Often, produced water pH is 5–6 rather than a pH of 2–3 for many scale inhibitor solutions. Brine compatibility is also important for continuously dosed inhibitors: in an extreme case, it is possible to form a scale consisting of Ca/Mg salts of the inhibitor at a point immediately downstream of where the inhibitor mixes with the produced fluids.

For squeeze treatments, three further studies are needed. First, compatibility tests are needed to check that the scale inhibitor does not precipitate when mixed with formation brines causing formation damage. Second, and especially for high-temperature reservoirs, thermal ageing tests are needed to make sure the inhibitor is

Scale Control

stable at the reservoir temperature for the expected squeeze lifetime. The inhibitor solution is aged in a static bottle and then performance-tested against a nonaged sample. It has long been believed that many classes of scale inhibitors were not thermally stable for high-temperature reservoirs based on static bottle-test thermal ageing studies followed by dynamic tube-blocking tests. For example, vinyl sulfonate polymers appeared to be thermally far more stable than aminophosphonates at high temperatures. However, it has been pointed out that in the field, the scale inhibitor is deposited on the formation rock, which can slow down the thermal degradation process. Consequently, laboratory thermal ageing studies should be carried out on scale inhibitors adsorbed to rock samples.[110]

The second type of test that can be performed for evaluating squeeze treatments is the static adsorption test.[248] This bottle test determines the adsorption (not precipitation) ability of a scale inhibitor on crushed formation rock. The final test, and the most important, is a dynamic core flood. The inhibitor is flooded into a core sample at reservoir conditions and back-produced using formation water. The concentration of inhibitor in the produced water is monitored until it drops below the MIC. From these data, one can plot an adsorption isotherm and determine the expected squeeze lifetime compared with other products. Permeability changes can also be monitored with a coreflood.

A few studies have focused on deposits formed on the surface of metals. One study demonstrated that bulk precipitation and surface deposition have different dependencies on the index of supersaturation, and so, to completely understand an industrial scaling system, both processes should be studied.[249]

3.9 CHEMICAL SCALE REMOVAL

This section reviews chemicals for metal carbonate, sulfate, sulfide, and lead scale removal. Soft halite (NaCl) scale is simply removed by washing or jetting with low-salinity water. If seawater is used, a sulfate scale inhibitor may need to be added. Mechanical methods of scale removal have been reviewed.[1,250] Abrasive jetting (water with sharp-edged sand particles) is better than water jetting alone but can damage tubulars. Rounded beads can reduce the damage. They are also acid-soluble, making clean-up operations simple. Viscosifying the water with a polymer such as xanthan helps reduce friction.[251–252]

3.9.1 Carbonate Scale Removal

Chapter 5 on acid stimulation reviews, among other topics, methods of stimulating carbonate reservoirs. The chemical solutions are very much the same for carbonate scale removal, which is also a form of stimulation. Therefore, this section will only give a brief discussion related to carbonate scale removal.

Calcium carbonate (calcite) is by far the most common carbonate scale, followed by iron carbonate (siderite).[253] Carbonates are dissolved by all number of acids, the easiest and cheapest being hydrochloric acid (HCl):

$$2HCl + CO_3^{2-} \rightarrow H_2O + CO_2 + 2Cl^-$$

The exothermicity of the reaction and the release of CO_2 gas can both speed up the rate of reaction. As the acid spends, the concentration of metal ions (e.g., calcium ions) in solution increases and acts to retard the reaction. An additive that must be used in an acid dissolver package is a corrosion inhibitor. Two other common additives are

- iron control agent
- water-wetting surfactant

These additives and others, such as those to counteract sludging problems are reviewed in Chapter 5 on acid stimulation. Organic acids have also been used as carbonate scale dissolvers. Use of organic acids such as acetic acid (CH_3COOH) will slow down the reaction with carbonates compared with HCl, and they are less corrosive than HCl. The use of long-chain organic acids has also been reported.[254] The reducing acid, formic acid, and buffered blends with formate ions can be useful to reduce corrosion problems in carbonate scale removal at high temperatures.[255–256] Gelled organic acid have been used for improved $CaCO_3$ removal in horizontal openhole wells.[257] Citric acid has also been investigated.[258]

One way to avoid corrosion and sludging problems in carbonate scale removal is to use a chelate dissolver instead of acids. Salts of EDTA, such as Na_2EDTA and other aminocarboxylate chelating agents, also known as sequestrants, have been used to some extent to remove carbonate scale formation damage, especially in high-temperature environments (Figure 3.23).[259–260] Otherwise, they are rarely used commercially as carbonate dissolvers probably because HCl and organic acids are less expensive. Phosphonate scale inhibitors with carboxylic acid groups, such as 2-phosphonobutane-1,2,4-tricarboxylic acid, can also be used.[261]

Other more biodegradable chelating agents have been investigated.[262] These include salts of hydroxyethyliminodiacetic acid (HEIDA) and hydroxyethylethylenediamine triacetic acid (HEDTA) and glutamic acid diacetic acid (GLDA) (Figure 3.24). A salt of HEIDA has the dissolver capability of 7.5% HCl and is biodegradable whereas EDTA is not. However HEIDA did not perform as well as HEDTA against gypsum scale.[263] These hydroalkyl chelates have a benefit over EDTA in that they are more soluble in acid solutions if they must be used for iron control in acid stimulation.[264] Another biodegradable chelate dissolver that is now commercially available is trisodium ethylenediaminodisuccinate (Figure 3.24).[265]

A viscoelastic surfactant such as *N*-erucyl-*N*,*N*-bis(2-hydroxyethyl)-*N*-methyl ammonium chloride can be used to obtain better placement of the dissolver. In use,

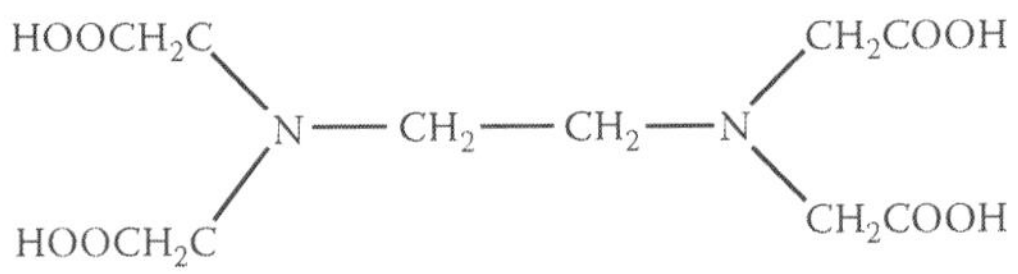

FIGURE 3.23 Ethylenediaminetetraacetic acid.

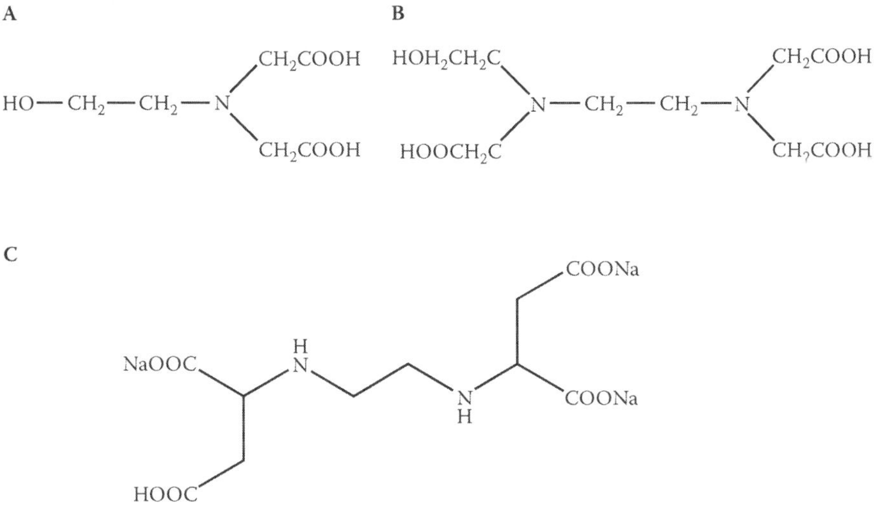

A

HO—CH₂—CH₂—N(CH₂COOH)(CH₂COOH)

B

(HOH₂CH₂C)(HOOCH₂C)N—CH₂—CH₂—N(CH₂COOH)(CH₂COOH)

C

[structure of trisodium ethylenediaminodisuccinate with NaOOC, HOOC, COONa, COONa groups and two NH centers connected by ethylene bridge]

FIGURE 3.24 Hydroxyethyliminodiacetic acid (A), hydroxyethylethylenediamine triacetic acid (B), and trisodium ethylenediaminodisuccinate (C).

formation hydrocarbons act on the surfactant to reduce the viscosity of the fluid so that the fluid selectively invades a hydrocarbon-bearing zone of the formation.[266]

Iron carbonate (siderite) scale can be removed using the usual chelates mentioned above. Iron oxides are harder to remove with high pH chelants. Amino acid/hydroxyaromatic chelates and sulfonated hydroxyaromatics are better at dissolving iron oxides at pH 7 or higher, while biodegradable citric acid can be used at low pH. Phosphonates also show good performance.[267]

3.9.2 Sulfate Scale Removal

Calcium sulfate (gypsum or anhydrite, $CaSO_4 \times H_2O$) is the easiest sulfate scale to remove chemically and barium sulfate (barite, $BaSO_4$), the hardest. Calcium sulfate is not soluble in acids but will dissolve in high-pH chelate solutions, the same products discussed above for carbonate scale removal.[268] The standard products for many years have been sodium salts of EDTA (Figure 3.23), but there is a move to more biodegradable products in environmentally strict areas such as the North Sea basin.[263] Other biodegradable chelate dissolvers such as those discussed for calcium carbonate scale dissolution and also glutamic acid N,N-diacetic acid, 2-(1,2-dicarboxyethylimino)-3-hydroxybutane diacid, carboxymethylimino-3-hydroxybutane diacid, 2-(1,2-dicarboxyethylimino) butanediacid, and (S)-aspartic acid-N-monoacetic acid have also been claimed.[269–271] An organosilane can be added to the chelate dissolver to stabilize fines migration.[272] Addition of chelating aminophosphonates (which are also scale inhibitors) to EDTA formulations has been claimed to enhance calcium sulfate dissolution.[273] A product that is claimed to enhance disintegration of gypsum scale better than EDTA is hydroxamic acid and/or certain salts thereof.[274] Chelate dissolver treatments can also remove other scales/damage such as iron oxides, calcium, and aluminum silicates, and bromides.[291]

An alternative method of removing gypsum scale (and maybe even barite scale) is to convert the calcium sulfate to calcium carbonate by pumping in aqueous CO_2/bicarbonate mix at pH < 7. The acidic solution then dissolves the carbonate.[275] The use of soda ash (sodium carbonate) to "convert" the gypsum to calcium carbonate and subsequently the use of an HCl wash to dissolve this salt is a low-cost two-stage process that has often been used in the United States for gypsum remediation.[1]

Barium sulfate or mixed strontium/barium sulfate scale is a notoriously hard scale. In severe scaling cases, it may be best to remove it by mechanical means such as milling or jet blasting.

A good review of mechanical scale-removal methods has been published.[250] A new mechanical method using plasma channel drilling has been patented.[276] There is only one chemical class capable that has been consistently used in the field for dissolving barium sulfate scale at an appreciable rate and that is salts of diethylenetriamine pentaacetic acid (DTPA) at pH higher than 12 (Figure 3.25).[277-278] Nearly all good commercial barium sulfate dissolvers contain at least some DTPA with performance enhancers (synergists).[279] However, a new class of barite dissolver has been reported with similar performance to DTPA-based dissolvers.[280] Little information on the structure of the dissolver has been published except to say that it is a diester and works at pH 5–6. The diester dissolver is also readily biodegradable and is, therefore, of particular use in environmentally sensitive areas. A barite dissolver with 41% 28-day biodegradation in the OECD306 test has also been claimed, but the structure was not reported.[281] Crown ethers and cryptates were researched in the late seventies as potential barium sulfate dissolvers, but even today, they are still too expensive for commercial use.[282-283]

Returning to the well-known DTPA-based dissolvers, DTPA salts are octadentate chelating ligands that can bind to Ba^{2+} ions via oxygen atoms on the five carboxylate groups as well through the lone pairs on the three nitrogen atoms to form soluble $[Ba(DTPA)]^{3-}$ species.[284] Studies with noncontact atomic force microscopy suggest that the initial reaction is that active sites of one DTPA molecule binds to two or three Ba^{2+} cations exposed on the (001) barite surface.[285] More expensive sequestrants such as triethylenetetraamine hexaacetic acid and DIOCTA have also been shown to be good dissolvers for barite scale.[286-288] A method of regenerating the chelate (EDTA, DTPA, etc.) from the dissolved scale solution has been claimed.[289] A completely different class of barite dissolver are ionic liquids such as trimethylamine dialuminum

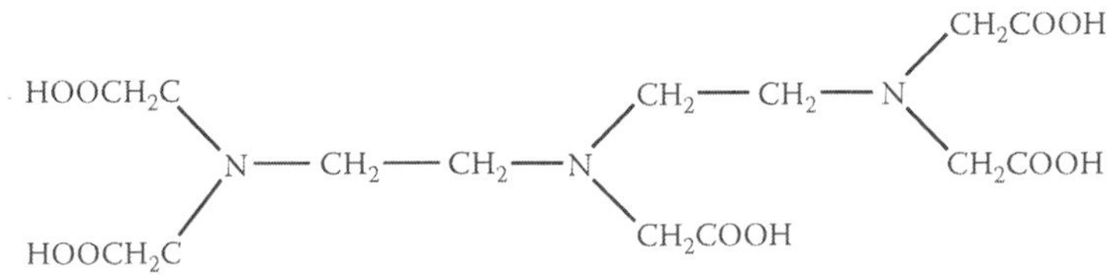

FIGURE 3.25 Diethylenetriamine pentaacetic acid.

heptachloride. These were shown to perform better than EDTA salts. However, their cost is, at present, considered too high for field applications.[290]

The reaction of barium sulfate with DTPA dissolvers is fairly slow at room temperature. The dissolution rate is improved at downhole temperatures and under good agitation. The potassium salt of DTPA, K_5DTPA appears to work better than the sodium salt.[291] Addition of a little excess base such as sodium hydroxide can be useful in case the dissolver reacts with CO_2, H_2S, or organic acids downhole, lowering the pH and rendering the dissolver less effective.[292] It has been shown in laboratory studies that solutions containing inorganic group I alkalis (K and Na) as pH-controlling agents deteriorate the permeability of artificially consolidated sandstone cores, independent of the fact that they were chelate containing or chelate-free. Using basic ethylamines gave the same effect but not with methylamines.[293]

Several enhancers can be used to catalyze the rate of reaction of the chelate with barite scale. For example, carbonate[294] and formate ions are commonly used today and oxalate ions have been used in the past.[278,295] Citrate, thiosulfate, nitriloacetate, mercaptoacetate, hydroxyacetate, or aminoacetate ions have also been proposed as synergists.[296] Sodium glucoheptonate has been claimed as a synergist for EDTA- or DTPA-based dissolvers[297] as well as 2-phosphonobutane-1,2,4-tricarboxylic acid.[298]

None of the best-performing chelates for barite scale (i.e., DTPA and EDTA) are seawater biodegradable according to the OECD 306 test. However, these molecules can be degraded under more realistic marine conditions, that is, photodegradation and subsequent biodegradation of the photodegradation products.[299]

Strontium sulfate can be dissolved using various chelates mentioned above such as EDTA or HEIDA salts.[300] Strontium sulfate is often found together coprecipitated with barium sulfate and, hence, some DTPA is then needed in the formulation. Calcite-barite mixed scales also need to be dissolved with sequestrants containing at least some DTPA.[301]

The dissolution rate of scales can also be accelerated by adding a small NIA, such as butyl triglycol ether. The NIA appears to work by cleaning the barite surface free from hydrocarbons so that it can react with the dissolver, and altering its wettability.[302] Barium and strontium sulfate scales contain small amount of radioactive radium ions in the lattice called NORM (naturally occurring radioactive material) or LSA (low specific activity) scale. Although the radioactivity levels are low, solids or solutions of this scale require specific handling and disposal regulations.

3.9.3 SULFIDE SCALE REMOVAL

Chemicals that can remove iron sulfide scales include:

- hydrochloric and organic acids
- acrolein (also as a biocide)
- tetrakishydroxymethylphosphonium salts (also biocides)
- chelates (sequestrants)

Iron sulfide, the most common sulfide scale, is generally soluble in both hydrochloric acid (HCl) and organic acids such as formic acid (HCOOH).[303] In general, iron sulfide deposits with low sulfur content have higher solubility in acids.[304] However, old samples of iron sulfide may have changed somewhat in their composition and are not so easy to dissolve in acids. In one study, pyrite and marcasite iron sulfides were found to be acid insoluble, pyrrhotite soluble at a low pace and mackinawite highly soluble.[305] Maleic acid has been claimed as a ferrous sulfide dissolver giving minimal H_2S production.[306]

Hydrochloric acid (HCl) will also remove zinc sulfide scales but not lead sulfide. HCl is corrosive and, thus, a corrosion inhibitor needs to be added to the formulation. However, one study showed that film-forming corrosion inhibitors reduced the rate of iron sulfide dissolution.[307] The same study showed that surfactants enhanced the dissolution rate but mutual solvent did not. However, a mutual solvent such as monobutyl glycol ether can be beneficial to remove hydrocarbons from the surface of the iron sulfide deposits.[308] A H_2S scavenger needs to be added to remove toxic H_2S gas produced by the reaction of HCl with iron sulfide:[309–310]

$$FeS + 2HCl \rightarrow Fe^{2+} + 2Cl^- + H_2S$$

However, aldehyde-based sulfide scavengers reduced the rate of dissolution of iron sulfide probably due to a polymeric deposit formed on the scale surface. Triazine scavengers can be used instead. As HCl acid spends and the pH increases, it is possible that insoluble iron salts can be formed. Hence, an iron-control agent such as citric acid can be added to prevent this. However, addition of citric acid to HCl was shown to reduce the amount of iron sulfide dissolved. Other chelates such as EDTA could be used but will probably have the same effect. Aminocarboxylate chelates at pH 8–10 have been claimed to remove iron sulfide by themselves, although the reaction is slow.[311]

Less corrosive acids than HCl can be used to remove sulfide scales at high temperatures, such as formic acid, thioglycolic acid, glyoxylic acid, and maleic acid.[312–314]

Acrolein (2-propenal) has also been shown to dissolve iron sulfide scales besides being a H_2S scavenger and biocide.[315–318] However, it should be noted that acrolein has high acute toxicity and is a suspected carcinogen, so it needs to be handled carefully.

Tetrakis hydroxymethylphosphonium salts such as the sulfate (THPS) have also been shown to remove iron sulfide scale both downhole and topside when coinjected with a small alkylamine or preferably an ammonium salt (see also Chapter 14).[319–321] These formulations can also be used in blends with a solution of a strong acid such as HCl.[322] The iron ends up being chelated to a nitrogen-phosphorus ligand giving the water a red color. In one field, use of THPS with a surfactant killed the SRB, removed iron sulfide scale, and, surprisingly, increased well production in several wells by as much as 300%.[323] In another field application, THPS was used to remove iron sulfide and control the growth of bacteria. Then, in the second-stage, formic acid was added to remove residual iron sulfide and some polymer residue in the near–well bore area.[321] THPS formulation with ammonium salts can be corrosive, especially at high temperatures. A corrosion inhibitor such as an acetylenic alcohol (e.g., propargyl alcohol) or a thio-compound such as thioglycolic acid can be added to alleviate this problem.[324–325]

It is also possible to use oxidizing agents such as chlorites/chlorine dioxide or permanganates for FeS removal, by conversion to soluble iron(III) ions. These oxidizing agents are so strong that corrosion would be expected. Chlorine dioxide has been used for the removal of iron sulfide sludge from water flood injection distribution systems and in acid stimulation treatments.[326,327]

3.9.4 Lead Scale Removal

Lead metal scale will only dissolve in very hot concentrated oxidizing acids such as nitric acid. However, this acid poses too much of a risk due to corrosion. It has been found that lead scale can be removed by treating it with a blend of acetic acid (CH_3COOH) and hydrogen peroxide (H_2O_2).[328] This generates the more oxidizing peracetic acid (CH_3COOOH) in situ, which is probably the active species. Similar lead dissolution results were obtained using potassium permanganate ($KMnO_4$) as the oxidant, but this led to MnO_2 precipitates. The MnO_2 precipitate could be removed using solutions of citric acid or other chelates.

REFERENCES

1. (a) W. Frenier and M. Ziauddin, *Formation, Removal, and Inhibition of Scale in the Oilfield Environment*, eds. N. Wolf and R. L. Hartman, Society of Petroleum Engineers, 2008. (b) J. C. Cowan and D. J. Weintritt, *Water-Formed Scale Deposits*, Gulf Publishing, 1976. (c) M. Davies and P. J. B. Scott, *Oilfield Water Technology*, Houston, TX: National Association of Corrosion Engineers (NACE), 2006.
2. I. R. Collins, "A New Model for Mineral Scale Adhesion," SPE 74655 (paper presented at the SPE International Symposium on Oilfield Scale, Aberdeen, UK, 30–31 January 2002).
3. M. M. Jordan, K. Sjursaether, I. R. Collins, N. D. Feasey, and D. Emmons, "Life Cycle Management of Scale Control within Subsea Fields and its Impact on Flow Assurance, Gulf of Mexico and North Sea Basin," *Chemistry in the Oil Industry VII*, Manchester, UK: Royal Society of Chemistry, 2002, 223.
4. (a) M. B. Tomson, A. T. Kan, and G. Fu, "Inhibition of Barite Scale in the Presence of Hydrate Inhibitors," SPE 87437, *SPE Journal* 10(3) (2005): 256. (b) R. Masoudi, B. Tohidi, A. Danesh, A. C. Todd, and J. Yang, "Measurement and Prediction of Salt Solubility in the Presence of Hydrate Organic Inhibitors," SPE 87468, *SPE Production & Operations* 21(2) (2006): 182.
5. (a) M. M. Jordan and E. J. Mackay, "Scale Control in Deepwater Fields, *World Oil* 2005, 226(9): 75-80. (b) G. M. Graham, E. J. Mackay, S. J. Dyer, and H. M. Bourne, "The Challenges for Scale Control in Deepwater Production Systems—Chemical Inhibition and Placement Challenges," Paper No. 02316 (paper presented at the Annual Spring Meeting of NACE International, CORROSION/2002, Denver, CO, 7–14 April 2002).
6. K. Davis, G. Wooodward, Y. Mottot, and D. Joubert, International Patent Application WO/2006/103203, 2006.
7. C. Sitz, J. Shumway, and C. Miller, "An Unconventional Scale from an Unconventional Reservoir," SPE 87434 (paper presented at the SPE International Symposium on Oilfield Scale, Aberdeen, UK, 26–27 May 2004).

8. D. L. Gallup and C. J. Hinrichsen, "Control of Silicate Scales in Steam Flood Operations," SPE 114042 (paper presented at the SPE International Oilfield Scale Conference, Aberdeen, UK, 28–29 May 2008).
9. G. Atkinson and M. Mecik, "The Chemistry of Scale Prediction," *Journal of Petroleum Science & Engineering* 17 (1997): 113.
10. G. Rousseau, C. Hurtevent, M. Azaroual, C. Kervevan, and M.-V. Durance, "Application of a Thermo-Kinetic Model to the Prediction of Scale in Angola Block 3 Field," SPE 80387 (paper presented at the International Symposium on Oilfield Scale, Aberdeen, UK, 29–30 January 2003).
11. H. K. Kotlar, S. Jacobsen, and E. Vollen, "An Integrated Approach for Evaluating Matrix Stimulation Effectiveness and Improving Future Design in the Gullfaks Field," SPE 50616 (paper presented at the SPE European Petroleum, Conference, The Hague, The Netherlands, 20–22 October 1998).
12. (a) G. M. Graham, R. Stalker, and R. McIntosh, "The Impact of Dissolved Iron on the Performance of Scale Inhibitors under Carbonate Scaling Conditions," SPE 80254 (paper presented at the SPE International Symposium on Oilfield Scale, Houston, TX, 5–8 February 2003). (b) S. Yean, H. Al Saiari, A. T. Kan, and M. B. Tomson, "Ferrous Carbonate Nucleation and Inhibition," SPE 114124 (paper presented at the SPE International Oilfield Scale Conference, Aberdeen, UK, 28–29 May 2008). (c) K. Chokshi, W. Sun, and S. Nesic, "Iron Carbonate Scale Growth and the Effect of Inhibition in CO_2 Corrosion of Mild Steel," Paper 05285 (paper presented at the NACE International Corrosion Conference, Houston, TX, 2005).
13. I. R. Collins, "Predicting the Location of Barium Sulfate Scale Formation in Production Systems," SPE 94366 (paper presented at the SPE International Symposium on Oilfield Scale, Aberdeen, UK, 11–12 May 2005).
14. (a) P. J. Webb and O. Kuhn, "Enhanced Scale Management through the Application of Inorganic Geochemistry and Statistics," SPE 87458 (paper presented at the SPE International Symposium on Oilfield Scale, Aberdeen, UK, 26–27 May 2004). (b) J. J. Tyrie, International Patent Application WO/2007/144562.
15. M. Brown, "Full Scale Attack, Review," *BP Technology Magazine*, Oct.–Dec. 1998, 30–32.
16. I. R. Collins and M. M. Jordan, Occurrence, "Prediction and Prevention of Zinc Sulfide Scale within Gulf Coast and North Sea High Temperature/High Salinity Production Wells," SPE 68317 (paper presented at the SPE International Symposium on Oilfield Scale, Aberdeen, UK, 30–31 January 2001).
17. M. M. Jordan, K. Sjursaether, M. C. Edgerton, and R. Bruce, "Inhibition of Lead and Zinc Sulfide Scale Deposits Formed during Production from High Temperature Oil and Condensate Reservoirs," SPE 64427 (paper presented at the SPE Asia Pacific Oil and Gas Conference and Exhibition, Brisbane, Australia, 16–18 October 2000).
18. F. A. Hartog, G. Jonkers, A. P. Schmidt, and R. D. Schuiling, "Lead Deposits in Dutch Natural Gas Systems," *SPE Production & Facilities* 17 (2002): 122.
19. R. Jasinski and D. Frigo, "The Modelling and Prediction of Halite Scale," Proceedings IBC International Conference, *Advances in Solving Oilfield Scaling*, Aberdeen, UK, 22–23 January 1996.
20. J. K. Smith and J. L. Przybylinski, "The Effect of Common Brine Constituents on the Effficacy of Halite Precipitation Inhibitors," Paper No. 06837 (paper presented at the NACE CORROSION Conference, 2006).
21. V. Khoi Vu, C. Hurtevent, and R. A. Davis, "Eliminating the Need for Scale Inhibition Treatments for Elf Exploration Angola's Girassol Field," SPE 60220 (paper presented at the SPE International Symposium on Oilfield Scale, Aberdeen, UK, 26–27 January 2000).

22. E. Mackay and M. M. Jordan, "Natural Sulfate Ion Stripping during Seawater Flooding in Chalk Reservoirs," *Chemistry in the Oil Industry VIII*, Manchester, UK: Royal Society of Chemistry, 2003, 133.
23. E. Mackay, K. Sorbie, V. Kavle, E. Sørhaug, K. Melvin, K. Sjursæther, and M. M. Jordan, "Impact of In-Situ Sulfate Stripping on Scale Management in the Gyda Field," 100516-MS (paper presented at the SPE International Oilfield Scale Symposium, Aberdeen, UK, 31 May–1 June 2006).
24. I. R. Collins and P. A. Sermon, International Patent Application WO/2006/008506, 2006.
25. E. Acton and G. J. Morris, International Patent Application WO/2000/079095, 2000.
26. (*a*) L. Rzeznik, M. Juenke, D. Stefanini, M. Clark, and P. Lauretti, "Two Year Results of a Breakthrough Physical Water Treating System for the Control of Scale in Oilfield Applications," SPE 114072 (paper presented at the SPE International Oilfield Scale Conference, Aberdeen, UK, 28–29 May 2008). (*b*) D. Stefanini, International Patent Application WO/2008/017849.
27. (*a*) J. Groenenboom, S.-W. Wong, and G. Nitters, U.S. Patent Application 20040195187, 2004. (*b*) B. Wang and C. Wang, International Patent Application WO/2009/000177.
28. (*a*) O. G. Maxson and G. D. Achenbach, U.S. Patent 4115606, 1978. (*b*) S. Brand, A. Dierdorf, H. Liebe, F. Osterod, G. Motz, and M. Günthner, International Patent Application WO/2007/096070.
29. (*a*) I. R. Collins, "A New Model for Mineral Scale Adhesion," SPE 74655 (paper presented at the International Symposium on Oilfield Scale, Aberdeen, UK, 30–31 January 2002). (*b*) W. C. Cheong, A. Neville, P. H. Gaskell, and S. Abbott, "Using Nature to Provide Solutions to Calcareous Scale Deposition," SPE 114082 (paper presented at the SPE International Oilfield Scale Conference, Aberdeen, UK, 28–29 May 2008).
30. (*a*) M. G. Mwaba, M. R. Golriz, and J. Gu, *International Journal of Heat Exchangers* VI (2005): 235. (*b*) R. E. Herzog, Q. Shi, J. N. Patil and J. L. Katz, "Magnetic Water Treatment: The Effect of Iron on Calcium Carbonate Nucleation and Growth," *Langmuir* 5 (1989): 861.
31. (*a*) E. Chibowski, L. Holysz, A. Szczes, and M. Chibowski, "Precipitation of Calcium Carbonate from Magnetically Treated Sodium Carbonate Solution," *Colloids Surface A: Physicochemical and Engineering Aspects* 225 (2003): 63. (*b*) F. F. Farshad and S. M. Vargas, "Scale Prevention, a Magnetic Treatment Approach," SPE 77850 (paper presented at the SPE Asia Pacific Oil and Gas Conference and Exhibition, Melbourne, Australia, 8–10 October 2002).
32. J. D. Donaldson and S. M. Grimes, "Control of Scale in Sea Water Applications by Magnetic Treatment of Fluids," SPE 16540 (paper presented at the SPE Offshore Europe, Aberdeen, UK, 8–11 September 1987).
33. J. S. Baker and S. A. Parsons, "Anti-Scale Magnetic Treatment," *Water and Waste Treatment* 39 (1996): 36–38.
34. S. A. Parsons, B. L. Wang, S. J. Judd, and T. Stephenson, "Magnetic Treatment of Calcium Carbonate Scale—Effect of pH Control," *Water Research* 31 (1997): 339.
35. G. N. Jefferson, U.S. Patent 5738766, 1998.
36. D. Walker, U.S. Patent 6145542, 2000.
37. J. S. Gill, "Development of Scale Inhibitors," Paper 229 (paper presented at the NACE CORROSION Conference, 1996).
38. G. M. Graham and A. J. B. Hennessey, "Scale Inhibitor Surface Interactions Using Synchroton Radiation Techniques" (paper presented at the Chemistry in the Oil Industry VIII, Royal Society of Chemistry, Manchester, UK, November 2003).
39. G. M. Graham, L. S. Boak, and K. S. Sorbie, "The Influence of Formation Calcium and Magnesium on the Effectiveness of Generically Different Barium Sulphate Oilfield Scale Inhibitors," SPE 81825, *SPE Production & Facilities* 18(1) (2003): 28.

40. M. D. Yuan, E. Jamieson, and P. Hammonds, "Investigation of Scaling and Inhibition Mechanisms and the Influencing Factors in Static and Dynamic Inhibition Tests," Paper 98067 (paper presented at the NACE CORROSION Conference, San Diego, 22–27 March 1998).
41. W. H. Leung and G. H. Nancollas, "Nitrolotri (Methylenephosphonic Acid) Adsorption on Barium Sulfate Crystals and Its Influence on Crystal Growth," *Journal of Crystal Growth* 44 (1978): 163.
42. M. B. Tomson, G. Fu, M. A. Watson, and A. T. Kan, "Mechanisms of Mineral Scale Inhibition," *SPE Production & Facilities* 18(3) (2003): 192.
43. S. He, A. T. Kan, and M. B. Tomson, "Mathematical Inhibitor Model for Barium Sulfate Scale Control," *Langmuir* 12 (1996): 1901.
44. K. S. Sorbie and N. Laing, "How Scale Inhibitors Work: Mechanisms of Selected Barium Sulfate Scale inhibitors Across a Wide Temperature Range," SPE 87470 (paper presented at the SPE International Symposium on Oilfield Scale, Aberdeen, UK, 26–27 May 2004).
45. P. Wilkie, S. Heath, and C. Strachan, "An Overview of Scale Inhibitor Detection Techniques and Some Recent Advances in Detection Methods for Mixtures of Polymeric Scale Inhibitors in Produced Brines that Enable Improved Scale Management in Sub Sea and Deep Water Fields" (paper presented at the 19th Oilfield Chemical Symposium, Geilo, Norway, 9–12 March 2008).
46. J. D. Sallis, W. Juckes, and M. E. Anderson, "Phosphocitrate: Potential to Influence Deposition of Scaling Salts and Corrosion," in *Mineral Scale Formation and Inhibition*, ed. Zahid Amjad, New York: Plenum Press, 1995, 87.
47. H. A. El Dahan and H. S. Hegazy, "Gypsum Scale Control by Phosphate Ester," *Desalination* 127 (2000): 111.
48. (*a*) M. M. Jordan, N. Feasey, C. Johnston, D. Marlow, and M. Elrick, "Biodegradable Scale Inhibitors. Laboratory and Field Evaluation of a 'Green' Carbonate and Sulfate Scale Inhibitor with Deployment Histories in the North Sea," *RSC Chemistry in the Oil Industry X*, Manchester, UK, 5–7 November 2007. (*b*) A. F. Miles, S. H. Bodnar, H. C. Fisher, S. Sidoe, C. D. Sitz, Progress Towards Biodegradable Phosphonate Scale Inhibitors, International Oilfield Chemistry Symposium, Geilo, Norway, 23–25 March 2009.
49. W. R. Hollingshad, U.S. Patent 3932303, 1976.
50. M. Crossman and S. P. R. Holt, U.S. Patent 6995120, 2006.
51. C. Holzner, W. Ohlendorf, H.-D. Block, H. Bertram, R. Kleinstuck, and H.-H. Moretto, U.S. Patent 5639909, 1997.
52. J. L. Hen, U.S. Patent 5059333, 1991.
53. G. Woodward, C. R. Jones, and K. P. Davis, International Patent Application, WO/2004/002994.
54. K. P. Davis, G. F. Docherty, and G. Woodward, International Patent Application WO/2000/018695.
55. K. P. Davis, G. P. Otter, and G. Woodward, International Patent Application WO/2004/078662.
56. N. J. Stewart and P. A. M. Walker, U.S. Patent 6527983, 2003.
57. J. L. Przybylinski, G. T. Rivers, and T. H. Lopez, U.S. Patent Application 20060113505, 2006.
58. T. Johnson, C. Roggelin, C. Simpson, and R. Stalker, "Phosphonate Based Scale Inhibitors for High Iron and High Salinity Environments," *Proceedings of Chemistry in the Oil Industry IX*, Manchester, UK: Royal Society of Chemistry, 2005, 368.
59. F. A. Devaux, J. H. Van Bree, T. N. Johnson, and P. P. Notte, International Patent Application WO/2008/017338.
60. P. P. Notte and F. A. Devaux, International Patent Application WO/2008/017339.
61. G. Woodward, G. P. Otter, K. P. Davis, and R. E. Talbot, International Patent Application WO/2001/085616.

62. D. Redmore, B. Dhawan, and J. L. Przybylinski, U.S. Patent 4857205, 1989.
63. G. E. Jackson, K. Mclaughlin, N. Poynton, and J. L. Przybylinski, WO 97/21905.
64. (a) G. E. Jackson, G. Salters, P. R. Stead, B. Dahwan, and J. Przybylinski, "Using Statistical Experimental Design to Optimise the Performance and Secondary Properties of Scale Inhibitors for Downhole Application," *Recent Advances in Oilfield Chemistry V*, Manchester, UK: Royal Society of Chemistry, 1994, 164. (b) M. A. Singleton, I. A. Collins, N. Poynton, and H. J. Formston, "Developments in PhosphonoMethylated PolyAmine (PMPA) Scale Inhibitor Chemistry for Severe $BaSO_4$ Scaling Conditions," SPE 60216 (paper presented at the SPE International Symposium on Oilfield Scale, Aberdeen, UK, 26–27 January 2000).
65. T. L. Herrera, M. Guzmann, K. Neubecker, and A. Göthlich, International Patent Application WO/2008/095945.
66. E. A. Kerr and J. Rideout, U.S. Patent 5604291, 1997.
67. (a) D. H. Emmons, D. W. Fong, and M. A. Kinsella, U.S. Patent 5213691, 1993. (b) G.B. Clubley and J. Rideout, European Patent EP0479465, 1994.
68. J. F. Pardue and J. F. Kneller, U.S. Patent 5018577, 1991.
69. M. J. Smith, P. Miles, N. Richardson, and M. A. Finan, UK Patent Application GB 1458235, 1976.
70. K. P. Davis, G. Otter, and G. Woodward, International Patent Application WO/2001/057050.
71. K. Davis, G. Woodward, J. Hardy, K. Carmichael, and G. Otter, International Patent Application WO/2005/023904.
72. G. Woodward, G. Otter, S. Zafar, D. Bendejaco, and O. Anthony, International Patent Application WO/2007/017647.
73. G. Otter, S. Zafar, G. Woodward, International Patent Application WO/2006/032896.
74. G. Woodward, G. P. Otter, K. P. Davis, and K. Huan, International Patent Application WO/2004/056886.
75. (a) K. P. Davis, D. R. E. Walker, G. Woodward, and A. C. Smith, European Patent Application EP0 861846, 1998. (b) K. P. Davis, S. D. Fidoe, G. P. Otter, R. E. Talbot, and M. A. Veale, "Novel Scale Inhibitor Polymers with Enhanced Adsorption Properties," SPE 80381 (paper presented at the SPE International Symposium on Oilfield Scale, Aberdeen, UK, 29–30 January 2003).
76. http://www.wateradditives.com.
77. J. E. LoSasso, U.S. Patent 6322708, 2001.
78. C. Gancet, R. Pirri, B. Boutevin, C. Loubat, and J. Lepetit, U.S. Patent 6900171, 2005.
79. M. Guzmann, J. Rieger, T. L. Herrera, and K.-H. Buechner, International Patent Application WO2007000398, 2007.
80. M. Guzmann, Y. Liu, R. Konrad, and D. Franz, International Patent Application WO/2007/125073.
81. C. A. Costello and G. F. Matz, U.S. Patent 4,460,477, 1984.
82. H. Trabitzsch, J. Frieser, A. Koschik, and H. Plainer, U.S. Patent 4271058, 1981.
83. P. Chen, T. Hagen, H. Montgomerie, R. Matheson, T. Haaland, B. Juliussen, and R. Benvie, "A Scale Inhibitor Chemistry Developed for Downhole Squeeze Treatments in a Water Sensitive and HTHP Reservoir" (paper presented at the 19th Oilfield Chemical Symposium, Geilo, Norway, 9–12 March 2008).
84. P. Chen, T. Hagen, H. Montgomerie, R. Wat, and O. M. Selle, International Patent Application WO/2008/020220.
85. (a) N. Kohler, G. Courbinm, C. Estievenart, and F. Ropital, "Polyaspartates: Biodegradable Alternatives to Polyacrylates or Noteworthy Multifunctional Inhibitors?," Paper 02411 (paper presented at the NACE CORROSION Conference, 2002). (b) Z. Quan, Y. Chen, X. Wang, C. Shi, Y. Liu, and C. Ma, "Experimental Study on Scale Inhibition Performance of a Green Scale Inhibitor Polyaspartic Acid," *Science in China Series B: Chemistry* 51(7) (2008): 695–699. (c) L. P. Kosan and K. C. Low, U.S. Patent 511651, 1992.

86. (a) R. J. Ross, K. C. Low, and J. E. Shannon, "Polyaspartate Scale Inhibitors: Biodegradable Alternatives to Polyacrylates," *Materials Performance* 36 (1997): 53. (b) A. P. Wheeler and L. P. Koskan, Large Scale Thermally Synthesized Polyaspartate as a Substitute in Polymer Applications, *Mat. Res. Soc. Symp. Proc.*, 1993, 292, 277. (c) K. C. Low, A. P. Wheeler, and L. P. Koskan, Commercial Poly(aspartic acid) and Its Uses, Advances in Chemistry Series 248, 1996, Am. Chem. Society, Washington, D.C.
87. I. R. Collins, B. Hedges, L. M. Harris, BP Exploration; J. C. Fan, and T. D. G. Fan, "The Development of a Novel Environmentally Friendly Dual Function Corrosion and Scale Inhibitor," SPE 65005 (paper presented at the SPE International Symposium on Oilfield Chemistry, Houston, TX, 13–16 February 2001).
88. R. Kleinstuck, H. Sicius, T. Groth, and W. Joentgen, U.S. Patent 5525257, 1996.
89. P. Chen, T. Hagen, A. Maclean, O. M. Selle, K. Stene, and R. Wat, "Meeting the Challenges of Downhole Scale Inhibitor Selection for an Environmentally Sensitive Part of the Norwegian North Sea," SPE 76452 (paper presented at the SPE Oilfield Scale Symposium, Aberdeen, UK, 30–31 January 2002).
90. H. T. R. Montgomerie, P. Chen, T. Hagen, R. Wat, O. M. Selle, and H. K. Kotlar, International Patent Application WO/2004/011771, 2004.
91. H. K. Kotlar and J. A. Haugen, International Patent Application WO2002/095187.
92. (a) J. Tang and R. V. Davis, U.S. Patent 5776875, 1998. (b) P. Y. Zhang, L. P. Zheng, and Y. J. Chai, *Polymer Materials* 25 (2008): 259. (c) Z. H. Quan, Y. C. Chen, and X. R. Wang, *Science in China Series B Chemistry* 51(7) (2008): 695.
93. D. I. Bain, G. Fan, J. Fan, and R. J. Ross, "Scale and Corrosion Inhibition by Thermal Polyaspartates," Paper No. 120 (paper presented at the NACE CORROSION Conference, 1999).
94. D. I. Bain, G. Fan, J. Fan, H. Brugman, and K. Enoch, "Laboratory and Field Development of a Novel Environmentally Acceptable Scale and Corrosion Inhibitor," Paper No. 2230 (paper presented at the NACE CORROSION Conference, 2002).
95. J. Tang, R. T. Cunningham, and B. Yang, U.S. Patent 5750070, 1998.
96. C. G. Carter, L.-D. G. Fan, J. C. Fan, R. P. Kreh, and V. Jovancicevic, U.S. Patent 5344590, 1994.
97. (a) T. Saeki, A. Kanzaki, J. Nakamura, G. Fukii, and S. Yyamaguchi, Japanese Patent JP2001163941, 2001. (b) T. Saeki, H. Nishibayashi, T. Hirata, and S. Yamaguchi, U.S. Patent 5856288, 1999.
98. (a) J. M. Brown and G. F. Brock, U.S. Patent 5409062, 1995. (b) R. C. Xiong, Q. Zhou, and G. Wei, "Corrosion Inhibition of a Green Scale Inhibitor Polyepoxysuccinic Acid," *Chinese Chemical Letters* 14(9) (2003): 955–957.
99. D. L. Verraest, J. P. Peters, H. van Bekkum, and G. M. van Rosmalen, "Carboxymethyl Inulin: A New Inhibitor for Calcium Carbonate Precipitation," *Journal of the American Oil Chemists Society* 73(1) (1996): 55.
100. H. C. Kuzee and H. W. C. Raaijmakers, U.S. Patent 6613899, 2003.
101. D. L. Verraest, J. G. Batelaan, J. A. Peters, and H. van Bekkum, U.S. Patent 5777090, 1998.
102. (a) B. Bazin, N. Kohler, A. Zaitoun, T. Johnson, and H. Raaijmakers, "A New Class of Green Mineral Scale Inhibitors for Squeeze Treatments," SPE 87453 (paper presented at the SPE International Symposium on Oilfield Scale, Aberdeen, UK, 26–27 May 2004). (b) S. Baraka-Lokmane, K. S. Sorbie, and N. Poisson, "The Use of Green Scale Inhibitors for Squeeze Treatments, Carbonate Coreflooding Experiments," *Geophysical Research Abstracts* 9 (2007): 02444. (c) S. Baraka-Lokmane, K. Sorbie, N. Poisson, N. Kohler, *Petr. Sci. Techn.*, 2009, 27, 427.

103. (*a*) K. A. Rodrigues, J. S. Thomaides, A. L. Cimecioglu, and M. Crossman, U.S. Patent Application 20070015678, 2007. (*b*) F. Decampo, S. Kesavan, and G. Woodward, International Patent Application WO/2008/140729. (*c*) S. Baraka-Lokmane, K. Sorbie, N. Poisson, N. Kohler, *Petr. Sci. Techn.*, 2009, 27, 427.
104. (*a*) A. Viloria, L. Castillo, J. A. Garcia, and J. Biomorgi, U.S. Patent Application 20070281866, 2007. (*b*) S. P. Holt, J. Sanders, K. A. Rodrigues, M. Vanderhoof, Biodegradable Alternatives for Scale Control in Oil Field Applications, SPE 121723, SPE International Symposium on Oilfield Chemistry, The Woodlands, TX, April 20–22, 2009.
105. M. H. Salimi, K. C. Petty, and C. L. Emmett, U.S. Patent 5263539, 1993.
106. L. J. Persinski, P. H. Ralston, and R. C. Gordon, Jr., U.S. Patent 3928196, 1975.
107. G. F. Matz, U.S. Patent 4536292, 1985.
108. D. O. Falk, U.S. Patent 5360065, 1994.
109. R. Pirri, C. Hurtevent, and P. Leconte, "New Scale Inhibitor for Harsh Field Conditions," SPE 60218 (paper presented at the International Symposium on Oilfield Scale, Aberdeen, UK, 26–27 January 2000).
110. R.Wat, L.-E. Hauge, K. Solbakken, K. E. Wennberg, L. M. Sivertsen, and B. Gjersvold, "Squeeze Chemical for HT Applications—Have We Discarded Promising Products by Performing Unrepresentative Thermal Aging Tests?," SPE 105505 (paper presented at the SPE International Symposium on Oilfield Chemistry, Houston, TX, 28 February–2 March 2007).
111. M. M. Jordan, K. Mackin, C. J. Johnston, and N. D. Feasey, "Control of Hydrogen Sulphide Scavenger Induced Scale and the Associated Challenge of Sulphide Scale Formation Within A North Sea High Temperature/High Salinity Field Production Wells: Laboratory Evaluation to Field Application," SPE 87433 (paper presented at the SPE 6th International Symposium on Oilfield Scale, Aberdeen, UK, 26–27 May 2004).
112. K. C. Taylor, H. A. Nasr-El-Din, and J. A. Saleem, "Laboratory Evaluation of Iron-Control Chemicals for High-Temperature Sour-Gas Wells," SPE 65010 (paper presented at the SPE International Symposium on Oilfield Chemistry, Houston, TX, 13–16 February 2001).
113. J.A Billman, "Antibiofoulants: A Practical Methodology for Control of Corrosion Caused by Sulfate-Reducing Bacteria," *Materials Performance* 36 (1997): 43.
114. S. D. Fidoe, R. E. Talbot, C. R. Jones, and R. Gabriel, U.S. Patent 6926836, 2005.
115. M. A. Mattox and E. J. Valente, U.S. Patent 6986358, 2006.
116. (*a*) M. Ke and Q. Qu, U.S. Patent 7159655, 2007. (*b*) T. Chen, H. Montgomerie, P. Chen, T. H. Hagen, and S. Kegg, Development of Environmentally Friendly Iron Sulfide Inhibitors and Field Application SPE International Symposium on Oil Field Chemistry, The Woodlands, TX, April 20–22, 2009.
117. C. Okocha, K. S. Sorbie, and L. S. Boak, "Inhibition Mechanisms for Sulfide Scales," SPE 112538 (paper presented at the SPE International Symposium and Exhibition on Formation Damage Control, Lafayette, LA, 13–15 February 2008).
118. M. Lehmann and F. Firouzkouhi, "A New Chemical Treatment to Inhibit Iron Sulfide Deposition," SPE 114065 (paper presented at the SPE International Oilfield Scale Conference, Aberdeen, UK, 28–29 May 2008).
119. D. Emmons and G. R. Chesnut, U.S. Patent 4762626, 1988.
120. I. R. Collins and M. M. Jordan, "Occurrence, Prediction, and Prevention of Zinc Sulfide Scale within Gulf Coast and North Sea High-Temperature and High-Salinity Fields," SPE 84963, *SPE Production & Facilities* 18(3) (2003): 200–209.
121. T. H. Lopez, M. Yuan, D. A. Williamson, and J. L. Przybylinski, "Comparing Efficacy of Scale Inhibitors for Inhibition of Zinc Sulfide and Lead Sulfide Scales," SPE 95097 (paper presented at the SPE International Symposium on Oilfield Scale, Aberdeen, UK, 11–12 May 2005).

122. J. L. Przybylinski, "Iron Sulfide Scale Deposit Formation and Prevention under Anaerobic Conditions Typically Found in the Oil Field," SPE 65030 (paper presented at the SPE International Symposium on Oilfield Chemistry, Houston, TX, 13–16 February 2001).
123. S. Dyer, K. Orski, C. Menezes, S. Heath, C. MacPherson, C. Simpson, and G. Graham, "Development of Appropriate Test Methodologies for the Selection and Application of Lead and Zinc Sulfide Inhibitors for the Elgin/Franklin Field," SPE 100627 (paper presented at the SPE International Oilfield Scale Symposium, Aberdeen, UK, 31 May–1 June 2006).
124. W. Kleinitz, M. Koehler, and G. Dietzsch, "The Precipitation of Salt in Gas Producing Wells," SPE 68953 (paper presented at the SPE European Formation Damage Conference, The Hague, The Netherlands, 21–22 May 2001).
125. (a) H. Guan, R. Keatch, C. Benson, N. Grainger, and L. Morris, "Mechanistic Study of Chemicals Providing Improved Halite Inhibition," SPE 114058 (paper presented at the SPE 9th International Conference on Oilfield Scale, Aberdeen, UK, 28–29 May 2008). (b) H. Guan, R. Keatch, N. Grainger, and C. Benson, "Development of a Novel Salt Inhibitor" (paper presented at the 19th Oil Field Chemistry Symposium, Geilo, Norway, 9–12 March 2008).
126. (a) T. Chen, H. Montgomerie, P. Chen, O. Vikane, and T. Jackson, Development of Halite Test Methodology, Inhibitors, and Field Application, International Oilfield Chemistry Symposium, Geilo, Norway, 23–25 March 2009. (b) T. Chen, H. Montgomerie, P. Chen, O. Vikane, and T. Jackson, Understanding the Mechanisms of Halite Inhibition and Evaluation of Halite Scale Inhibitor by Static and Dynamic Tests, SPE 121458 (poster presented at SPE International Symposium on Oilfield Chemistry, 20–22 April 2009, The Woodlands, Texas).
127. M. A. Damme-van Weele, Ph.D. thesis, Twente University of Technology, The Netherlands, 1965.
128. M. C. Van der Leeden and G. M. van Rosmalen, *Chemicals in the Oil Industry*, Manchester, UK: Royal Society of Chemistry, 1988, 68.
129. C. Rodriguez-Navarro, L. Linares-Fernandez, E. Doehne, and E. Sebastian, "Effects of Ferrocyanide Ions on NaCl Crystallization in Porous Stone," *Journal of Crystal Growth* 243 (2002): 503.
130. D. M. Frigo, L. A. Jackson, S. M. Dora, and R. A. Trompert, SPE 60191 (paper presented at the SPE Second International Symposium on Oilfield Scale, Aberdeen, UK, 26–27 January 2000).
131. N. H. Zaid, U.S. Patent 5396958, 1995.
132. D. A. Smith, S. Sucheck, S. Cramer, and D. Baker, "Nitrilotriacetamide: Synthesis in Concentrated Sulfuric Acid and Stability in Water," *Synthetic Communications* 25(24) (1995): 4123.
133. J. W. Kirk, U.S. Patent 7028776, 2006.
134. R. Ralston et al., U.S. Patent 3367416, 1968.
135. S. Sarig, A. Glasner, and J. A. Epstein, "Crystal Habit Modifiers," *Journal of Crystal Growth* 28(3) (1975): 295.
136. S. Sarig and F. Tartakovsky, "Crystal Habit Modifiers: II. The Effect of Supersaturation on Dendritic Growth," *Journal of Crystal Growth* 28(3) (1975): 300.
137. J. W. Kirk and J. B. Dobbs, "A Protocol to Inhibit the Formation of Natrium Chloride Salt Blocks," SPE 74662 (paper presented at the SPE Oilfield Scale Symposium, Aberdeen UK, 30–31 January 2002).
138. S. Szymczak, R. Perkins, M. McBryde, and M. El-Sedaway, "Salt Free: A Case History of a Chemical Application to Inhibit Salt Formation in a North African Field," SPE 102627 (paper presented at the SPE International Symposium on Oilfield Chemistry, Houston, TX, 28 February–2 March 2007).

139. S. M. Heath, A. R. Thornton, E. K. McAra, M. Sim, A. Arefjord, E. Samuelsen, and R. Frederiksen, "Downhole Scale Control on Amerada Hess South Arne through Continuous Injection of Scale Inhibitor in the Water Injection System" (paper presented at the International Oilfield Chemistry Symposium, Geilo, Norway, 23–25 March 2009).
140. (a) G. Poggesi, C. Hurtevent, and D. Buchart, "Multifunctional Chemicals for West African Deep Offshore Fields," SPE 74649 (paper presented at the International Symposium on Oilfield Scale, Aberdeen, UK, 30–31 January 2002). (b) N. Fleming, K. Ramstad, S. H. Eriksen, E. Moldrheim, and T. R. Johansen, "Development and Implementation of a Scale-Management Strategy for Oseberg Soer," SPE 100371 (paper presented at the SPE International Oilfield Scale Symposium, Aberdeen, UK, 31 May–1 June 2006).
141. I. Fjelde, "Scale Inhibitor Injection in Gas-Lift Systems" (paper presented at the 11th International Oil Field Chemicals Symposium, Fagernes, Norway, 19–22 March 2000).
142. N. Fleming, J. A. Stokkan, A. M. Mathisen, K. Ramstad, and T. Tydal, "Maintaining Well Productivity through Deployment of a Gas Lift Scale Inhibitor: Laboratory and Field Challenges," SPE 80374 (paper presented at the International Symposium on Oilfield Scale, Aberdeen, UK, 29–30 January 2003).
143. G. Poggesi, C. Hurtevent, and J. L. Brazy, Scale "Inhibitor Injection Via the Gas Lift System in High Temperature Block 3 Fields in Angola," SPE 68301 (paper presented at the SPE International Symposium on Oilfield Scale, Aberdeen, UK, 30–31 January 2001).
144. M. N. Kelly, J. S. James, D. M. Frigo, D. W. Driessen, and A. D. Waldie, "Application of Scale Dissolver and Inhibitor Squeeze through the Gas Lift Line in a Sub-Sea Field," SPE 95100 (paper presented at the SPE International Symposium on Oilfield Scale, Aberdeen, UK, 11–12 May 2005).
145. A. Daminov, V. Ragulin, SPE, and A. Voloshin, "Mechanism Formations of Corrosion Damage of Inter Equipment in Wells by Continuous Scale-Inhibitor Dosing Utilizing Surface Dosing Systems: Testing Scale and Corrosion Inhibitors," SPE 100476 (paper presented at the SPE International Oilfield Corrosion Symposium, Aberdeen, UK, 30 May 2006).
146. K. S. Sorbie and R. D. Gdanski, "A Complete Theory of Scale-Inhibitor Transport and Adsorption/Desorption in Squeeze Treatments," SPE 95088 (paper presented at the SPE International Symposium on Oilfield Scale, Aberdeen, UK, 11–12 May 2005).
147. E. J. Mackay and K. S. Sorbie, "An Evaluation of Simulation Techniques for Modelling Squeeze Treatments," SPE 56775 (paper presented at the SPE Annual Technical Conference and Exhibition, Houston, TX, 3–6 October 1999).
148. A. T. Kan, G. Fu, M. Al-Thubaiti, J. Xiao, and M. B. Tomson, "A New Approach to Inhibitor Squeeze Design," SPE 80230 (paper presented at the SPE International Symposium on Oilfield Chemistry, Houston, TX, 5–7 February 2003).
149. M. B. Tomson, A. T. Kan, and G. Fu, "Control of Inhibitor Squeeze through Mechanistic Understanding of Inhibitor Chemistry," *SPE Journal* 11(3) (2006): 283–293.
150. (a) E. Hills, S. Touzet, and B. Langlois, International Patent Application WO/2005/001241. (b) K. Du, Y. M. Zhou, and L. Y. Dai, *International Journal of Polymer Materials* 57 (2008): 785.
151. J. E. Pardue, "A New Inhibitor for Scale Squeeze Applications," SPE 21023 (paper presented at the SPE SPE International Symposium on Oilfield Chemistry, Anaheim, CA, 20–22 February 1991).
152. P. A. Read and T. Schmidt, U.S. Patent 5090479, 1992.

153. M. M. Jordan, K. S. Sorbie, G. M. Graham, K. Taylor, K. E. Hourston, and S. Hennessey, "The Correct Selection and Application Methods for Adsorption and Precipitation Scale Inhibitors for Squeeze Treatments in North Sea Oilfields," SPE 31125 (paper presented at the SPE SPE Formation Damage Control Symposium, Lafayette, LA, 14–15 February 1996).
154. M. M. Jordan, K. S. Sorbie, P. Chen, P. Armitage, P. Hammond, and K. Taylor, "The Design of Polymer and Phosphonate Scale Inhibitor Precipitation Treatments and the Importance of Precipitate Solubility in Extending Squeeze Lifetime," SPE 37275 (paper presented at the SPE International Symposium on Oilfield Chemistry, Houston, TX, 18–21 February 1997).
155. D. C. Berkshire, J. B. Lawson, and E. A. Richardson, U.S. Patent 4357248, 1982.
156. (a) M. R. Rabaioli and T. P. Lockhart, *Journal of Petroleum Science and Engineering* 15 (1996): 1156. (b) M. B. Tomson, A. T. Kan, G. Fu, D. Shen, H. A. Nasr-El-Din, H. Al-Saiari, and M. Al-Thubaiti, "Mechanistic Understanding of the Rock/Phosphonate Interactions and the Effect of Metal Ions on Inhibitor Retention," *SPE Journal* 13 (2008): 325.
157. E. S. Snavely Jr. and J. Hen, U.S. Patent 4787455, 1988.
158. A. T. Kan, G. Fu, D. Shen, M. B. Tomson, and H. A. AlSaiari, "Enhanced Inhibitor Treatments with the Aid of Transition Metal Ions," SPE 114060 (paper presented at the SPE 9th International Conference on Oilfield Scale, Aberdeen, UK, 28–29 May 2008).
159. F. F. D. Rosario, C. N. Khalil, M. C. Bezerra, and S. B. Rondinini, U.S. Patent 5840658, 1998.
160. J. Hen, International Patent Application WO 9403706.
161. J. Hen, A. Brunger, B. K. Peterson, M. D. Yuan, and J. P. Renwick, "A Novel Scale Inhibitor Chemistry for Downhole Squeeze Application in High Water Producing North Sea Wells," SPE 30410 (paper presented at the SPE Offshore Europe, Aberdeen, UK, 5–8 September 1995).
162. H. M. Bourne, G. D. M. Williams, J. Ray, and A. Morgan, "Extending Squeeze Lifetime through In-Situ pH Modification—Laboratory and Field Experience" (paper presented at the 8th International Oil Field Chemical Symposium, Geilo, Norway, 2–5 March 1997).
163. R. J. Faircloth and J. B. Lawson, U.S. Patent 5211237, 1993.
164. J. D. Lynn and H. A. Nasr-El-Din, "A Novel Low-Temperature, Forced Precipitation Phosphonate Squeeze for Water Sensitive, Non-Carbonate Bearing Formations," SPE 84404 (paper presented at the SPE Annual Technical Conference and Exhibition, Denver, CO, 5–8 October 2003).
165. J. A. McRae, S. M. Heath, C. Strachan, L. Matthews, and R. Harris, "Development of an Enzyme Activated, Low Temperature, Scale Inhibitor Precipitation Squeeze System," SPE 87441 (paper presented at the SPE International Symposium on Oilfield Scale, Aberdeen, UK, 26–27 May 2004).
166. I. R. Collins, International Patent Application WO 9830783, 1998.
167. R. G. Chapman, I. R. Collins, S. P. Goodwin, A. R. Lucy, and N. J. Stewart, U.S. Patent 5690174, 1997.
168. I. R. Collins, L. G. Cowie, M. Nicol, and N. J. Stewart, "The Field Application of a Scale Inhibitor Squeeze Enhancing Additive," SPE 38765 (paper presented at the SPE Annual Technical Conference and Exhibition, San Antonio, TX, 5–8 October 1997).
169. M. M. Jordan, C. J. Graff, and K. N. Cooper, "Development and Deployment of a Scale Squeeze Enhancer and Oil-Soluble Scale Inhibitor to Avoid Deferred Oil Production Losses During Squeezing Low-Water Cut Wells, North Slope, Alaska," SPE 58725 (paper presented at the SPE International Symposium on Formation Damage Control, Lafayette, LA, 23–24 February 2000).

170. I. R. Collins, L. G. Cowie, M. Nicol, and N. J. Stewart, "Field Application of a Scale Inhibitor Squeeze Enhancer Additive," SPE 54525 (paper presented at the SPE Annual Technical Conference, San Antonio, TX, 5–8 October 1997).
171. I. R. Collins, S. P. Goodwin, J. C. Morgan, and N. J. Stewart, U.S. Patent 6225263, 2001.
172. H. M. Bourne, S. L. Booth, and A. Brunger, "Combining Innovative Technologies to Maximize Scale Squeeze Cost Reduction," SPE 50718 (paper presented at the SPE International Symposium on Oilfield Chemistry, Houston, TX, 16–19 February 1999).
173. H. T. R. Montgomerie, P. Chen, T. Hagen, R. Wat, O. M. Selle, and H. K. Kotlar, International Patent Application WO/2004/011772, 2004.
174. O. M. Selle, R. M. S. Wat, O. Vikane, H. Nasvik, P. Chen, T. Hagen, H. Montgomerie, and H. Bourne, "A Way Beyond Scale Inhibitors, Extending Scale Inhibitor Squeeze Life through Bridging," SPE 80377 (paper presented at the SPE 5th International Symposium on Oilfield Scale, Aberdeen, Scotland, 29–30 May 2003).
175. P. Chen, X. Yan, T. Hagen, and H. T. R. Montgomerie, International Patent Application WO/2007/015090, 2007.
176. H. Montgomerie, P. Chen, T. Hagen, O. Vikane, R. Matheson, V. Leirvik, C. Froytlog, and J. O. Saeten, "Development of a New Polymer Inhibitor Chemistry for Downhole Squeeze Applications," SPE 113926 (paper presented at the SPE International Oilfield Scale Conference, Aberdeen, UK, 28–29 May 2008).
177. J. Hen, U.S. Patent 5089150, 1992.
178. T.-Y. Yan, International Patent Application WO9305270, 1993.
179. O. Bache and S. Nilsson, "Ester Cross-Linking of Polycarboxylic Acid Scale Inhibitors as a Possible Means to Increase Inhibitor Squeeze Lifetime," SPE 60190 (paper presented at the SPE International Symposium on Oilfield Scale, Aberdeen, UK, 26–27 January 2000).
180. N. Fleming, K. Ramstad, A.-M. Mathisen, A. Nelson, and S. Kidd, "Innovative Use of Kaolinite in Downhole Scale Management: Squeeze Life Enhancement and Water Shutoff," SPE 113656 (paper presented at the SPE 9th International Conference on Oilfield Scale, Aberdeen, UK, 28–29 May 2008).
181. (a) D. R. Watkins, J. J. Clemens, J. C. Smith, S. N. Sharma, and H. G. Edwards, U.S. Patent 5224543, 1993. (b) M. Garcia-Lopez de Victoria, J. W. Still, and T. Bui, U.S. Patent Application 20090025933.
182. K. Cheremisov, D. Oussoltsev, K. K. Butula, A. Gaifullin, I. Faizullin, and D. Senchenko, "First Application of Scale Inhibitor during Hydraulic Fracturing Treatments in Western Siberia," SPE 114255 (paper presented at the SPE International Oilfield Scale Conference, Aberdeen, UK, 28–29 May 2008).
183. A. M. Fitzgerald and L. G. Cowie, "A History of Frac Pack Scale Inhibitor Deployment," SPE 112474 (paper presented at the SPE International Symposium and Exhibition on Formation Damage Control, Lafayette, LA, 13–15 February 2008).
184. O. J. Vetter, S. Lankford, and T. Nilssen, "Well Stimulations and Scale Inhibitors," SPE 17284 (paper presented at the SPE Permian Basin Oil and Gas Recovery Conference, Midland, TX, 10–11 March 1988).
185. M. Norris, D. Perez, H. M. Bourne, and S. M. Heath, "Maintaining Fracture Performance through Active Scale Control," SPE 68300 (paper presented at the SPE International Symposium on Oilfield Scale, Aberdeen, UK, 30–31 January 2001).
186. S. Szymczak, G. Brock, and J. M. Brown, D. Daulton, and Brian Ward, "Beyond the Frac: Using the Fracture Process as the Delivery System for Production Chemicals Designed to Perform for Prolonged Periods of Time," SPE 107707 (paper presented at the SPE Rocky Mountain Oil & Gas Technology Symposium, Denver, CO, 16–18 April 2007).

187. S. Szymczak, J. M. Brown, S. Noe, and G. Gallup, "Long-Term Scale Inhibition Using a Solid Scale Inhibitor in a Fracture Fluid," SPE 102720 (paper presented at the SPE Annual Technical Conference and Exhibition, San Antonio, TX, 24–27 September 2006).
188. R. J. Powell, R. D. Gdanski, M. A. McCabe, and D. C. Buster, "Controlled-Release Scale Inhibitor for Use in Fracturing Treatments," SPE 28999 (paper presented at the SPE International Symposium on Oilfield Chemistry, San Antonio, TX, 14–17 February 1995).
189. D. V. S. Gupta, J. M. Brown, and S. Szymczak, "Multi-Year Scale Inhibition from a Solid Inhibitor Applied during Stimulation," SPE 115655 (paper presented at the SPE Annual Technical Conference and Exhibition, Denver, CO, 21–24 September 2008).
190. S. M. Heath and H. M. Bourne, U.S. Patent 7196040, 2007.
191. M. Tyndall, L. Maschio, B. C. O. Bustos, and B. Lungwitz, "Simultaneous Fracturing and Inhibitor Squeeze Treatments in Gas Reservoirs: A New Approach to Tailor Chemistry to Reservoir Conditions," *Chemistry in the Oil Industry X: Oilfield Chemistry*, Manchester, UK, RSC, 5–7 November 2007.
192. P. S. Smith, "Combined Scale Removal and Scale Inhibition Treatments," SPE 60222 (paper presented at the SPE 2nd International Symposium on Oilfield Scale, Aberdeen, Scotland, 26–27 January 2000).
193. P. S. Smith, L. G. Cowie, H. M. Bourne, M. Grainger, and S. M. Heath, "Field Experiences with a Combined Acid Stimulation and Scale Inhibition Treatment," SPE 68312 (paper presented at the SPE International Symposium on Oilfield Scale, Aberdeen, UK, 30–31 January 2001).
194. P. S. Smith, C. C. Clement Jr., and A. Mendoza Rojas, "Combined Scale Removal and Scale Inhibition Treatments," SPE 60222 (paper presented at the SPE 2nd International Symposium on Oilfield Scale, Aberdeen, Scotland, 26–27 Jan. 2000).
195. R. T. Barthorpe, "The Impairment of Scale Inhibitor Function by Commonly Used Organic Anions," SPE 25158 (paper presented at the SPE International Symposium on Oilfield Chemistry, New Orleans, LA, 2–5 March 1993).
196. R. Castano, J. Villamizar, O. Diaz, S. A. Hocol, M. Avila, S. Gonzalez, E. D. Dalrymple, S. Milson, and D. Everett, "Relative Permeability Modifier and Scale Inhibitor Combination in Fracturing Process at San Francisco Field in Colombia, South America," SPE 77412 (paper presented at the SPE Annual Technical Conference and Exhibition, San Antonio, TX, 29 September–2 October 2002).
197. P. Powell, M. A. Singleton, and K. S. Sorbie, U.S. Patent 6913081, 2005.
198. I. R. Collins, T. Jones, and C. G. Osborne, International Patent Application WO/2004/016906, 2004.
199. A. F. Miles, O. Vikane, D. S. Healey, I. R. Collins, J. Saeten, H. M. Bourne, and R. G. Smith, "Field Experiences Using 'Oil Soluble' Non-Aqueous Scale Inhibitor Delivery Systems," SPE 87431 (paper presented at the SPE International Symposium on Oilfield Scale, Aberdeen, UK, 26–27 May 2004).
200. M. M. Jordan, I. R. Collins, A. Gyani, and G. M. Graham, "Coreflood Studies Examine New Technologies that Minimize Intervention throughout Well Life Cycle," SPE 74666, *SPE Production & Operations* 21(2) (2006): 161–173.
201. H. Guan, K. S. Sorbie, and E. J. Mackay, "The Comparison of Non-aqueous and Aqueous Scale-Inhibitor Treatments: Experimental and Modelling Studies," SPE 87445, *SPE Production & Operations* 21(4) (2006): 419.
202. J. M. Reizer, M. G. Rudel, C. D. Sitz, R. M. S. Wat, and H. Montgomerie, U.S. Patent 6379612, 2002.
203. J. M. Reizer, M. G. Rudel, C. D. Sitz, R. M. S. Wat, H. Montgomerie, and A. F. Miles, U.S. Patent Application 20020150499, 2002.

204. R. Wat, H. Montgomerie, T. Hagen, R. Boereng, H. K. Kotlar, and O. Vikane, "Development of an Oil-Soluble Scale Inhibitor for a Subsea Satellite Field," SPE 50706 (paper presented at the SPE International Symposium on Oilfield Chemistry, Houston, TX, 16–19 February 1999).
205. N. J. Jenvey, A. F. MacLean, A. F. Miles, and H. T. R. Montgomerie, "The Application of Oil Soluble Scale Inhibitors into the Texaco Galley Reservoir: A Comparison with Traditional Squeeze Techniques to Avoid Problems Associated with Wettability Modification in Low Water-Cut Wells," SPE 60197 (paper presented at the SPE International Symposium on Oilfield Scale, Aberdeen, UK, 26–27 January 2000).
206. S. M. Heath, J. J. Wylde, M. Archibald, M. Sim, and I. R. Collins, "Development of Oil Soluble Precipitation Squeeze Technology for Application in Low and High Water Cut Wells," SPE 87451 (paper presented at the SPE International Symposium on Oilfield Scale, Aberdeen, UK, 26–27 May 2004).
207. J. J. Wylde and E. K. McAra, "Optimization of an Oil Soluble Scale Inhibitor for Minimizing Formation Damage: Laboratory and Field Studies," SPE 86477 (paper presented at the SPE International Symposium and Exhibition on Formation Damage Control, Lafayette, LA, 18–20 February 2004).
208. T. A. Lawless and R. N. Smith, "New Technology, Invert Emulsion Scale Inhibitor Squeeze Design," SPE 50705 (paper presented at the SPE International Symposium on Oilfield Chemistry, Houston, TX, 16–19 February 1999).
209. S. Beare, "The Development and Laboratory Evaluation of an Emulsified Scale Inhibitor Squeeze Treatment" (paper presented at the NIF 11th International Oilfield Chemical Symposium, Fagernes, Norway, March 2000).
210. C. Romero, B. Bazin, and A. Zaitoun and F. Leal-Calderon, "Behavior of a Scale Inhibitor Water-in-Oil Emulsion in Porous Media," SPE 98275 (paper presented at the SPE International Symposium and Exhibition on Formation Damage Control, Lafayette, LA, 15–17 February 2006).
211. M. M. Jordan, F. Murray, A. Kelly, and K. Stevens, "Deployment of Emulsified Scale—Inhibitor Squeeze to Control Sulfate/Carbonate Scales within Subsea Facilities in the North Sea Basin," SPE 80249 (paper presented at the International Symposium on Oilfield Chemistry, Houston, TX, 5–7 February 2003).
212. I. R. Collins, International Patent Application WO/2001/046553, 2001.
213. A. F. Miles, H. M. Bourne, R. G. Smith, and I. R. Collins, "Development of a Novel Water in Oil Microemulsion Based Scale Inhibitor Delivery System," 80390 (paper presented at the International Symposium on Oilfield Scale, Aberdeen, UK, 29–30 January 2003).
214. R. Collins and I. Vervoort, U.S. Patent 6581687, 2003.
215. L. Del Gaudio, R. Bortolo, and T. P. Lockhart, "Nanoemulsions: A New Vehicle for Chemical Additive Delivery," SPE 106016 (paper presented at the International Symposium on Oilfield Chemistry, Houston, TX, 28 February–2 March 2007).
216. I. R. Collins, M. M. Jordan, and S. E. Taylor, "The Development and Application of a Novel Scale Inhibitor Deployment System," SPE 80286, *SPE Production & Facilities* 17(4) (2002): 221.
217. I. R. Collins, M. M. Jordan, and S. E. Taylor, "The Development and Application of a Novel Scale Inhibitor for Deployment in Low Water Cut, Water Sensitive or Low Pressure Reservoirs," SPE 60192 (paper presented at the SPE International Symposium on Oilfield Scale, Aberdeen, UK, 26–27 January 2000).
218. I. R. Collins, M. M. Jordan, N. Feasey, and G. D. Williams, "The Development of Emulsion-Based Production Chemical Deployment Systems," SPE 65026 (paper presented at the SPE International Symposium on Oilfield Chemistry, Houston, TX, 13–16 February 2001).

219. H. A. Nasr-El-Din, J. D. Lynn, M. K. Hashem, and G. Bitar, "Field Application of a Novel Emulsified Scale Inhibitor System to Mitigate Calcium Carbonate Scale in a Low Temperature, Low Pressure Sandstone Reservoir in Saudi Arabia," SPE 77768 (paper presented at the SPE Annual Technical Conference and Exhibition, San Antonio, TX, 29 September–2 October 2002).
220. I. R. Collins, P. D. Ravenscroft, and C. I. Bates, International Patent Application WO97/45625, 1997.
221. I. R. Collins, S. D. Duncum, M. M. Jordan, and N. D. Feasey, "The Development of a Revolutionary Scale-Control Product for the Control of Near-Well Bore Sulfate Scale within Production Wells by the Treatment of Injection Seawater," SPE 100357 (paper presented at the SPE International Oilfield Scale Symposium, Aberdeen, UK, 31 May–1 June 2006).
222. H. K. Kotlar, O. M. Selle, O. A. Aune, L. Kilaas, and A. D. Dyrli, International Patent Application WO/2002/040827.
223. H. K. Kotlar, O. M. Selle, O. A. Aune, L. Kilaas, and A. D. Dyrli, International Patent Application WO/2002/040826.
224. C. R. Clark, D. L. Whitfill, D. P. Cords, E. F. McBride, and H. E. Bellis, U.S. Patent 4986353, 1991.
225. L. A. McDougall, J. C. Newlove, and J. A. Haslegrave, U.S. Patent 4670166, 1987.
226. H. K. Kotlar, O. M. Selle, O. A. Aune, L. Kilaas, and A. D. Dyrli, International Patent Application WO/2002/040828.
227. I. R. Collins and S. N. Duncum, International Patent Application WO/2003/106810, 2003.
228. P. J. C. Webb, T. A. Nistad, B. Knapstad, P. D. Ravenscroft, and I. R. Collins, "Economic and Technical Features of a Revolutionary Chemical Scale Inhibitor Delivery Method for Fractured and Gravel Packed Wells: Comparative Analysis of Onshore and Offshore Subsea Applications," SPE 39451 (paper presented at the SPE Formation Damage Control Conference, Lafayette, LA, 18–19 February 1998).
229. H. M. Bourne, S. M. Heath, S. McKay, J. Fraser, L. Stott, and S. Muller, "Effective Treatment of Subsea Wells with a Solid Scale Inhibitor System," SPE 60207 (paper presented at the International Symposium on Oilfield Scale, Aberdeen, UK, 26–27 January 2000).
230. P. A. Read, European Patent EP656459, 1995.
231. H. M. Bourne and P. A. Read, International Patent Application WO/1996/027070.
232. H. K. Kotlar, O. M. Selle, O. A. Aune, L. Kilaas, Lars, and A. D. Dyrli, U.S. Patent Application 20040060702.
233. D. Shen, P. Zhang, A. T. Kan, G. Fu, J. Farrell, and M. B. Tomson, "Control Placement of Scale Inhibitors in the Formation with Stable Ca-DTPMP Nanoparticle Suspension and its Transport Porous Media," SPE 114063 (paper presented at the SPE International Oilfield Scale Conference, Aberdeen, UK, 28–29 May 2008).
234. E. J. Mackay and K. S. Sorbie, "Modelling Scale Inhibitor Squeeze Treatments in High Crossflow Horizontal Wells," SPE 50418 (paper presented at the SPE 3rd International Conference on Horizontal Well Technology, Calgary, Alberta, 1–4 November 1998).
235. M. M. Jordan, M. C. Edgerton, Cole-Hamilton; and K. Mackin, "The Application of Novel Wax Diverter Technology to Allow Successful Scale Inhibitor Squeeze Treatment into a Sub Sea Horizontal Well, North Sea Basin," SPE 49196 (paper presented at the SPE Annual Technical Conference and Exhibition, New Orleans, LA, 27–30 September 1998).
236. M. M. Jordan, C. J. Tomlinson, A. R. P. Pritchard, and M. Lewis, "Laboratory Testing and Field Implementation of Scale Inhibitor Squeeze Treatments to Subsea and Platform Horizontal Wells, North Sea Basin," SPE 50436 (paper presented at the SPE International Conference on Horizontal Well Technology, Calgary, Alberta, Canada, 1–4 November 1998).

237. M. M. Jordan, K. Sjursaether, and I. R. Collins, "Scale Control within the North Sea Chalk/Limestone Reservoirs—The Challenge of Understanding and Optimizing Chemical-Placement Methods and Retention Mechanisms: Laboratory to Field," SPE 86476, *SPE Production & Facilities* 20(4) (2005): 262.
238. J. S. James, D. M. Frigo, M. M. Townsend, G. M. Graham, F. Wahid, and S. M. Heath, "Application of a Fully Viscosified Scale Squeeze for Improved Placement in Horizontal Wells," SPE 94593 (paper presented at the SPE International Symposium on Oilfield Scale, Aberdeen, UK, 11–12 May 2005).
239. R. Stalker, G. M. Graham, D. Oliphant, and M. Smillie, "Potential Application of Viscosified Treatments For Improved Bullhead Scale Inhibitor Placement in Long Horizontal Wells—A Theoretical and Laboratory Examination," SPE 87439 (paper presented at the SPE International Symposium on Oilfield Scale, Aberdeen, United Kingdom, 26–27 May 2004).
240. R. Stalker, K. Butler, and G. Graham and R. Wat, L.-E. Hauge, and K. E. Wennberg, "Evaluation of Polymer Gel Diverters for a High-Temperature Field with Special Focus on the Formation Shape Factor—An Important Parameter for Enhancing Matrix Placement of Stimulation Chemicals," SPE 107806 (paper presented at the European Formation Damage Conference, Scheveningen, The Netherlands, 30 May–1 June 2007).
241. S. M. Heath, M. Sim, M. Archibald, and R. Stalker, "Development of a Viscosified Scale Inhibitor for Improving Placement in Acid Fractured Limestone Reservoirs," SPE 114046 (paper presented at the SPE International Oilfield Scale Conference, Aberdeen, UK, 28–29 May 2008).
242. O. M. Selle, M. Springer, I. H. Auflem, P. Chen, R. Matheson, A. A. Mebratu, and G. Glasbergen, "Gelled Scale Inhibitor Treatment for Improved Placement in Long Horizontal Wells at Norne and Heidrun fields," SPE 112464 (paper presented at the SPE International Symposium and Exhibition on Formation Damage Control, Lafayette, LA, 13–15 February 2008).
243. N. D. Feasey, M. M. Jordan, E. J. Mackay, and I. R. Collins, "The Challenge that Completion Types Present to Scale Inhibitor Squeeze Chemical Placement: A Novel Solution Using a Self-Diverting Scale Inhibitor Squeeze Process," SPE 86478 (paper presented at the SPE International Symposium and Exhibition on Formation Damage Control, Lafayette, LA, 18–20 February 2004).
244. F. De Campo, A. Colaco, and S. Kesavan, International Patent Application WO/2008/066918.
245. I. Drela, P. Falewicz, and S. Kuczkowska, "New Rapid Test for Evaluation of Scale Inhibitors," *Water Research* 32(10) (1998): 3188.
246. (*a*) S. J. Dyer and G. M. Graham, *Journal of Petroleum Science and Engineering* 35 (2002): 95. (*b*) G. M. Graham, I. R. Collins, R. Stalker, and I. J. Littlehales, "The Importance of Appropriate Laboratory Procedures for the Determination of Scale Inhibitor Performance," SPE 74679 (paper presented at the International Symposium on Oilfield Scale, Aberdeen, UK, 30–31 January 2002); (*c*) E. N. Halvorsen, A. K. Halvorsen, K. Reiersølmeon, T. R. Andersen, and C. Bjornstad, New Method for Scale Inhibitor Testing, SPE 121663, SPE International Symposium on Oilfield Chemistry, The Woodlands, TX, April 20–22, 2009.
247. A. M. Pritchard, L. Cowie, J. R. Goulding, G. Graham, A. C. Greig, B. M. Hamblin, A. Hunton, and S. Terry, *Chemicals in the Oil Industry III*, Manchester, UK: Royal Society of Chemistry, 1988, 140.
248. M. D. Yuan, K. S. Sorbie, P. Jiang, P. Chen, M. M. Jordan, A. C. Todd, K. E. Hourston, and K. Ramstad, "Phosphonate Scale Inhibitor Adsorption on Outcrop and Reservoir Rock Substrates—The 'Static' and 'Dynamic' Adsorption Isotherms," *Recent Advances in Oilfield Chemistry V*, Manchester, UK: Royal Society of Chemistry, 1994, 164.

249. T. Chen, A. Neville, and M. Yuan, "Calcium Carbonate Scale Formation—Assessing the Initial Stages of Precipitation and Deposition," *Journal of Petroleum Science and Engineering* 46 (2005): 185.
250. M. Crabtree, D. Eslinger, P. Fletcher, A. Johnson, and G. King, "Fighting Scale—Removal and Prevention," *Oilfield Review (Schlumberger)* 11 (1999): 30.
251. A. Johnson, D. Eslinger, and H. Larsen, "An Abrasive Jetting Removal System, SPE 46026 (paper presented at the SPE/IcoTA Coiled Tubing Roundtable," Houston, TX, 15–16 April 1998).
252. R. J. Tailby, C. B. Amor, and A. McDonough, "Scale Removal from the Recesses of Side-pocket Mandrels," SPE 54477 (paper presented at the SPE/IcoTA Coiled Tubing Roundtable, Houston, TX, 25–26 May 1999).
253. M. S. Mirza and V. Prasad, "Scale Removal in Khuff Gas Wells," SPE 53345 (paper presented at the Middle East Oil Show and Conference, Bahrain, 20–23 February 1999).
254. T. Huang, P. M. McElfresh, and A. D. Gabrysch, "Acid Removal of Scale and Fines at High Temperatures," SPE 74678 (paper presented at the SPE Oilfield Scale Symposium, Aberdeen, UK, 30–31 January 2002).
255. S. M. Proctor, "Scale Dissolver Development and Testing for HP/HT Systems," SPE 60221 (paper presented at the SPE International Symposium on Oilfield Scale, Aberdeen, UK, 26–27 January 2000).
256. H. Williams, R. Wat, P. Chen, T. Hagen, K. Wenneberg, V. Viken, and G. M. Graham, "Scale Dissolver Application under HPHT Conditions—Use of an HPHT 'Stirred Reactor' for in-situ Scale Dissolver Evaluations," SPE 95127 (paper presented at the SPE International Symposium on Oilfield Scale, Aberdeen, UK, 11–12 May 2005).
257. O. M. Selle, R. M. S. Wat, H. Nasvik, and A. Mebratu, "Gelled Organic Acid System for Improved $CaCO_3$ Removal in Horizontal Openhole Wells at the Heidrun Field," SPE 90359 (paper presented at the SPE Annual Technical Conference and Exhibition, Houston, TX, 26–29 September 2004).
258. M. Al-Khaldi, H. A. Nasr-El-Din, S. Metha, and A. Aamri, "Reaction of Citric Acid with Calcite," *Chemical Engineering Science* 62 (2007): 5880.
259. C. M. Shaughnessy and W. E. Kline, "EDTA Removes Formation Damage at Prudhoe Bay," *Journal of Petroleum Technology* 35 (1983): 1783.
260. J. S. Rhudy, "Removal of Mineral Scale from Reservoir Core by Scale Dissolver," SPE 25161 (paper presented at the SPE International Symposium on Oilfield Chemistry, New Orleans, LA, 2–5 March 1993).
261. K. D. Demadis, P. Lykoudis, R. G. Raptis, and G. Mezei, "Phosphonopolycarboxylates as Chemical Additives for Calcites Scale Dissolution and Metallic Corrosion Inhibition Based on a Calcium-Phosphonotricarboxylate Organic-Inorganic Hybrid," *Crystal Growth & Design* 6(5) (2006): 1064.
262. (*a*) R. S. Boethling, E. Sommer, and D. DiFiore, "Designing Small Molecules for Biodegradability," *Chemical Reviews* 107 (2007): 2207. (*b*) T. O. Boonstra, M. Heus, A. Carstens, and J. LePage, International Patent Application WO/2009/024518. (*c*) J. LePage, C. DeWolf, J. Bemelaar, and H. A. Nasr-El-Din, An Environmentally Friendly Stimulation Fluid for High-Temperature Applications, SPE 121709, SPE International Symposium on Oilfield Chemistry, The Woodlands, TX, April 20–22, 2009.
263. W. W. Frenier, "Novel Scale Removers Are Developed for Dissolving Alkaline Earth Deposits," SPE 65027 (paper presented at the SPE International Symposium on Oilfield Chemistry, Houston, TX, 13–16 February 2001).
264. W. W. Frenier and D. Wilson, "Use of Highly Acid-Soluble Chelating Agents in Well Stimulation Services," SPE 63242 (paper presented at the SPE Annual Technical Conference and Exhibition, Dallas, TX, 1–4 October 2000).
265. K. B. Charkhutian, B. L. Libutti, and M. A. Murphy, International Patent Application WO/2006/028917.

266. T. G. J. Jones, G. J. Tustin, P. Fletcher, and C.-W. Lee, U.S. Patent Application 20070119593, 2007.
267. R. P. Kreh, W. H. Henry, J. Richardson, and V. R. Kuhn, in *Mineral Scale Formation and Inhibition*, ed. Zahid Amjad, New York: Plenum Press, 1995.
268. A. F. Clemmit, D. C. Balance, and A. G. Hunton, "The Dissolution of Scales in Oilfield Systems," SPE 140190, Offshore Europe, Aberdeen, UK, 10–13 September 1985.
269. M. Asakawa, Y. Sumida, M. Shimomura, S. Okuno, T. Morimoto, M. Morita, and H. Suenaga, U.S. Patent 6103686, 2000.
270. H. Yamamoto, Y. Takayanagi, K. Takahashi, and T. Nakahama, U.S. Patent 6221834, 2001.
271. K. B. Charkhutian, B. L. Libutti, F. L. M. De Cordt, and J. S. Ruffini, U.S. Patent 6797177, 2004.
272. L. J. Kalfayan, D. R. Watkins, and G. S. Hewgill, U.S. Patent 4992182, 1991.
273. G. H. Zaid and B. A. Wolf, U.S. Patent 6494218, 2002.
274. M. B. Lawson, U.S. Patent 4096869, 1978.
275. J. M. Paul, U.S. Patent 5146988, 1992.
276. S. J. Mac Gregor and S. M. Turnbull, International Patent Application WO/2003/069110.
277. J. Hen, U.S. Patent 5068042, 1991.
278. A. Putnis, C. C. Putnis, and J. Paul, "The Efficiency of a DTPA-Based Solvent in the Dissolution of Barium Sulfate Scale Deposits," SPE 29094 (paper presented at the SPE International Symposium on Oilfield Chemistry, San Antonio, TX, 14–17 February 1995).
279. M. M. Jordan, G. M. Graham, K. S. Sorbie, A. Matharu, R. Tomlins, and J. Bunney, "Scale Dissolver Application: Production Enhancement and Formation Damage Potential," SPE 66565, *SPE Production & Facilities* 14(4) (2000): 288.
280. J. Rebeschini, C. Jones, G. Collins, S. Edmunds, and A. Archer, "The Development and Performance Testing of a Novel Biodegradable Barium Sulfate Scale Dissolver" (paper presented at the 19th Oil Field Chemistry Symposium, Geilo, Norway, 9–12 March 2008).
281. R. Boereng, P. Chen, T. Hagen, C. Sitz, R. Thoraval, and H. K. Kotlar, "Creating Value with Green Barium Sulfate Scale Dissolvers—Development and Field Deployment on Statfjord Unit," SPE 87438, (paper presented at the SPE International Symposium on Oilfield Scale, Aberdeen, UK, 26–27 May 2004).
282. F. De Jong, D. N. Reinhoudt, and G. Torny-Schutte, U.S. Patents 4215000 and 4190462, 1980.
283. A. Van Zon, F. De Jong, and G. Torny-Schutte, U.S. Patent 4288333, 1981.
284. H. A. Nasr-El-Din, S. H. Mutairim, H. H. Al-Hajji, and J. D. Lynn, "Evaluation of a New Barite Dissolver: Lab Studies," SPE 86501 (paper presented at the SPE International Symposium and Exhibition on Formation Damage Control, Lafayette, LA, 18–20 February 2004).
285. K.-S. Wang, R. Resch, K. Dunn, P. Shuler, Y. Tang, B. E. Koel, and T. H. Yen, "Dissolution of the Barite (001) Surface by the Chelating Agent DTPA as Studied with Non-Contact Atomic Force Microscopy," *Colloids Surfaces A, Physical and Engineering Aspects* 160(3) (1999): 217.
286. I. Lahkatos and J. Lakatos-Szabo, "Potential of Different Polyamino Carboxylic Acids as Barium and Sulfate Dissolvers," SPE 94633 (paper presented at the SPE European Formation Damage Conference, Scheveningen, The Netherlands, 25–27 May 2005).
287. I. Lakatos, J. Lakatos-Szabo, and B. Kosztin, "Optimization of Barite Dissolvers by Organic Acids and pH Regulation" (paper presented at the SPE Oilfield Scale Symposium, Aberdeen, UK, 30–31 January 2002).

288. I. Lakatos, J. Lakatos-Szabo, and B. Kosztin, "Comparative Study of Different Barite Dissolvers: Technical and Economic Aspects," SPE 73719-MS (paper presented at the SPE International Symposium and Exhibition on Formation Damage Control, Lafayette, LA, 20–21 February 2002).
289. R. Keatch, International Patent Application WO/2007/109798.
290. B. J. Palmer, D. Fu, R. Card, and M. J. Miller, U.S. Patent 6924253, 2005.
291. R. Boering, K. O. Bakken, O. Vikane, and A. Angelsen, "A Stimulation Treatment of a Sub Sea Well Using a Scale Dissolver," SPE 50736 (paper presented at the SPE International Symposium on Oilfield Chemistry, Houston, TX, 16–19 February 1999).
292. H. K. Kotlar, O. M. Selle, and F. Haavind, "A 'Standardized' Method for Ranking of Scale Dissolver Efficiency: A Case Study from the Heidrun Field," SPE 74668 (paper presented at the SPE Oilfield Scale Symposium, Aberdeen, UK, 30–31 January 2002).
293. I. J. Lakatos, J. Lakatos-Szabo, J. Toth, and T. Bodi, "Improvement of Placement Efficiency of $BaSO_4$ and $SrSO_4$ Dissolvers Using Organic Alkalis as pH Controlling Agents," SPE 106015 (paper presented at the European Formation Damage Conference, Scheveningen, The Netherlands, 30 May–1 June 2007).
294. F. J. Quattrini, U.S. Patent 3660287, 1972.
295. (a) R. L. Morris and J. M. Paul, U.S. Patent 4980077, 1990. (b) J. M. Paul and R. L. Morris, U.S. Patent 5282995, 1994.
296. (a) T. F. D'Muhala, U.S. Patent 4708805. (b) R. L. Morris and J. M. Paul, U.S. Patent 5049297, 1991. (c) R. L. Morris and J. M. Paul, U.S. Patent 5084105, 1991.
297. R. D. Tate, U.S. Patent 5685918, 1997.
298. J. Hen, U.S. Patent 5059333, 1991.
299. C. A. de Wolf, I. C. M. Huybens, K. van Ginkel, and R. Geerts, "Chemistry Biodegradable Solvents for Barium Sulfate Dissolution in Offshore Oilfields" (paper presented at the Chemistry in the Oil Industry X: Oilfield Chemistry, Royal Society of Chemistry, Manchester, UK, 5–7 November 2007).
300. J. Al-Ashhab, H. Al-Matar, and S. Mokhtar, "Techniques Used to Monitor and Remove Strontium Sulfate Scale in UZ Producing Wells," SPE 101401 (paper presented at the SPE Abu Dhabi International Petroleum Exhibition and Conference, Abu Dhabi, UAE, 5–8 November 2006).
301. M. M. Jordan, D. Marlow and C. Johnston, "The Evaluation of Enhanced (Carbonate/Sulfate) Scale Dissolver Treatments for Near-Wellbore Stimulation in Subsea Production Wells, Gulf of Mexico," SPE 100356 (paper presented at the SPE Oilfield Scale Symposium, Aberdeen, UK, 31 May–1 June 2006).
302. G. D. Williams and I. R. Collins, "Enhancing Mineral Scale Dissolution in the Near-Wellbore Region," SPE 56774 (paper presented at the SPE Annual Technical Conference and Exhibition, Houston, TX, 3–6 October 1999).
303. H. A. Nasr-El-Din and A. Y. Al-Humaidan, "Iron Sulfide Scale: Formation, Removal and Prevention," SPE 68315 (paper presented at the SPE International Symposium on Oilfield Scale, Aberdeen, UK, 30–31 January 2001).
304. W. G. F. Ford, M. L. Walker, M. P. Halterman, D. L. Parker, D. G. Brawley, and R. G. Fulton, "Removing a Typical Iron Sulfide Scale: The Scientific Approach," SPE 24327 (paper presented at the SPE Rocky Mountain Regional Meeting, Casper, WY, 18–21 May 1992).
305. J. Leal, J. R. Solares, H. A. Nasr-El-Din, C. Franco, F. Garzon, H. M. Marri, S. A. Aqeel, and G. Izquierdo, "A Systematic Approach to Remove Iron Sulfide Scale: A Case History," SPE 105607 (paper presented at the SPE Middle East Oil and Gas Show and Conference, Kingdom of Bahrain, 11–14 March 2007).
306. M. B. Laswon, U.S. Patent 4351673, 1982.
307. H. A. Nasr-El-Din, B. A. Fadhel, A. Y. Al-Humaidan, W. W. Frenier, and D. Hill, "An Experimental Study of Removing Iron Sulfide Scale from Well Tubulars," SPE 60205

(paper presented at the SPE International Symposium on Oilfield Scale, Aberdeen, UK, 26–27 January 2000).
308. A. R. Miller, U.S. Patent 6887840, 2005.
309. W. W. Frenier, M. D. Coffey, J. D. Huffines, and D. C. Smith, U.S. Patent 4220550, 1980.
310. W. W. Frenier, U.S. Patent 4310435, 1982.
311. L. D. Martin, U.S. Patent 4276185, 1981.
312. G. R. Buske, U.S. Patent 4289639, 1981.
313. M. B. Lawson, U.S. Patent 4351673, 1982.
314. G. H. Zaid and B. A. Wolf, U.S. Patent 6774090, 2004.
315. R. M. Jorda, SPE 280 (paper presented at the SPE Production Research Symposium, Tulsa, OK, 12–13 April 1962).
316. C. Reed, J. Foshee, J. E. Penkala, and M. Roberson, "Acrolein Application to Mitigate Biogenic Sulfides and Remediate Injection Well Damage in a Gas Plant Water Disposal System," SPE 93602 (paper presented at the SPE International Symposium on Oilfield Chemistry, The Woodlands, TX, 2–4 February 2005).
317. J. Penkala, M. D. Law, D. D. Horaska, and A. L. Dickinson, "Acrolein 2-Propenal: A Versatile Microbiocide for Control of Bacteria in Oilfield Systems," Paper 04749 (paper presented at the NACE CORROSION Conference, 2004).
318. T. Salma, "Cost Effective Removal of Iron Sulfide and Hydrogen Sulfide from Water Using Acrolein," SPE 59708 (paper presented at the SPE Permian Basin Oil and Gas Recovery Conference, Midland, TX, 21–23 March 2000).
319. (*a*) M. A. Mattox and E. J. Valente, U.S. Patent 6986358, 2006. (*b*) M. A. Mattox, U.S. Patent 6866048, 2005.
320. P. D. Gilbert, J. M. Grech, R. E. Talbot, M. A. Veale, and K. A. Hernandez, "Tetrakishydroxymethylphosphonium sulfate (THPS), for Dissolving Iron Sulfides Downhole and Topside—A Study of the Chemistry Influencing Dissolution," Paper 02030 (paper presented at the NACE CORROSION Conference, 2002).
321. H. A. Nasr-El-Din, A. M. Al-Mohammad, M. A. Al-Hajri and J. B. Chesson, "A New Chemical Treatment to Remove Multiple Damages in a Water Supply Well," SPE 95001 (paper presented at the SPE European Formation Damage Conference, Sheveningen, The Netherlands, 25–27 May 2005).
322. R. E. Talbot, C. R. Jones, and J. M. Grech, International Patent Application WO/2005/026065.
323. P. R. Rincon, J. P. McKee, C. E. Tarazon, L. A. Guevara, and B. Vinccler, "Biocide Stimulation in Oilwells for Downhole Corrosion Control and Increasing Production," SPE 87562 (paper presented at the SPE International Symposium on Oilfield Corrosion, Aberdeen, UK, 28 May 2004).
324. R. E. Talbot and J. M. Grech, International Patent Application WO/2005/040050.
325. C. R. Jones and J. M. Grech, International Patent Application WO/2004/083131.
326. J. Romaine, T. G. Strawser, and M. L. Knippers, "Application of Chlorine Dioxide as an Oilfield-Facilities-Treatment Fluid," SPE 29017, *SPE Production & Facilities* 11(1) (1996): 18.
327. J. F. McCafferty, E. W. Tate, and D. A. Williams, "Field Performance in the Practical Application of Chlorine Dioxide as a Stimulation Enhancement Fluid," SPE 20626, *SPE Production & Facilities* 8(1) (1993): 9.
328. J. G. R. Eylander, D. M. Frigo, F. A. Hartog, and G. Jonkers, "A Novel Methodology for In-Situ Removal of NORM from E & P Production Facilities," SPE 46791 (paper presented at the SPE International Conference on Health, Safety and Environment in Oil and Gas Exploration and Production, Caracas, Venezuela, 7–10 June 1998).
329. O. M Selle, A. Mebratu, H. Montgomerie, P. Chen, and T. Hagen, International Patent Application WO/2008/152419.

4 Asphaltene Control

4.1 INTRODUCTION

Asphaltene deposition is a major problem both upstream and downstream in the petroleum industry.[1-5] Asphaltenes can block reservoir pores in the near-well area and deposit in the production tubing and downstream pipeline and facilities.[6] Many reservoirs with crudes containing asphaltenes produce without evidence of asphaltene deposition until the oil stability is disturbed or destabilized. Asphaltenes are destabilized by gas breakout (pressure drop), condensate treatments, gas or gas liquid injection (CO_2, NGL floods), acid stimulation, low-pH scale-inhibitor squeeze treatments, crude blending, and high shear or streaming potential. If this destabilization occurs in a formation that contains charged minerals, the asphaltenes can adsorb and alter wettability and permeability.[7] A high concentration of asphaltenes in a crude does not necessarily result in an asphaltene deposition problem. In fact, crudes with as much as 20% asphaltenes may present no deposition problems whereas crudes with asphaltene contents as low as 0.2 wt.% have been shown to cause asphaltene deposits.[8]

Asphaltenes are generally defined as the fraction of oil that is insoluble in light aliphatic hydrocarbons such as pentane and heptane but soluble in aromatic solvents such as toluene. However, it is possible to encounter asphaltene problems in the field even with a crude oil that, in the laboratory, does not deposit asphaltenes from addition of excess heptane. Asphaltenes are considered to be among the heaviest components of crude oil. Related to the asphaltenes are the lower molecular weight maltenes or simple "resins," which also have polycyclic polar groups but more aliphatic side-chain character and are heptane-soluble. Although maltenes and resins do not form damaging precipitates, it is generally believed that they help stabilize asphaltenes in solution.

Asphaltenes are organic solids consisting of various polyaromatic structures with aliphatic chains as well as containing heteroatoms such as sulfur, nitrogen, and oxygen, and metals such as nickel, vanadium, and iron. The metals form complexes and impart electrical charge, which in turn may influence asphaltene deposition. The percentage of each element in asphaltenes varies from oil to oil but average values are 76–86 wt.% C, 7.3–8.5 wt.% H, 5.0–9.0 wt.% S, 0.7–1.2 wt.% O, 1.3–1.4 wt.% N, 0.1–0.2 wt.% metals (mostly Ni, V, Fe).[9-12] Various functional groups can be found in asphaltenes. Sulfur is present mostly as sulfidic and thiophenic groups and, to a lesser extent, as sulfoxide.[13] Asphaltene nitrogen is virtually all-present in aromatic groups such as pyrrolic, and to a lesser extent, pyridinic, groups and only very occasionally as tertiary amines.[14] Oxygen is present mainly in carbonyl and hydroxyl/phenolic groups. This includes ketones and carboxylic acids.

To understand how asphaltene dispersants (ADs) and inhibitors (AIs) work, it is necessary to determine the structure, molecular weight, and mechanism of aggregation of asphaltenes. The structure of asphaltene monomers and the size of the aromatic ring system in crude oil asphaltenes has been the subject of much discussion. Two models have been proposed in the literature. The first is the "continental" or "Island" model, which proposes a monomer molecular asphaltene structure with a molecular weight in the range of ~500–1,000 Da, with a maximum ~750 Da, consisting of, on average, a core of about six to seven fused aromatic rings surrounded by several aliphatic groups with some heteroatoms (Figure 4.1).[10,15–19] The second model is the "archipelago" or "rosary-type" model, which proposes that individual asphaltene monomers are comprised of clusters of polycondensed groups consisting of five to seven aromatic rings each connected by short aliphatic side chains, possibly containing polar heteroatom bridges (Figure 4.2).[20–27] Asphaltenes are present as a large range of structures, so only representative structures for the two models have been drawn in Figures 4.1 and 4.2 to illustrate the basic differences.

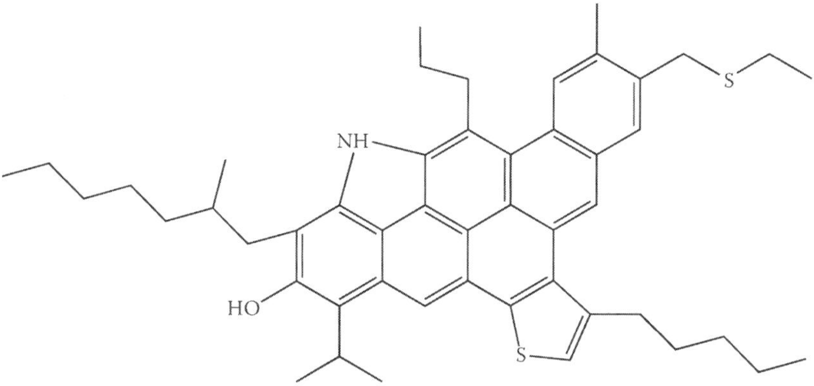

FIGURE 4.1 Representative structure and typical molecular weight of a proposed "continental" asphaltene molecule.

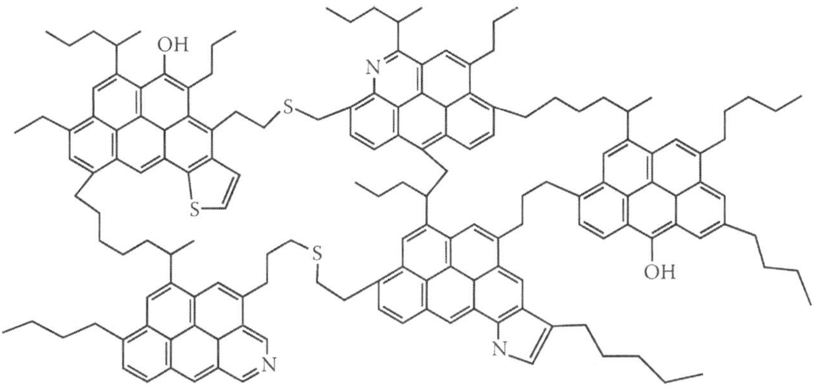

FIGURE 4.2 Representative structure of a proposed "archipelago" asphaltene molecule.

An article published in 2008 shows that earlier conclusions that asphaltenes have continental structures, derived from measurements of fluorescence decay and depolarization kinetic times, are wrong and that therefore asphaltenes do not have single condensed ring core architectures.[28] Another paper from 2007 claims that asphaltene molecular weights are bimodal with one component in the roughly megadalton range and a second component in the roughly 5-kDa range.[29] However, a later paper disputes these results and concludes that asphaltene molecular weights are monomodal with molecular weights as low as 750–1,000 Da, fitting the continental model.[30] The evidence gathered in this paper is very comprehensive, being obtained from four molecular diffusion techniques and seven mass spectral techniques from many groups around the world. An example is asphaltene mass spectra recorded with two-step laser mass spectrometry (L^2MS), in which desorption and ionization are decoupled and no plasma is produced.[31] Thus, the presence of archipelago structures in asphaltenes now seems very unlikely.

At high pressure in the reservoir, asphaltenes exist as individual molecules or, at most, nanoaggregates in the crude oil. It has been assumed for about 70 years that polar resins, acting as surfactants, stabilize asphaltenes in solution. However, evidence by measuring asphaltene gravitational gradient and by high-Q ultrasonics and NMR diffusion measurements suggests that resins are not associated with asphaltenes at all in the reservoir.[32] However, two key parameters that have been shown to control the stability of asphaltene aggregates (micelles) in a crude oil are the ratio of aromatics to saturates and that of resins (or maltenes) to asphaltenes. When these ratios decrease, asphaltene monomers or aggregates will flocculate and form larger aggregates that may deposit in the system.[8,33–34] It is known that some monomeric surfactant AIs (to be discussed later), designed to act as artificial resins or resin enhancers, are able to keep asphaltenes dispersed. Thus, it may be that small asphaltene clusters or aggregates, which do not precipitate from solution, are interacted by resins only at higher aggregation numbers and not in the reservoir.

When the pressure drops during oil production but remains above the bubble point, the density of the oil decreases and the volumes occupied by the lighter components (C6–) increases more rapidly than the heavier less-compressible components (C7+ including aromatics).[35] Thus, the polarity of the oil decreases and the asphaltenes may begin to associate and eventually flocculate. This often occurs across the perforations downhole but can occur anywhere in the system, as far as the processing facilities.[36] Thus, minimization of pressure drops during production is one way of controlling asphaltene deposition.

Asphaltene self-association is a topic of much discussion.[37] The size of the first petroleum asphaltene aggregates (or micelles) appears to be about 20 Å or 8–10 molecules of asphaltene.[10,38] Others describe the first asphaltene aggregates as polydisperse oblate cylinders[25] or loose and nonspherical, fractallike particles.[39] The size of the aggregates is dependent on the polarity of the solvent. However, in mixtures with resins or synthetic ADs and AIs, these aggregates do not necessarily aggregate further or deposit on pipe walls. In fact, macroscopic asphaltene particles visible to the naked eye may be present, yet no deposition may occur.[40] Asphaltene flocculation has been shown to be reversible indicating that addition of some resins or dispersants to flocculated asphaltenes can prevent their deposition.[39] However, asphaltene deposition is generally considered irreversible.

It has been suggested that the aggregated asphaltene molecular structure is governed principally by the balance between the capacity of fused aromatic ring systems to stack via π-bonding, reducing solubility, and the steric disruption of stacking due to alkyl groups, increasing solubility.[41-42] Other interactions of importance between the polar parts of asphaltene molecules will also occur such as acid-base (electron donor-acceptor) and hydrogen bonding interactions.[43-44] One study obtained results that point toward the existence of alkyl layers surrounding precipitated asphaltene solids more in line with the "continental" model for asphaltene monomers.[45] A micellar model for asphaltenes was proposed, which explained many of the experimental determinations. The model considers the existence of alkyl moieties surrounding the inner aromatic cores. In another study, the least soluble fractions of asphaltene appear to form aggregates that agglomerate to insoluble particles with a highly porous nature.[46]

Nonchemical techniques that have been recommended for asphaltene control include the following:

- Avoid mixing certain crude streams. The blending of crude feedstocks is a common cause for asphaltene precipitation. Light, nonasphaltic crude is a possible precipitant for heavier crudes.
- Operate outside the AFE (asphaltene formation envelope). By manipulating temperature, pressure, or flow, it may be possible to minimize the occurrence of conditions that promote asphaltene deposition and thus extend the onstream efficiency of well production and equipment.
- Mechanical cleaning of the wells and surface equipment: this includes the use of wireline methods and the opening up of vessels, for example, separators, and literally digging out the accumulated material.
- Use higher flows to erode deposits.

There are two chemical methods of controlling asphaltenes in production operations, which will be discussed in this chapter:

- prevention with ADs and AIs
- remedial treatment with asphaltene dissolvers (solvents or deasphalted oil)

4.2 ASPHALTENE DISPERSANTS AND INHIBITORS

There are clearly two classes of additives that can prevent asphaltene deposition. They are ADs and AIs.[47] AIs provide real inhibition in that they prevent the aggregation of asphaltene molecules. Thus, an AI can shift the onset of asphaltene flocculation pressure. Hence, it can move asphaltene precipitation and subsequent deposition out of the wellbore to a point in the production system where it could be dealt with much more easily. AIs can be ranked by studying the asphaltene flocculation point determined, for example, by light transmission with a fiber-optic sensor. ADs do not affect the asphaltene flocculation point but reduce the particle size of flocculated asphaltenes keeping them in suspension in the oil.[36,48–49] ADs disperse preformed asphaltene flocculates. Many AIs can also function as ADs, but ADs do not generally function as AIs.

It is well known that AIs and ADs can be oil-specific. For example, a polymeric AI with protic polar heads and aliphatic tails prevented asphaltene deposition in oil A but not in oil B.[50] In contrast, a nonpolymeric amine performed well in oil B but not oil A. The reason for these observations was explained in terms of acid-base chemistry. Thus, the nonpolymeric basic amine interacts preferably with organic acids in oil A and not with asphaltene molecules, whereas the protic polymeric inhibitor does not react with acidic species in oil A and is free to interact with asphaltenes. Oil B was also found to contain a higher abundance of basic nitrogen species in its asphaltenes, which would favor interaction with the protic polymeric inhibitor.

In general, AIs are polymers (or resins) whereas ADs are usually nonpolymeric surfactants, although many polymeric surfactant AIs function as ADs. To prevent the aggregation of asphaltene molecules, AIs need several molecular points of interaction for good inhibition, hence, the need for polymers. If, in addition, the AI contains alkyl long chains these can help disperse any formed asphaltene aggregates. This is assumed to be occurring also with nonpolymeric ADs. The polar and/or aromatic headgroups in AD surfactants interact with aggregated asphaltenes and the long alkyl chains on the periphery of the asphaltene aggregate help change the polarity of the outside of the aggregate and make it more similar to and dispersible in the crude oil. Some studies have shown that increasing the dosage of an AD can actually have a detrimental effect on asphaltene aggregation, possibly because of self-association of the AD surfactant molecules.[51]

For AIs, one study on asphaltene precipitation with heptane showed that until you reach a critical AI concentration, you do not see any effect; at the critical concentration and above, however, you see a dramatic effect since it actually stops the asphaltenes from flocculating.[52] The ADs in the same study, however, showed no critical concentration effects and act almost proportionately to concentration. A study on some commercial AIs showed that some products did not reduce the onset pressure for asphaltene flocculation or the average particle size.[53] However, the AIs did reduce the cumulative particle count relative to an untreated sample. A thermodynamic model has been proposed for asphaltene precipitation inhibition that treats asphaltenes as micelles.[54] The adsorption interaction between an asphaltene and an amphiphile molecule (natural resin or synthetic additive) is considered the most important parameter for the stabilization of the asphaltene micelles in crude oil. The authors further proposed that the adsorption enthalpy could be used as the most important criterion to search for efficient amphiphiles.

At least two research groups have shown that some polymeric AIs can actually increase the amount of flocculated asphaltene relative to an untreated system.[55–56] In another study, a polymeric AI actually had a negative impact on the AD test at low concentrations (100 ppm) for a medium weight percent asphaltene Gulf of Mexico crude, whereas at higher concentrations (> 500 ppm), the product was very effective.[57]

Since AIs can prevent asphaltene flocculation, they are best applied upstream of the bubble-point pressure, which is commonly downhole. ADs may be used further downstream, for example, where the AI treatment may not have been sufficient and the flocculated asphaltenes need dispersing to prevent them from depositing. It is general field practice to use either an AI or an AD, but not both, as an AI often has dispersant properties of its own. An AI is usually injected as a continual or batch treatment downhole either through a capillary string or gas lift system or can be injected further downstream such as at the wellhead. Treatment levels of good commercial AIs are often in the range of 20–100 ppm. A few reports of squeeze treatments with AIs into the near-wellbore have been published, but this deployment technique is not very common.[58–60] In this case, the key is getting the oil-soluble AI to have good adsorption to the formation rock and not be produced back too quickly. Otherwise, treatment lifetimes become uneconomically short. A laboratory study highlighted that selection of the best-performing AI will not necessarily be the optimum choice for squeezing if it has poor rock adsorption.[61] Today, squeeze treatment lifetimes of perhaps 2–6 months are possible depending on the severity of the asphaltene problem. To aid adsorption of the AI onto the rock, very polar groups such as carboxylic acids can be introduced into the polymer. Phosphonate groups could also be introduced into AI polymers, as they are known to adsorb more strongly than carboxylic acid groups.

Use of live and dead oils in testing AIs and ADs can give different results. Tests with dead oils can be useful for rough screening of classes of ADs and AIs[58,62–68] but high-pressure live oil deposition tests are recommended, either in pressure cells or pressure-drop pipe-loops.[8,53,69–76] Some laboratory studies used precipitated asphaltenes that had been redissolved in an aromatic solvent. These solutions will lack the natural resins, meaning AIs and ADs that depend on the presence of natural resins for optimum performance will not perform as well. One study showed that tests using live oils that had not been depressurized needed a lower concentration of AI for asphaltene stabilization than using standard laboratory precipitation methods with dead oils.[71]

In this review on asphaltene control additives, monomeric surfactants that generally perform as ADs will be discussed first. Polymers that are AIs, many of which are also ADs, are discussed afterward. Polymers are most often used commercially both as AIs and ADs, although some monomeric surfactants have been used in the field with some success. One class of commercial AI polymer is used in the field both downhole and topside.[57,77] They can act as synergists with demulsifiers for oil-water separation by removing the asphaltenes from the oil-water interface. They are mentioned here in this section because no structural details of the polymers are available except that the polymers contain only carbon, hydrogen, and oxygen. The polymers have also been formulated with biodegradable solvents to give them a better environmental profile.[78]

4.3 LOW MOLECULAR WEIGHT, NONPOLYMERIC ASPHALTENE DISPERSANTS

The various classes of low molecular weight, monomeric ADs can be summarized as follows:

Asphaltene Control

- very low polarity alkylaromatics
- alkylaryl sulfonic acids
- phosphoric esters and phosphonocarboxylic acids
- sarcosinates
- ethercarboxylic acids
- aminoalkylenecarboxylic acids
- alkylphenols and their ethoxylates
- imidazolines and alkylamide-imidazolines
- alkylsuccinimides
- alkylpyrrolidones
- fatty acid amides and their ethoxylates
- fatty esters of polyhydric alcohols
- ion-pair salts of imines and organic acids
- ionic liquids

Alkylaryl sulfonic acids, such as dodecylbenzenesulfonic acid (DDBSA), have been used for many years in the field as ADs, both upstream and in refineries. Other surfactants such as alkylphenols (which are endocrine disrupters) and some amide/imide products have also been used in the field.

4.3.1 Low-Polarity Nonpolymeric Aromatic Amphiphiles

Of all the classes of asphaltene control additives that will be discussed, we begin with the least polar. Hexadecylnaphthalene and hexadecylnapthoxide have been claimed as AIs rather than ADs (Figure 4.3).[79–80] It is claimed that these chemicals prevent precipitation of asphaltenes in the first place, rather than involving dispersion of the precipitate.

These low-polarity molecules will interact with asphaltenes by π–π interactions between the aromatic rings of naphthalene and the asphaltene monomers as well as via the polar groups. This prevents the asphaltene monomers from stacking and aggregating while the aliphatic tails of the additives will interact with the hydrocarbon solvent, with which they are compatible. The two-ring naphthyl group appears to give better π–π interactions than a single phenyl group. However, a more polar head group than naphthyl may give even better interactions with asphaltene monomers.

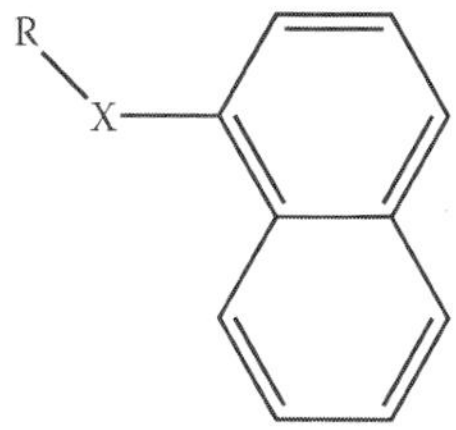

FIGURE 4.3 Naphthalene-based AIs. R = long alkyl chain and X is optionally a spacer group such as ether, ester, or amide group.

For example, one group measured neutron and X-ray scattering intensities from asphaltenic aggregates in several solvents. They found that aggregate sizes were two to four times smaller in high-polarity solvents such as pyridine and tetrahydrofuran than in benzene.[81] Another group found that mixtures of hydrocarbon aromatic solvents (e.g., toluene, xylene) and quinoline or alkylquinoline were better asphaltene dissolvers than hydrocarbon aromatic solvents alone.[82] Dimethyl formamide and N-methyl pyrrolidone (NMP) are also better asphaltene dissolvers than aromatic solvents.[50] This means that asphaltene control additives with more polar head groups such as pyridinyl, quinolinyl, tetrahydrofuryl, and dimethylamidyl should interact better with asphaltenes than less polar head groups such as phenyl or naphthyl. This is further supported by molecular simulation studies. Simulations showed that for asphaltene aggregates in quinoline, some stacking interactions could be disrupted, while in 1-methylnaphthalene, it was not observed.[43] Thus, hexadecylnaphthalene and hexadecylnapthoxide may not have a polar-enough head group for optimum asphaltene interaction.

4.3.2 Sulfonic Acid-Based Nonpolymeric Surfactant ADs

One way to make the aromatic head group of the ADs (or AIs) in Section 4.3.1 more polar is to add a sulfonic acid group. The most common AD in this class is DDBSA, which is cheap and has been used commercially with some success (Figure 4.4).[62,83] Longer-chain alkylarylsulfonic acids have also been patented.[84–85]

Several research groups have investigated the asphaltene-dispersing power of DDBSA as well as its asphaltene solubilizing power.[86–90] For example, it was found that dodecylbenzene sulfonic acid is a flocculant at low concentration but a dispersant at higher concentrations.[91] Another group examined the properties of a number of monomeric additives.[92] They found that increasing the polarity on the headgroup gave better asphaltene stabilization through stronger acid-base interactions. Thus, the performance of dodecylbenzene sulfonic acid as an AD was better than nonylphenol (NP), which in turn was better than nonylbenzene. Dodecylbenzene sulfonic acid was also shown by Fourier transform infrared (FTIR) spectroscopy to interact via hydrogen bonds with asphaltenes. Besides acid-base interactions, this group also proposed that the sulfonic acid group could donate its proton to C=C bonds in asphaltenes.[93] However, they noted that using too many polar groups, or one group that is too polar, can reduce the solubility of the surfactant in oil, rendering it ineffective as an AD. To support this claim, another group found that the sodium salt of dodecylbenzene sulfonic acid, which has a very polar ionic head group, was a fairly poor AD.[94] Other workers have found similar results.[95] For example, in order of decreasing performance as an AD, it was found that dodecyl resorcinol > dodecylbenzene sulfonic acid > nonyl phenol > toluene. The workers proposed that the effect

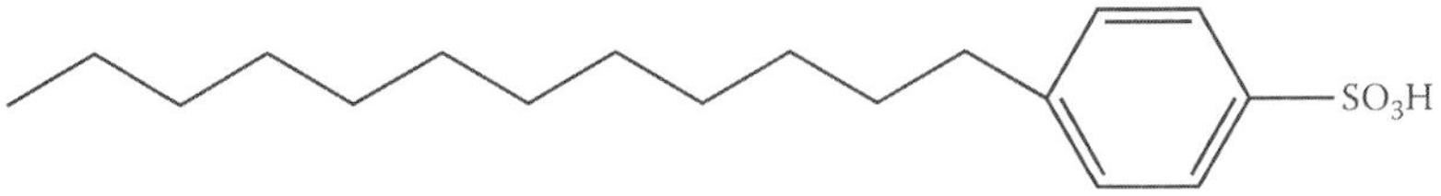

FIGURE 4.4 Structure of the 4-isomer of DDBSA.

Asphaltene Control   119

of the surfactants is due to the interaction between the acidic head of these molecules and the asphaltene. The mechanism of inhibition was explained in terms of a micellization model.[96–98]

Another study gave results showing that a more polar group attached to an aryl head performs better as an AD.[99] In order of decreasing performance as an AD, it was found that dodecylbenzene sulfonic > nonyl phenol > dodecylphenol bis-ethoxylate > nonylbenzene. The only anomaly in the order is that dodecylphenol bis-ethoxylate performed worse than nonyl phenol even though it has a more polar head group (higher HLB value). This may suggest that simple alcohols or alcohol polyethoxylates are not good head groups for ADs as they give poor interactions with polar groups in asphaltenes. The patent literature seems to bear this out as there are no good examples of any class of polyethoxylate surfactant claimed as ADs. The reason is that the highly acidic proton in a sulfonic acid group can hydrogen-bond better to amine residues in asphaltenes than a proton in a lone hydroxyl group. The same group also determined that the above amphiphiles adsorb by a two-step adsorption mechanism (LS or S type) onto asphaltene particles.[99] The first step is adsorption of the amphiphile onto asphaltene, the second step is adsorption of amphiphiles onto adsorbed amphiphiles forming a double layer.

Related sulfonic acid-based improvements on dodecylbenzene sulfonic acid as an AD appear to have gone in two directions. In one patent, the aryl head group has been removed leaving only aliphatic alkyl chains directly bonded to the sulfonate. However, the inventors have claimed only the use of secondary alkanesulfonic acids having chain lengths of 8–22 carbons, preferably 11–18 carbons, as ADs. (Figure 4.5).[100] The alkanesulfonic acid is preferably formulated as a solution or microemulsion and can further contain optional alkyl-formaldehyde resin, oxyalkylated amines, or wax-dispersing agents. The alkanesulfonic acids provide reduced precipitate amounts, slow the rate of precipitate formation, form a more finely divided precipitate, and reduce the tendency of the precipitate to be deposited on surfaces. The only example of a secondary alkanesulfonic acid tested was not compared with any other AD. The reason why secondary rather than primary alkanesulfonic acids are claimed may be because two alkyl tails give better interaction with the hydrocarbon solvent.

This is also seen in a second patented improvement on DDBSA, which is to use a branched alkyl polyaromatic sulfonic acid. Preferred structures are sulfonated alkyl-naphthalenes with at least one branch in the tail and preferably 30+ carbons in the tail.[101–103] An example of a good AD in this class with C15 and iso-C15 tails is given in Figure 4.6. One sulfonic acid group was determined to be the most effective head attached to the aromatic structure, better than carboxylic acid, hydroxyl, and amine groups. A straight-chain paraffinic tail was found to be ineffective above 16 carbons. This was because of decreased solubility in the oil caused by crystallization with

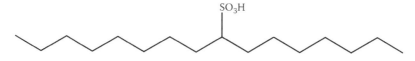

FIGURE 4.5   Structure of a preferred secondary alkanesulfonic acid.

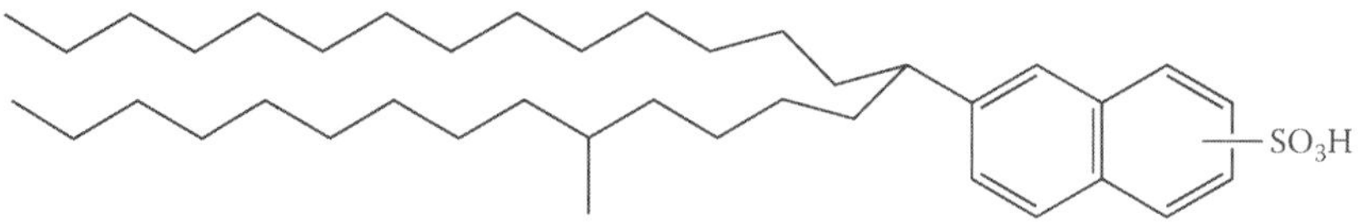

FIGURE 4.6. A preferred branched alkyl (*n*-C15 and *iso*-C15) naphthalene sulfonic acid AD.

other AD tails and with waxes in the oil. Further, if the aromatic part of the AD was one ring or two connected rings (biphenyl), the performance was worse than two fused-ring systems (e.g., naphthalene). In addition, the researchers found that *n*-alkylaryl sulfonic acids, such as DDBSA, lose their ability to disperse asphaltenes with time. Both of these problems were solved by using two branched tails of varying length. As a result, the effectiveness of the dispersant increased with total tail length, well above 30 carbons, and it remained effective with time. These sulfonated alkylnaphthalenes appear to be the best sulfonic acid-based monomeric surfactant ADs investigated.

4.3.3 Other Nonpolymeric Surfactant ADs with Acidic Head Groups

There are several further studies on the use of AD surfactants with acidic head groups that are not sulfonic acid. One group studied small alkylcarboxylic acid amphiphiles as well as alkylamines.[104] They proposed that the effectiveness of an amphiphile in stabilizing asphaltenes is dependent on its ability to adsorb onto the asphaltene surface. For example, they found that hexylbenzoic acid adsorbed better to asphaltene surfaces than hexylamine. However, neither additive adsorbed as well as didocecylbenzene sulfonic acid and DDBSA. The poor results with alkylamines, which are basic, compared with acid groups such as sulfonic or carboxylic acid suggest that hydrogen bonding from an acid group in an amphiphile to basic sites in an asphaltene molecule (e.g., amines, hydroxyl groups) is more pronounced than that of amine groups in amphiphiles bonding to acidic protons in asphaltene molecules. This is also reflected in the relatively lower amount of oxygen-containing groups (hydroxyl, carboxyl) in asphaltenes compared with nitrogen groups (amines).

The use of a highly acidic group in the head of an AD amphiphile, such as sulfonic or carboxylic acid seems to have further preference in the patent literature. Although ordinary fatty acids have not been claimed in patents, claims have been made for both ether carboxylic acids and phosphoric esters as ADs (Figure 4.7). Both contain highly acidic protons, which can form hydrogen bonds to amines or hydroxyl groups in the asphaltenes or interact with metal ions destabilizing the asphaltene aggregation process. It is not clear why the introduction of a polyether (polyalkoxylate) chain into ether carboxylic acids is an advantage over a simple fatty acid. The polyether chain makes the head of the amphiphile more polar and may possibly interact via the oxygen atoms with either acidic protons or metals ions.[105]

Like the ether carboxylic acids, most of the examples of phosphoric esters investigated do not have aromatic groups that could aid interaction via $\pi-\pi$ overlap with

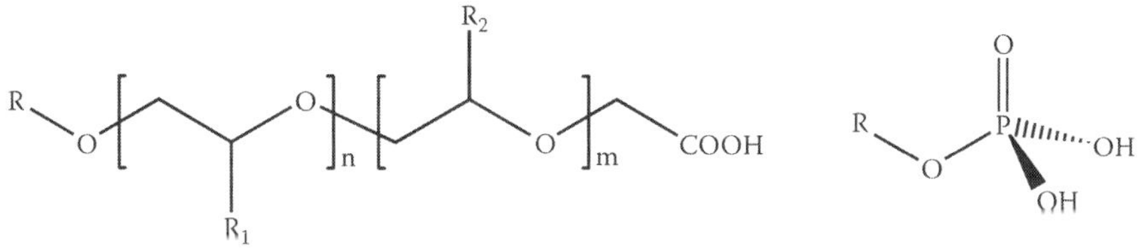

Asphaltene Control

FIGURE 4.7 Structures of ether carboxylic acids and phosphoric monoesters (R = alkyl or alkylaryl and R_1 and R_2 are H or Me).

asphaltenes.[106–107] A preferred nonaromatic example is isooctyl acid phosphate. A blend with dodecylbenzene sulfonic acid gave an improved performance as an AD over the phosphoric acid ester alone.[108] One group reported results showing that phosphoric esters of alkylphenylethoxylates blended with fatty acid diethanolamides performed well as ADs.[109] Besides hydrogen bonding from the phosphoric groups to asphaltenes, the phenyl group in the phosphoric ester of alkylphenylethoxylates can also interact via π–π overlap with asphaltenes (Figure 4.8). Synergistic blends of phosphoric esters with carboxylic acids or carboxylic acid derivatives have also been patented.[110] These apparently performed better than nonylphenolformaldehyde resins, which will be discussed later. Another patent claims phosphonocarboxylic acid esters as ADs, made from esterification of phosphono or phosphino acids such as phosphonosuccinic acid, with alcohols with a C6–25 alkyl, alkylaryl, or alkenyl group.[111]

The asphaltene-dispersing power of natural naphthenic acids as well as some synthetic ones has been examined using near infrared spectroscopy.[112] It was found that naphthenic acids adsorb to asphaltenes and disperse them, but the performance was lower than a commercial AD based on fatty acids and amines. The best synthetic naphthenic acid was decahydro-1-naphthalenepentanoic acid (Figure 4.9). The rings

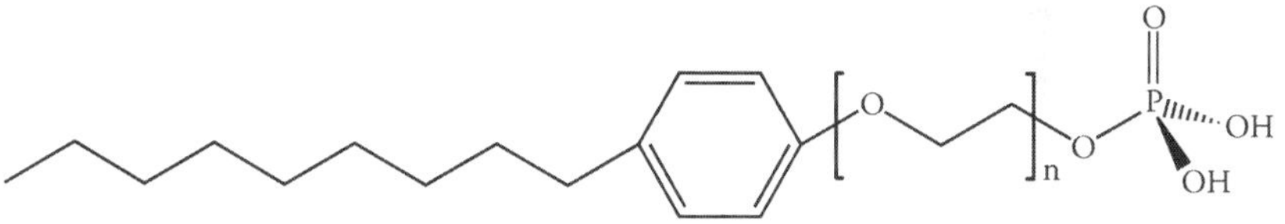

FIGURE 4.8 A preferred AD class of phosphoric ester of alkylphenylethoxylates.

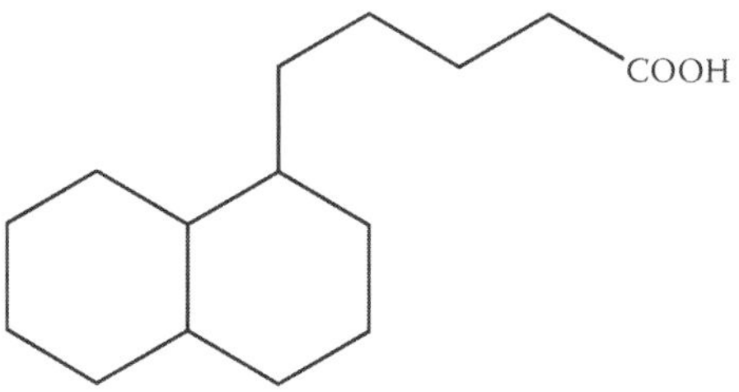

FIGURE 4.9 The structure of a synthetic naphthenic acid, decahydro 1-naphthalenepentanoic acid.

of this naphthenic acid are not aromatic and therefore cannot interact by π–π overlap with asphaltene molecules. For this reason, it is suggested that alkylphenylcarboxylic acids might perform better than naphthenic acids.

Other classes of ADs containing carboxylic acids in the head of the amphiphile have also been patented. For example, sarcosinate surfactants have been shown to be useful ADs in laboratory tests.[113] This class has the structure shown in Figure 4.10. This class can form hydrogen bonds to asphaltenes via both the amide and carboxylic acid groups.

Another class of amphiphile-containing acidic head groups that have been claimed as ADs are the reaction products of amines and unsaturated organic acids (Figure 4.11).[114]

A preferred example of an AD in this class is the reaction product of oleyl amine with acrylic acid in a 1:2 ratio. The structure should be a Michael addition product as shown in Figure 4.12. It may exist as a zwitterion on contact with water but this is unlikely to occur in a nonpolar solvent such as a hydrocarbon.

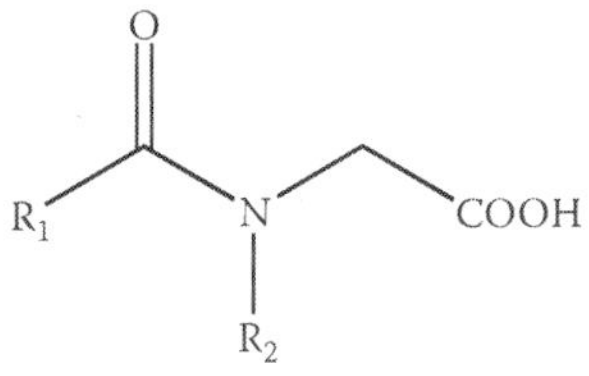

FIGURE 4.10 The structure of sarcosinate ADs: R_1 and R_2 are preferably long alkyl or alkenyl chains.

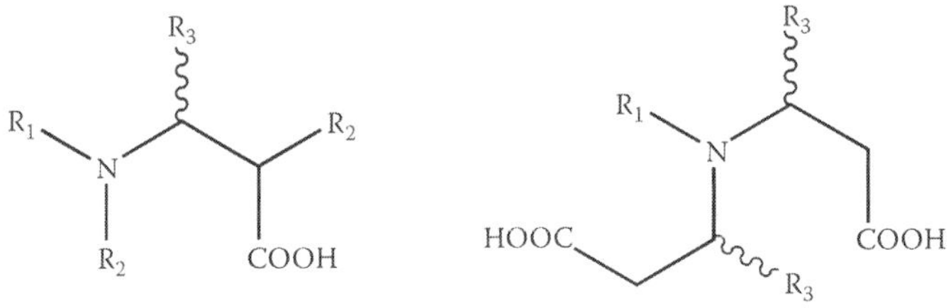

FIGURE 4.11 General structures of the reaction product of amines and unsaturated organic acids.

FIGURE 4.12 The tautomeric structure of the reaction product of oleylamine with acrylic acid in a 1:2 ratio.

Asphaltene Control

4.3.4 AMIDE AND IMIDE NONPOLYMERIC SURFACTANT ADs

A number of nonionic amphiphiles with amide or imide groups have been claimed as ADs, some of which are in commercial use. The simplest amide classes that have been investigated contain a single amide group and no other functionality. They include long-chain N,N-dialkylamides and alkylpyrrolidones (Figure 4.13).[50,115] The alkylpyrrolidones contain a 5-ring resembling the pyrrolic groups found in asphaltenes.

Polyisobutylene succinimide, which can be considered a small amphiphile, has also been claimed as an AD.[116] This molecule contains an imide head group as shown in Figure 4.14, which can form hydrogen bonds to asphaltenes via either carbonyl group. The authors of the patent also claim that blends of polyisobutylene succinimide with other amphiphiles can repeptize precipitated asphaltenes. Related to this is another AD patent claiming the reaction product of an amine with polyisobutylene succinic anhydride (which has C50–C70 carbons in the chain).[117] The amines can be fatty amines, tertiary alkylamines, or polyamines such as diethylenetriamine. The reaction product should contain both carboxyl and amide groups for best effect plus one or more alkyl tails. These additives are also claimed to reduce the viscosity of crudes by preventing asphaltene agglomeration.[118]

Experiments have been carried out showing that blends of fatty acid diethanolamides and phosphoric esters of alkylphenylethoxylates are good ADs.[109] Phosphoric esters were discussed earlier in this section. The structure of the diethanolamides is given in Figure 4.15. This class has three functional groups that can form hydrogen bonds to asphaltenes.

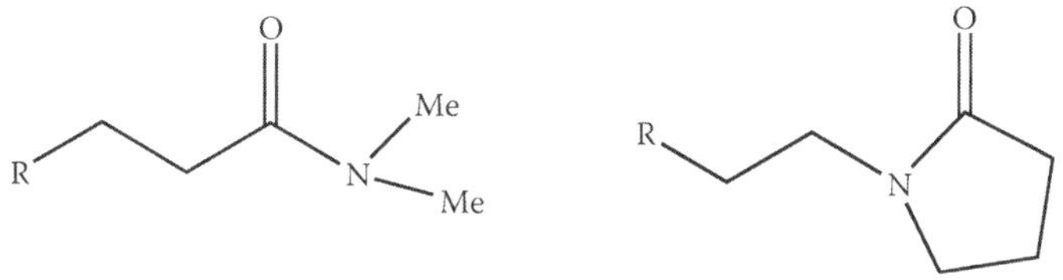

FIGURE 4.13 The structures of N,N-dialkylamides and alkylpyrrolidone ADs. R is preferably greater than eight carbons.

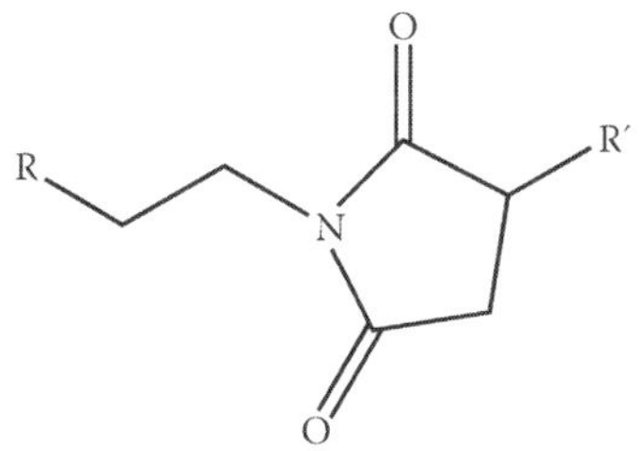

FIGURE 4.14 The general structure of alkylsuccinimides such as polyisobutylene succinimide.

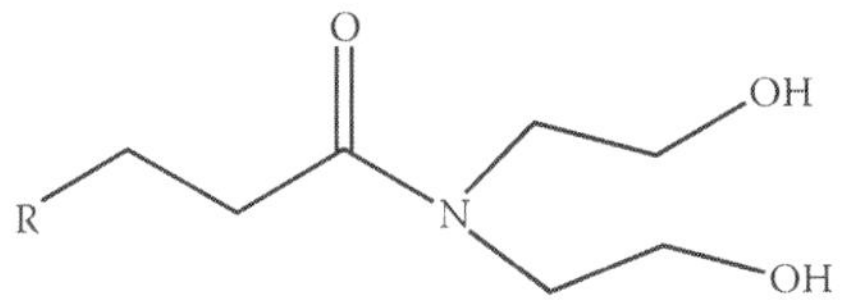

FIGURE 4.15 The structure of fatty acid diethanolamides.

More complicated amide products have been claimed as ADs, blended with alkylaryl sulfonic acids and emulsifiers.[119] These are derived from the condensation of a fatty acid with a polyamine of formula $H_2N-[(CH_2)_n-NH]_m-R$ where $n = 1–4$, $m = 1–6$, and R = H or alkyl. A preferred example of a polyamine is diethylenetriamine. Reaction of this triamine with 2 mol of fatty acid such as tall acid gives acyclic amides as well as cyclic imidazoline products as shown in Figure 4.16. Again, we have cyclic structures that resemble pyrrolic groups in asphaltenes. The ring imine and amide groups in these ADs can hydrogen-bond to acidic protons in the asphaltenes, while there may also be some π–π overlap between the unsaturated imidazoline ring and aromatic rings in the asphaltenes. Similar condensation products, blended with solvents such as NMP, N-ethyl pyrrolidone, dimethylformamide, and aromatics, have also been patented.[120] These blends are claimed to both inhibit and dissolve asphaltenes.

AD blends of two components, one of which is an amide-based surfactant, have also been patented.[121] The amide surfactant results from the condensation of a linear N-alkyl polyamine with a cyclic anhydride. Examples are the reaction product of maleic anhydride and N-oleyl-diamino-1,3-propane or the reaction product of phthalic anhydride and N-stearyl methyl-1-diamino-1,3-propane (Figure 4.17). Both structures contain 5-rings and imide groups, which may adsorb by acid-base interactions to pyrrolic groups in asphaltenes. The phthalimide structure contains, additionally, an aromatic ring, which can adsorb via π–π interactions with aromatic rings in the asphaltenes. The second component in the blend results from the reaction of an ethoxylated amine with a carboxylic acid of eight to 30 carbons. An example is the compound obtained by diesterification of triethanolamine by a tallow fatty acid in a molecular ratio of 1:2. These blends are claimed to delay the flocculation of asphaltenes in fuels.

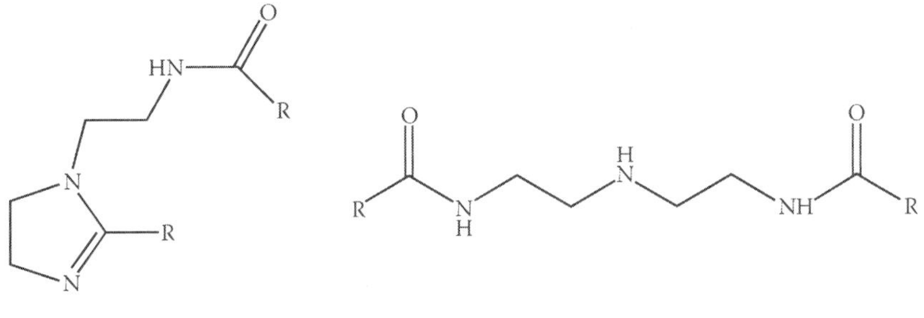

FIGURE 4.16 Possible products from the condensation of diethylenetriamine with 2 mol of a fatty acid. R is preferably C16–18.

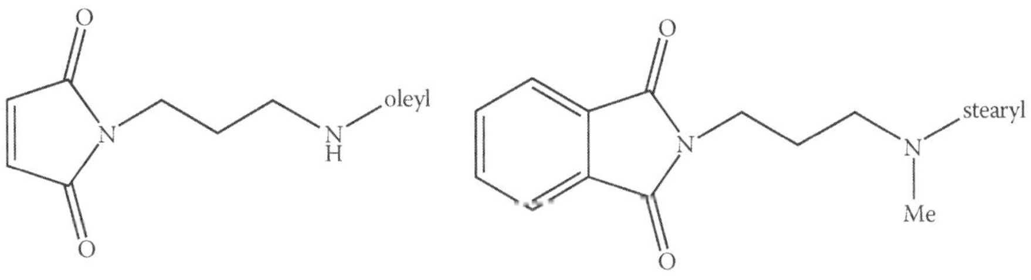

FIGURE 4.17 The structures of the reaction product of maleic anhydride and N-oleyl-diamino-1,3-propane (left) or the reaction product of phthalic anhydride and N-stearyl methyl-1-diamino-1,3-propane (right).

4.3.5 ALKYLPHENOLS AND RELATED ADs

Alkylphenols have been sold commercially as ADs for downstream applications, although they are known marine endocrine disrupters and cannot be used on environmental grounds in certain regions (see Chapter 1). Several groups have investigated monomeric alkylphenols as ADs in which the mildly acidic phenolic head group is assumed to resemble part of the asphaltene structure (Figure 4.18). Some of the results with monomeric alkylphenols were compared with alkylaryl sulfonic acids in Section 4.3.2. One study found that dodecylbenzene sulfonic acid and its sodium salt performed better than several less polar alkylphenols, where the alkyl chain has 2–12 carbon atoms.[122] The performance of the alkylphenols increased with increasing size of the alkyl tail, dodecylphenol (DDP) being the best AD in this class. Another group also carried out studies on alkylphenols with varying aliphatic chain lengths.[88–89] It was proposed that surfactants with short aliphatic chain lengths cannot peptize asphaltenes by forming a steric stabilization layer but get imbedded in or coprecipitate with the asphaltene. Conversely, too long an alkyl chain may lead to a poorer interaction of the surfactant with asphaltene.

Studies on the stabilization of asphaltenes using NP, DDP, and a nonylphenol-ethoxylate (NPE) have been carried out.[123] In order of decreasing performance as an AI, it was found that NP > DDP > NPE. Thus, in this study, making the head of the amphiphile more polar by ethoxylation gave a worse performance than the unethoxylated alkylphenols, although conflicting results have also been published.[95] Another group carried out studies on the asphaltene-peptizing ability of NPEs and found that they performed better than aliphatic amines[124] and alcohols.[125] Clearly, the use of a surfactant with an aromatic ring was better than using an aliphatic-based surfactant. NP also performed better than NPE phosphoric ester.

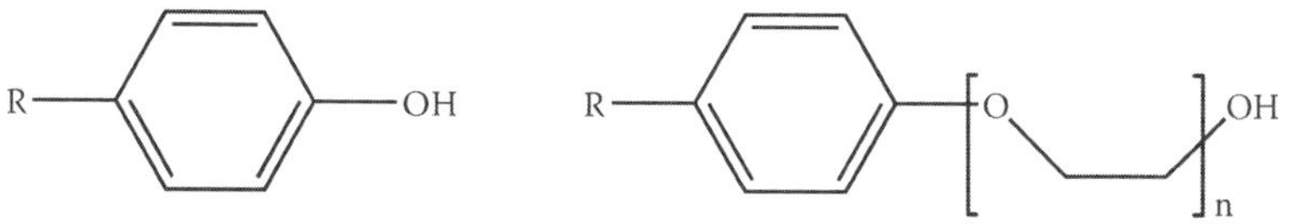

FIGURE 4.18 Structure of 4-alkylphenols and 4-alkylphenyl ethoxylates.

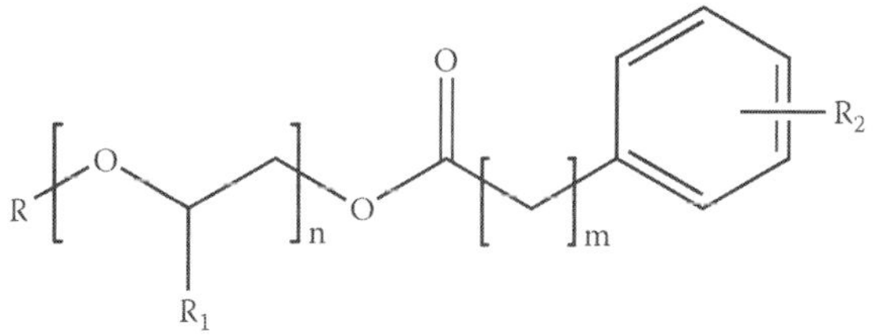

FIGURE 4.19 Preferred structures of ether carboxylic phenyl esters as ADs. R = H or CH$_3$, R$_2$ = C1–4 alkyl.

A study on the stabilization of asphaltenes by phenolic compounds extracted from cashew-nut shell liquid (CNSL) has been carried out.[56] The liquid extracted from cashew-nut shell is composed almost completely of phenolic compounds containing 15 carbon chains with variable unsaturation degrees, *meta*-substituted in the aromatic ring. The results showed that CNSL and cardanol, with C15 alkyl chains, have a performance comparable to NP as asphaltene-stabilizing agents.

Alkylphenols have activity as ADs as they contain a π-interacting aromatic ring and a polar hydrogen bonding group, phenol. These two features have been incorporated into a new class of single-headed amphiphiles that inhibited as well as helped dissolve asphaltenes.[126] These amphiphiles are ether carboxylic phenyl esters incorporating a phenyl ring with a polar chain made up of an ester group and one or more alkoxylate chains (Figure 4.19).

4.3.6 Ion-Pair Surfactant ADs

Oil-soluble ion-pair surfactants were first claimed as ADs in the early nineties. Besides surfactant-asphaltene interaction discussed earlier, it is possible that these ionic ADs selectively bind to metals in the asphaltenes, improving their adsorption. An example is a mixture of an alkylarylsulfonic acid and an alkylimidazoline. The acid proton is lost to the imidazoline and an anion-cation ion pair is formed.[127] Several patent applications have been filed on other surfactant ion pairs as ADs. For example, oil-soluble salt reaction products of amines and organic acids with the preferred formula R^{1}R^{2}R^{3}N$^+$R^{4}COO$^-$ have been claimed, where the molecule may also contain a further polar group 2–10 carbons away from the carboxylate group.[128] Examples are the reaction products of fatty amines with 2-hydroxybutyric acid or salicylic acid (Figure 4.20). These salts are oil-soluble where they will exist as ion pairs.

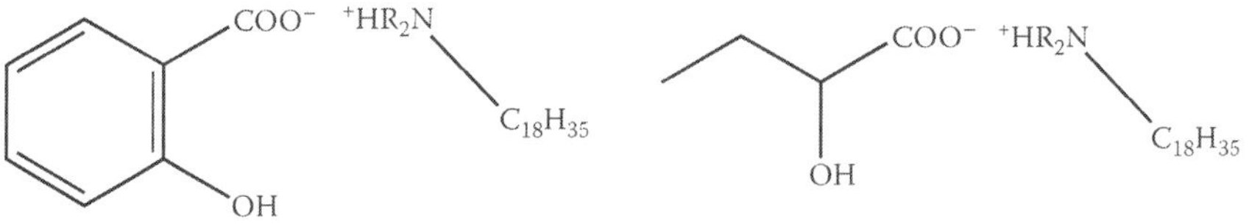

FIGURE 4.20 The structures of ion pair salt products of alkylamines and organic hydroxyacids.

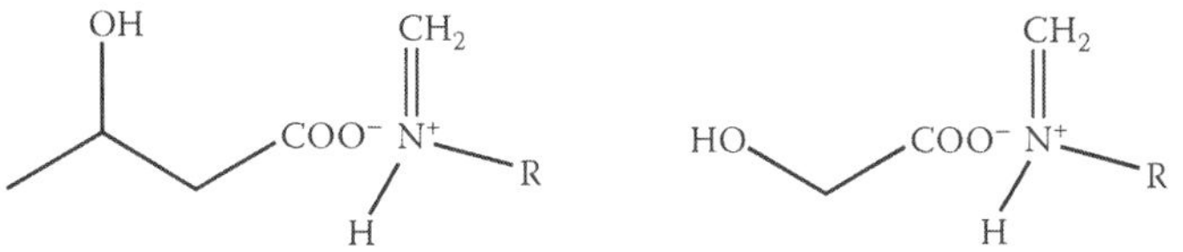

FIGURE 4.21 Structures of ion-pair salt reaction products of imines and organic acids.

In a related patent from the same author, the reaction product of an imine and an organic acid is claimed as an AD.[129] A preferred structure is a salt formed from the reaction of long-chain tertiary alkyl methyleneimines and a carboxylic acid such as glycolic acid (Figure 4.21). These salts are also oil-soluble and also probably exist as ion pairs.

Other ion pair ADs are formed between long-chain unsaturated alkylimidazolines and a C2–10 organic acid having at least one hydroxy group or at least one additional carboxyl group such as ascorbic or oxalic acid.[130] Blends of alkylimidazolines with EDTA-tertiary alkyl primary amine complex and 10–80% of an alkylbis(2-hydroxyethyl)amide have also been claimed.[131]

4.3.7 Miscellaneous Nonpolymeric ADs

Asphaltene precipitation studies have been carried out on a range of surfactant ADs including ionic liquids with cationic surfactants.[122] For example, *N*-butylisoquinolinium cations gave almost as good a performance as the *p*-alkylbenzenesulfonic acids. Long alkyl chains on the cationic surfactant worsened the performance. The best ionic liquids have an anion with high charge density, in connection with cations with sufficiently low charge densities. The mechanism proposed is that the ionic liquids can effectively prevent asphaltene precipitation from the reservoir oils by breaking the asphaltene associations, which are due to the local nonneutrality of the charge densities of the cation and the anion.

AD experiments with esters of polyhydric alcohols with carboxylic acids have been reported, although claims also include ethers formed from reacting glycidyl ethers or epoxides with polyhydric alcohols, and esters formed by reacting glycidyl ethers or epoxides with carboxylic acids.[132] Preferred AD examples are decaglycerol tetraoleate and sorbitan monooleate, which are used in some emulsifier formulations (Figure 4.22). Both molecules contain several hydroxyl groups in the polar head making them much more polar than alkylphenols and better ADs.

A range of nonionic amphiphiles for stabilizing asphaltenes in the process of making fuel oils has been claimed.[133] For example, in asphaltene precipitation tests, a blend of an alkylglycol, *o*-dichlorobenzene and naphtha solvent was found to give less precipitate than the well-known dodecylbenzene sulfonic acid and phosphoric esters. Another blend that performed well was tetrahydroxy-*p*-benzoquinone dissolved in a glycol or methanol. The patent also claims that these additive blends repeptize precipitated asphaltenes. Oxazoline derivatives of polyalkyl or polyalkenyl *N*-hydroxyalkyl succinimides have also been claimed as ADs.[134]

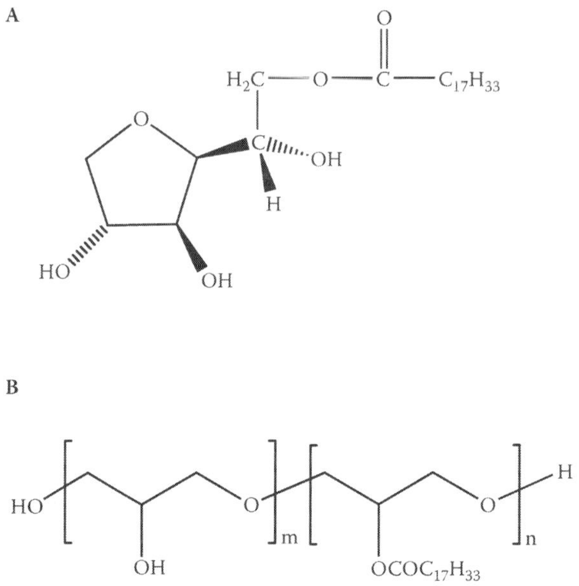

FIGURE 4.22 The structures of sorbitan monooleate (A) and polyglycerol polyoleates (B).

4.4 OLIGOMERIC (RESINOUS) AND POLYMERIC AIs

Most commercial AIs for upstream deployment appear to be based on polymeric surfactants. Some of them also function as ADs.[66–67,70,135–136] Polymeric surfactant AIs contain many polar groups each of which can in theory interact with an asphaltene monomer. Strong binding between the AI and asphaltene molecules is assumed to be one requisite for good performance.[137] This section will review various categories of oligomeric (2–12 monomer units) and polymeric (>12 monomer units) AIs. Most AI treatments are continuous injection although there have been reports of AI squeeze treatments.[58–60] For squeeze purposes, the presence of acid groups in the AI, which can bind well to the rock, will increase the amount of AI that is adsorbed. This will give the AI squeeze treatment a longer lifetime. Typical acid groups used in polymeric AIs are often carboxylic acid groups. This functionality cannot easily be introduced into all classes of AIs.

The various classes of polymeric surfactant AIs or ADs can be summarized as follows:

- alkylphenol/aldehyde resins and similar sulfonated resins
- polyolefin esters, amides, or imides with alkyl, alkylenephenyl, or alkylenepyridyl functional groups
- alkenyl/vinyl pyrrolidone copolymers
- graft polymers of polyolefins with maleic anhydride or vinyl imidazole
- hyperbranched polyester amides
- lignosulfonates
- polyalkoxylated asphaltenes

Asphaltene Control

4.4.1 ALKYLPHENOL-ALDEHYDE RESIN OLIGOMERS

There are several reports on the efficiency of polyalkylphenol resins as AIs. The performance of these additives appears to depend on the polymerization procedure for making the additives. One of the classes of polymeric AIs that have been most investigated and finds regular use in the oil industry is the alkylphenol-aldehyde resin oligomers. Reaction of an alkylphenol with, for example, formaldehyde, gives an oligomer of usually 2–12 alkyphenol groups connected by methylene bridges (Figure 4.23). Polyalkoxylates of these oligomers are commonly used as demulsifiers.

Adsorption isotherms on asphaltenes with NP, nonylphenolic formaldehyde resin (NPR), and native resins (NRs) have been reported.[138] In increasing order of effectiveness as AIs, it was found that NP < NR < NPR. The adsorption isotherms fitted an LS (Langmuir-S) shape that was explained using a two-step adsorption mechanism. In the first step, the surfactants are adsorbed individually on the asphaltene surface. In the second, the interactions between adsorbed surfactants become predominant and the formation of amphiphile aggregates in the surface begins. In an experimental theoretical approach to the activity of amphiphiles as asphaltene stabilizers, it was found that there is a balance between the polarizability of the amphiphile and its dipole moment on its ability to adsorb to asphaltenes.[139] If the amphiphile is too polar, then it makes it insoluble also.

Studies on the stabilization of asphaltenes using cardanol and polycardanol have been reported[56] (it should be noted that the polycardanol was made by *cationic* polymerization of the monomer, usually, more alkylphenol resins are polymerized with aldehydes). The alkyl group of 15 carbons is *meta* to the phenolic group in cardanol, whereas the alkyl group in NPE is *para* to the phenolic group. It was found that polycardanol was not only less efficient than its monomer, but, instead, enhanced the precipitation of asphaltenes. This effect was ascribed to the large number of phenol groups present in the polymer that may flocculate the asphaltene particles or increase its polarity, reducing its solubility in aliphatic solvents. Thus, polycardanol either interacts with several asphaltene molecules aggregating them or the hydroxyl groups in the polycardanol stick out into the solvent raising the polarity and lowering the solubility of the adsorbed asphaltenes. In contrast, it has been shown that polycardanol, made by the condensation of formaldehyde with cardanol, gave good

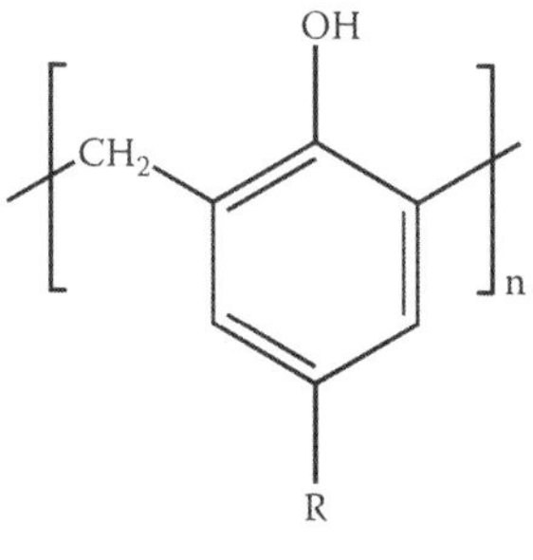

FIGURE 4.23 A typical structure of an alkylphenol formaldehyde resin AI where R = C3–24 and $n = 2–12$.

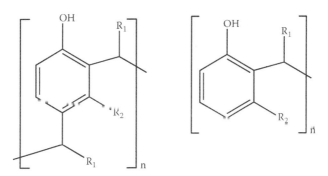

FIGURE 4.24 Two possible structures found in cardanol-aldehyde resins where R_1 is very variable and R_2 has approximately 15 carbons as a mix of at least four isomers.

results as an AI.[140] Examples of possible structures of cardanol-aldehyde resins are given in Figure 4.24.

Several improvements on the basic alkylphenol-formaldehyde resin AIs have been patented. For example, it has been found that sulfonated alkyl phenol formaldehyde resins performed better than nonsulfonated resins.[141–142] Interestingly, the structure of these resins resembles dodecylbenzene sulfonic acid units joined together. Thus, the polymers have several groups for acid-base and hydrogen bonding interactions with asphaltenes and several alkyl tails, which are compatible with the hydrocarbon solvent. Another group found that alkylphenol formaldehyde resins treated with polyamines such as triethylenetetraamine gave better performance in caustic-treated petroleum crude oil than DDBSA and esters of polyolefin/maleic anhydride copolymers.[143]

There are also several patents on the use of a second additive that can act synergistically with alkylphenol formaldehyde resins to increase the AI performance. For example, it has been found that the performance of alkylphenolformaldehyde resins could be improved synergistically by the addition of oxyalkylated amines such as ethoxylated triethylenetetraamine.[144] It has also been found that alkylphenol formaldehyde resins in combination with hydrophilic-lipophilic vinylic polymers gave improved results as AIs compared with the resins alone.[145–146] Other synergy patents mention specific hydrophilic-lipophilic vinylic polymers as amphiphilic ester copolymers such as lauryl methacrylate-hydroxyethylmethacrylate copolymer. Ester copolymers will be discussed in the next section.[147–149]

There are a few further studies that have compared the performance as ADs of alkylphenol formaldehyde resins and monomeric amphiphiles and other polymeric amphiphiles. For example, it was found that synergistic blends of phosphoric esters with ether carboxylic acids (discussed earlier) performed better than nonylphenolformaldehyde resins.[110] Another study found that dodecylphenolformaldehyde resin (DPR, MW = 200–2,000) was a better peptizer of asphaltenes than octadecene-maleic anhydride copolymer (POM, MW = 10,000; Figure 4.25).[150] The molecular weight of the amphiphilic polymer was considered critical. Too high a molecular weight could lead to several polymer-asphaltene particle interactions leading to unwanted coagulation. POM was found to associate more strongly with asphaltenes than DPR and monomeric amphiphiles, indicating that POM should be

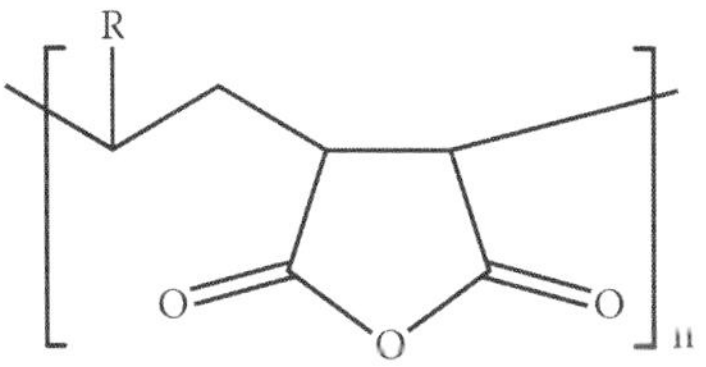

Asphaltene Control

FIGURE 4.25 The structure of alkene-maleic anhydride copolymer AIs. R is preferably a long alkyl group.

the best inhibitor. Hydrogen bonding between the polymers and asphaltenes was considered the primary mechanism for the association. The authors believe POM's anhydride groups give better hydrogen bonding to asphaltenes than DPR's phenolic hydroxy groups. The possibility of π–π interactions between DPR aromatic rings and asphaltenes was not considered. In contrast to this study, another group found that POM actually increased the amount of asphaltene flocculation relative to an untreated sample.[55] It was concluded from FTIR studies that the anhydride group was hydrogen bonding to OH groups in the asphaltenes. The contrasting results with POM are hard to understand, although it should be noted that in the latter study, they used fuel oil, whereas the first study used crude oil. The latter results can also be rationalized if the asphaltene sample had already aggregated to some degree before chemical treatment. The aggregates will have many aliphatic alkyl groups sticking out into the solvent. The polymer, which cannot penetrate to the polar part of the asphaltene aggregates, has also many long aliphatic alkyl groups. These groups can interact with several asphaltene aggregates with the polar anhydride groups facing out into the hydrocarbon solvent, causing them to flocculate to larger particles. Thus, the polar anhydride groups are facing out into the solvent and not the alkyl groups. In line with the above study in crude oil, it was found that C28 α-olefin (octacocene)/maleic anhydride copolymer was an excellent dispersant for asphaltenes in hydrocarbon refinery streams.[151]

4.4.2 Polyester and Polyamide/Imide AIs

Polyester or polyamide/imide-based AIs have been used for some time in commercial applications by several service companies.[70] The ester groups in polyesters and the amide groups in polyamides are usually provided by reactions with acrylic and/or maleic anhydride monomers (vinyl alkanoates could also be used). Unesterified monomers of this type also allow for incorporation of free carboxylic groups for better rock adsorption in squeeze applications. Typical ester examples are (meth)acrylate copolymers, styrene/maleate ester, and alkene,maleate ester copolymers shown in Figure 4.26. The ester groups can be exchanged with amide groups to make polyamides. Olefin-maleic copolymers can be reacted with amines to provide succinimide groups (Figure 4.27). Such polymers are commercially available AIs.

A copolymer made from a lipophilic monomer, as the major component, and a hydrophilic monomer has been claimed as a useful AI. An example is lauryl methacrylate/hydroxyethylmethacrylate copolymer. The use of a hydrophilic

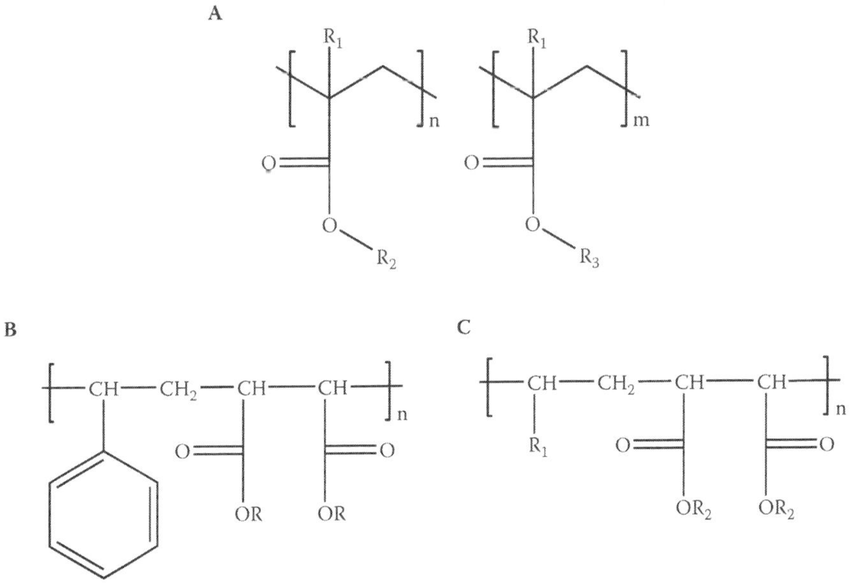

FIGURE 4.26 Structures of (meth)acrylate (R_1 = H or CH_3), styrene/maleate diester and alkene/maleate diester copolymers.

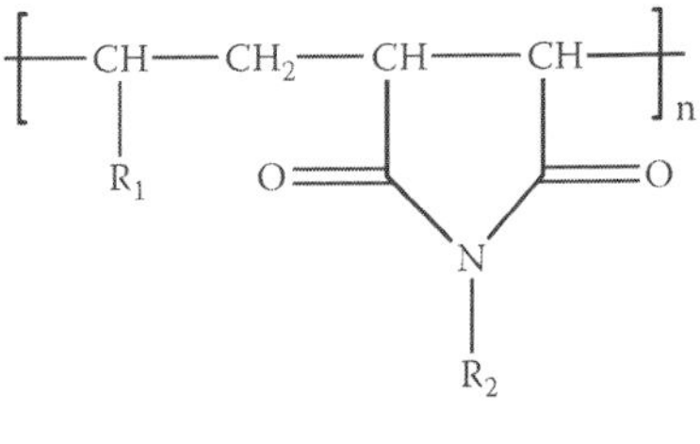

FIGURE 4.27 Polyalkylenesuccinimide copolymers.

monomer in the copolymer will raise the polarity of the end of the pendant groups giving stronger hydrogen bonding interactions with asphaltene particles.[147,149,152] In line with this rationalization, the authors mention that a commercial product with only lipophilic monomers, a diester of alkylene-maleic anhydride reacted with a fatty alcohol, was a worse product. Further, they found that α-olefin/maleic anhydride copolymers that were not esterified performed better as AIs than the lipophilic copolymers.

Instead of using alkyl chains in the side groups of polyesters and polyamides, it would seem advantageous to use aromatic rings that can make π–π interactions with asphaltenes. Examples of this class have been patented.[153] Preferred examples are transesterified products of methyl methacrylate and hydroxymethylpyridines, which are subsequently polymerized (Figure 4.28). The hydroxymethylpyridines could be the 2-, 3-, or 4-isomers. They could also be reacted with alkene-maleic anhydride

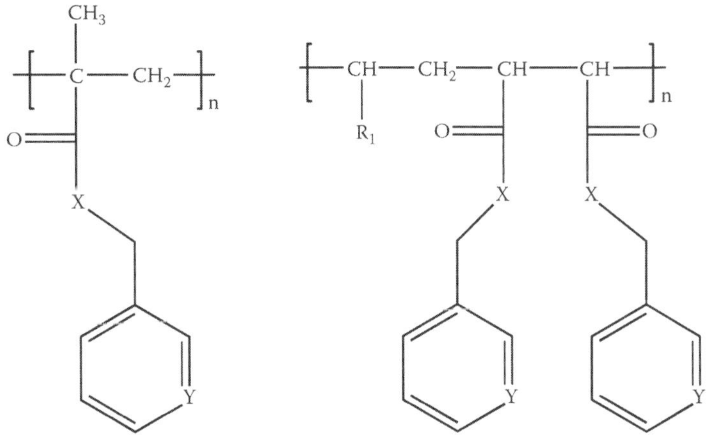

Asphaltene Control

FIGURE 4.28 The structure of two classes of AI polymers with pendant aromatic rings. R_1 = alkyl or alkenyl, X = O or NH, and Y = CH or N.

copolymers instead of meth(acrylates). All these polymers will have pendant pyridine rings with neighboring polar ester groups. Other aromatic rings such as phenyl or naphthyl groups could also be incorporated into the copolymers with ester or amide spacer groups.

In a related patent by the same group, the use of polyesters or polyamides as AIs has been claimed, made by partially derivatizing polycarboxylic acids with amines or alcohols containing ring structures.[154] Examples of ring structures are aromatic and heterocyclic rings. Specific examples of polymers are 4-alkylphenylmethacrylate where the alkyl group R has nine to 12 carbons (Figure 4.29). Esters of maleic anhydride copolymers could also be used.

Another method of incorporating amide, imide, or ester groups into a polymer is to first graft an anhydride onto a polyvinyl backbone then react the product with an amine or alcohol.[155] They prepared and tested polyester amides based on polyisobutylene reacted first with maleic anhydride and then with an amine such as

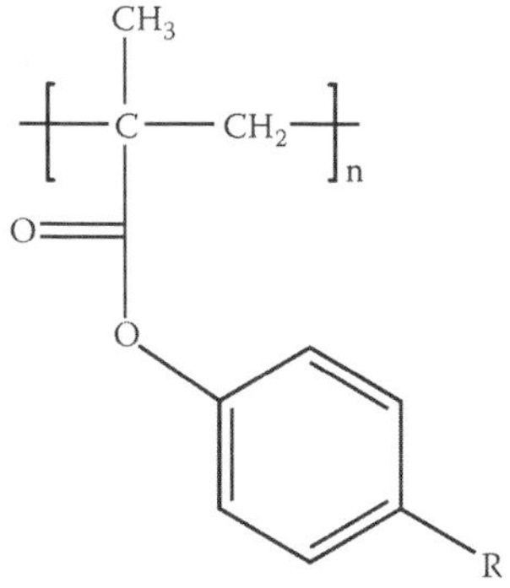

FIGURE 4.29 The structure of 4-alkylphenylmethacrylate polymer AIs. R has nine to 12 carbons.

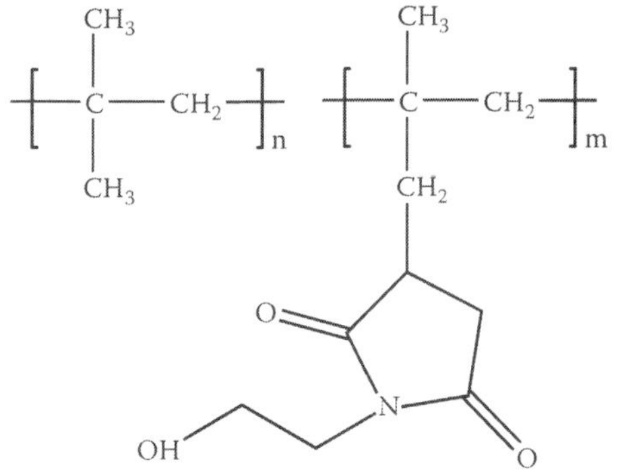

FIGURE 4.30 The structure of graft copolymer AIs with pendant *N*-hydroxyethylsuccinimide groups.

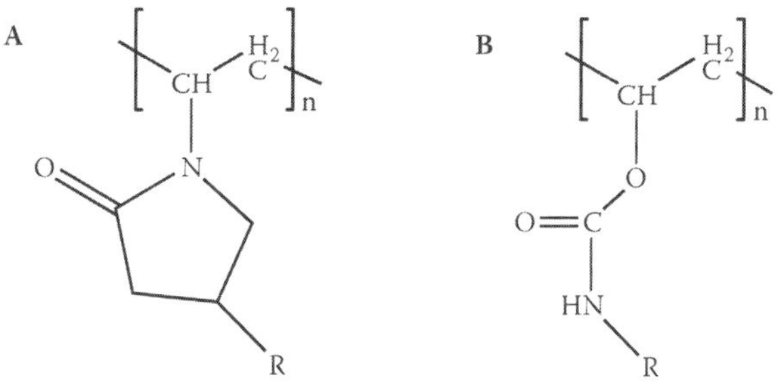

FIGURE 4.31 The structures of (A) 1-vinyl-4-alkyl-2-pyrrolidone polymers and (B) polyvinylalkyl carbamates. R is a long chain alkyl.

monoethanolamine (Figure 4.30). The copolymers have low saponification number, that is, high polyisobutylene/maleic anhydride ratio.

Another class of polymeric amide AI is the 1-vinyl-4-alkyl-2-pyrrolidone polymers (Figure 4.31). The pyrrolidone group has a strong hydrogen-bonding carbonyl group and shows structural similarities to the five-ring pyrroles found in asphaltenes.[50] Polyvinylalkyl carbamates are another related class of AI.

All the ester and amide polymers discussed so far have a linear polyvinyl backbone. A new class of hyperbranched polyesteramides has been found to be good solubilizing additives for asphaltenes (Figure 4.32).[156] These polymers have a dendrimeric three-dimensional structure with alkyl groups pointing out in all directions (a full hyperbranched polyesteramide structure is illustrated in Section 9.2.2.3). The polyesteramides are made by condensing a cyclic acid anhydride with a dialkanolamine in a ratio of $n:n + 1$ where n is an integer (by varying n one can vary the molecular weight of the polymer). This gives a polymer with hydroxyl groups at the tips. The hyperbranching is caused by the dialkanolamine, which has three reactive groups. By reacting a carboxylic acid, such as a fatty acid, with the hydroxylated hyperbranched polymer, the tips of the polymer can be modified to become less

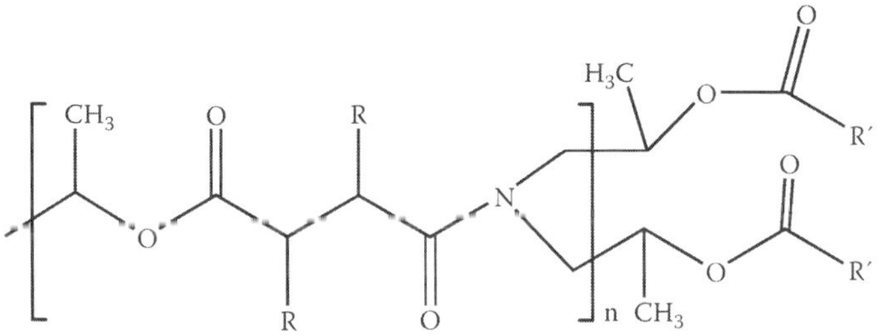

FIGURE 4.32 Structural units and end groups in hyperbranched polyesteramide AIs. R can be H or hydrocarbyl, R' is a long alkyl chain.

hydrophilic. Preferred examples of asphaltene-solubilizing additives are composed of succinic anhydride and diisopropanolamine in a molecular ratio of about 5:6, in which part of the functional hydroxyl groups have been modified by esterification with coco fatty acid and polyisobutenyl succinic anhydride (containing ca. 22 isobutenyl monomers). A second preferred polyesteramide is based on structural units composed of succinic anhydride and di-isopropanolamine wherein part of the functional hydroxyl groups have been modified by reaction with polyisobutenyl succinic anhydride alone.

Structurally complicated mixtures of polyesteramides preferably formed by reaction of polymerized long-chain hydroxyacids with polyamines have been claimed as AIs with dispersant properties.[157] Typical polyhydroxyacids are poly(ricinoleic acid) or poly(12-hydroxy stearic acid) and typical polyamines are alkylenepolyamine bottoms or polyethyleneimine.

4.4.3 OTHER POLYMERIC ASPHALTENE INHIBITORS

The use of graft copolymers containing pendant polar rings as AIs have been claimed.[158–160] The graft copolymers are made by grafting a vinylic monomer with nitrogen and/or oxygen atom(s) with a polyolefin. Examples of vinylic monomers are *N*-vinylimidazole and 4-vinylpyridine, but also *N*-vinyllactams can be used (Figure 4.33).

A related class to monomeric arylsulfonates is the lignosulfonate polymers. This class has been claimed for use as asphaltene-deposition inhibitors in squeeze treatments.[161] The structure of a lignosulfonate is given in Figure 4.34. The polar head group contains phenolic as well as sulfonic acid groups. There is some resemblance to the alkyarylsulfonic acids, but now with other functional groups and a more complex polymeric structure of interconnected phenolic groups.

A brief mention should be made of a novel method to prepare semisynthetic AIs. It has been shown that one can derivatize asphaltenes themselves by phosphoalkoxylation to produce a product that can inhibit formation of the same asphaltenes.[162] Reacting acidic groups in asphaltenes with PCl_3 then with polypropyleneglycols produces a mixture of structures, two of which are shown in Figure 4.35. These compounds probably have less ability to self-associate than the original asphaltenes and have a more resin-like character.

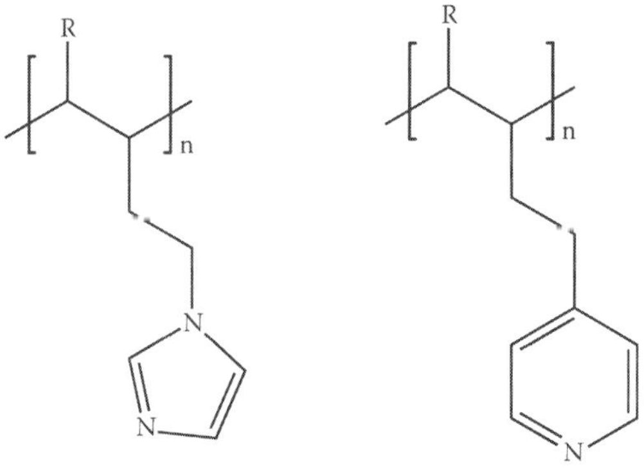

FIGURE 4.33 Graft polymer AIs with pendant imidazole (left) or pyridine rings (right). R is preferably an alkyl chain.

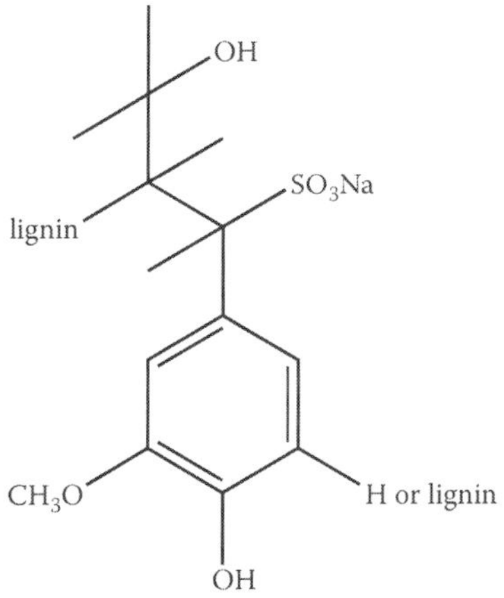

FIGURE 4.34 The partial structure of a lignosulfonate showing the aromatic and polar groups.

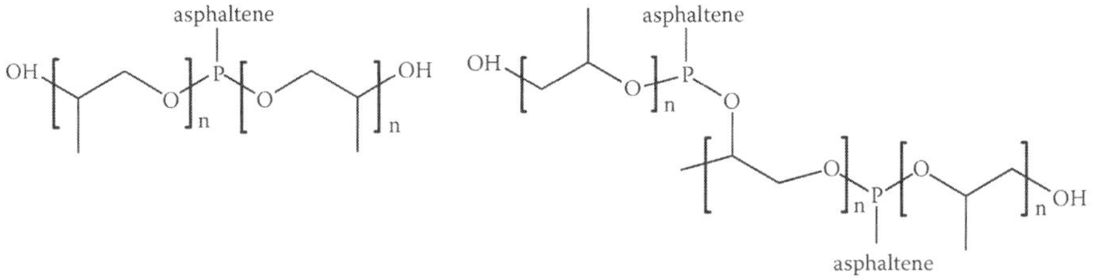

FIGURE 4.35 AIs based on the phosphoalkoxylation of asphaltenes.

Another patent claims the use of biodegradable molecules having tetrapyrrolitic patterns as stabilizers for asphaltenes.[163] These molecules can be chlorophyll-based molecules extracted from plant leaves and then derivatized.

4.5 SUMMARY OF ADs AND AIs

As a generalization, monomeric surfactants behave as dispersants, whereas polymeric surfactants are true inhibitors of asphaltene flocculation but may also function as dispersants. The mechanisms by which these ADs and AIs interact with asphaltenes can be summarized as follows:

- π–π interactions between asphaltenes and unsaturated or aromatic groups
- acid-base interactions
- hydrogen bonding
- dipole-dipole interactions
- complexing of metal ions

A key feature with nearly all ADs and AIs seems to be one or more polar functional groups that interact with the asphaltene monomers or aggregates, plus one or more alkyl chains that cooperatively form a less polar steric stabilization layer of alkyl tails around the asphaltenes solubilizing them in the crude oil. The alkyl tails are compatible with the crude oil, which has aliphatic hydrocarbons as its major component. Ring structures in ADs and AIs, particularly those of an aromatic nature, are also quite common. Branched alkyl groups in surfactants, of the correct length, appears to be of better help in dispersing asphaltenes than straight chain alkyl groups. Molecular modeling may help deepen our understanding of some of these issues. Good solvents to use with AIs and ADs are aromatic or polar solvents that can help peptize the asphaltenes (see Section 4.6 on asphaltene dissolvers); examples are toluene, xylenes, trimethylbenzenes, 1-methylnaphthalene, tetrahydronaphthalene (tetralin), quinoline, isoquinoline, dimethylformamide, NMP, alkylglycols, and blends thereof.

4.6 ASPHALTENE DISSOLVERS

Asphaltene deposits can occur anywhere from downhole to the process facilities and beyond. Severe deposits in the processing facilities may need to be mechanically dug out. If there is a reservoir asphaltene deposition problem and it is very severe, reperforating or refracturing may be necessary. Enzymes have been used to good effect to remove asphaltene and resin formation damage in China.[164] The only good chemical method to remove asphaltene deposits is to use asphaltene dissolvers or solvents. To remove asphaltic deposits in flowlines, dissolvers are either batch-treated or recirculated to the affected zone for a number of hours. For treatment of tubulars, the product is used neat or diluted with crude and recirculated via the annulus or bullheaded. Effective dissolvers can dissolve up to their own weight of asphaltenes at downhole temperatures in just a few hours. Asphaltic sludge formed after acid stimulation jobs are often more difficult to remove with solvents since the asphaltenes are chemically bonded through water layers onto charged mineral surfaces.[165–166] Compatibility of the asphaltene solvent with elastomers and plastics materials that are contained within the system must be checked before use.

Most asphaltene dissolvers are based on aromatic solvents, sometimes with added enhancers.[167] Deasphalted oils with high aromatic content have also been used as a low-cost alternative to chemical solvents.[168–169] Xylene (a mixture of 1,2- 1,3-, and 1,4-dimethylbenzene), with a low flash point of 28°C (82°F), is probably the most common aromatic solvent.[170] The even more volatile toluene (flash point 5°C [41°F]) is also used but is somewhat less effective.[171] Most substituted aromatic solvents are now classed as marine pollutants. The aromatic solvents interact via π–π orbital overlap with the aromatic components of asphaltene aggregates, replacing asphaltene-asphaltene π–π interactions and thereby solubilizing them. Studies have shown that the presence of paraffinics (acyclic or cyclic nonaromatic hydrocarbons) in asphaltene dissolvers is detrimental to the performance.[172] The same study showed that monocyclic and bicyclic aromatic solvents performed better than tricyclic or polycyclic aromatic chemicals. Another study showed that the bicyclic molecules tetralin (1,2,3,4-tetrahydronaphthalene) and 1-methylnaphthalene performed better than mono-ring solvents such as n-propylbenzene, toluene, and xylene in both the amount of asphaltenes dissolved and the rate of dissolution (Figure 4.36).[173] Higher temperatures and agitation helps to improve the asphaltene dissolution rate. Most commercial asphaltene dissolvers are formulated to contain a large percentage of mono-ring aromatic chemicals and optionally minor percentages of bicyclic aromatic chemicals due to the higher cost. A high–flash point commercial dissolver based on distillates, coal tar, naphthalene oils, and methylnaphthalene fractions has been used in Italy. Terpene-based solvents (containing for example d-limonene), with improved health, safety, and environmental profiles compared with aromatic solvents, have also been used in the field but their dissolving capacity is limited (Figure 4.36).[174] Terpene solvents are also useful for dissolving waxes.[175]

Since asphaltenes contain heteroatoms that give the structures some polarity, aromatic solvents with heteroatoms and polar groups should also make good asphaltene dissolvers. Carbon disulphide, which has a very low flash point, is a good asphaltene dissolver. Pyridines are good asphaltene dissolvers, but they are toxic and incompatible with many elastomers.[176] Alternatively, a hydrocarbon aromatic solvent can be used in combination with a polar cosolvent. By combining aromatic solvents and additives with polar functional groups, it has been shown that the overall cosolvent polarity can be matched to the asphaltene type of the field. Such blending enhanced the dissolution power of xylene, used as basic solvent, and increased asphaltene desorption from mineral surfaces of reservoir rock.[177] One patent claims

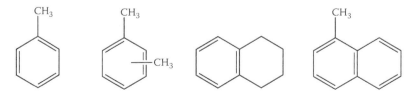

FIGURE 4.36 Toluene, xylenes, tetralin, and 1-methylnaphthalene.

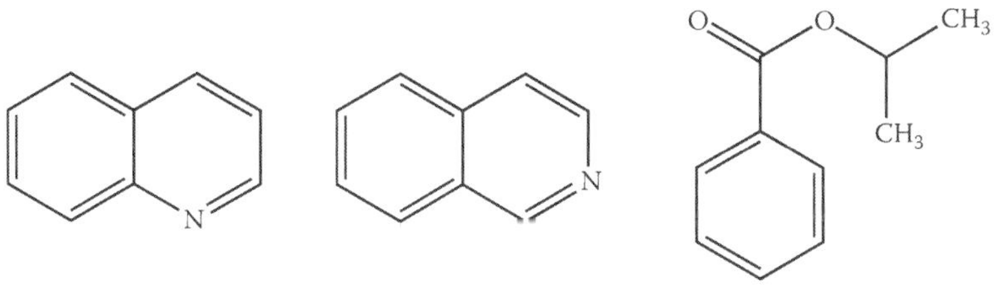

Asphaltene Control

[Structures of quinoline, isoquinoline, and isopropylbenzoate]

FIGURE 4.37 Quinoline and isoquinoline and isopropylbenzoate.

that aromatic solvents with preferably 3–10% additional quinoline and isoquinoline (or C1–4 alkyl substituted), which have higher polarity compared with normal aromatics, have improved asphaltene dissolution rates (Figure 4.37).[178] Another patent claims that addition of benzotriazole to the aromatic solvent improves the performance.[179] Alkyl or alkenyl esters of certain aromatic carboxylic acids, preferably isopropyl benzoate are also good asphaltene dissolvers as well as having attractive toxicological and environmental profiles.[180] These solvents can be the basis of the external phase in retarded emulsified acid-stimulation fluids. The emulsion removes both inorganic and organic formation damage.[181] Solvents with heteroatoms do not necessarily have to be aromatic to dissolve asphaltenes. For example, solvents containing 1–15% N-substituted imidazolines, and preferably NMP (or N-ethyl pyrrolidone), are also good dissolvers (Figure 4.38).[182] Compositions comprising of kerosene or an aromatic solvent with at least one C4–C30 olefin or oxidation product thereof have been claimed as improved asphaltene dissolvers.[183] Other "greener" solvents such as glycol ethers, alkanolamines, and esters have been screened and may find increasing use in the future to improve environmental impact.

Enhancers are sometimes added to the asphaltene solvent to improve its performance but the choice of enhancer appears to depend on whether the asphaltenes are adsorbed onto the rock formation or elsewhere in the system. Thus, one study concluded that toluene with 2 wt.% chemicals bearing polar groups (such as polymers) but not proton-donor groups (such as alkylbenzene sulfonic acids) performed well at enhancing the dissolution of bulk asphaltene solids compared with toluene alone.[136] However, both polymers and the alkylbenzene sulfonic acids increased the dissolution of asphaltenes adsorbed on rock in coreflooding experiments, the sulfonic acid

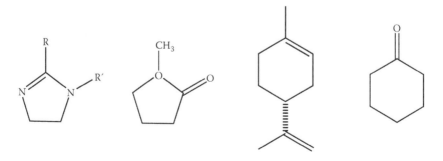

FIGURE 4.38 Imidazolines, NMP, limonene (a terpene), and cyclohexanone.

performing best. Competitive interaction of an additive with the rock active sites was suggested as the mechanism by which the asphaltenes are desorbed from the rock. One service company found that addition of a range of ADs to the asphaltene dissolver did not enhance the amount or rate of asphaltene dissolution.

There have also been developed high–flash point emulsion-based asphaltene dissolvers in which an aromatic solvent but no BETX (benzene, ethylbenzene, toluene, or xylenes) is used. These emulsion-based dissolvers can sometimes perform better than straight aromatic solvents such as xylene[184] (the same publication reports a useful list of keys to a long-lasting asphaltene removal treatment). The water-external emulsion compositions can comprise (1) water; (2) an organic solvent blend further comprising (a) a nonpolar organic solvent such as a terpene blend or crude oil from which light aromatic solvents have been previously distilled and (b) a polar organic solvent such as NMP or cyclohexanone; and (3) a surfactant adapted for forming an emulsion of the organic solvent blend and the water and for helping break down asphaltene deposits (Figure 38).[185] The combination of these two polar organic solvents unexpectedly resulted in better dissolution of asphaltenes than either of the two solvents alone in the composition. The emulsion-based dissolver has since been optimized and tested on asphaltenes from various parts of the globe.[186] Acids emulsified in aromatic solvents such as xylene have been used to enhance well productivity by acid stimulation and dissolving asphaltenes downhole.[187] Microemulsions have also been claimed removing formation damage such as asphaltene deposits.[188]

REFERENCES

1. K. J. Leontaritis, *Oil & Gas Journal* 1 (1998): 122.
2. E. Y. Sheu and O. C. Mullins, eds., *Asphaltenes: Fundamentals and Applications*, New York: Plenum Press, 1995.
3. J. G. Speight, *The Chemistry and Technology of Petroleum*, 3rd ed., New York: Marcel Decker, 1999, 412–467.
4. E. Y. Sheu and O. C. Mullins, eds., *Asphaltenes*, New York: Springer, 1996.
5. O. C. Mullins and E. Y. Sheu, eds., *Structures and Dynamics of Asphaltenes*, New York: Springer, 1999.
6. K. J. Leontaritis and G. A. Mansoori, "Asphaltene Deposition: A Survey of Field Experiences and Research Approaches," *Journal of Petroleum Science and Engineering* 1 (1988): 229.
7. K. Barker and M. E. Newberry, "Inhibition and Removal of Low-pH-Fluid–Induced Asphaltic Sludge Fouling of Formations in Oil and Gas Wells," SPE 102738 (paper presented at the SPE International Symposium on Oilfield Chemistry, Houston, TX, 28 February–2 March 2007).
8. L. C. C. Marques, J. B. Monteiro, and G. González, *Journal of Dispersion Science and Technology* 28 (2007): 391.
9. F. Trejo, G. Centeno, and J. Ancheyta, "Precipitation, Fractionation and Characterization of Asphaltenes from Heavy and Light Crude Oils," *Fuel* 83 (2004): 2169.
10. O. C. Mullins, "Molecular Structure and Aggregation of Asphaltenes and Petroleomics," SPE 95801 (paper presented at the SPE Annual Technical Conference and Exhibition, Dallas, TX, 9–12 October 2005).

11. J. P. Maclean and P. Kilpatrick, "Comparison of Precipitation and Extrography in the Fractionation of Crude Oil Residua," *Energy & Fuels* 11 (1997): 570.
12. S. Asomaning, "Test Methods for Determining Asphaltene Stability in Crude Oils," *Petroleum Science and Technology* 21 (2003): 581.
13. N. G. Graham and M. L. Gorbaty, "Sulfur K-Edge X-ray Absorption Spectroscopy of Petroleum Asphaltenes and Model Compounds," *Journal of the American Chemical Society* 111 (1989): 3182.
14. S. Mitra-Kirtley, O. C. Mullins, J. van Elp, S. J. George, J. Chen, and S. P. Cramer, "Determination of the Nitrogen Chemical Structures in Petroleum Asphaltenes Using XANES Spectroscopy," *Journal of the American Chemical Society* 115 (1993): 252.
15. J. Castillo, A. Fernandez, M. A. Ranaudo, and S. Acevedo, "New Techniques and Methods for the Study of Aggregation, Adsorption, and Solubility of Asphaltenes. Impact of These Properties on Colloidal Structure and Flocculation," *Petroleum Science and Technology* 19 (2001): 75.
16. O. C. Mullins, E. Y. Sheu, A, Hammami, and A. G. Marshall, *Asphaltenes, Heavy Oils, and Petroleomics*, New York: Springer, 2006.
17. G. W. Zajac, N. K. Sethi, and J. T. Joseph, "Molecular Imaging of Petroleum Asphaltenes by Scanning Tunneling Microscopy," *Scanning Microscopy* 8 (1994): 463.
18. A. Sharma, H. Groenzin, A. Tomita, and O. C. Mullins, "Probing Order in Asphaltenes and Aromatic Ring Systems by HRTEM," *Energy & Fuels* 16 (2002): 490.
19. (*a*) H. Groenzin and O. C. Mullins, "Molecular Size and Structure of Asphaltenes from Various Sources," *Energy & Fuels* 14 (2000): 677. (*b*) A. E. Pomerantz, M. R. Hammond, A. L. Morrow, O. C. Mullins, and R. N. Zare, *Energy Fuels*, 2009, 23, 1162.
20. O. P. Strausz, T. W. Mojelsky, F. Faraji, and E. M. Lown, P. Peng, "Additional Structural Details on Athabasca Asphaltene and Their Ramifications," *Energy & Fuels* 13(2) (1999): 207.
21. J. Murgich, "Molecular Simulation and the Aggregation of the Heavy Fractions in Crude Oils," *Molecular Simulation* 29 (2003): 451.
22. O. P. Strausz, T. W. Mojelsky, and E. M. Lown, "The Molecular Structure of Asphaltene: An Unfolding Story," *Fuel* 71 (1992): 1355.
23. O. P. Strausz, P. Peng, and J. Murgich, "About the Colloidal Nature of Asphaltenes and the MW of Covalent Monomeric Units," *Energy & Fuels* 16 (2002): 809.
24. K. L. Gawrys, Ph.D. thesis, North Carolina State University, 2005.
25. K. L. Gawrys and P. K. Kilpatrick, "Asphaltenic Aggregates are Polydisperse Oblate Cylinders," *Journal of Colloid Interface Science* 288 (2005): 325.
26. V. Calemma, R. Rausa, P. D'Antona, and L. Montanari, "Characterization of Asphaltenes Molecular Structure," *Energy & Fuels* 12 (1998): 422.
27. S. A. A. Castro, J. G. Negrin, A. Fernandez, G. Escobar, V. Piscitelli, F. Delolme, and G. Dessalces, "Relations Between Asphaltene Structures and Their Physical and Chemical Properties: The Rosary-Type Structure" *Energy & Fuels* 21(4) (2007): 2165.
28. O. P. Strausz, I. Safarik, E. M. Lown, and A. Morales-Izquierdo, "A Critique of Asphaltene Fluorescence Decay and Depolarization-Based Claims about Molecular Weight and Molecular Architecture," *Energy & Fuels*, 22 (2008): 1156–1166.
29. A. A. Herod, K. D. Bartle, and R. Kandiyoti, "Characterization of Heavy Hydrocarbons by Chromatographic and Mass Spectrometric Methods: An Overview," *Energy & Fuels* 21 (2007): 2176.
30. O. C. Mullins, B. Martínez-Haya, and A. G. Marshall, "Contrasting Perspective on Asphaltene Molecular Weight: This Comment vs. the Overview of A. A. Herod, K. D. Bartle, and R. Kandiyoti," *Energy & Fuels* 22 (2008): 1765.
31. A. E. Pomerantz, M. R. Hammond, A. L. Morrow, O. C. Mullins, and R. N. Zare, "Two-Step Laser Mass Spectrometry of Asphaltenes," *Journal of the American Chemical Society* 130 (2008): 7216.

32. O. C. Mullins, S. S. Betancourt, M. E. Cribbs, F. X. Dubost, J. L. Creek, A. Ballard, and L. Venkataramanan, "The Colloidal Structure of Crude Oil and the Structure of Oil Reservoirs," *Energy & Fuels* 21(5) (2007): 2785.
33. M. S. Diallo, T. Cagin, J. L. Faulon, and W. A. Goddard III, "Asphaltenes and Asphalts," in *Developments in Petroleum Science 40B*, eds. T. F. Yen and G. V. Chilingarian, Elsevier, Chap. 5, 103, 2000.
34. J. X. Wang, J. S. Buckley, N. E. Burke, and J. L. Creek, "A Practical Method for Anticipating Asphaltene Problems," SPE 87638, *SPE Production & Facilities* 19(3) (2004): 152.
35. J. X. Wang, J. S. Buckley, N. E. Burke, and J. L. Creek, "A Practical Method for Anticipating Asphaltene Problems," *SPE Production & Facilities* 19(3) (2004): 152–160.
36. S. Asomaning and A. Yen, "Prediction and Solution of Asphaltene Related Problems in the Field," *Chemistry in the Oil Industry VII*, Royal Society of Chemistry, 2002, 277.
37. H. W. Yarranton, "Asphaltene Self-Association," *Journal of Dispersion Science and Technology* 26 (2005): 5.
38. R. E. Guerra, K. Ladavac, A. B. Andrews, O. C. Mullins, and P. N. Sen, "Diffusivity of Coal and Petroleum Asphaltene Monomers by Fluorescence Correlation Spectroscopy," *Fuel* 86 (2007): 2016.
39. K. Rastegari, W. Y. Svrcek, and H. W. Yarranton, "Kinetics of Asphaltene Flocculation," *Industrial and Engineering Chemistry Research* 43 (2004): 6861.
40. J. Wang, J. S. Buckley, and J. L. Creek, "Asphaltene Deposition on Metallic Surfaces," *Journal of Dispersion Science and Technology* 25 (2004): 287.
41. E. Buenrostro-Gonzalez, H. Groenzin, C. Lira-Galeana, and O. C. Mullins, "The Overriding Chemical Principles that Define Asphaltenes," *Energy & Fuels* 15(4) (2001): 972.
42. J. Murgich, J. Rodriguez, and Y. Aray, "Molecular Recognition and Molecular Mechanics of Micelles of Some Model Asphaltenes and Resins," *Energy & Fuels* 10 (1996): 68.
43. T. Takanohashi, S. Sato, and R. Tanaka, "Molecular Dynamics Simulation of Structural Relaxation of Asphaltene Aggregates," *Petroleum Science and Technology* 21 (2003): 491.
44. T. Takanohashi, S. Sato, and R. Tanaka, "Structural Relaxation Behaviors of Three Different Asphaltenes Using MD Calculations," *Petroleum Science and Technology* 22 (2004): 901.
45. L. Carbognani and E. Rogel, "Solid Petroleum Asphaltenes Seem Surrounded by Alkyl Layers," *Petroleum Science and Technology* 21(3–4) (2003): 537.
46. K. L. Gawrys, M. Spiecker, and P. K. Kilpatrick, "The Role of Asphaltene Solubility and Chemistry on Asphaltene Aggregation," *Petroleum Science and Technology* 21 (2003): 461.
47. H.-J. Oschmann, "New Methods for the Selection of Asphaltene Inhibitors in the Field," *Chemistry in the Oil Industry VII*, Royal Society of Chemistry, Manchester, 13–14 2001, 254.
48. L. C. C. Marques, G. Gonzalez, and J. B. Monteiro, "A Chemical Approach to Prevent Asphaltenes Flocculation in Light Crude Oils: State-of-the-Art," SPE 91019 (paper presented at tyhe SPE Annual Technical Conference and Exhibition, Houston, TX, 26–29 September, 2004).
49. D. F. Smith, G. C. Klein, A. T. Yen, M. P. Squicciarini, R. P. Rodgers, and A. G. Marshall, "Crude Oil Polar Chemical Composition Derived from FT–ICR Mass Spectrometry Accounts for Asphaltene Inhibitor Specificity," *Energy & Fuels* 22 (2008): 3112.
50. J. Marugán, "Characterization of the Onset of Asphaltenes by Focus-Beamed Laser Reflectance: A Tool for Chemical Additives Screening" (paper presented at the 9th International Conference on Petroleum Phase Behavior and Fouling, Victoria, BC, 15–19 June 2008).

51. M. Barcenas, P. Orea, E. Buenrostro-González, L. S. Zamudio-Rivera, and Y. Duda, "Study of Medium Effect on Asphaltene Agglomeration Inhibitor Efficiency," *Energy & Fuels* 22 (2008): 1917.
52. S. Asomaning, "Methods for Selecting Asphaltene Inhibitors and New Insights into Inhibitor Mechanisms," Abstr. No. GEOC 142, 225th ACS National Meeting, New Orleans, LA, 23–27 March 2003.
53. K. Karan, A. Hammami, M. Flannery, and B. A. Stankiewicz, "Evaluation of Asphaltene Instability and Chemical Control during Production of Live Oils," *Petroleum Science and Engineering* 21 (2003): 629.
54. H. Pan and A. Firoozabadi, "Thermodynamic Micellization Model for Asphaltene Precipitation Inhibition," *AIChE Journal* 46 (2000): 416.
55. L. Z. Pillon, "Effect of Dispersants and Flocculants on the Colloidal Stability of Asphaltene Constituents," *Petroleum Science and Engineering* 19 (2001): 863.
56. L. F. Bandeira Moreira, E. F. Lucas, and G. González, "Stabilization of Asphaltenes by Phenolic Compounds Extracted from Cashew-Nut Shell Liquid," *Journal of Applied Polymer Science*, 73(1) (1999): 29.
57. J. Dunlop, "Novel High Performance Dispersants for Oil Industry Applications" (paper presented at the 4th International Conference on Petroleum Phase Behaviour and Fouling, Trondheim, Norway, 23–26 June 2003).
58. S. J. Allenson and M. A. Walsh, "A Novel Way to Treat Asphaltene Deposition Problems Found in Oil Production," SPE 37286 (paper presented at the SPE International Symposium on Oilfield Chemistry, Houston, TX, 18–21 February 1997).
59. L. M. Cenegy, "Survey of Successful World-Wide Asphaltene Inhibitor Treatments in Oil Production Fields," SPE 71542 (paper presented at the SPE Annual Technical Conference and Exhibition, New Orleans, LA, 30 September–3 October 2001).
60. O. Niemeyer, "New Squeeze Applications for Asphaltene Blocked Wells" (paper presented at the 7th International Conference on Petroleum Phase Behaviour and Fouling, Asheville, NC, 25–29 June 2006).
61. K. Allan, "Asphaltene Inhibitor Squeezing—What Can We Learn from Scale Inhibitor Squeezing" (paper presented at the 9th International Conference on Petroleum Phase Behavior and Fouling, Victoria, BC, June 15–19, 2008).
62. R. B. De Boer, K. Leerlooyer, M. R. P. Eigner, and A. R. D. van Bergen, "Screening of Crude Oils for Asphalt Precipitation: Theory, Practice and the Selection of Inhibitors," *SPE Production & Facilities* 10 (1995): 55–61.
63. J. K. Borchardt, "Chemicals Used in Oil-Field Operations," in *Oil-Field Chemistry*, eds. J. K. Borchardt and T. F. Yen, *American Chemical Society Symposium Series 396*, Washington, DC: ACS, 1989.
64. S. Takhar, "A Fast and Effective Chemical Screening Technique for Identifying Asphaltene Inhibitors for Field Deployment," Proceedings of the Second International Conference on Fluid and Thermal Energy Conversion, eds. A. Mansoori and A. Suwono, 1997, S83–S90.
65. H. J. Oschmann, "New Methods for the Selection of Asphaltene Inhibitors in the Field," Special Publication, *Royal Society of Chemistry* 280 (2002): 254–263.
66. M. N. Bouts, R. J. Wiersma, H. M. Muijs, and A. J. Samuel, "An Evaluation of New Asphaltene Inhibitors: Laboratory Study and Field Testing," *Journal of Petroleum Technology* 782 (1995): 782.
67. M. B. Manek, "Asphaltene Dispersants as Demulsification Aids," SPE 28972 (paper presented at the SPE International Symposium on Oilfield Chemistry, San Antonio, TX, 14–17 February, 1995.

68. A. K. M. Jamaluddin, J. Nighswander, N. B. Joshi, D. Calder, and B. Ross, "Asphaltene Characterization: A Key to Deepwater Developments" SPE 77936 (paper presented at the SPE Asia Pacific Oil and Gas Conference and Exhibition, Melbourne, Australia, 8–10 October 2002).
69. A. Yen, R. Yin, and S. Asomaning, "Evaluating Asphaltene Inhibitors: Laboratory Tests and Field Studies," SPE 65376 (paper presented at the SPE International Symposium on Oilfield Chemistry, Houston, TX, 13–16 February 1995).
70. H. Anfindsen, P. Fotland, and A. M. Mathisen (paper presented at the 9th International Oilfield Chemical Symposium, Geilo, Norway, 22–25 March 1995).
71. M. A. Aquino-Olivos, E. Buenrostro-Gonzalez, S. I. Andersen, and C. Lira-Galeana, "Investigations of Inhibition of Asphaltene Precipitation at High Pressure Using Bottomhole Samples," *Energy & Fuels* 15(1) (2001): 236.
72. T. Maqbool, I. A. Hussein, and H. S. Fogler (paper presented at the 9th International Conference on Petroleum Phase Behaviour and Fouling, Victoria, BC, Canada, 15–19 June 2008).
73. A. Hammami, C. H. Phelps, T. Monger-McClure, and T. M. Little, "Asphaltene Precipitation from Live Oils: An Experimental Investigation of Onset Conditions and Reversibility," *Energy & Fuels* 14 (2000): 14.
74. S. Asomaning, "Assessing the Performance of Asphaltene Inhibitors Using High Pressure Methods: The Deepwater Gulf of Mexico Experience," RSC/EOSCA Proceedings of Chemicals in the Oil Industry, Manchester, UK, 3–5 November 2003, 35.
75. S. Asomaning and C. Gallagher, "High Pressure Asphaltene Deposition Technique for Evaluating the Deposition Tendency of Live Oil and Evaluating Inhibitor Performance, Preprints" (paper presented at the Second International Conference on Petroleum and Gas Phase Behaviour and Fouling, Copenhagen, Denmark, August 26–31 2000).
76. H. Alboudwarej, W. Y. Svrcek, A. Kantzas, and H. W. Yarranton, "A Pipe Loop Apparatus to Investigate Asphaltene Deposition," *Petroleum Science and Engineering* 22 (2005): 799.
77. J. Dunlop, "Novel High Performance Additives for Asphaltene Control in Oil Production Operations," RSC/EOSCA Chemistry in the Oil Industry VIII, Manchester, UK, 3–5 November 2003.
78. J. Dunlop, "Low Environmental Asphaltene Inhibitors" (paper presented at the Tekna 19th International Oilfield Chemistry Symposium, Geilo, Norway, 9–12 March 2008).
79. R. J. Gochin, U.S. Patent Application 20050082231, 2005.
80. R. J. Gochin and A. Smith, U.S. Patent 6270653, 2005.
81. D. Espinat, J. C. Ravey, V. Guille, J. Lambard, T. Zemb, and J. P. Cotton, "Colloidal Macrostructure of Crude Oil Studied by Neutron and X-ray Small Angle Scattering Techniques," *Journal De Physique IV* 3(C8) (1993): 181.
82. A. Del Bianco and F. Stroppa, European Patent EP0737798, 1996.
83. M. E. Newberry and K. M. Barker, U.S. Patent 4414035 1983.
84. N. Feldman, U.S. Patent 4441890, 1984.
85. W. H. Stover and S. A. Hunter, U.S. Patent 4182613, 1980.
86. L. C. Rocha Junior, M. S. Ferreira, and A. C. S. Ramos, "Inhibition of Asphaltene Precipitation in Brazilian Crude Oils Using New Oil Soluble Amphiphiles," *Journal of Petroleum Science* 51 (2006): 26.
87. C. L. Chang and H. S. Fogler, "Stabilization of Asphaltenes in Aliphatic Solvents Using Alkylbenzene Derived Amphiphiles. 1. Effect of the Chemical Structure of Amphiphiles on Asphaltenes Stabilization," *Langmuir* 10 (1994): 1749.
88. C. L. Chang and H. S. Fogler, "Asphaltene Stabilization in Alkyl Solvents Using Oil-Soluble Amphiphiles," SPE 25185 (paper presented at the SPE International Symposium on Oilfield Chemistry, New Orleans, LA, 2–5 March, 1993).

89. H. Pan and A. Firoozabadi, "Thermodynamic Modeling of Asphaltene Precipitation Inhibition," *AIChE Journal* (2000): 416.
90. H. H. Ibrahim and R. O. Idem, "Interrelationships Between Asphaltene Precipitation Inhibitor Effectiveness, Asphaltenes Characteristics, and Precipitation Behavior during n-Heptane (Light Paraffin Hydrocarbon)-Induced Asphaltene Precipitation," *Energy & Fuels* 18(4) (2004): 1038.
91. L. Goual and A. Firoozabadi, "Effect of Resins and DBSA on Asphaltene Precipitation from Petroleum Fluids," *AIChE Journal* 50 (2004): 470.
92. C. L. Chang and H. S. Fogler, "Stabilization of Asphaltenes in Aliphatic Solvents Using Alkylbenzene-Derived Amphiphiles. 2. Study of the Asphaltene-Amphiphile Interactions and Structures Using Fourier Transform Infrared Spectroscopy and Small-Angle X-ray Scattering Techniques," *Langmuir* 10 (1994): 1758.
93. P. Permsukarome, C. L. Chang, and H. S. Fogler, "Kinetic Study of Asphaltene Dissolution in Amphiphile/Alkane Solutions," *Industrial and Engineering Chemistry Research* 36 (1997): 3960.
94. A. C. S. Ramos, C. C. Delgado, R. S. Mohamed, V. R. Almeida, and W. Loh, "Reversibility and Inhibition of Asphaltene Precipitation in Brazilian Crude Oils," SPE 38967 (paper presented in the Latin American and Caribbean Petroleum Engineering Conference, 30 August–3 September 1997).
95. T. A. Al-Sahhaf, A. F. Mohammed, and A. S. Elkilani, "Retardation of Asphaltene Precipitation by Addition of Toluene, Resins, Deasphalted Oil and Surfactants," *Fluid Phase Equilibria* 194 (2002): 1045.
96. A. I. Victorov and A. Firoozabadi, "Thermodynamics of Asphaltene Deposition Using a Micellization Model," *AIChE Journal* 42 (1996): 1753.
97. H. Pan and A. Firoozabadi, "Thermodynamic Micellization Model for Asphaltene Precipitation: Micellar Growth and Precipitation," *SPE Production and Facilities* 13(2) (1998): 118.
98. H. Pan and A. Firoozabadi, "A Thermodynamic Micellization Model for Asphaltene Precipitation from Reservoir Crudes at High Pressures and Temperatures," *SPE Production and Facilities* 15 (2000): 58.
99. O. Leon, E. Rogel, A. Urbina, A. Andujar, and A. Lucas, "Study of the Adsorption of Alkyl Benzene-Derived Amphiphiles on Asphaltene Particles," *Langmuir* 15(22) (1999): 7653.
100. D. Miller, A. Vollmer, and M. Feustel, U.S. Patent 5925233, 1999.
101. I. Wiehe, R. Varadaraj, T. Jermansen, R. J. Kennedy, and C. H. Brons, International Patent Application WO/00/32546, 2000.
102. R. Varadaraj and C. H. Brons, U.S. Patent Application 20040072361, 2004.
103. I. Wiehe and T. G. Jermansen, "Design of Synthetic Dispersants for Asphaltene Constituents," *Petroleum Science and Engineering* 21 (2003): 527.
104. J. A. Ostlund, M. Nyden, H. S. Fogler, and K. Holmberg, "Functional Groups in Fractionated Asphaltenes and the Adsorption of Amphiphilic Molecules," *Colloids and Surfaces A* 234 (2004): 95.
105. D. Miller, A. Vollmer, M. Feustel, and P. Klug, U.S. Patent 6063146, 2000.
106. N. Ikenaga, Y. Watanabe, and S. Hayashi, Japanese Patent JP63023991, 1988.
107. C. A. Stout, Canadian Patent CA1142114, 1983.
108. R. F. Miller, U.S. Patent 4425223, 1984.
109. S. V. Tapavicza, W. Zoellner, C. P. Herold, J. Groffe, and J. Rouet, U.S. Patent 6344431, 2002.
110. D. Miller, A. Vollmer, M. Feustel, and P. Klug, U.S. Patent 6204420, 2001.
111. G. Woodward, C. R. Jones, and K. P. Davis, International Patent Application WO/2004/002994.

112. I. H. Auflem, T. E. Havre, and J. Sjoblom, "Near Infrared Study on the Dispersive Effects of Amphiphiles and Naphthenic Acids on Asphaltenes in Model Heptane-Toluene Mixtures," *Colloid Polymer Science* 280 (2002): 695.
113. D. Miller, A. Vollmer, M. Feustel, and P. Klug, International Patent Application, WO 98/16595.
114. M. Ravindranath, European Patent Application EP 1359206, 2003.
115. J. M. Romocki, International Patent Application, WO 94/18430, 1994.
116. M. Ferrara, International Patent Application WO95/20637.
117. M. Ravindranath and R. M. Banavali, U.S. Patent Application US2004238404.
118. R. Banavali, "Reducing Viscosity of Asphaltenic Crudes Via Chemical Additives," Chemistry in the Oil Industry VIII, Royal Society of Chemistry, 2003, 1.
119. A. Lesimple, C. P. Herold, D. Groffe, and W. Breuer, International Patent Application WO01/27438, 2001.
120. H. L. Becker and B. W. Wolf, U.S. Patent 5504063, 1996.
121. C. Bernasconi, A. Faure, and B. Thibonnet, U.S. Patent 4622047, 1986.
122. Y. F. Hu and T. M. Guo, "Effect of the Structures of Ionic Liquids and Alkylbenzene-Derived Amphiphiles on the Inhibition of Asphaltene Precipitation from CO_2-Injected Reservoir Oils," *Langmuir* 21(18) (2005): 8168.
123. O. León, E. Contreras, and E. Rogel, "Amphiphile Adsorption on Asphaltene Particles: Adsorption Isotherms and Asphaltene Stabilization," *Colloids and Surfaces A* 189 (2001): 123.
124. G. D. Sutton, U.S. Patent 3914132, 1975.
125. G. Gonzalez and A. Middea, "Peptization of Asphaltene by Various Oil Soluble Amphiphiles," *Colloids and Surfaces* 52 (1991): 207.
126. T. Cox, N. Grainger, and E. G. Scovell, Canadian Patent CA2404316 (TW546370B), 2003.
127. J. Groffe, J. Rouet, and D. Chauvie, French Patent, FR2679151, 1993.
128. M. Ravindranath, U.S. Patent Application US2004232042.
129. M. Ravindranath, U.S. Patent Application US2004232044.
130. B. D. Chheda, U.S. Patent Application 20070124990.
131. R. M. Banavali, B. D. Cheda, and G. M. Manari, U.S. Patent Application 20060079434.
132. P. J. Breen, International Patent Application, WO 01/55281, 2001.
133. M. Ferrara, International Patent Application WO95/20637.
134. V. J. Mena Cervates, L. S. Zamudio-Rivera, M. Lozada y Casso, H. Beltrán Conde, E. Buenrostro-González, S. López Ramirez, Y. Douda, A. Morales Pacheco, R. Hernández Altamirano, and M. Barcenas Castañeda, Mexican Patent Request, File No. MX/E/2007/084388.
135. S. S. Schantz and W. K. Stephenson, "Asphaltene Deposition: Development and Application of Polymeric Asphaltene Dispersants," SPE 22783 (paper presented at the SPE Annual Technical Conference and Exhibition, Dallas, TX, 6–9 October, 1991).
136. L. Barberis Canonico, A. Del. Bianco, G. Piro, F. Stroppa, C. Carniani, and E. I. Mazzolini, "A Comprehensive Approach for the Evaluation of Chemicals for Asphaltene Deposit Removal," *Recent Advances in Oilfield Chemistry*, The Royal Society of Chemistry, 1994, 220.
137. A. L. Soldan, L. C. F. Barbosa, R. L. A. Santos, J. C. C. B. R. Moreira, S. C. Menezes, M. A. G. Teixeira, C. R. Souza, R. B. Haag, L. C. C. Marques, and A. N. Sanmartin, "1st SPE Brazil Sect. Colloid Chemistry in Oil Production," Proceedings Asphaltenes and Wax Deposition International Symposium, Rio de Janeiro, Brazil, 26–29 November, 1995, 51–55.
138. O. León, E. Contreras, E. Rogel, G. Dambakli, J. Espidel, and S. Acevedo, "The Influence of the Adsorption of Amphiphiles and Resins in Controlling Asphaltene Flocculation," *Energy & Fuels* 15 (2001): 1028.

139. E. Rogel, E. Contreras, and O. León, "An Experimental Theoretical Approach to the Activity of Amphiphiles as Asphaltene Stabilizers," *Petroleum Science and Engineering* 20 (2002): 725.
140. D. Leinweber, M. Feustel, E. Wasmund, and H. Grundner, International Patent Application WO/02/64706, 2002.
141. A. Behler, W. Breuer, and M. Hof, International Patent Application WO3/054348, 2003.
142. A. Behler, U.S. Patent Application 20050091915.
143. M. B. Manek and N. K. Sawney, U.S. Patent 5494607, 1996.
144. D. Miller, M. Feustel, A. Vollmer, R. Vybiral, and D. Hoffmann, U.S. Patent 6180683, 2001.
145. W. K. Stephenson, B. D. Mercer, and D. G. Comer, U.S. Patent 5143594, 1992.
146. W. K. Stephenson, B. D. Mercer, and D. G. Comer, U.S. Patent 5100531, 1992.
147. W. Stephenson and M. Kaplan, U.S. Patent 5021498, 1991.
148. W. Stephenson, J. Walker, B. Krupay, and S. Wolsey-Iverson, Canadian Patent CA2075749, 1993.
149. W. Stephenson and M. Kaplan, U.S. Patent 5073248, 1991.
150. C. L. Chang and H. S. Fogler, "Peptization and Coagulation of Asphaltenes in Apolar Media Using Oil-Soluble Polymers," *Petroleum Science and Engineering* 14 (1996): 75.
151. D. G. Comer and W. K. Stephenson, U.S. Patent 5214224 1993.
152. W. Stephenson, J. Walker, B. Krupay, and S. Wolsey-Iverson, Canadian Patent CA2075749, 1993.
153. S. Handa, P. Hodgson, International Patent Application WO9858580, 1998.
154. S. Handa, P. K. G. Hodgson, and W. J. Ferguson, U.K. Patent Application GB 2337522A. 1999.
155. W. Breuer, P. Birnbrich, D. Groffe, S. Von Tapavicza, C.-P. Herold, and M. Hof, International Patent Application WO 02/18454, 2002.
156. P. M. W. Cornelisse, International Patent Application WO02/102928, 2002.
157. M. Wilkes and M. Davies, International Patent Application, WO/2006/047745.
158. F. J. Boden, R. P. Sauer, I. L. Goldblatt, and M. E. McHenry, U.S. Patent 6686321.
159. F. J. Boden, R. P. Sauer, I. L. Goldblatt, and M. E. McHenry, U.S. Patent 5663126.
160. F. J. Boden, S. P. Sauer, I. L. Goldblatt, and M. E. McHenry, U.S. Patent 5873389.
161. D. M. Bilden and V. E. Jones, U.S. Patent 6051535, 2000.
162. R. L. Sung, T. F. DeRosa, D. A. Storm, and B. J. Kaufman, U.S. Patent 5207891, 1993.
163. J. Rouet, D. Groffe, and M. Salaun, International Patent Application WO/2008/084178.
164. Y. Wang, A. Kantzas, B. Li, Z. Li, Q. Wang, and M. Zhao, "New Agent for Formation-Damage Mitigation in Heavy-Oil Reservoir: Mechanism and Application," SPE 112355 (paper presented at the SPE International Symposium and Exhibition on Formation Damage Control, Lafayette, LA, 13–15 February 2008).
165. S. T. Dubey and M. H. Waxman, "Asphaltene Adsorption and Desorption from Mineral Surfaces," SPE 18462, *SPE Reservoir Engineering* 6(3) (1991): 389.
166. G. Piro, M. R. Rabianoli, and L. B. Canonico, "Evaluation of Asphaltene Removal Chemicals: A New Testing Method," SPE 27386, Lafayette, LA, 1994.
167. A. Del Bianco and F. Stroppa, U.S. Patent 5382728, 1995.
168. A. K. M. Jamaluddin and T. W. Nazarko, U.S. Patent 5425422, 1995.
169. S. R. King and C. R. Cotney, "Development and Application of Unique Natural Solvents for Treating Paraffin and Asphaltene Related Problems," SPE 35265 (paper presented at the SPE Mid-Continent Gas Symposium, Amarillo, TX, 28–30 April 1996).
170. M. G. Trbovich and G. E. King, "Asphaltene Deposit Removal: Long-Lasting Treatment with a Co-Solvent," SPE 21038 (paper presented at the SPE International Symposium on Oilfield Chemistry, Anaheim, CA, 20–22 February 1991).

171. M. Galoppini, "Asphaltene Deposition Monitoring and Removal Treatments: An Experience in Ultra Deep Wells," SPE 27622 (paper presented at the European Production Operations Conference and Exhibition, Aberdeen, UK, 15–17 March 1994).
172. C. W. Benson, R. A. S. Simcox, and I. C. Huldal, "Tailoring Aromatic Hydrocarbons for Asphaltene Removal," Royal Society of Chemistry, *Chemicals in the Oil Industry IV* 1991, 215.
173. L. Barberis Canonica, A. Del Bianco, G. Piro, F. Stroppa, C. Carniani, and F. J. Mazzolini, "A Comprehensive Approach for the Evaluation of Chemicals for Asphaltene Deposit Removal," Royal Society of Chemistry, *Chemicals in the Oil Industry V*, 13–15 April 1994, Ambleside, Cumbria, UK, 220.
174. J. Curtis, "Environmentally Favorable Terpene Solvents Find Diverse Applications in Stimulation, Sand Control and Cementing Operations," SPE 84124 (paper presented at the SPE Annual Technical Conference and Exhibition, Denver, CO, 5–8 October 2003).
175. C. J. Bushman, International Patent Application WO/2008/024488.
176. W. Kleinitz, "Asphaltene Precipitates in Oil Production Wells" (paper presented at the 8th Oil Field Chemical Symposium, NIF, Geilo, Norway, 1997).
177. L. Minssieux, "Removal of Asphalt Deposits by Cosolvent Squeeze: Mechanisms and Screening," SPE 69672, *SPE Journal* 6(1) (1001): 39.
178. A. Del Bianco and F. Stroppa, U.S. Patent 5690176, 1997.
179. M. B. Lawson and K. J. Snyder, U.S. Patent 4,033,784, 1977.
180. E. G. Scovell, N. Grainger, and T. Cox, International Patent Application WO/2001/074966.
181. E. G. Scovell, N. Grainger, and T. Cox, International Patent Application WO/2001/088333.
182. H. L. Becker and B. W. Wolf, U.S. Patent 1996.
183. M. L. Trimble, M. A. Fleming, B. L. Andrew, G. A. Tomusiak, P. M. Digiacinto, and L. M. Heymans, International Patent Application, WO2008/010923.
184. S. Lightford, E. Pitoni, F. Armesi, and L. Mauri, "Development and Field Use of a Novel Solvent-Water Emulsion for the Removal of Asphaltene Deposits in Fractured Carbonate Formations," SPE 101022 (paper presented at the SPE Annual Technical Conference and Exhibition, San Antonio, TX, 24–27 September 2006).
185. S. Lightford and F. Armesi, International Patent Application WO/2007/129348.
186. K. A. Frost, R. D. Daussin, and M. S. van Domelen, "New, Highly Effective Asphaltene Removal System with Favorable HSE Characteristics," SPE 112420 (paper presented at the 2008 SPE International Symposium and Exhibition on Formation Damage Control, Lafayette, LA, 13–15 February 2008).
187. W. A. Fatah and H. A. Nasr-El-Din, "Acid Emulsified in Xylene: A Cost Effective Treatment to Remove Asphaltene Deposition and Enhance Well Productivity," SPE 117251 (paper presented at the Eastern Regional/AAPG Eastern Section Joint Meeting, Pittsburgh, PA, 11–15 October 2008).
188. L. Quintero, T. A. Jones, D. E. Clark, A. D. Gabrysch, A. Forgiarini, and J-L. Salager, U.S. Patent Application 20090008091.

5 Acid Stimulation

5.1 INTRODUCTION

About 50% of all oil reservoirs worldwide are based on carbonate minerals (limestone/chalk/dolomite) and about 50% are sandstone (quartz, feldspar etc.), although they can contain a small percentage carbonate minerals also. Acid stimulation is used to increase permeability both in production and injector wells, carbonate or sandstone, by dissolving various acid-soluble solids naturally present in the rock matrix or as formation damage. There are many types of formation damage only some of which can be treated with acids, for example, organic deposits such as wax and asphaltenes cannot be treated. This section on acidizing should also be read together with the section on scale dissolvers in Chapter 3 on scale control since low-pH acidizing will also remove carbonate and sulfide scale deposits. To chemically remove sulfate scale, high pH chelates such as salts of polyaminocarboxylic acids are used.

Stimulation by acidizing is an old production enhancement technique dating as far back as the nineteenth century.[1] Several books describe the fundamentals of acid stimulation, including a good introduction to modern techniques.[2] Acid stimulation needs to be carried out with a full knowledge of the history of the well to determine the best course of action since there have been many cases of acid stimulation causing temporary or permanent formation damage, including turning oil-producing wells into 100% water producers. This probably stems from the complex, heterogeneous nature of formation minerals and the unpredictability of their response to conventional oilfield acid formulations. Here, we will give just a brief summary of the techniques used and concentrate on the chemicals involved in the various acidizing treatment strategies. The various books and articles in the list of references at the end of this chapter should be consulted for a more comprehensive understanding of acid stimulation.

There are two basic methods of using acids to stimulate production:

- fracture acidizing
- matrix acidizing

5.2 FRACTURE ACIDIZING OF CARBONATE FORMATIONS

Fracturing can be done hydraulically with proppants or with acids. In both cases, the goal is to create long, open channels from the wellbore penetrating deep into the formation. In fracture, acidizing some or all of the acid treatment is pumped in above the fracturing pressure. Fracture acidizing is usually carried out on carbonate reservoirs, which have lower permeability than sandstone reservoirs. Fracture acidizing of carbonate formations (chalk, limestone, and dolomite) can be used to either remove formation damage or stimulate undamaged formations. Once fractures have been formed by the overpressure, the acid is needed to etch out the fractures leaving high and low points along the channel. This produces a conductive channel within the fracture where oil or gas can migrate. The acids used are the same as in carbonate matrix acidizing discussed below.

A problem with fracture acidizing is that as the acid is injected, it tends to react with the most reactive rock and/or the rock with which it first comes into contact. Thus, much of the acid is used up near the wellbore and is not available for etching of the fracture faces farther from the wellbore. Furthermore, the acidic fluid follows the paths of least resistance, which are, for example, either natural fractures in the rock or areas of more permeable or more acid-soluble rock. This process creates typically long-branched passageways in the fracture faces leading away from the fracture, usually near the wellbore. These highly conductive microchannels are called "wormholes" and are very deleterious because later-injected fracturing fluid tends to leak off into the wormholes rather than lengthening the desired fracture. To block the wormholes, techniques called leak-off control techniques have been developed. This blockage should be temporary, because the wormholes are preferably open to flow after the fracturing treatment; fluid production through the wormholes adds to total production. Commonly, the same methods may be used for leak-off control in acid fracturing and for "diversion" in matrix acidizing. Thus, an acid-etched fracture can be created using either viscous fingering (pad-acid) or viscous acid fracturing. With viscous fingering, a fracture is first formed using a viscous gelled water pad. Acid with lower viscosity is then injected, which fingers through the viscous pad in the fracture, etching out an uneven pattern as it goes. Viscous acid fracturing uses viscous acid systems such as gelled, emulsified, or foamed systems, or chemically retarded acids. These chemical methods are discussed later in this section.

5.3 MATRIX ACIDIZING

In matrix acidizing, the acid treatment is pumped into the production well at or below the formation-fracturing pressure. Matrix acidizing is useful for stimulating both sandstone and carbonate reservoirs. A useful state-of-the-art review was published in 2003.[3]

In carbonate matrix, acidizing the objective is to allow the acid to dissolve channels called wormholes in the near-wellbore region, reaching as far as possible into the formation. If the formation is undamaged, the production rate can be doubled at best, but with damaged formations, higher production rates can be obtained. It should be noted that carbonate matrix acid stimulations are also useful for treating carbonate-cemented sandstones and damage from acid-soluble species such as calcium carbonate ($CaCO_3$), lost circulation material, or carbonate or sulfide scales.

5.4 ACIDS USED IN ACIDIZING

5.4.1 ACIDS FOR CARBONATE FORMATIONS

The most common acid used in carbonate fracture or matrix acidizing is hydrochloric acid (HCl). Organic acids such as acetic acid (CH_3COOH) and formic acid

(HCOOH) are sometimes used particularly for high-temperature applications. Concentrations of HCl used in the field vary: 15 wt.% is common, but a concentration as high as 28 wt.% may be used (commercial HCl is usually sold as a 37 wt.% aqueous solution). Lower concentrations can be used as pickling acids to clean up the well in a preflush (to remove scale and rust) or an afterflush.

Calcium carbonate rock (limestone or chalk) dissolves in the acid to release carbon dioxide and form a calcium chloride solution. The reaction with HCl is given below.

$$CaCO_3 + 2HCl \rightarrow CaCl_2 + CO_2 + H_2O$$

Dolomite rock, which contains magnesium and calcium, will release both cations on treatment with acids. Strong acids such as hydrochloric acid form predominantly unbranched wormholes, whereas weaker organic acids and so-called retarded acids form more branched wormholes.

In some high-temperature applications, HCl does not produce acceptable stimulation results due to lack of penetration or surface reactions. Organic acids, like formic acid and acetic acid, were introduced to offer a slower-reacting and thus deeper-stimulating acid.[4] These "retarded" acids have shortcomings due to solubility limitations of acetate or formate salts and also corrosivity at high temperatures.[4,5] However, corrosion is less than with HCl. High-pH chelating agents for matrix acidizing, such as salts of EDTA or hydroxyaminocarboxylic acids have also been proposed.[5,7] By adjusting the flow rate and pH of the fluid, it becomes possible to tailor the slower-reacting chelate solutions to the well conditions and achieve maximum wormhole formation with a minimum amount of solvent. In addition, use of high-pH solvents significantly reduces corrosion problems.[8,9] Chelates such as EDTA are considerably more expensive than HCl and organic acids. Long-chained carboxylic acids have also been investigated, offering low-corrosion rates, good dissolving power at high temperature, high biodegradability, and easy and safe handling.[10]

5.4.2 Acids for Sandstone Formations

In sandstone matrix acidizing, the primary purpose is to remove acid-soluble damage in the well and near-wellbore area, thus providing a better pathway for the flow of oil or gas. Treating an undamaged sandstone well with matrix acids does not usually lead to stimulation unless the reservoir is naturally fractured. Some carbonate-based damage can be removed with the same acids used in carbonate matrix acidizing. But in a sandstone reservoir, which is composed mainly of quartz and aluminosilicates (such as feldspars), migration of particles (fines) into the pores of the near-wellbore area can cause reduced production. These fines will not dissolve in strong acids such as hydrochloric acid but will dissolve in hydrofluoric acid (HF).

Although highly corrosive, HF is classed as a weak acid due to its low ionization in water. HF is also very toxic. HF, or more usually HF-releasing chemicals such as ammonium bifluoride (NH_4HF_2), is used for sandstone matrix acidizing combined with hydrochloric acid (HCl) or organic acids. HF will also dissolve clays left behind after drilling operations such as bentonite. An aqueous HF/HCl blend is often called a "mud acid." A preflush and overflush of an ammonium salt is often used to remove incompatible ions such as Na^+, K^+, and Ca^{2+} that could lead to precipitation of insoluble fluorosilicate salts (e.g., Na_2SiF_6). The concentrations of the acids used in sandstone and carbonate matrix acidizing treatments vary somewhat according to the service companies who carry out such operations. Guidelines to the concentrations have been documented.[11] Generally, an HF concentration with a maximum of 3% has been used due to the fear of deconsolidation of the near-wellbore of sandstone reservoirs. HCl/HF ratios usually vary from 4:1 to 9:1.

In sandstone acidizing, one has to be particularly careful of reprecipitation of reaction products, which could cause new formation damage.[12] They occur mostly if the well is shut-in for a long period. The chemistry is complicated, but basically, HF reacts first with aluminosilicates to generate fluorosilicates, which react further with clays to form insoluble sodium or potassium fluorosilicates. Overflushes of dilute HCl or NH_4Cl can be used to push these potentially precipitous solutions away from the critical near-well area and deeper into the formation. Another method to control this precipitation problem is by using delayed acid formulations that generate HF slowly. Examples are clay acid (tetrafluoroboric acid, HBF_4)[13] and self-generating acids, which can be esters that hydrolyze to acids at elevated temperatures.[14] Buffered acid systems that allow for deeper penetration can also be used. For example, a buffered HF acid solution of pH 1.9–4 containing organic acids mixed with salts of organic acids and a phosphonate to alleviate the formation of siliceous precipitates can be used.[15] High-pH–buffered systems have been used successfully in single-stage sandstone-acidizing treatments, eliminating the need for preflushes and overflushes.[16]

The aluminum in the clays reacts later on with HF after the silicates have reacted. Aluminum fluoride salts are soluble in the spent acid unless diluted or the pH is raised by postflushes. Chelating agents such as polyaminocarboxylic acids or buffered organic acids can be added to the acid itself to prevent this precipitation from happening.[17] Such blends can be used in single-stage treatments, compared with multistage mud acid treatments, and are especially useful for high-temperature wells.[18,19] In addition, insoluble calcium fluoride can precipitate in the spent acid if too much calcium carbonate is present in the sandstone reservoir. In such cases, HCl treatment alone will suffice. Insoluble iron (III) salts can also cause problems if the pH of the spent acid is raised above approximately 2. It has been proposed that the HCl-based preflush used in sandstone acidizing may be sufficient to remove much of the formation damage in certain cases, for example, calcium carbonate scales. This avoids potential formation damage caused by products of HF acidizing.[20] A phosphonate scale inhibitor can also be added to avoid reprecipitation of carbonate scales.

5.5 POTENTIAL FORMATION DAMAGE FROM ACIDIZING

There are several other ways that acidizing, both for sandstone and carbonate reservoirs, can lead to formation damage if not carried out correctly. These include:

- loss of near-wellbore compressive strength due to using too much HF either in volume or concentration;
- formation of emulsions or asphaltic sludge due to incompatibility between the acid and formation fluids;

- water-blocking and wettability-alteration damage (this can be repaired with mutual solvent treatments (mixed with water or hydrocarbon solvent) containing surfactants);
- fines migration after acidizing (this is fairly common in sandstone acidizing; bringing the well on slowly after treatment can minimize this damage).

Experience has shown that for sandstone formations, oil wells respond to matrix acidizing in a different manner as compared with gas wells. For oil wells, the improvement in permeability resulting from the stimulation treatment peaks at a certain acid volume and then drops as the volume of acid injected increases. For gas wells, however, the resulting improvement in permeability is roughly proportional to the volume of acid injected, and is normally better than that obtained with oil wells. It is, therefore, expected that stimulation of oil wells in sandstone formations could be improved by displacing the oil in the zone to be treated with gas. Gas injection prior to acidizing is sought to minimize the formation of emulsions or sludge resulting from reactions between the spent acid products and the oil that otherwise would be contacted.[21]

5.6 ACIDIZING ADDITIVES

The acid main flush contains several additives to bring control to the treatment. These almost always include:

- corrosion inhibitor
- iron control agent
- water-wetting surfactant

Many other additives can also be used. The three classes listed above will be discussed first followed by other classes of additives.

5.6.1 CORROSION INHIBITORS FOR ACIDIZING

5.6.1.1 General Discussion

The use of highly acidic stimulation fluids can lead to severe metal corrosion and hydrogen and chloride stress cracking. Chapter 8 entitled corrosion control discusses various classes of inhibitors injected primarily into production streams. Corrosion rates of carbon or chrome steels (corrosion resistant alloys) during acid stimulation are higher than under normal production conditions. Corrosion inhibitors for protection of carbon and chrome steels (such as duplex steel) during acidizing are mostly different from those used to treat production fluids and usually dosed at higher concentrations.[24c,25] For example, the well-known imidazoline and phosphate ester surfactants used in production pipelines are not usually used in acidizing operations, although unsaturated acid (e.g., acrylic acid) derivatives of amines and imidazolines have been claimed.[22] Acidizing corrosion inhibitors must be able to prevent reaction of corrosive acids with steels and be stable in high-concentration acid solutions at low pH and high reservoir temperatures. They must also be cost-effective since the concentration of corrosion inhibitor in the acid solution needs to be fairly high. The inhibitor dosage usually increases with increasing well temperature. The choice of corrosion inhibitor is also limited in regions with strict environmental requirements. Progress has been made in finding greener acid-corrosion inhibitors, particularly for high-temperature wells, but sometimes at the expense of performance.[23–25]

Historically, arsenic compounds were used as corrosion inhibitors in acid stimulation packages, but due to their high toxicity, they have been phased out. Many commercial acid inhibitors used to be based on Mannich condensation products. They were made by reacting a reactive ketone, formaldehyde, and an amine to produce a Mannich base.[26,27] Because this reaction rarely goes to completion, some formaldehyde, an unwanted carcinogen, remains in the reaction product that is formulated as the commercial inhibitor. Amine-based inhibitors, such as quaternary ammonium surfactants described below, can be used as replacements but many of them are also toxic. Imines (Schiff bases) have also been investigated as corrosion inhibitors for acid solutions.[27] Acetylenic alcohols and α,β-unsaturated aldehydes, such as cinnamaldehyde and related derivatives, have been used as more environmentally-friendly corrosion inhibitors.[28,29]

These are also discussed below in more detail. Solvents and wetting surfactants are also used in acidizing corrosion inhibitor packages. In general, most organic corrosion inhibitors for acid stimulation possess electronegative atoms, a high degree of conjugated double or triple bonds and aromatic rings and a high degree of planarity. These features have been described as giving them the ability to hold tightly to the ferrous surface of the tubulars by mingling electrons with the "electron fog" of the steel.[29]

An easy way to help control corrosion is to use a percentage of a thermally stable, reducing acid in the HCl (carbonate) or HCl/HF (sandstone) formulation to prevent iron oxidation. Formic acid (HCOOH) is the simplest and cheapest example. In addition, reducing agents such as iodide ions can be added, which forms hydriodic acid in situ, a strong reducing agent.[30] Elemental iodine has also been claimed as a corrosion inhibitor intensifier.[31] Organic acid blends containing reducing acids such as formic acid are particularly useful for high-temperature applications where corrosion issues are paramount.[32,33]

5.6.1.2 Nitrogen-Based Corrosion Inhibitors

To achieve satisfactory performance, most film-forming corrosion inhibitors contain a hetero atom (e.g., nitrogen, oxygen, sulfur) having a nonbonding pair of electrons available for interaction with a metal surface. Monomeric quaternary ammonium surfactants are an exception to this rule, but they do tolerate highly acidic solutions. However, their performance may not be good enough especially when used alone for high-temperature acidizing jobs. Most corrosion inhibitors packages for mineral acid stimulation (HCl or HCl/HF blends) contain quaternary ammonium compounds or amines mixed with unsaturated oxygen compounds. Some gemini quaternary surfactants (two quaternary head groups and two hydrophobic tails) appear to show good corrosion inhibition.[34] Aromatic quaternary surfactants (e.g., pyridinium or quinolinium) appear to be better than alkyl quaternary surfactants for acidizing jobs (Figure 5.1).[35a] Quaternary naphthylmethyl quinolium chloride with antimony

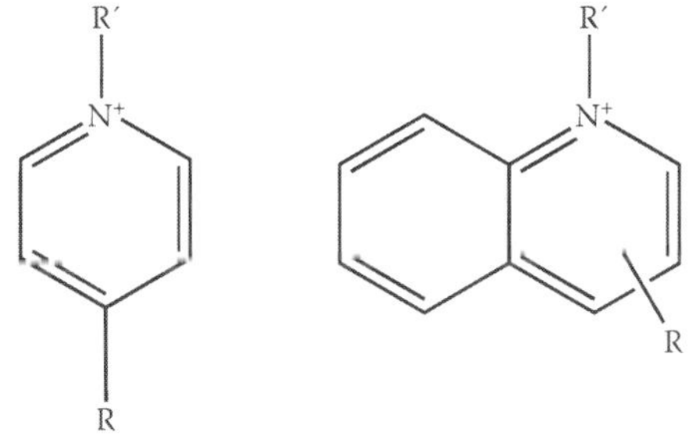

FIGURE 5.1 Pyridinium and quinolinium ions.

chloride as an inhibitor aid has also been claimed.[35b] 2-Aminopyridine and other aminopyridines, which, like all amines, will be protonated in acid solution, has been shown to be a good corrosion inhibitor of mild steel.[36] The performance of some quaternary surfactants such as benzylquinolinium chloride can be improved by adding synergists such as long-chain carboxylic acids.[37] Sulfur compounds such as thiosulfate salts (e.g., $Na_2S_2O_3$) or thioglycolic acid ($HSCH_2COOH$) can be used as intensifiers in organic acid compositions.[38] The performance of quaternary ammonium surfactants can also be improved by adding enhancers or intensifiers such as molybdate ions[39] or reducing agents; for example, iodide ions is common, and cuprous, antimony, bismuth salts, and formic acid have also been proposed.[40,41] Bismuth salts are less toxic than antimony salts. Enhancers often have to be used in high temperature applications. Hypophosphate or hypophosphite salts have been proposed as corrosion inhibitors for weak acid-stimulation formulations.[42] Yet another method to improve the performance is to use a polymeric or oligomeric amine/quaternary ammonium salt. Oligomerized aromatic amines or quaternary derivatives containing several benzylquinolinium moieties have been proposed as acid corrosion inhibitors well as poly(toluidine) and quaternized polyethyleneimines.[42,43] A family of amine surfactants with only partly ethoxylated amines have been proposed for use in production operations or acidizing jobs. Examples are fatty alkylamines, alkyletheramines, or alkylamidopropylamines reacted with only a mole of ethylene oxide, e.g., $R'C(O)NHCH_2CH_2CH_2NHCH_2CH_2OH$.[45] A new family of mono/diamine-based corrosion inhibitors has been reported.[46] Polyalkoxylatediamines have been claimed as inhibitors for acid treatment of sour wells.[47] Amines that quaternize in acid solution and alkyl quaternary ammonium salts are not allowed in environmentally sensitive areas such as parts of the North Sea basin. Therefore, other acidizing corrosion inhibitors are needed here.[48]

5.6.1.3 Oxygen-Containing Corrosion Inhibitors Including Those with Unsaturated Linkages

Unlike corrosion inhibitors for production operations, quite a few acidizing corrosion inhibitors contain oxygen in the head group and an unsaturated linkage. The unsaturated oxygen compounds polymerize on the metal surface. Quaternary ammonium or amine compound are usually added to help hold the polymers to the surface as well as providing their own surface inhibition. Some commercial acid inhibitor packages

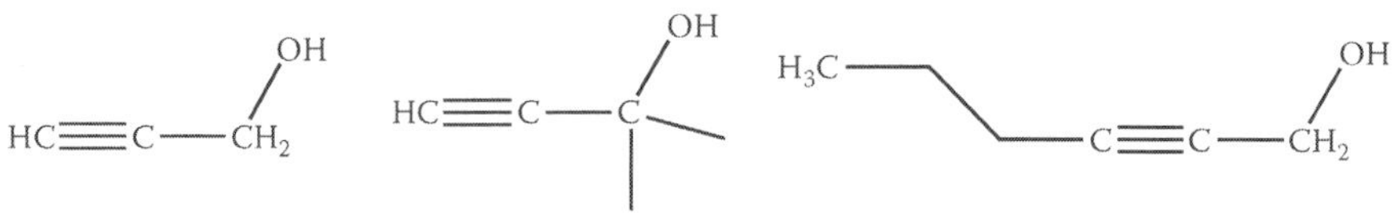

FIGURE 5.2 Propargyl alcohol, 2-methyl-3-butynol, and 1-hexyn-3-ol.

are formulated with acetylenic alcohols. Examples are 1-octyn-3-ol, 1-hexyn-3-ol, 2-methyl-3-butynol, and 1-propyn-3-ol (propargyl alcohol; Figure 5.2). Diacetylenic diols such as 3,8-dihydroxy-1,9-decadiyne have also been studied.[49] They react in acid at elevated temperatures to form oligomers that have film-forming properties on steel.[30] Their performance can be improved by adding intensifiers mentioned above or elemental iodine, quaternary ammonium surfactants, or amines such as hexamethylenetetramine, which can also scavenge any hydrogen sulfide released by reaction of the acids with metal sulfides.[50] However, some small acetylenic alcohols are fairly toxic and can present handling problems such as in the back production of hot, spent acid. A new twist on acetylenic alcohols is the use of 2,3-di-iodo-2-propen-1-ol (used in excess propargyl alcohol), made from the reaction of propargyl alcohol and iodine.[51] The compound provides iodine to the media in a stable form that does not appear to degrade over time. Acetylenic sulfides can also be employed in this invention in lieu of acetylenic alcohols. Examples of these are dipropargyl sulfide, bis-(1-methyl-2-propynyl)sulfide, and bis-(2-ethynyl-2-propyl)sulfide.

Cinnamaldehyde, another oxygenated molecule, and related derivatives are useful additives that have been used for some years in acidizing corrosion inhibitor packages (Figure 5.3).[48,52,53] They are generally more environment-friendly than quaternary ammonium products and Mannich condensation compounds. They are also less toxic than most acetylenic alcohols.[56] The key feature of their structure is an aldehyde and a conjugated double bond. In acid solution, cinnamaldehyde forms oligomeric species that form a film on the metal surface via various organometallic and metal-oxygen interactions. By themselves, cinnamaldehydes have limited corrosion inhibition performance, but they can be blended with other inhibitors and intensifiers such as surfactants, additional aldehydes, acetylenic compounds, alkyl or arylphosphines, or antimony or iodide salts for improved performance.[54] Cinnamaldehydes blended with aliphatic aldehydes such as glyoxylic acid or glyoxal have also been claimed.[55] Related cinnamyl chemistry is actively pursued for improved corrosion inhibition. For example, cinnamaldehyde performs well when combined with an organosulfur compound such as thioethanol (to form a thioacetal) and optionally a quaternary ammonium surfactant.[29,56,57] Urea has also been shown to work synergistically with cinnamaldehyde and related derivatives to improve acidizing corrosion inhibition.[58]

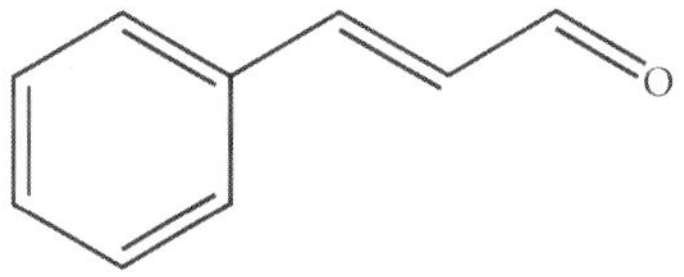

FIGURE 5.3 Cinnamaldehyde.

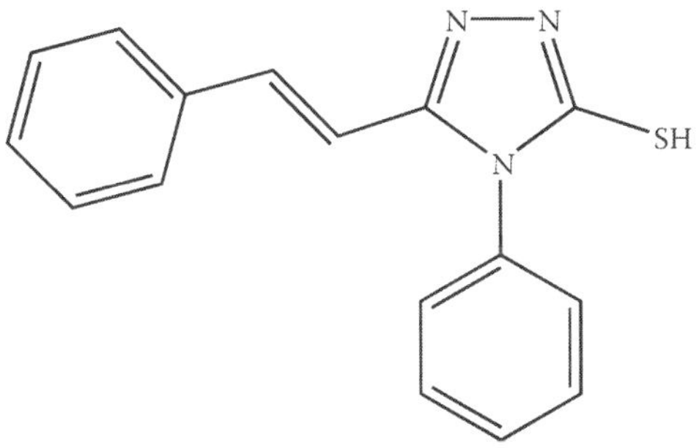

FIGURE 5.4 3-Cinnamyl-4-phenyl-5-mercapto-1,2,4-triazole. The cinnamyl group can be replaced with various substituted phenyl groups.

The reaction product of an α,β-unsaturated aldehyde or ketone such as cinnamaldehyde with primary or secondary amines and alkanolamines also gives improved acid corrosion inhibitors.[59] A composition comprising of cinnamaldehyde or a substituted cinnamaldehyde together with a reaction product of a C3–9 ketone such as acetophenone, thiourea, formaldehyde, and hydrochloric acid has been claimed as an environment-friendly acid corrosion inhibitor.[60] Biodegradable ester quaternary surfactants of cinnamyl alcohol, such as [bis(2-hydroxyethyl)coco betaine] cinnamyl alcohol esterquat, can be used as corrosion inhibitors: they also have antisludging and demulsifier properties.[61] Cinnamyl thiotriazoles also performed well in up to 3 M mineral acids (Figure 5.4).[62]

Conjugation in corrosion inhibitors such as cinnamaldehyde appears to give good interaction with the steel surface as this gives planar molecules. For example, a Schiff base derivative of cinnamaldehyde with phenylenediamine, 2,4-dicinnamyledene iminophenylene, gave good steel corrosion inhibition in acid media (Figure 5.5).[63] Another class of corrosion inhibitor with conjugation is the α-alkenylphenones (Figure 5.6). Several inhibitors in this class formed 2-benzoylallyl alcohol as an intermediate in 15% HCl. This molecule polymerizes to form poly(phenylvinylketone), which was identified as the film-forming component on the steel coupons. Blends of the α-alkenylphenones with surfactants gave improved performance.[64]

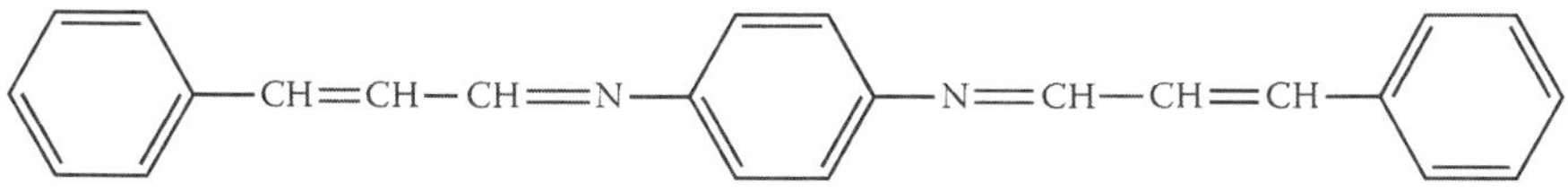

FIGURE 5.5 2,4-Dicinnamyledene iminophenylene.

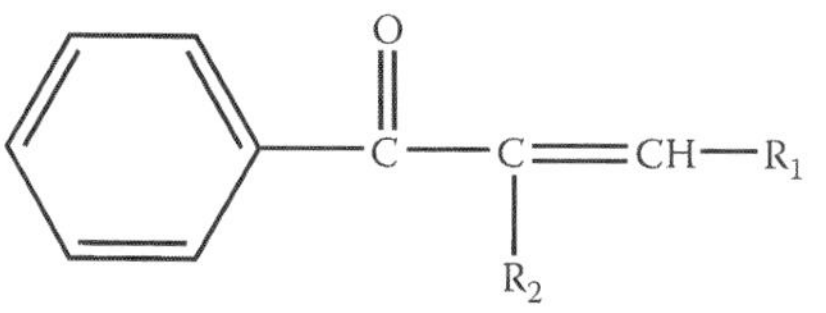

FIGURE 5.6 α-Alkenylphenones. For example, R_1 = H and R_2 = H, CH_2OH, or CH_2OCH_3.

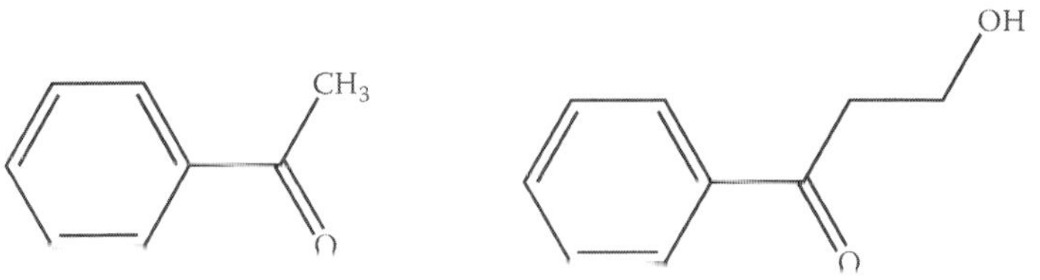

FIGURE 5.7 Acetophenone and 3-hydroxy-1-phenyl-1-propanone.

Ketones such as acetophenone or 3-hydroxy-1-phenyl-1-propanone, as additives containing oxygen, have been studied for use in acid corrosion inhibitor packages (Figure 5.7). For example, a composition of acetophenone with a pyridinium or quinolium quaternary salt and antimonium or bismuth ions is a good acid corrosion inhibitor for up to 120°C (248°F).[65] Condensation products formed by reacting an aldehyde (e.g., formaldehyde) in the presence of a nitrogen-containing compound (e.g., an alkylamine) and a carbonyl compound (e.g., acetophenone) have also been proposed,[66] as well as aldol base-catalyzed condensation products of ketones, aldehydes, and fatty acids with antimony and bismuth compounds as synergists.[67] Rosin amine components (e.g., dehydroabietylamine), a ketone component (e.g., acetophenone, hydroxyacetophenone, and/or di-acetophenone), one or more carboxylic acid components (e.g., formic, glycolic, citric acid), and paraformaldehyde have been shown to be good corrosion inhibitors for acidizing with lower toxicity.[68] Terpenes and hydroxypropionic acid have been proposed as corrosion inhibitor intensifiers.[69,70]

5.6.1.4 Corrosion Inhibitors Containing Sulfur

Besides nitrogen and oxygen, the other common heteroatom found particularly in a number of organic acidizing corrosion inhibitors is sulfur. The sulfur-containing inhibitor is usually used in a formulation with other classes of corrosion inhibitors, such as acetylenic alcohols, cinnamaldehydes, and quaternary nitrogen compounds, and is particularly useful for reducing pitting corrosion.

A common class of sulfur-containing corrosion inhibitors are the 1,3-dialkylthioureas such as 1,3-dibutylthiourea,[70] although more water-soluble molecules such as the parent thiourea can be used. A whole range of other thio compounds have been proposed for use in formulations with cinnamaldehydes.[29–56] If the compound contains a thiol group (–SH) it can react with an aldehyde, such as cinnamaldehyde or crotonaldehyde, to form a new compound that can also be used as a corrosion inhibitor.[57] The reaction product of thiourea (or alternatively a primary amine), with formaldehyde, and an aromatic ketone such as acetophenone in the presence of an organic acid (e.g., acetic acid) and a mineral acid has been found effective as an acidizing corrosion inhibitor.[60]

There has been a fair amount of research on acidizing corrosion inhibitors containing both nitrogen and sulfur, especially in heterocyclic rings. For example, oxadiazole (X = O) and thiadiazole (X = S) compounds shown in Figure 5.8 were shown to exhibit good cathodic corrosion inhibition in 1 M HCl. Higher concentrations were not investigated.[71]

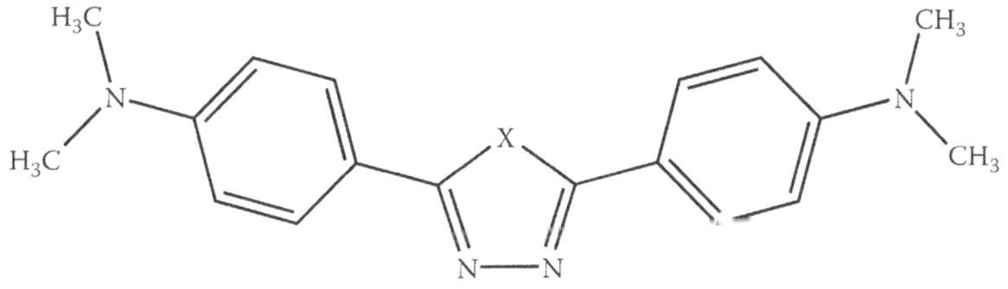

Acid Stimulation

FIGURE 5.8 Oxadiazole (X = O) and thiadiazole (X = S) compounds with good corrosion inhibition in acid media.

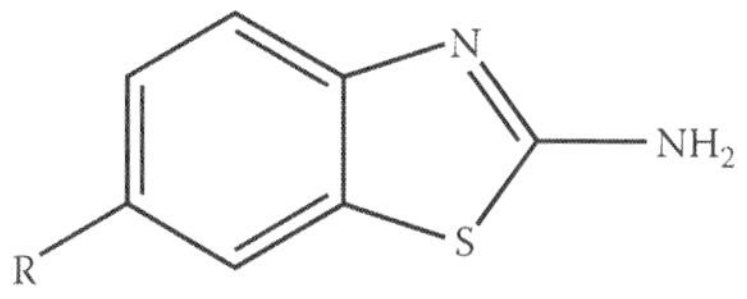

FIGURE 5.9 6-Substituted aminobenzothiazoles.

Aminobenzothiazoles also gave good corrosion inhibition in acidic media (Figure 5.9). The positive charge on the amine at low pH may also contribute to the performance besides interaction of the ring nitrogen and sulfur atoms with the steel surface. Inhibition was better for R = Cl than R = H due to a bigger dipole moment in the chlorinated molecule. Interestingly, a cinnamaldehyde derivative (probably a Schiff base) performed even better than the aminobenzothiazoles. Potassium iodide intensified the corrosion inhibition performance.[72]

Other heterocyclic compounds containing both nitrogen and sulfur have been found to be good corrosion inhibitors for steel in acid solutions. A product with three nitrogen and one sulfur atom in a ring structure with alkene (C=C) and methane (C=N) conjugation performed significantly better than propargyl alcohol in boiling 15% HCl.[73] Other sulfur-nitrogen compounds with good corrosion inhibition properties include thiotriazoles, thiobiuret compounds, and thiosemihydrazides, as well as sulfoximines ($R_1R_2S(=NH)O$) and sulfur-substituted (iso)thioureas.[62,63]

5.6.2 IRON CONTROL AGENTS

There is a tendency for iron sulfide scale to form in the well and near-wellbore area especially in sour wells. Acids used to treat the well can dissolve the iron sulfide, but in the process, hydrogen sulfide is generated, which is toxic and stimulates corrosion. In addition, the dissolved iron tends to precipitate, in the form of ferric hydroxide or ferric sulfide, as the acid in the treatment fluid becomes spent and the pH of the fluid increases. Precipitation of iron compounds can cause damage to the permeability of the formation. In addition, the release of iron(III) ions into solution can exacerbate asphaltic sludging formation damage. Sludging is formed by reaction of the iron ions with polar groups in asphaltenes in the reservoir oil. Control of sludging with iron control agents and demulsifiers can also alleviate emulsion block problems.[74]

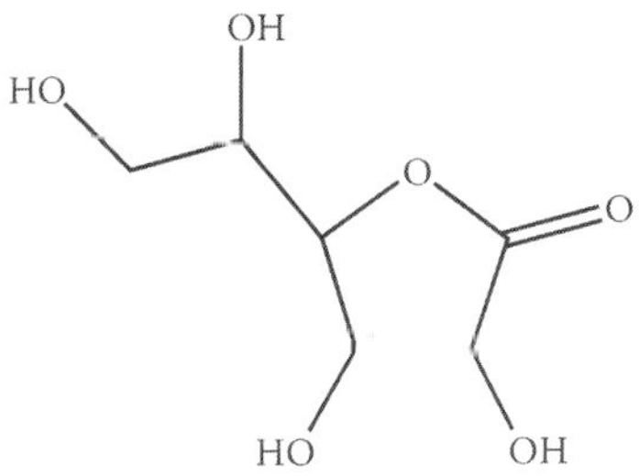

FIGURE 5.10 Erythorbic acid.

There are two classes of iron control chemicals designed to prevent these effects from happening:

- reducing agents to reduce iron(III) to iron(II) ions (ferric → ferrous)
- complexing agents (also called "sequestering agents" or "chelates")

Reducing agents that can be used include:

- Metal ions in reduced oxidation states, for example, tin(II) or copper(I) ions
- Iodide salts or iodine
- Reducing acids such as formic acid, hypophosphorous acid, or a hypophosphorous acid precursor such as a metal phosphinate catalyzed by antimony(V) or copper(II) ions.[75]
- Erythorbic acid (isoascorbic acid) or ascorbic acid (Figure 5.10).
- Reducing thioacids (e.g., thioglycolic acid, $HSCH_2COOH$) with catalysts such as copper(II) ions and iodide ions[76] or ketones that react with sulfides.[77]

The reducing agents such as iodide and the inorganic acids also help to prevent corrosion. Other transition metal ions with several easily accessible oxidation states, such as vanadium or rhenium ions, combined with reducing agents can also be used. Reducing agents such as iodide ions first reduce the metal ions to a lower oxidation state, which can then reduce iron(III) to iron(II) ions, keeping them soluble. In effect, the metal ions are electron-transfer agents and catalyst precursors.[78–80]

Complexing agents are commonly used for iron control. Complexing agents are chelates such as citric acid, EDTA, or nitrilotriacetic acid, the same products discussed in Section 3.9 on scale dissolvers in Chapter 3 on scale control. The first two chelates are the most used commercially for iron control. EDTA gives additional iron control by becoming a reducing agent above 120°C (248°F).[81] In addition, a sulfide modifier such as a ketone, aldehyde, or acetal can be added to combine with dissolved sulfides.[82,83] Dithiocarbamate compositions to sequester iron have also been proposed.[84] Another method claimed to reduce the precipitation of metal sulfides from acid solutions is to add an oxime such as acetaldoxime. The oxime preferentially reacts with sulfide ions in the solution and thereby prevents the sulfide ions from reacting with metal ions therein.[85]

5.6.3 WATER-WETTING AGENTS

Water-wetting agents are needed in the acid stimulation treatment to remove any oily film from the rock or scale so that the aqueous acid has good contact. Water-wetting agents are also necessary to clean up the well and leave the formation water-wet, thereby enhancing the flow of oil or gas. Water-wetting agents can be simple monomeric surfactants but with a high HLB (hydrophilic-lipophilic balance). Examples are nonionic surfactants such as the alkyl ethoxylates and alkylphenyl ethoxylates. A mutual solvent ("musol") can also be used. A mutual solvent is miscible with both water and oil and aids in water-wetting the formation. The most common solvent is monobutyl glycol ether also called butyl glycol. Another is dipropylene glycol methyl ether. Mutual solvents can also remove trapped water (water blocks) caused by an all-aqueous treatment, thus providing stimulation.[97] The film-forming corrosion inhibitor in the acid solution may also provide water-wetting.

5.6.4 OTHER OPTIONAL CHEMICALS IN ACIDIZING TREATMENTS

There are several other optional chemicals needed depending mainly on the formation fluids and type of damage. These include:

- Clay stabilizer: This is useful to prevent migration or swelling of clays caused by the sandstone acid treatment. Approximately 5 wt.% ammonium salts in the preflush is useful. Improved clay stabilizers are polyamines and polyquaternary amines such as polydimethyldiallylammonium chloride, again preferably added to the preflush to protect clays during the initial ion exchange.[86,87]
- Fines fixing agent: Most fines in sandstone formations are not clay but quartz, feldspar, and other minerals. Clay stabilizers will not control these fines. An organosilane agent, such as 3-aminopropyltriethoxy silane, has been used to fix these fines. It reacts in situ with water to form polysiloxanes, which bind siliceous fines.[88,89]
- Antisludging agent
  - Iron-control agents, discussed above, can prevent asphaltic sludging, if they avoid formation of uncomplexed ferric ions.
  - Asphaltene solvents, inhibitors, and dispersants can also be used, especially in emulsified acids (see Section 5.8.9 below on emulsified acids or Chapter 4 on asphaltene control).[90] Examples are xylene as solvent and the alkylaryl sulfonic acids as dispersants.[91] Other antisludging surfactants that can be used in acid treatments include ester quaternary ammonium surfactants, ethoxylated alkyl phenols, alkoxylated alkyl-substituted phenol sulfonates and ester, or salts of sulfonated fatty acids.[61,92] Treatment to stabilize the asphaltenes previous to acid stimulation or low-pH scale inhibitor squeeze treatments was most effective in one field.[93]
  - Biodegradable ester quaternary surfactants, which also have corrosion inhibition properties.[94]

- An invention to prevent sludging combines HCl, a quaternary corrosion inhibitor (preferably aromatic) and a conjugate ion pair of a cationic amine oxide surfactant and an anionic surfactant that does not react with the corrosion inhibitor. A preferable conjugate ion pair is dimethylcocoalkylamine oxide and dodecyl sodium sulfate.[95]
- Demulsifier: The demulsifier may be added as a spearhead in a hydrocarbon solvent before the main acid solution. Some emulsion problems can be treated using iron control additives in the acid solution as discussed above. Some demulsifiers, such as N-alkylated polyhydroxyetheramines, have been specifically designed for use in acid-stimulation operations[96] (see also Chapter 11 on demulsifiers).
- Alcohols: Small alcohols can be beneficial in gas well stimulation by helping release spent acid from the formation by lower the surface tension. Methanol or glycol may be needed for preventing gas hydrates plugs in cold gas wells in deepwater.
- Calcium sulfate scale inhibitor: Necessary if sulfate concentrations are high in the formation water (see Chapter 3 on scale control).
- H_2S scavenger: If the acidizing treatment contacts sulfide scales toxic H_2S can be released. An aldehyde or other H_2S scavenger can be used (see also Chapter 15 on hydrogen sulfide scavengers).[98]
- Foaming agent: These are surfactants used together with nitrogen gas to help remove spent acid and lift a gas well back on stream.
- Drag reducer: These are usually water-soluble polymers such as high molecular weight polyacrylamides. A drag reducer is useful in deep-well treatments or when you need a high treatment rate. They provide fluids with low friction pressures.[99]
- Surfactants: In gas wells, reduction of the surface tension is required to reduce the capillary forces that trap the aqueous phase in the formation thus avoiding water blockage.[100]

5.7 AXIAL PLACEMENT OF ACID TREATMENTS

In addition to determining the most effective combination of acid blends and volumes for each particular reservoir, treatment design and planning are often performed to ensure that the acid is placed across the entire interval. The successful acid placement in matrix treatments of open-hole horizontal wells is even more difficult due to the length of the zone treated and potential variation of the formation properties. A successful diversion technique is critical to place the acid to the location where damage exists.

To improve contact of the acid solutions with the interval to be treated, one can use either mechanical or chemical placement techniques. Mechanical placement can be done by packer systems, ball sealers, or coiled tubing. Chemical placement uses so-called diverters. Diverters are useful in a number of other well operations besides acidizing, for example, scale-inhibitor squeeze treatments. The diverter is usually applied in a preflush and temporarily plugs the zone or zones of highest permeability allowing the main flush to react with other less permeable or more damaged zones. On back production, the diverter is removed from the perforations. There have also been developed self-diverting viscoelastic fluids, which are discussed later.

There are at least four categories of diverters for acidizing:

- solid particles that degrade, dissolve, or melt in hot-produced water or oil
- polymer gels
- foams
- viscoelastic surfactants (VESs)

Solid particles are not usually used in HCl carbonate acidizing because of the high solubility of the formation in the acid and the formation of channels. The other categories given above have all been used successfully in carbonate acidizing.

In addition to the above methods, relative permeability modifiers (RPMs) have been used for improving the diversion of acidizing treatments[101] (see Chapter 2 on water and gas control). In one carbonate field, the use of a hydrophobically modified RPM in matrix-acidizing treatments gave better production results than the use of foam or polymer gel diverters.[102] Emulsified acids are used to achieve better penetration but the viscosity of these formulations can also give them diversion capabilities. These are discussed in Section 9. Another method of improving the axial placement of HF acid treatments is the so-called maximum pressure differential and injection rate (MADPIR) technique. By maintaining maximum injection rate, while always increasing it to maintain the maximum allowable matrix injection pressure (i.e., below the fracturing pressure), the need for diverter is removed.[103]

5.7.1 Solid Particle Diverters

In HCl and non-HF acidizing treatments, the most common diverter is rock salt (NaCl). Rock salt can be added to the main acid flush. Benzoic acid flakes, oil-soluble resins, and wax beads are good particle diverters that can be used in sandstone acidizing treatments. The resins and wax particles cannot be used in low-temperature wells such as gas wells or injection wells as they may not melt at the well temperature. Other suitable solid additives include polyesters, polycarbonates, polyacetals, polymelamines, polyvinyl chlorides, and polyvinyl acetates. When combined with a viscosified fluid they are claimed to give a deeper barrier in the near-wellbore area.[104]

Solid diverters have generally been superseded by other diverter techniques in the last 20 years. However, a new diverter, solid particles of a polyester, that degrades by aqueous hydrolysis at elevated temperatures, has been shown to have superior performance in acid stimulation jobs.[105] Successful field trials involving the use of a degradable (hydrolyzable) polymeric fiber technology designed to achieve effective acid diversion during acid fracturing have also been reported. The significant viscosity increase achieved in the pad by the addition of fibers, and its particulate bridging mechanism plug off just the stimulated zones effectively.[106]

5.7.2 Polymer Gel Diverters

Polymer gels have higher viscosity than normal acidizing solutions, which means they will penetrate low-permeability intervals first. Thus, there are two ways to use a viscous solution such as a polymer gel. First, one can inject a polymer gel viscous pill (or preflush). This technique relies upon the viscosity of the preflush to influence the injection pressure of the interval it enters. As the preflush enters the formation, the viscosity of the pill will restrict the injection of other fluids into this area. As the injection pressure increases within this portion of the interval, other sections of the interval will break down and begin accepting fluid. A low-viscosity acid main flush will then be able to penetrate the low-permeable damaged zones. Second, one can gel the acid solution itself. Gelled acid diverters give two benefits to an acidizing treatment:

1. Reaction rates are significantly slowed allowing better acid penetration (see radial axial placement of acidizing treatments below).
2. The higher viscosity decreases the leak-off due to increased flow resistance.

These two benefits combined increases the tendency of the subsequent treating acid to be diverted elsewhere. By pumping alternately gelled and ungelled acid stages, the whole zone of interest can be acidized. This is especially useful in carbonate matrix acidizing.[107] Gel breakers can also be used to reduce the viscosity of the gel after the acidizing treatment. Polymer gelled acids are also used as leak-off control agents in fracture acidizing.

A typical chemical diverting fluid is a gelled hydroxyethylcellulose (HEC) pill. This method is severely limited if the temperature of the gelled HEC exceeds approximately 95°C (203°F). Above this temperature, the base viscosity and life of the pill are greatly decreased. Other typical polymers that have been used are acrylamide polymers and other natural or polysaccharide polymers such as guar, xanthan, scleroglucan, or succinoglycan.[108] Biodegradable scleroglucan or diutan when partially cross-linked have better thermal stability than most natural polymers and are easier to disperse than poorly biodegradable synthetic polymers such as polyacrylamides.[109] HEC, the natural polymers, and polyacrylamide are not so stable in acids at high well temperatures or tolerant of high salinities. Thus, various other synthetic polymers have been proposed for such applications including polymers and copolymers of vinyl lactam, N,N-dimethyl acrylamide, acrylamidopropanesulfonic acid, and quaternary monomers such as acrylamidoethyltrimethylammonium chloride and dimethylaminoethyl methacrylate, although these are generally more expensive and poorly biodegradable materials.[110–112] Clarified xanthans made by genetic engineering or bacteria selection have been claimed for use in gelled acids up to 150°C.[113]

The main flush acidic solution can be injected with a polymer and a second chemical, which cross-links the polymer at raised temperature in situ in the wellbore, increasing the viscosity of the solution further. Tetravalent and trivalent metal salts (zirconium(IV), titanium(IV)) are commonly used to cross-link polysaccharides, such as guar, in fracture acidizing. Metal salts such as aluminum salts, iron(III) salts, phenolic compounds and/or small monoaldehydes, and polyamines have been used as cross-linking agents with anionic polymers such as partially hydrolyzed acrylamide polymers, some of them similar in composition to chemicals used in water shut-off treatments (see Chapter 2 on water and gas control).[114,115] Some polymer solutions naturally thicken at raised temperatures in the wellbore. These include polyoxyalkylenes such as ethylenediamine reacted with propylene oxide and then ethylene oxide to a molecular weight of up to 30,000.[116,117] Viscous oil-in-water emulsions have also been proposed (see Section 9).

Self-diversion of polymer gelled acids can also be accomplished by a change in pH due to the acid being spent and the resultant pH increase. The pH increase can activate a highly valent metallic reagent, such as Zr(IV), that cross-links the polymer chains. The subsequent viscosity increase causes a higher flow resistance and diversion of the unspent acid into other zones. Further increase of pH deactivates the metallic cross-linker using a reducing agent and breaks down the fluid to the original polymer chains.[118] Self-gelling acid treatments are called "low leak-off control acids." They have become very popular in field operations for generating deeply penetrating wormholes in carbonate formations. Cross-linking xanthan can be especially difficult and/or impractical, but an oxidized form of cross-linked xanthan, has been claimed to be more stable and useful in self-diverting acids.[119]

A problem, which is sometimes encountered when using polymer-thickened compositions in treating formations, is the ease of removal of the treating composition after the operation is completed. Some thickened or highly viscous polymeric solutions are difficult to remove from the pores of the formation or the fracture after the operation is complete.[120] Sometimes, a clogging residue can be left in the pores of the formation, or in the fracture. This can inhibit the production of fluids from the formation and can require costly cleanup operations.[121] One method to avoid this is to delay the thickening of the solution. For example, chromium(VI) salts with a reducing agent such as sulfite ions or thioacetamide have been proposed. Such compositions will form chromium(III) ions in situ, which can then cross-link with the polymer.[122] Another method that relies on aldehydes to cross-link the polymers is to use an aldehyde precursor such as an acetal, which decomposes to form the aldehyde in situ at raised temperatures in the well bore.[123] Breakers such as enzymes and oxidizing agents such as sodium persulfate can be injected to break down polysaccharide-based fluids. This is very common in hydraulic fracturing. If the polysaccharide polymer is cross-linked with metal ions (e.g., titanium or zirconium), the viscosity can be reduced after a delay period by coinjecting the breaker coated with a water-soluble resin. Examples of breakers for metal salts are ionic materials that will preferentially complex with the metal ions such as fluoride ions (e.g., fluorspar, cryolite), sulfates, phosphates, phosphonates, and carboxylates (e.g., EDTA).[124]

One method for leak-off control in fracture acidizing is to incorporate into the acidic fluid first a chemical or chemicals that will form a barrier to fluid flow after a substantial amount of the acid is consumed ("spent") and the pH increases, and second, another chemical or chemicals that will destroy the barrier as more acid is spent and the pH increases further. That initially strongly acidic system initially has low viscosity but includes a soluble ferric ion source and a polymeric gelling agent that is cross-linked by ferric ions at a pH of about 2 or higher but not at lower pHs. Typical polymers are anionic polyacrylamides. However, the polymer is not cross-linked by ferrous ions. Therefore, the system includes a reducing agent that reduces ferric ions to ferrous ions, but only at a pH above ~3–3.5. Consequently, as the acid spends, for example, in a wormhole, and the pH increases to ~2 or higher, the polymer crosslinks and a very viscous gel forms that inhibits further flow of fresh acid into the wormhole. As the acid spends further (after the treatment) and the pH continues to rise, the reducing agent converts the ferric ions to ferrous ions and the gel reverts to a more fluid waterlike state. Hydrazine salts and hydroxylamine salts have been used as the reducing agents in the past but due to their toxic and carcinogenic properties, carbohydrazides, semicarbohydrazides, ketoximes, and aldoximes have been proposed as less hazardous replacements.[125]

5.7.3 Foam Diverters

Foamed acids were proposed as early as the seventies for fracture acidizing but is also used for diversion and better penetration in matrix acidizing.[126] Foam can aid acid-well stimulation by diverting acid into damaged or low-permeability layers near the wellbore. The foam is made using a water-soluble surfactant and a gas such as nitrogen, carbon monoxide, carbon dioxide, or a natural gas.[127] Optionally, a viscosifying polymer can be used such as HEC, carboxymethylcellulose polymers, polyacrylamides, polyacrylates, and polysaccharides, or a VES.[128,129]

Foam-forming surfactants include anionic, cationic, amphoteric, and nonionic surfactants, in increasing order of performance. Anionic surfactants are adversely affected by the presence of crude oils and severely deteriorate in the presence of strong acids such as hydrochloric acid. Cationic surfactants are moderately good foamers in the presence of acid, but produce unstable foams in the presence of crude oils. The same is true of amphoteric surfactants. Nonionic surfactants foam well initially but the life of these foams in the presence of acids and crude oils remains too short for acidizing operations. Improved foams can be made using a nonionic primary surfactant (e.g., ethoxylated alcohols or polyglycosides) together with a cationic cosurfactant (e.g., fluorinated quaternary ammonium chloride). Polymers (e.g., polysaccharides or partially hydrolyzed polyacrylamide) may be added to the surfactant solution to enhance the foam mobility reduction.[130] Another class of foam-forming surfactants are the ethoxylated fatty amines, which become quaternized in situ by the acid.[131,132] Guidelines for the use of foamed acids have been published.[133,134] High-temperature foam acid stimulation studies have been carried out with a number of surfactants.[135]

The foams must have several properties. Foams must be stable in the presence of the acids and the reservoir hydrocarbons. The foam must be capable of producing a stiff foam that is substantially less mobile (e.g., at least 100 times) than gas to block the flow of the acid. Lastly, the foam must maintain its flow-blocking capabilities (i.e., stiffness) during the injection period of the acid, after which, the foam should inherently break down to once again allow flow through the higher-permeable zones.

Foamed fluids have been shown to be able to block a formation not just by their viscosity but also by the mechanism of breaking and reforming under dynamic flow conditions. Furthermore, foamed fluids will block a formation more effectively the greater the bubble size in the foam relative to the pore size. When there is stratification (layers of varying permeability), diversion is achieved by generating and maintaining a stable foam in the higher permeability zone or zones during the entire treatment.

Acid Stimulation

When there is a long zone to be treated, diversion is achieved by treating part of the zone with acid, then placing a foam to block entry of subsequently injected acid into that part of the zone, and then injecting more acid. These alternating steps may be repeated. The result is complete zonal coverage by the treating fluid and effective damage removal by the acid, even from severely damaged zones. Depending upon the type and concentration of the surfactants used and the foam quality, foams can generate different levels of yield stress. Foamed fluids have also been known to support solid particles and enhance the stability and viscous flow behavior of fluids. Foamed fluids have also been recognized as one of the best diversion fluids for acid stimulation. Other benefits of foamed fluids are that they are inherently cleaner than nonfoamed fluids, even if they contain polymers, because they contain less liquid and that they help kick off flowback and cleanup because they provide energy to the system to help overcome resistance, for example, the hydrostatic head, to flow back; that they are "energized" is particularly important in depleted reservoirs.

A combination of a foamed acid and viscoelastic acid system has been proposed (see next section). The resistance to flow of the gelled foamed VES system is greater than expected from a foam or viscoelastic gel system alone. Betaine surfactants[129] or alkylamidoamine oxide surfactants are used to make the viscoelastic foams.[136]

5.7.4 VISCOELASTIC SURFACTANTS

Another type of viscous fluid diverting agent used to assist in formation stimulation (both fracturing and matrix acidizing) is a VES or surfactant mixture. These are not the same surfactants that are used to make foams.[127] Viscoelasticity, also known as anelasticity, describes materials that exhibit both viscous and elastic characteristics. The rheology of VES fluids, in particular the increase in viscosity of the solution, is attributed to the three-dimensional structure formed by the components in the fluids. When the surfactant concentration significantly exceeds a critical level, the surfactant molecules aggregate and form structures such as micelles or vesicles that can interact to form a network exhibiting viscoelastic behavior.[137] Although generally more expensive than polymer-gelled acids, modern VES acid stimulation has the benefit that it usually leaves little or no residue (formation damage) after treatment.[138] Surfactants as gelling agents require fairly high concentrations to create the necessary viscosity, higher than the concentrations needed to make polymeric gels.

VES solutions are usually formed by the addition of certain reagents to concentrated solutions of surfactants, frequently consisting of long-chain amphoteric or quaternary ammonium salts such as cetyltrimethylammonium bromide (CTAB). Many common reagents or cosurfactants can be added to generate extra viscoelasticity and stability to the surfactant solutions depending on their ionicity. Salts such as ammonium chloride, potassium chloride, magnesium chloride, sodium salicylate, and sodium isocyanate and nonionic organic molecules such as chloroform can be used. Certain cationic surfactant/anionic surfactant blends with a nonaqueous solvent also form viscoelastic solutions. Salicylic acid or phthalic acid can be used with amphoterics. The electrolyte content of many surfactant solutions is also an important control on their viscoelastic behavior.[139,140]

Individual surfactants that have been shown to form viscoelastic solutions come from many classes. Some of those that have been proposed for use in the oilfield include:

- Cationic surfactants[141]
  - Examples are erucyl methyl bis(2-hydroxyethyl) ammonium chloride or 1 erucamidopropyl-1,1,1-trimethyl ammonium chloride (Figure 5.11)[142]
  - Erucyl methyl bis(2-hydroxyethyl) ammonium chloride with sodium salicylate (0.06 wt.%) or erucyl amine cosurfactants give improved performance being stable over a wide pH range.[144]
  - Gemini and nongemini bis-quaternary and other polycationic surfactants.[144]
- Zwitterionic/amphoteric surfactants
  - Examples are betaine surfactants such as oleylamidopropyl betaine or erucylamidopropyl betaine (Figure 5.12).[145,146]
- Anionic surfactants
  - Alkyl taurate anionic surfactants[147]
  - Methyl ester sulfonates[148]
  - Sulfosuccinates
- Amine oxides and amidoamine oxides
  - Dimethylaminopropyltallowamide oxide (Figure 5.13).[143,149,150]
- Ethoxylated fatty amines

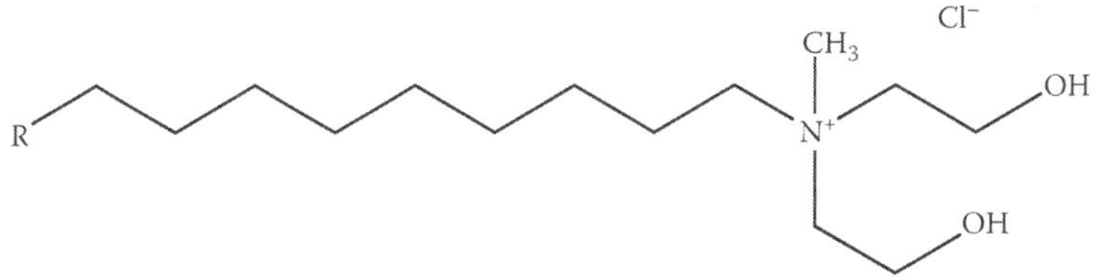

FIGURE 5.11 Long-chain alkyl methyl bis(2-hydroxyethyl) ammonium chloride surfactants.

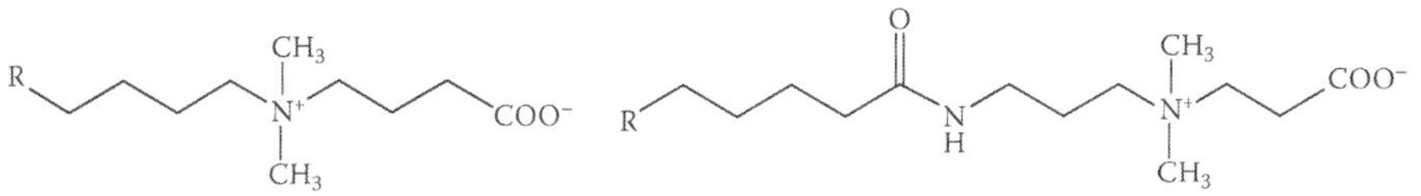

FIGURE 5.12 Zwitterionic/amphoteric surfactants.

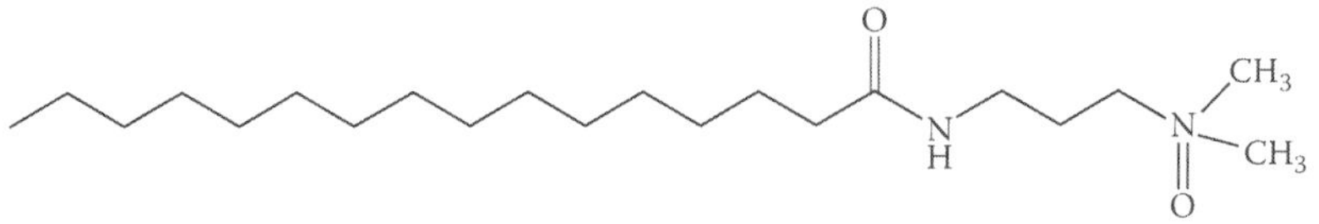

FIGURE 5.13 Dimethylaminopropyltallowamide oxide.

There are two types of application for acid diversion with surfactant micellar fluids. They are:

- self-diverting surfactant fluid
- surfactant micellar fluids as diversion pills

A self-diverting fluid is composed of fresh acid and surfactant.[151] The initial fresh acid system has low, waterlike viscosity during mixing and pumping. When the fresh acid fluid system contacts and reacts with carbonate in the formation, the fresh acid system loses its acidity, the fluid pH increases, and due to the presence of divalent ions like Ca^{2+} and/or Mg^{2+}, the fluid starts to become viscous (or gelled). As the local viscosity increases, the fluid system then effectively diverts the trailing acid fluid. The following fresh acid will be diverted to other areas to stimulate the formation. This is a continuous diverting process during acid injection.

The surfactant micellar diversion pill used in acidizing treatments is composed of surfactant and brine and is viscous (or gelled) when pumped. The diversion pills are pumped with fresh acid in different stages. During a pumping treatment, spacers are usually pumped to separate the diversion pills and acid fluid. The surfactant diversion pill is pH-sensitive, as in the self-diverting surfactant acid system, and the pills lose viscosity at low pH. Erucyl methyl bis(2-hydroxyethyl) ammonium chloride, a commercial VES used in surfactant micellar diversion pills, does not require a large counterion to form viscoelastic micelles. It can also be used at a concentration too low to viscosify the fluid, but that is concentrated in the formation so that the fluid system gels.[143] Cationic and amphoteric/zwitterionic VESs, in particular, those comprising a betaine moiety (such as $R(Me)_2N^+CH_2CH_2CH_2COO^-$), are useful at temperatures up to about 160°C (320°F). They are, therefore, of particular interest for medium- to high-temperature wells. However, like the cationic VESs mentioned above, they are not compatible with high brine concentration. Cosurfactants useful in extending the brine tolerance, increasing the gel strength, and reducing the shear sensitivity of the betaine VES-fluid include sodium dodecylbenzene sulfonate.[152] Other suitable cosurfactants for betaine VES are certain chelating agents such as trisodium hydroxyethylethylenediamine triacetate.[153]

The wormlike micelles formed by many VESs are sensitive to hydrocarbons. By utilizing this sensitivity, the fluid may selectively block water-bearing zones while the hydrocarbon-bearing zone is unaffected. Anionic surfactants have been reported to have this effect.[154] However, some VES fluids such as betaines cannot discriminate between zones with various permeabilities as long as the zones are hydrocarbon-bearing. Further, unlike polymer-based fluids, which rely upon filter cake deposition to control leak-off to the formation, VES diverting agents control fluid leak-off into the formation through the structure size of the micelles. The micellar-based viscoelastic betaine surfactant fluids usually have high leak-off rates to the formation due to the small size of the worm-like micelles.

Betaine surfactants have been shown to be applicable for self-diverting acid systems (VDAs or SDVAs) at formation temperatures up to 160°C (320°F).[155,156] Zwitterionic amine oxide surfactants can be used up to about 125°C (257°F).[157] The viscosity increase is due to a drop in pH and an increase in the aqueous concentration of Group II metal cations, leading transformation of spherical micelles to wormlike micelles. The viscous barrier forces the following acid into other zones that have lower injectivity so that those zones can also be effectively stimulated. After the acidizing treatment, the viscous barrier breaks down upon contacting produced hydrocarbon, lower salinity produced water, produced preflush fluids, or postflush fluids, leaving no solid residue to cause formation damage to the rock, because the VES system contains no solids or polymer. VDA zwitterionic diverters have also been used in carbonate gas wells. They are stable in 28% HCl and to reaction products after acid is spent after reaction with carbonates.[158] A review of lessons learned from matrix acidizing with VES diverters states that the surfactant loading must not be too high or else formation damage may occur. However, the damage can be removed using mutual solvent. Further, high corrosion inhibitor or iron salt concentrations can adversely affect the VES performance.[157]

To avoid formation damage, the breaking of the gel of the aqueous viscoelastic treating fluid can be accomplished by several mechanisms including contact with hydrocarbons in the formation, change in pH, dilution/change in salinity, contact with alkoxylated alcohol solvents, and contact with a reactive agent. Relying on the formation of hydrocarbons or aqueous dilution, especially for dry gas reservoirs, to break down the viscosity of the gel will extend the treatment time and delay production. Many operators that have experience with VES have learned that remedial VES cleanup often is necessary. Thus, for more assured and complete fluid cleanup, chemical breakers have been developed that destroy the micellar structure of some VES fluids used as diverters.[159] For example, compounds that have charges opposite to the VESs head group can act as breakers by destroying the micellar structure of the VES fluid. These were designed particularly for cleanup after drilling and hydraulic fracturing operations.[160] Other breakers include chemicals that degrade the surfactants at elevated temperature or compounds that degrade at elevated temperature to generate a micelle breaker in situ.[161] Internal breakers include bacteria, transition metal ion sources, saponified fatty acids, mineral oils, hydrogenated polyalphaolefin oils, saturated fatty acids, unsaturated fatty acids such as polyenoic acids and monoenoic acids and combinations thereof.[162,163] A VES fluid containing a metal ion having at least two oxidation states and a redox reagent has been proposed. The metal ions, such as iron(II) ions may be encapsulated. They are released in the formation after the acid has done its job and are oxidized to iron(III) ions, which reduce the gel viscosity.[164] Break-enabling ammonium salts can also be used in the self-diverting acid main flush.[165] However, iron(III) ions have also been shown to form strong gels with several VESs, leading to a new form of formation damage.[166] Internal breakers for carbonate self-diverting surfactant/organic acids have also been developed.[151]

There have been a number of variations or improvements on the basic use of VESs either as leak-off control or diverting agents. For example, the benefits of both foam and VES acid treatments can be obtained by combining both properties in one solution.[167] Another variation is the use of certain anionic surfactant VES fluid systems mixed with a small percentage of a lower alcohol such as methanol. The surfactant concentration as injected is insufficient to divert fluid flow in the formation. The fluid only develops the ability to divert fluids as it flows through the formation via loss of the lower alcohol from solution.[168]

Acid Stimulation

The use of a viscoelastic amidoamine oxide surfactant in HF-acidizing compositions has been shown to avoid deconsolidation problems in the near-wellbore area of sandstone reservoirs. An example of such a surfactant is shown in Figure 5.13.[169,170] This self-diverting surfactant solution viscosifies as the acid spends and then decreases in viscosity as the acid spends even more.

An improved VES diverting pill solution is to use a viscoelastic amphoteric surfactant, such as lecithin, with a quaternary amine polyelectrolyte capable of reacting with the amphoteric surfactant to increase the viscosity, and a nonaqueous solvent.[171,172] The proposed polymer is a polyquaternary derivative of HEC. This has an advantage over many other micellar VES fluids because this mixture forms aggregated vesicles in solution, which are much larger than micelles. Thus, fluid leak-off is better controlled. Introducing a polyelectrolyte also reduced the surfactant loading and stabilizes the solution up to about 170°C (338°F).[137] The fluids break at low pH or the fluids can contain internal breakers such as persulfate salts. Other polymer/surfactant blends have been proposed to have superior viscoelastic properties than the surfactants alone.[142]

There are some surfactant/cosurfactant systems that are thermally sensitive, increasing in viscosity with temperature. The simplest system reported is CTAB with 5-methyl salicylic acid.[173] It does not appear that these systems have been exploited for use in oilfield chemicals.

A surfactant-based gelling acid stimulation package has been used to stimulate high–water-cut heavy oil wells.[174] The treatment temporarily plugs water zones and effectively stimulates oil zones with chemical diversion. In addition, the acid system is claimed to have a unique inherent property to limit acid penetration in high–water-saturation zones, while enhancing deeper penetration in high–oil-saturation layers.

5.8 RADIAL PLACEMENT OF ACIDIZING TREATMENTS

One problem encountered during acidizing of all formations, especially high-temperature formations, is that the acid is rapidly consumed by the reactive material immediately adjacent to the borehole before the acid can penetrate any significant distance into the formation. A hypothetical system that is so highly retarded that no reaction takes place while the acid is pumped into the reservoir has been proposed as the ultimate for a matrix treatment.[175]

"Good penetration" in matrix acidizing is usually considered to be a distance of a few meters, and in fracture-acidizing operations, a distance of 100–200 m. But there are several other ways to improve the penetration using so-called retarded acids:

- oil-wetting surfactants
- weak organic acids for carbonate acidizing
- weak sandstone acidizing fluorinated agents
- buffered organic acids
- gelled or viscous acids
- foamed acids
- temperature-sensitive acid-generating chemicals and enzymes
- emulsified acids

5.8.1 Oil-Wetting Surfactants

Oil-wetting surfactants coat the pore surfaces slowing the rate of attack of acid. Some corrosion inhibitors and surfactant antisludging additives, discussed in Section 5.6, can also have this function. These systems are simple and applicable even in high-temperature wells.[176,177] Emulsified acids are also oil-wetting systems (see Section 5.8.9).

5.8.2 Weak Organic Acids

Weaker but more expensive acids than HCl such as acetic acid, formic acid, citric acid, ethylenediaminetetraacetic acid, glycolic acid, and tartaric or sulfamic acid can be used in carbonate acidizing to retard the acidizing process, especially in high-temperature wells where HCl may be spent very quickly. They are also less corrosive than inorganic acids. Blends of organic and inorganic acids can also be used. One study in high-temperature wells showed that organic acids did not show enough retardation, as only part of the interval was treated. Therefore, other retardation methods must be used in addition. For example, besides the methods discussed in this chapter, coated solid organic acid particles can be used, but the ion of the acid may not always be compatible with cations dissolved from the formation, e.g., Ca^{2+} with citrate ions in carbonate acidizing.[205] HF, citric acid, and a phosphonate have been proposed for use in sandstone acidizing.[178] A review of the advantages of the most commonly used organic acids, formic and acetic acid, has been published.[4]

5.8.3 Weak Sandstone-Acidizing Fluorinated Agents

Deeper acid stimulation treatments for sandstones will avoid a rapid decline in production by stabilizing fines and precipitation of acidization products near the well bore. Weaker sandstone-acidizing fluorinated agents have been proposed such as clay acid, HBF_4, mixed with chelating agents to chelate aluminum species.[179] HBF_4 is generated by reacting boric acid (H_3BO_3) with HF. Reaction of HBF_4 with sandstone is slower than with HF allowing the acid to penetrate deeper into the formation and avoid near-wellbore damage. Another system uses aluminum chloride that reacts with HF to generate aluminum fluoride species with lower acidizing rates than free HF. Ammonium bifluoride with a pentaphosphonic acid has also been proposed as a retarded sandstone acid with lower corrosivity than normal mud acids.[180] By slowing down the kinetics, one can use these retarded acids for high-temperature applications. A review of sandstone-retarded acid systems has been published.[181]

5.8.4 Buffered Acids

It has been proposed to use buffered acid solutions of pH 1.9–4 containing organic acids mixed with salts of organic acids and a phosphonate for sandstone acidizing. The main acid flush has a similar pH and works as a retarded acid giving deeper formation penetration.[15] High pH-buffered systems have been used successfully in single stage sandstone acidizing treatments, eliminating the need for preflushes and overflushes.[16] For carbonate formations, the use of aminopolycarboxylic anionic

Acid Stimulation

species gives a slower reaction than HCl when the pH is in the preferred range of about 4–9.5.[182]

5.8.5 Gelled or Viscous Acids

Polymer gels or VESs described earlier in this chapter can be used for better acid penetration besides their use as diverters or self-diverters.[120] Thus, their chemistry will not be repeated here. Such agents serve to thicken the acid solution and thus increase the viscosity thereof. These work as retarded acid solutions allowing deeper penetration of the acid. Higher viscosities are also advantageous in carbonate-fracturing acidizing operations in that the more viscous acidic solutions produce wider and longer fractures. Once penetration has been attained, the surfactant or polymer stability must be sufficient to permit the contact of the acidizing solution with the formation for a period sufficient for the acid in the composition to significantly react with the acid-soluble components of the formation and stimulate the production of fluids by creating new passageways or enlarging existing passageways through the formation. Gelled acids have about one-tenth of the diffusion rate of ungelled acids.

In one laboratory study, polyacrylamide used with HCl, formic, or acetic acids did not show any retardation whereas addition of xanthan to the acids did show acidizing retardation.[183]

A novel temperature-controlled gelled retarded acid has been developed for carbonate fracture acidizing based on using a specific polyquaternary polymer. This system is viscous during injection and penetration but loses its viscosity as it reaches the maximum reservoir temperature.[184]

5.8.6 Foamed Acids

Besides their use as diverters, foamed acid treatments can be used as retarded acids, particularly in fracture acidizing (see Section 5.7.3 on foam diverters above). Many types of surfactants can be used to create foams, the best category are the nonionic surfactants such as ethoxylated alcohols. Fluorinated quaternary ammonium salts can help stabilize foams as can addition of polymers.[185]

5.8.7 Temperature-Sensitive Acid-Generating Chemicals and Enzymes

One retardation approach for carbonate acidizing has involved the use of a treatment fluid comprising an aqueous solution of an ester of acetic acid and, optionally, an enzyme that may facilitate cleaving the ester and releasing acetic acid so as to acidize the subterranean formation.[186–188] Enzymes such as a hydrolase, a lipase, or an esterase have also been used to generate acetic acid from esters such as methyl acetate, ethyl acetate, methyl formate, or 1,2-ethanediol diacetate. Efficient delivery of the enzyme-generated acid to the open-hole section of a long horizontal well gave superior cleanup than HCl treatments.[189] However, the use of such treatment fluids may be problematic for reasons such as the price of the enzyme, the cost and difficulty in providing storage for the enzyme (particularly in tropical and desert locations), the formation temperature, and that only a fraction of the theoretical acid is generated.

Ester-based systems that hydrolyze downhole without the need for enzymes can also be used.[206] An example is a blend of 1,2,3-propanetriol diacetate and ammonium or potassium formate.

Formate esters that hydrolyze in aqueous media at reservoir temperatures to generate formic acid, a stronger acid than acetic acid, have been proposed as superior acid-generating products. For example, formate esters such as ethylene glycol monoformate or ethylene glycol diformate can be used to delay the formation of formic acid in carbonate acidizing allowing for greater penetration. This method also prevents corrosion problems during injection.[190]

Another acid-generating system for acid fracturing of high-temperature carbonate reservoirs is to use a solid polylactide and/or polyglycolic acid. These decompose in water at raised temperatures to give lactic and glycolic acids. The solid acid precursor may be mixed with a solid acid-reactive material to accelerate the hydrolysis and/or coated to slow the hydrolysis. Water-soluble liquid compounds are also given that accelerate the hydrolysis. The method ensures that the acid contacts fracture faces far from the wellbore.[191,192]

5.8.8 EMULSIFIED ACIDS

Emulsified acids have been around as long as gelled acids as retarded acids to increase penetration. They can be used in fracture or matrix acidizing.[193,194] The emulsion is usually acid in oil and is created by using an oil-soluble surfactant with a low HLB value (3–6). The emulsion is often designed to break at the reservoir temperature, releasing the aqueous acid phase. As emulsions have higher viscosities than the aqueous acid, they also have diverting properties.

Various surfactants have been proposed for making the emulsions, varying from cationic, anionic, and nonionic types. One early, commercially successful emulsified acid is based on a saturated hydrocarbon sulfonate and an alkylarylsulfonic acid or water-soluble salt thereof.[195] Another system uses a long-chain amine, which becomes a quaternized, emulsifying surfactant only when blended with HCl.[196] Blends have also been proposed such as a fatty amine and the diethanolamide of oleic acid.[197] The viscosity of emulsified acids also gives them diverting properties besides being retarded acids. Acid-internal emulsions also have the advantage of avoiding corrosion problems during injection.

Acid stimulation emulsions can also be used to dissolve asphaltene deposits, another cause of formation damage, if the organic solvent phase is made up of asphaltene solvents such as xylenes. Alkyl or alkenyl esters of certain aromatic carboxylic acids, preferably isopropyl benzoate, can be used in the emulsion to give the product a more attractive toxicological and environmental profile.[198]

Emulsified acids can be formulated in such a way that the emulsion breaks when the acids spend.[199] In this way, excessively high treating pressures, caused by flow of high viscosity emulsions in the formation, can be avoided. Emulsified acids with 14–19 times the retardation of normal carbonate HCl-based acids have been reported. Emulsified acids are therefore also useful for high-temperature applications.[200,201] Emulsified 15% HCl was found to give good penetration in low-permeability dolomite wells.[202]

Microemulsions of acids have been claimed as improved retarded acids compared with normal macroemulsions. Microemulsions are here meant thermodynamically stable, isotropic solutions. An example of a surfactant that makes an acid-in-oil microemulsion is made by reacting 2 mol of a 2-ethylhexyl epoxide with 1 mol of ethanolamine and then quaternizing the product.[203]

Certain acid-external emulsions have also been shown to work as retarded acids. Mixtures of aqueous HCl, formic/acetic acids, a hydrocarbon, and an organic diphosphonate amine surfactant were shown to give long reaction times with carbonate cores.[204]

REFERENCES

1. B. B. Williams, J. L. Gidley, and R. S. Schecter, "Acidizing Fundamentals," *Monograph Series*, vol. 6, Dallas: Society of Petroleum Engineers, 1979.
2. L. Kalfayan, *Production Enhancement with Acid Stimulation*, 2nd ed., Tulsa, OK: PennWell Corporation, 2008.
3. P. Rae and G. di Lullo, "Matrix Acid Stimulation—A Review of the State-of-the-Art," SPE 82260 (paper presented at the SPE European Formation Damage Conference, The Hague, Netherlands, 13–14 May 2003).
4. M. Buijse, P. de Boer, M. Klos, and G. Burgos, "Organic Acids in Carbonate Acidizing," SPE 82211 (paper presented at the SPE European Formation Damage Conference, The Hague, Netherlands, 13–14 May 2003).
5. (*a*) W. W. Frenier, C. N. Fredd, and F. Chang, "Hydroxyaminocarboxylic Acids Produce Superior Formulations for Matrix Stimulation of Carbonates at High Temperatures," SPE 71696 (paper presented at the SPE Annual Technical Conference and Exhibition, New Orleans, LA, 30 September–3 October 2001). (*b*) T. Huang, P. M. McElfresh, and A. Gabrysch, "Acid Removal of Scale and Fines at High Temperatures," SPE 74678 (paper presented at the SPE Oilfield Scale Symposium, Aberdeen, UK, 30–31 January 2002).
6. W. W. Frenier, C. N. Fredd, and F. Chang, "Hydroxyaminocarboxylic Acids Produce Superior Formulations for Matrix Stimulation of Carbonates," SPE 68924 (paper presented at the SPE European Formation Damage Conference, The Hague, Netherlands, 21–22 May 2001).
7. A. Husen A. Ali, W. W. Frenier, Z. Xiao, and M. Ziauddin, "Chelating Agent-Based Fluids for Optimal Stimulation of High-Temperature Wells," SPE 77366 (paper presented at the SPE Annual Technical Conference and Exhibition, San Antonio, TX, 29 September–2 October 2002).
8. C. N. Fredd and H. S. Fogler, "The Influence of Transport and Reaction on Wormhole Formation in Porous Media," *Journal of American Institute of Chemical Engineers* 44 (1998): 1933.
9. C. N. Fredd and H. S. Fogler, *Journal of American Institute of Chemical Engineers* 44 (1998): 1949.
10. T. Huang, P. M. McElfresh, and A. D. Gabrysch, "Carbonate Matrix Acidizing Fluids at High Temperatures: Acetic acid, Chelating Agents, or Long-Chained Carboxylic Acids?," SPE 82268 (paper presented at the SPE European Formation Damage Conference held in The Hague, 13–14 May 2003).
11. H. O. McLeod, "Matrix Acidizing," *Journal of Petroleum Technology* (1984): 2055.

12. G. R. Coulter and A. R. Jennings, "A Contemporary Approach to Matrix Acidizing," SPE 38594 (paper presented at the SPE Annual Technical Conference and Exhibition, San Antonio, TX, 5–8 October 1997).
13. L. J. Kalfayan and D. R. Watkins, "A New Method for Stabilizing Fines and Controlling Dissolution during Sandstone Acidizing," SPE 20076 (paper presented at the SPE California Regional Meeting, Ventura, CA, 4–6 April 1990).
15. M. N. Al-Dahlan, H. A. Nasr-El-Din, and A. A. Al-Qahtani, "Evaluation of Retarded HF Acid Systems," SPE 65032 (paper presented at the SPE International Symposium on Oilfield Chemistry, Houston, TX, 13–16 February 2001).
15. P. J. Rae, G. Di. L. Arias, A. B. Ahmad, and L. J. Kalfayan, U.S. Patent Application 20050016731.
16. P. Rae and G. Di Lullo, "Single Step Matrix Acidizing with HF—Eliminating Preflushes Simplifies the Process, Improves the Results," SPE 107296 (paper presented at the European Formation Damage Conference, Scheveningen, The Netherlands, 30 May–1 June 2007).
17. W. Frenier, M. Ziauddin, S. Davies, and F. Chang, U.S. Patent 7192908, 2007.
18. F. E. Tuedor, Z. Xiao, M. J. Fuller, D. Fu, G. Salamat, S. N. Davies, and B. Lecerf, SPE 98314 (paper presented at the SPE International Symposium and Exhibition on Formation Damage, Lafayette, LA, 15–17 February 2006).
19. H. A. Nasr-El-Din, S. Kelkar, and M. M. Samuel, "Investigation of a New Single Stage Sandstone Acidizing Fluid for High Temperature Formations," SPE 107636 (paper presented at the European Formation Damage Conference, Scheveningen, The Netherlands, 30 May–1 June 2007).
20. A. N. Martin, "Stimulating Sandstone Formations with Non-HF Treatment Systems," SPE 90774 (paper presented at the SPE Annual Technical Conference and Exhibition, Houston, TX, 25–29 September 2004).
21. M. A. Aggour, M. Al-Muhareb, S. A. Abu-Khamsin, and A. A. Al-Majed, "Improving Sandstone Matrix Simulation of Oil Wells by Gas Preconditioning," *Petroleum Science and Technology* 20(3) (2002): 425.
22. D. A. Williams, J. R. Looney, D. S. Sullivan, B. I. Bourland, J. A. Haslegrave, P. J. Clewlow, N. Carruthers, and T. M. O'Brien, U.S. Patent 5322630, 1994.
23. J. Hall and S. Almond, "OSPAR Regulators Drive Design: How Regulations Have Impacted a Service Company's Product Development Program," Proceedings of the Chemistry in the Oil Industry IX, Manchester, UK: Royal Society of Chemistry, 2005, 61.
24. (*a*) W. W. Frenier, "Review of Green Chemistry Corrosion Inhibitors for Aqueous Systems (paper presented at the Proceedings of the 9th European Symposium on Corrosion Inhibitors, University of Ferrara, Ferrara, Italy, September 2000). (*b*) D. G. Hill and H. Romijn, "Reduction of Risk to the Marine Environment from Oilfield Chemicals: Environmentally Improved Acid Corrosion Inhibition for Well Stimulation," Paper 00342 (paper presented at the NACE CORROSION Conference, 2000). (*c*) A. Rostami, H. Nasr-El-Din, Review and Evaluation of Corrosion Inhibitors Used in Well Stimulation, SPE121726, SPE International Symposium on Oilfield Chemistry, The Woodlands, TX, April 20–22, 2009.
25. W. Frenier and M. Ziauddin, Formation, Removal, and Inhibition of Scale in the Oilfield Environment, Eds. N. Wolf and R. L. Hartman, Society of Petroleum Engineers, 2008.
26. (*a*) A. J. Saukaitis and G. S. Gardner, U.S. Patent 2758970, 1956. (*b*) R. C. Mansfield, J. G. Morrison, and C. J. Schmidle, U.S. Patent 2874119, 1959.
27. M. N. Desai, M. B. Desai, C. B. Shah, and S. M. Desai, "Schiff Bases as Corrosion Inhibitors for Mild Steel in Hydrochloric Acid Solutions," *Corrosion Science* 26 (1986): 827.

28. D. G. Hill, K. Dismuke, W. Shepherd, I. Witt, H. Romijn, W. Frenier, and M. Parris, Development Practices and Achievements for Reducing the Risk of Oilfield Chemicals, SPE 80593, SPE/EPA/DOE Exploration and Production Environmental Conf, 10–12 March 2003, San Antonio, TX.
29. M. Vorderbruggen and H. Kaarigstad, "Meeting the Environmental Challenge: A New Acid Corrosion Inhibitor for the Norwegian Sector of the North Sea," SPE 102908 (paper presented at the SPE Annual Technical Conference and Exhibition, San Antonio, TX, 24–27 September 2006).
30. W. W. Frenier and A. Iob, Paper 150 (paper presented at the NACE CORROSION Conference, 1988).
31. D. A. Williams, L. A. McDougall, and J. R. Looney, U.S. Patent 5543388, 1996.
32. E. P. da Motta, M. H. V. Quiroga, A. F. L. Aragao, and A. Pereira, "Acidizing Gas Wells in the Merluza Field Using an Acetic/Formic Acid Mixture and Foam Pigs," SPE 39424 (paper presented at the SPE International Symposium on Formation Damage Control, Lafayette, LA, 18–19 February 1998).
33. M. S. Van Domelen and A. R. Jennings Jr., "Alternate Acid Blends for HPHT Applications," SPE 30419, SPE Offshore Europe Conference, Aberdeen, UK, 5–8 September 1995.
34. V. Sharma, M. Borse, S. Jauhan, K. B. Pai, and S. Devi, *Surfarctants and Detergents* 42 (2005): 163.
35. (*a*) G. Schmitt and K. Bedbur, Wekst. Koros., 1985, 38, 575. (*b*) W. W. Frenier, U.S. Patent 5096618, 1992.
36. (*a*) O. O. Adeyemi and S. O. Oluwafemi, "2-Aminopyridine as an Effective Inhibitor for the Corrosion of Mild Steel in Acidic Solutions," *Bulletin of Electrochemistry* 22 (2006): 317. (*b*) S. N. Hettiarachchi, C. Subhash, and D. D. Macdonald, International Patent Application WO/1990/001478.
37. M. M. Brezinski, U.S. Patent 5763368, 1998.
38. M. M. Brezinski, U.S. Patent 5976416, 1999.
39. M. L. Walker, U.S. Patent 5441929, 1995.
40. M. M. Brezinski and B. Desai, U.S. Patent 5697443, 1997.
41. D. A. Williams, P. K. Holifield, J. R. Looney, and L. A. McDougall, U.S. Patent 5200096, 1993.
42. J. M. Cassidy, J. L. Lane, and C. E. Kiser, International Patent Application WO/2007/141524.
43. M. M. Brezinski, U.S. Patent 5756004, 1998.
44. (*a*) P. Manivel and G. Venkatachari, "The Inhibitive Effect of Poly(*p*-Toluidine) on Corrosion of Iron in 1M HCl Solutions," *Journal of Applied Polymer Science* 104 (2007): 2595. (*b*) B. Gao, X. Zhang, and Y. Sheng, "Studies on Preparing and Corrosion Inhibition Behavior of Quaternized Polyethyleneimine for Low Carbon Steel in Sulfuric Acid," *Materials Chemistry and Physics* 108 (2008): 375.
45. K. Overkempe, W. J. E. Parr, and J. C. Speelman, World Patent Application WO/2003/054251.
46. A. A. AlTaq, S. A. Ali, and H. A. Nasr-El-Din, "Inhibition Performance of a New Series of Mono/Diamine-Based Corrosion Inhibitors for HCl Acid Solutions," SPE 114087 (paper presented at the SPE 9th International Conference on Oilfield Scale, Aberdeen, UK, 28–29 May 2008).
47. R. L. Hoppe, R. L. Martin, M. K. Pakulski, and T. D. Schaffer, U.S. Patent Application 20070261853.
48. J. M. Cassidy, J. L. Lane, K. A. Frost, and C. E. Kiser, International Patent Application WO/2007/034155.
49. D. S. Sullivan III, C. E. Strubelt, and K. W. Becker, U.S. Patent 4039336, 1977.

50. G. P. Funkhouser, J. M. Cassidy, J. L. Lane, K. Frost, T. R. Gardner, and K. L. King, U.S. Patent 6192987, 2001.
51. A. Cizek and A. Hackerott, International Patent Application WO/01/79590.
52. W. W. Frenier, Paper 96154 (paper presented at the 51st NACE International Corrosion Forum, Denver, CO, March 1996).
53. W. Frenier and F. Growcock, U.S. Patent 4734259, 1988.
54. (a) M. A. Vorderbruggen and D. A. Williams, U.S. Patent 6117364, 2000. (b) J. M. Cassidy, C. E. Kiser, J. L. Lane, and K. A. Frost, U.S. Patent Application 20070071887, 2007.
55. J. M. Cassidy, C. E. Kiser, and J. L. Lane, U.S. Patent Application 20080139414.
56. W. W. Frenier and D. G. Hill, U.S. Patent 6399547, 2002.
57. T. D. Welton and J. M. Cassidy, World Patent Application, WO/2005/075707.
58. A. Punet Plensa and L. Lozano Salvatella, World Patent Application WO/2006/136262.
59. J. M. Cassidy and K. A. Frost, U.S. Patent Application 20050123437.
60. (a) G. A. Scherubel, R. Reid, A. L. Fauke, and K. Schwartz, U.S. Patent 5854180, 1998. (b) A. Cizek, International Patent Application WO/2002/103081.
61. J. M. Cassidy, J. L. Lane, and C. E. Kiser, U.S. Patent 7163056, 2007.
62. (a) M. A. Quraishi and R. Sardar, "Aromatic Triazoles as Corrosion Inhibitors for Mild Steel in Acidic Environments," NACE CORROSION Conference, September 2002, 748. (b) M. A. Quraishi, Paper 04421 (paper presented at the NACE CORROSION Conference, 2004).
63. K. C. Pilai and R. Narayan, *Corrosion Sci.*, 1983, 23, 151.
64. W. W. Frenier, D. G. Hill, F. B. Growcock, and V. R. Lopp, "α-Alkenylphenones, A New Class of Corrosion Inhibitors, Provide Improved Inhibition in Strong HCl" (paper presented at the RSC 3rd Chemicals in the Oil Industry Symposium, 19–29 April 1988).
65. R. J. Jasinski and W. W. Frenier, U.S. Patent 5120371, 1992.
66. M. D. Coffey, M. Y. Kelly, and W. C. Kennedy Jr., U.S. Patent 4493775, 1985.
67. M. L. Walker, U.S. Patent 5591381, 1997.
68. D. R. McCormick and J.P Bershas, U.S. Patent Application 20070018135, 2007.
69. (a) A. Penna, G. F. Di Lullo Arias, and P. J. Rae, U.S. Patent Application 20060264335, 2006. (b) T. D. Welton and J. M. Cassidy, U.S. Patent Application 20070010404, 2007.
70. (a) J. D. Anderson, E. S. Hayman Jr., and E. A. Rodzewich, U.S. Patent 3992313, 1976. (b) J. D. Nichols, R. Derby, G. T. Von dem Bussche, and D. A. Hannum, U.S. Patent 4557838, 1985.
71. F. Bentiss, M. Traisnel, H. Vezin, H. F. Hildebrand, and M. Lagrenee, "2,5-Bis(4-Dimethylaminophenyl)-1,3,4-Oxadiazole and 2,5-Bis(4-Dimethylaminophenyl)-1,3,4-Thiadiazole as Corrosion Inhibitors for Mild Steel in Acidic Media," *Corrosion Science* 46 (2004): 2781.
72. M. Ajmal, M. A. W. Khan, S. Ahmad, and M. A. Quraishi, Paper 217 (paper presented at the NACE CORROSION, 1996).
73. M. A. Quaraishi and D. Jamal, "CAHMT: A New Eco-Friendly Acidizing Corrosion Inhibitor," *Corrosion* 56(10) (2000): 983.
74. H.-N. Lin, R. D. Martin, J. M. Brown, and G. F. Brock, U.S. Patent 6132619, 2000.
75. M. Girgis-Ghaly and J. R. Delorey, U.S. Patent 6308778, 2001.
76. J. P. Feraud, H. Perthuis, and P. Dejeux, U.S. Patent 6306799, 2001.
77. M. M. Brezinski, T. R. Gardner, K. L. King, and J. L. Lane Jr., U.S. Patent 6225261, 2001.
78. M. M. Brezinski, U.S. Patent 6415865, 2002.
79. C. Smith, D. Oswald, D. Skibinski, and N. Sylvestre, U.S. Patent Application 20060281636, 2006.

80. M. M. Brezinski, U.S. Patent 6653260, 2002.
81. C. W. Crowe, U.S. Patent 4633949, 1987.
82. W. R. Dill and M. L. Walker, U.S. Patent 4888121, 1989.
83. C. D. Williamson, U.S. Patent 5126059, 1992.
84. I. C. Jacobs and N. E. S. Thompson, U.S. Patent 5112505, 1992.
85. M. M. Brezinski and R. D. Gdanski, U.S. Patent 5264141, 1993.
86. C. Smith, D. Oswald, and M. D. Daffin, U.S. Patent Application 20060289164, 2006.
87. R. E. Himes and E. F. Vinson, U.S. Patent 4842073, 1989.
88. D. R. Watkins, L. J. Kalfayan, and G. S. Hewgill, U.S. Patent 5039434, 1991.
89. F. O. Stanley, S. A. Ali, and J. L. Boles, "Laboratory and Field Evaluation of Organosilane as a Formation Fines Stabilizer," SPE Paper 29530 (paper presented at the SPE Production Operations Symposium, Oklahoma City, OK, April 1995).
90. B. S. Douglass and G. E. King, "A Comparison of Solvent/Acid Workovers in Embar Completions-Little Buffalo Basin Field," SPE 15167 (paper presented at the SPE Rocky Mountain Regional Meeting, Billings, MT, 19–21 May 1986).
91. (*a*) R. J. Dyer, U.S. Patent 5622921, 1997. (*b*) C. Smith and D. Skibinski, U.S. Patent Application 20070062698, 2007.
92. (*a*) A. R. Mokadam, C. E. Strubelt, D. A. Williams, and K. M. Webber, U.S. Patent 5543387, 1996. (*b*) W. G. F. Ford, U.S. Patent 4823874, 1989. (*c*) A. R. Mokadam, U.S. Patent 5797456, 1998.
93. K. M. Barker and M. E. Newberry, "Inhibition and Removal of Low-pH Fluid-Induced Asphaltic Sludge Fouling of Formations in Oil and Gas Wells," SPE 102738 (paper presented at the SPE International Symposium on Oilfield Chemistry, Houston, TX, 28 February–2 March 2007).
94. J. M. Cassidy, C. E. Kiser, and J. L. Lane, U.S. Patent Application 20060201676, 2006.
95. J. M. Cassidy, C. E. Kiser, and J. L. Lane, U.S. Patent Application 20060040831, 2006.
96. D. S. Treybig, D. Williams, and K. T. Chang, International Patent Application WO/2003/053536.
97. I. R. Collins, S. P. Goodwin, J. C. Morgan, and N. J. Stewart, U.S. Patent 6225263, 2001.
98. W. W. Frenier and D. G. Hill, U.S. Patent 6068056, 2000.
99. A. Ahrenst, B. Lungwitz, C. N. Fredd, C. Abad, N. Gurmen, Y. Chen, J. Lassek, P. Howard, W. T. Huey, Z. Azmi, D. Hodgson III, and O. Bustos, U.S. Patent Application 20080064614.
100. H. A. Nasr-El-Din, A. M. Al-Othman, K. C. Taylor, and A. H. Al-Ghamdi, "Surface Tension of Acid Stimulating Fluids at High Temperature," *Journal of Petroleum Science and Engineering* 43 (2004): 57.
101. D. Dalrymple, L. Eoff, B. R. Reddy, and J. Venditto, U.S. Patent 7182136, 2007.
102. (*a*) B. Garcia, E. Soriano, W. Chacon, and L. Eoff, "Novel Acid-Diversion Technique Increases Production in the Cantarell Field, Offshore Mexico," SPE 112413, (paper presented at the SPE International Symposium and Exhibition on Formation Damage Control, Lafayette, LA, 13–15 February 2008). (*b*) B. Garcia, E. Soriano, W. Chacon, and L. Eoff, "Novel Acid Diversion Technique Boosts Production," *World Oil* 229 (2001): 121.
103. G. Paccaloni, "A New, Effective Matrix Stimulation Diversion Technique," *SPE Production and Facilities* 10(3) (1995): 151.
104. C. Abad, J. C. Lee, P. F. Sullivan, E. Nelson, Y. Chen, B. Baser, and L. Lin, U.S. Patent Application 20070032386.
105. G. Glasbergen, B. Todd, M. Van Domelen, and M. Glover, "Design and Field Testing of a Truly Novel Diverting Agent," SPE 102606 (SPE Annual Technical Conference and Exhibition, San Antonio, TX, 24–27 September 2006).

106. J. R. Solares, J. J. Duenas, M. Al-Harbi, A. Al-Sagr, V. Ramanathan, and R. Hellman, "Field Trial of a New Non-Damaging Degradable Fiber-Diverting Agent Achieved Full Zonal Coverage during Acid Fracturing in a Deep Gas Producer in Saudi Arabia," SPE 115525 (paper presented at the SPE Annual Technical Conference and Exhibition, Denver, CO, 21–24 September 2008).
107. N. A. Menzies et al., "Modelling of Gel Diverter Placement in Horizontal Wells," SPE 56742 (paper presented at the SPE Annual Technical Conference and Exhibition, Houston, TX, 3–6 October 1999).
108. J. F. Tate, U.S. Patent 3749169, 1973.
109. (*a*) T. D. Welton, R. W. Pauls, and I. D. Robb, U.S. Patent Application 20060247135. (*b*) T. D. Welton, R. W. Pauls, L. Song, J. E. Bryant, S. R. Beach, and I. D. Robb, International Patent Application WO/2008/096164.
110. (*a*) B. L. Swanson and L. E. Roper; U.S. Patent 4205724, 1980. (*b*) L. D. Burns and G. A. Stahl, U.S. Patent 4690219, 1987.
111. L. R. Norman, M. W. Conway, and J. M. Wilson, "Temperature Stable Acid Gelling Polymers: Laboratories Evaluation and Field Results," SPE 10260 (paper presented at the SPE 56th Annual Fall Technical Conference, San Antonio, TX, 5–7 October 1981).
112. D. J. Poelker, J. McMahon, and D. Harkey, U.S. Patent 6855672, 2005.
113. R. W. Pauls and T. D. Welton, International Patent Application WO/2009/022107.
114. B. L. Swanson, U.S. Patent, 4103742, 1978.
115. W. Abdel Fatah, H. A. Nasr-El-Din, T. Moawad, and A. Elgibaly, "Effects of Crosslinker Type and Additives on the Performance of In-Situ Gelled Acids," SPE 112448 (paper presented at the SPE International Symposium and Exhibition on Formation Damage Control, Lafayette, LA, 13–15 February 2008).
116. C. G. Inks, U.S. Patent 4163727, 1979.
117. E. Clark Jr. and B. L. Swanson, U.S. Patent 4997582, 1991.
118. (*a*) S. Mukherjee and G. Gudney, SPE 25395, *Journal of Petroleum Technology* 45 (1993): 102. (*b*) A. Saxon, B. Chariag, and M. Rahman, SPE 37734 (paper presented at the SPE Middle East Oil Show, Bahrain, 15–18 March 1997).
119. T. D. Welton, International Patent Application WO/2008/102138
120. J. D. Lynn and H. A. Nasr-El-Din, "A Core Based Comparison of the Reaction Characteristics of Emulsified and In-Situ Gelled Acids in Low Permeability, High Temperature, Gas Bearing Carbonates," SPE 65386 (paper presented at the SPE International Symposium on Oilfield Chemistry, Houston, TX, 13–16 February 2001).
121. B. L. Swanson, U.S. Patent 4055502, 1977.
122. R. L. Clampitt and J. E. Hessert, U.S. Patent 4068719, 1978.
123. C. B. Josephson, U.S. Patent 4476033, 1984.
124. J. L. Boles, A. S. Metcalf, and J. Dawson, U.S. Patent 5497830, 1996.
125. D. G. Hill, U.S. Patent Application 20050065041, 2005.
126. R. Gdanski, "Experience and Research Show Best Designs for Foam-Diverting Acidizing," *Oil & Gas Journal* (1993): 85.
127. H. A. Nasr-El-Din, *Surfactant Use in Acid Stimulation, Surfactants: Fundamentals and Applications in the Petroleum Industry,* ed. L. L. Schramm, Cambridge, UK: Cambridge University Press, 2000, 329.
128. H. A. Volz, U.S. Patent 4044833, 1977.
129. P.-A. Francini, K. Chan, M. Brady, and C. Fredd, U.S. Patent Application, 20050020454, 2005.
130. S. Thach, U.S. Patent 5529122, 1996.
131. D. R. Watkins, U.S. Patent 4737296, 1988.
132. L. R. Norman and T. R. Gardner, U.S. Patent 4324669, 1982.
133. K. Thompson and R. Gdanski, "Laboratory Study Provides Guidelines for Diverting Acid with Foam," *SPE Production and Facilities* 8(4) (1993): 285.

134. S. Siddiqui, S. Talabani, J. Yang, S. T. Saleh, and M. R. Islam, "An Experimental Investigation of the Diversion Characteristics of Foam in Berea Sandstone Cores of Contrasting Permeabilities," *Chemical Engineering Science* 37 (2003): 51.
135. S. I. Kam, W. W. Frenier, S. N. Davies, and W. R. Rossen, "Experimental Study of High-Temperature Foam for Acid Diversion," *Journal of Petroleum Science and Engineering* 58 (2007): 138.
136. K. E. Cawiezel and J. C. Dawson, U.S. Patent Application 20050067165, 2005.
137. C. Zeiler, D. Alleman, and Q. Qu, "Use of Viscoelastic Surfactant-Based Diverting Agents for Acid Stimulation: Case Histories in GOM," SPE 90062 (paper presented at the SPE Annual Technical Conference, Houston, TX, 26–29 September 2004).
138. B. Lungwitz, C. Fredd, M. Brady, M. Miller, S. Ali, and K. Hughes, "Diversion and Cleanup Studies of Viscoelastic Surfactant-Based Self-Diverting Acid," SPE 86504, *SPE Production & Operations* 22(1) (2007): 121.
139. J. Yang, "Viscoelastic Wormlike Micelles and Their Applications," *Current Opinion in Colloid Interface Science* 7 (2002): 276.
140. D. P. Acharya and H. Kunieda, "Wormlike Micelles in Mixed Surfactant Solutions," *Advances in Colloid and Interface Science* 123–126 (2006): 401.
141. H. A. Nasr-El-Din, T. D. Welton, L. Sierra, and M. S. van Domelen, "Optimization of Surfactant-Based Fluids for Acid Diversion," SPE 107687 (paper presented at the European Formation Damage Conference, Scheveningen, The Netherlands, 30 May–1 June 2007).
142. (*a*) D. S. Treybig, G. N. Taylor, and D. K. Moss, U.S. Patent Application 20060025321, 2006. (*b*) R. Franklin, M. Hoey, and R. Pramachandran, "The Use of Surfactants to Generate Viscoelastic Fluids," Proceedings of the Chemistry in the Oil Industry VII, Manchester, UK: Royal Society of Chemistry, 2002.
143. D. Fu, M. Panga, S. Kefi, and M. Garcia-Lopez de Victoria, U.S. Patent 7237608, 2007.
144. (*a*) M. M. Samuel, K. I. Dismuke, R. J. Card, J. E. Brown, and K. England, U.S. Patent 6306800, 2001. (*b*) P. W. Knox, International Patent Application WO/2007/056393.
145. M. S. Dahayanake, J. Yang, J. H. Y. Niu, P.-D. Derian, R. Li, and D. Dino, U.S. Patent, 6258859, 2001.
146. D. Fu and F. Chang, U.S. Patent 6929070, 2005.
147. R. S. Hartshorne, T. L. Hughes, T. G. J. Jones, and G. J. Tustin, U.S. Patent Application 20050124525, 2005.
148. T. D. Welton, S. J. Lewis, and G. P. Funkhouser, U.S. Patent 7159659, 2007.
149. P. M. McElfresh and C. F. Williams, U.S. Patent 7216709, 2007.
150. R. E. Dobson Sr., D. K. Moss, and R. S. Premachandran, U.S. Patent 7060661, 2006.
151. T. Huang and J. B. Crews, "Do Viscoelastic-Surfactant Diverting Fluids for Acid Treatments Need Internal Breakers?," SPE 112484 (paper presented at the SPE International Symposium and Exhibition on Formation Damage Control, Lafayette, LA, 13–15 February 2008).
152. P. D. Berger and C. H. Berger, U.S. Patent Application 20060084579, 2006.
153. D. Fu, Diankui, Y. Chen, Z. Xiao, M. Samuel, and S. Daniel, U.S. Patent 7148185, 2006.
154. G. Di Lullo, A. Ahmad, P. Rae, L. Anaya, and R. Ariel Meli, "Toward Zero Damage: New Fluid Points the Way," SPE 69453 (paper presented at the SPE Latin American and Caribbean Petroleum Engineering Conference, Buenos Aires, Argentina, 25–28 March 2001).
155. (*a*) F. F. Chang; Q. Qu, and M. J. Miller, U.S. Patent 6399546, 2002. (*b*) F. F. Chang, Q. Qu, and W. Frenier, "A Novel Self-Diverting Acid Developed for Matrix Simulation of Carbonate Reservoirs," SPE 65033 (paper presented at the SPE International Symposium on Oilfield Chemistry, Houston, TX, 13–16 February 2001).

156. D. Taylor, P. Santhana Kumar, D. Fu, M. Jemmali, H. Helou, F. Chang, S. Davies, and M. Al-Mutawa, "Viscoelastic Surfactant Based Self-Diverting Acid for Enhanced Simulation in Carbonate Reservoirs," SPE 82263 (paper presented at the SPE European Formation Damage Conference, The Hague, Netherlands, 13–14 May 2003).
157. H. A. Nasr-El-Din, J. B. Chesson, K. E. Cawiezel, and C. S. Devine, SPE 102468 (paper presented at the SPE Annual Technical Conference and Exhibition, San Antonio, TX, 24–27 September 2006).
158. B. Lungwitz, C. Fredd, M. Brady, and T. Bui, "Application of Viscoelastic Surfactant Based Self-Diverting Acid in Gas Wells," Proceedings of the Chemistry in the Oil Industry IX, Manchester, UK: Royal Society of Chemistry, 31 October–2 November 2005, 214.
159. J. B. Crews, "Internal Phase Breaker Technology for Viscoelastic Surfactant Gelled Fluids," SPE 93449 (paper presented at the SPE International Symposium on Oilfield Chemistry, The Woodlands, TX, 2–4 February 2005).
160. E. B. Nelson, B. Lungwitz, K. Dismuke, M. Samuel, G. Salamat, T. Hughes, J. Lee, P. Fletcher, D. Fu, R. Hutchins, M. Parris, and G. J. Tustin, U.S. Patent Application 20020004464, 2002.
161. J. B. Crews and T. Huang, "Internal Breakers for Viscoelastic Surfactant Fracturing Fluids," SPE 106216 (paper presented at the SPE International Symposium on Oilfield Chemistry, Houston, TX, 28 February–2 March 2007).
162. J. B. Crews and T. Huang, U.S. Patent Application 20070151726.
163. J. B. Crews and T. Huang, U.S. Patent Application 20070299142.
164. T. D. Welton, R. D. Gdanski, and R. W. Pauls, U.S. Patent Application 20070060482.
165. D. Fu and M. Garcia-Lopez De Victoria, U.S. Patent 7341107, 2008.
166. A. R. Al-Nakhli, H. A. Nasr-El-Din, and A. A. Al-Baiyat, "Interactions of Iron and Viscoelastic Surfactants: A New Formation-Damage Mechanism," SPE 112465 (paper presented at the SPE International Symposium and Exhibition on Formation Damage Control, Lafayette, LA, 13–15 February 2008).
167. P.-A. Francini, K. Chan, M. Brady, and C. Fredd, U.S. Patent 7148184, 2006.
168. M. Garcia-Lopez De Victoria, Y. Christanti, G. Salamat, and Z. Xiao, U.S. Patent Application 20060131017, 2006.
169. K. E. Cawiezel and C. S. Devine, U.S. Patent Application 20050137095, 2005.
170. K. E. Cawiezel and C. S. Devine, "Nonpolymer Surfactant Enhances High-Strength Hydrofluoric Acid Treatments," SPE 95242 (paper presented at the SPE Annual Technical Conference and Exhibition, Dallas, TX, 9–12 October 2005).
171. Q. Qu and D. Alleman, U.S. Patent 7115546, 2006.
172. D. Alleman, Q. Qu, and R. Keck, "The Development and Successful Field Use of Viscoelastic Surfactant-Based Diverting Agents for Acid Stimulation," SPE 80222 (paper presented at the SPE International Symposium on Oilfield Chemistry, Houston, TX, 5–7 February 2003).
173. T. S. Davies, A. M. Ketner, and S. R. Raghavan, "Self-Assembly of Surfactant Vesicles that Transform into Viscoleastic Wormlike Micelles upon Heating," *Journal of the American Chemical Society* 128 (2006): 6669.
174. M. A. Samir, I. Elnashar, M. Samuel, and M. Jemmali, "Smart Chemical Systems for the Stimulation of High-Water-Cut Heavy Oil Wells," SPE 116746 (paper presented at the 2008 SPE Annual Technical Conference and Exhibition, Denver, CO, 21–24 September 2008).
175. B. B. Williams, J. L. Gidley, and R. R. Schechter, "Acidizing Fundamentals," SPE Monograph No. 6, New York: SPE, 1979.
176. C. Knox et al., U.S. Patent 3343602, 1967.
177. C. C. Bombardieri, U.S. Patent 3434545, 1969.
178. J. L. Boles and M. Usie, U.S. Patent 6443230, 2002.

179. W. Frenier and F. F. Chang, U.S. Patent 6806236, 2004.
180. G. Di Lullo and P. Rae, "A New Acid for True Stimulation of Sand Stone Reservoirs," SPE 37015 (paper presented at the SPE International 6th Asia Pacific Oil and Gas Conference, Adelaide, Australia, 28–31 October 1996).
181. M. N. Al-Dahlan, H. A. Nasr-El-Din, and A. A. Al-Qahtani, "Evaluation of Retarded HF Systems," SPE 65032 (paper presented at the SPE International Symposium on Oilfield Chemistry, Houston, TX, 13–16 February 2001).
182. W. Frenier, U.S. Patent Application 20020170715, 2002.
183. M. M. Amro, "Extended Matrix Acidizing Using Polymer-Acid Solutions," SPE 106360 (paper presented at the SPE Technical Symposium of Saudi Arabia Section, Dhahran, Saudi Arabia, 21–23 May 2006).
184. F. Zhou, Y. Liu, C. Xiong, J. Peng, X. Yang, X. Liu, Y. Lian, C. Qian, J. Yang, and F. Chen, SPE 104446 (paper presented at the SPE International Oil & Gas Conference and Exhibition, Beijing, China, 5–7 December 2006).
185. D. L. Holcombe, "Foamed Acid as a Means for Providing Extended Retardation," SPE 6376 (paper presented at the SPE Permian Basin Oil and Gas Recovery Conference, Midland, TX, 10–11 March 1977).
186. R. E. Harris and I. D. McKay, "New Applications for Enzymes in Oil and Gas Production," SPE 50621 (paper presented at the SPE European Petroleum Conference, The Hague, Netherlands, 20–22 October 1998).
187. V. Moses and R. E. Harris, U.S. Patent 5678632, 1997.
188. P. Leschi, G. Demarthon, E. Davidson, and D. Clinch, "Delayed-Release Acid System for Cleanup of Al Khalij Horizontal Openhole Drains," SPE 98164 (paper presented at the SPE International Symposium and Exhibition on Formation Damage Control, Lafayette, LA, 15–17 February 2006).
189. R. E. Harris, I. D. McKay, J. M. Mbala, and R. P. Schaaf, "Stimulation of a Producing Horizontal Well Using Enzymes that Generate Acid In-Situ—Case History," SPE 68911 (paper presented at the SPE European Formation Damage Conference, The Hague, Netherlands, 21–22 May 2001).
190. B. L. Todd and E. Davudson, U.S. Patent Application 20040163814.
191. J. W. Still, K. Dismuke, and W. Frenier, U.S. Patent Application 20040152601.
192. H. A. Nasr-El-Din, A. Al-Zahrani, J. Still, T. Lesko, and S. Kelkar, "Laboratory Evaluation of an Innovative System for Fracture Stimulation of High-Temperature Carbonate Reservoirs," SPE 106054 (paper presented at the SPE International Symposium on Oilfield Chemistry, Houston, TX, 28 February–2 March 2007).
193. M. A. Buijse and M. S. Van Domelen, "Novel Application of Emulsified Acids to Matrix Stimulation of Heterogeneous Formations," SPE 39583 (paper presented at the SPE International Symposium on Formation Damage Control, Lafayette, LA, 18–19 February 1998).
194. H. A. Nasr-El-Din and M. M. Samuel, "Development and Field Application of a New, Highly Stable Emulsified Acid," SPE 115926 (paper presented at the Annual Technical Conference and Exhibition, Denver, CO, 2008, 22–24 September 2008).
195. C. W. Crowe, U.S. Patent 3779916, 1973.
193. C. W. Crowe, U.S. Patent 3962102, 1976.
196. G. A. Scherubel, U.S. Patent 4140640, 1979.
198. E. G. Scovell, N. Grainger, and T. Cox, International Patent Application WO/2001/088333.
199. D. K. Sarma, P. Agarwal, E. Rao, and P. Kumar, "Development of a Deep-Penetrating Emulsified Acid and Its Application in a Carbonate Reservoir," SPE 105502 (paper presented at the 15th SPE Middle East Oil and Gas Conference, Bahrain, 11–14 March 2007).

200. R. C. Navarrete, B. A. Holms, S. B. McDonnell, and D. E. Linton, "Emulsified Acid Enhances Well Production in High-Temperature Carbonate Formations," SPE 50612 (paper presented at the SPE European Petroleum Conference, The Hague, Netherlands, 20–22 October 1998).
201. A. T. Jones, C. Rodenburg, D. G. Hill, A. H. Akbar Ali, and P. de Boer, "An Engineered Approach to Matrix Acidizing HTHP Sour Carbonate Reservoirs," SPE 68915 (paper presented at the SPE European Formation Damage Conference, The Hague, Netherlands, 21–22 May 2001).
202. P. Kasza, M. Dziadkiewicz, and M. Czupski, "From Laboratory Research to Successful Practice: A Case Study of Carbonate Formation Emulsified Acid Treatments," SPE 98261 (paper presented at the SPE International Symposium and Exhibition on Formation Damage Control, Lafayette, LA, 15–17 February 2006).
203. E. M. Andreasson, F. Egeli, K. A. Holmberg, B. Nystrom, K. G. Stridh, and E. M. Sterberg, U.S. Patent 4650000, 1987.
204. W. R. Dill, U.S. Patent 4322306, 1982.
205. M. Al-Khaldi, H. A. Nasr-El-Din, S. Metha, and A. Aamri, "Reactions of Citric Acid with Calcite," *Chemical Engineering Science* 62 (2007): 5880.
206. R. E. Harris and I. D. Mckay, International Patent Application WO/2004/007905.

6 Sand Control

6.1 INTRODUCTION

Sand (or "fines") production is common in many oil and gas wells throughout the world. The flow of abrasive sand through wells and production lines causes unwanted erosion of equipment, and its production may also exacerbate oil-water separation in the process facilities. There are a number of ways to reduce sand production mechanically, including the use of screens, gravel packing, frac-packing, and modification to the perforation technique usually carried out at the well-completion stage. For poorly consolidated reservoirs, which are still producing excessive sand, chemical sand control can be an option. This can be especially rewarding for subsea wells if expensive intervention costs can be avoided.

6.2 CHEMICAL SAND CONTROL

Chemical sand control has been carried out for many years with resins or epoxy, which harden unconsolidated sand. Typical systems are based on bisphenol A–epichlorohydrin resin, polyepoxide resin, polyester resin, phenol-aldehyde resin, urea-aldehyde resin, furan resin, urethane resin, and glycidyl ethers.[1] If the resin comprises bisphenol A–epichlorohydrin polymer, a preferred curing agent is 4,4-methylenedianiline. If the resin comprises a polyurethane, the curing agent is preferably a diisocyanate. The furan resin system is one of the most common: the key chemical is furfuryl alcohol and does not require a curing agent, as it is self-polymerizing in the presence of acid catalysts[2–3] (Figure 6.1). These systems are designed to maintain sufficient permeability of the formation to allow production. Self-diverting resin-based sand consolidation fluids have been claimed that allow a greater interval to be treated than conventional resin treatments.[4] Most resin-based chemicals are not considered to be very environmentally friendly.

Various aqueous and nonaqueous tackifying chemicals including silyl-modified polyamides that impart a sticky character to sand particles, hindering their movement, have been claimed.[5] Cross-linked polymer gels, similar to those used in water shut-off treatments have also been proposed for sand consolidation.[5–6] Polymer gel systems, such as those based on polyacrylamides, are claimed to impart a lower probability of failure to the formation compared to resin treatments.

Inorganic chemical systems for sand consolidation have also been developed. For example, a system based on an insoluble silica source and a source of calcium hydroxide (e.g., aqueous solutions of calcium chloride and sodium hydroxide) has been claimed.[7] The components of the aqueous system react to produce a calcium silicate hydrate gel having cementitious properties within the pores of the formation.

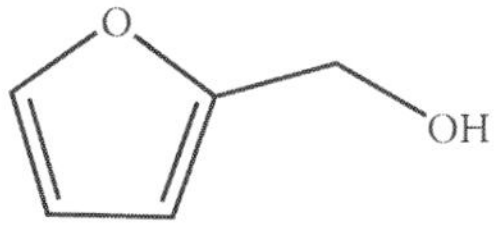

FIGURE 6.1 Furfuryl alcohol.

An enzyme-based process for consolidation of sand with calcium carbonate has also been proposed.[8] It requires calcium chloride, urea, and urease enzyme. The enzyme catalyzes the decomposition of urea to ammonia and carbon dioxide, raising the pH. In the presence of soluble calcium ions, insoluble calcium carbonate is formed that deposits on the sand and core, binding them together.

Since about 2005, a new sand-consolidation method based on organosilane chemistry has been developed and used in the field.[9–13] In comparison to other treatments, this method only increases the residual strength of the formation by a small amount. The treatment is oil-soluble and will therefore not alter the relative permeability in the oil-bearing zones, thereby reducing the risk of increased skin due to changes in saturation. This system is especially beneficial for fields with low reservoir pressure. The method is employed by simple bullheading and can have self-diverting properties. In laboratory studies, the organosilane treatment was shown to give better overall performance with regards sand consolidation and moderate permeability reduction compared to other treatments such as water-soluble gelling polymers and the $CaCl_2$/urea/enzyme system discussed earlier.

Preferred oil-soluble organosilanes that can be used are 3-aminopropyltriethoxysilane and bis-(triethoxy silylpropyl)amine or mixtures thereof (Figure 6.2). They are usually mixed in diesel and bullheaded into the well. The authors suggest that the presence of the amine function appears to result in better adsorption of the organosilane to the sand grains. It is also believed that the presence of an amine group may contribute to the formation of a gel-like structure having viscoelastic properties. The authors suggest that the organosilane compounds react with water and hydrolyze. The resulting chemicals then react with siliceous surfaces in the formation (e.g., the surface of silica sand), coat any sand particles, and bind them in place by the formation of silicate bridges restricting their movement. The advantage of bifunctional organosilanes, such as bis-(triethoxy silylpropyl)amine, is their ability to bind two particles together. The organosilanes are claimed to be environmentally acceptable with low bioaccumulation potential and high biodegradation.[14]

Several types of wells have been treated with the organosilane system at a chemical concentration in the range of 5–7 vol.% of the active components. The first results in terms of sand production reduction were mixed, with the subsea well responding

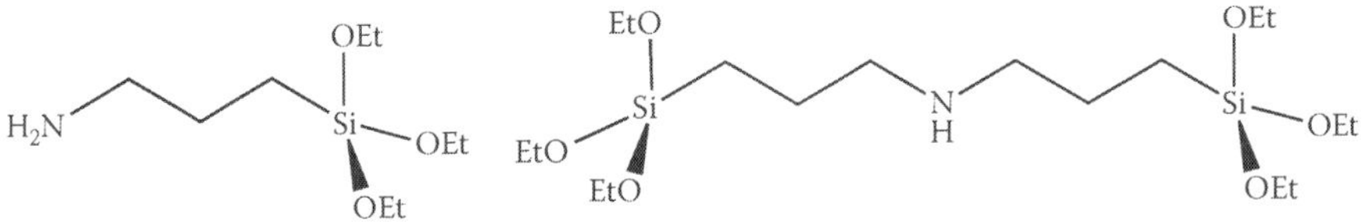

FIGURE 6.2 3-Aminopropyltriethoxysilane and bis-(triethoxy silylpropyl)amine.

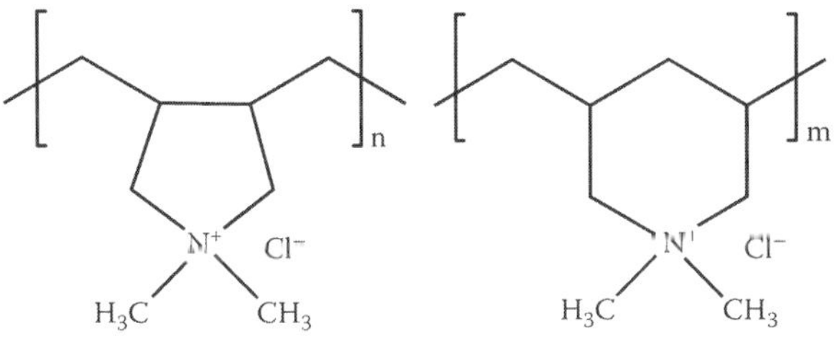

FIGURE 6.3 Polydiallyldimethyldiallylammonium chloride. The five-ring pyrolidinium monomer is the major component and the six-ring piperidinium monomer is the minor component.

best to the treatment. A moderate reduction in permeability was observed in some wells, which reduced the PI (production index) of the well by 10–15%. However, the production of the wells was limited by sand production levels, so such a reduction in PI was acceptable. Correct placement, especially in horizontal wells, was shown to be critical when it comes to performance with regards to increasing the maximum sand-free rate (MSFR). Organosilanes used at higher concentrations than those for sand consolidation have also been claimed as chemicals for water shut-off treatments.[15]

Another claimed sand consolidation method, which imparts small incremental forces or a relatively weak residual strength to the formation, is by using a positively charged water-soluble polymer. Examples are polyaminoacids, such as polyaspartate and copolymers comprising aspartic acid and proline and/or histidine, and poly(diallyl ammonium salts) such as polydimethyldiallylammonium chloride and mixtures thereof[16] (Figure 6.3). It is thought that by virtue of its length and multiple positive charges, the polymer may interact electrostatically with a number of different particles of the formation thereby holding or binding them together. In so doing, the polymer chain is likely to span the interstitial space between sand particles of the formation. The result is simply the formation of a "mesh-like" or "net-like" structure that does not impair fluid flow. Hence, the permeability of a subterranean formation treated according to the method described by the present invention is largely unchanged after treatment.

REFERENCES

1. P. D. Nguyen, J. A. Barton, and O. M. Isenberg, U.S. Patent 7013976, 2006.
2. B. W. Surles, P. D. Fader, R. H. Friedman, and C. W. Pardo, U.S. Patent 5199492, 1993.
3. M. Parlar, S. A. Ali, R. Hoss, D. J. Wagner, L. King, C. Zeiler, and R. Thomas, "New Chemistry and Improved Placement Practices Enhance Resin Consolidation: Case Histories from the Gulf of Mexico," SPE 39435 (paper presented at the SPE Formation Damage Control Conference, Lafayette, LA, 18–19 February 1998).
4. J. A. Ayoub, J. P. Crawshaw, and P. W. Way, U.S. Patent 6632778, 2003.
5. P. D. Nguyen, L. Sierra, E. D. Dalrymple, and L. S. Eoff, International Patent Application WO/2007/010190.
6. S. G. James, E. B. Nelson, and F. J. Guinot, U.S. Patent 6450260, 2002.
7. E. B. Nelson, S. Danican, and G. Salamat, U.S. Patent 7111683, 2006.

8. R. E. Harris and I. D. McKay, "New Applications for Enzymes in Oil and Gas Production," SPE 50621 (paper presented at the European Petroleum Conference, the Hague, Netherlands, 20–22 October 1998).
9. H. K. Kotlar and F. Haavind, International Patent Application WO/2005/124100.
10. H. K. Kotlar, F. Haavind, M. Springer, S. S. Bekkelund, and O. Torsaeter, "A New Concept of Chemical Sand Consolidation: From Idea and Laboratory Qualification to Field Application," SPE 95723 (paper presented at the SPE Annual Technical Conference and Exhibition, Dallas, TX, 9–12 October 2005).
11. H. K. Kotlar, F. Haavind, M. Springer, S. S. Bekkelund, A. Moen, and O. Torsaeter, "Encouraging Results with a New Environmentally Acceptable, Oil-Soluble Chemical for Sand Consolidation: From Laboratory Experiments to Field Application," SPE 98333 (paper presented at the International Symposium and Exhibition on Formation Damage Control, Lafayette, LA, 15–17 February 2006).
12. F. Haavind, S. S. Bekkelund, A. Moen, H. K. Kotlar, J. S. Andrews, and T. Haaland, "Experience with Chemical Sand Consolidation as a Remedial Sand-Control Option on the Heidrun Field," SPE 112397 (paper presented at the SPE International Symposium and Exhibition on Formation Damage Control, Lafayette, LA, 13–15 February 2008).
13. A. Jordan and B. Comeaux, "Keeping Fines in their Place to Maximise Inflow Performance," *World Oil* 228 (2007): 115–122.
14. H. K. Kotlar, A. Moen, T. Haaland, and T. Wood, "Field Experience with Chemical Sand Consolidation as a Remedial Sand-Control Option," OTC 19417 (paper presented at the Offshore Technology Conference, Houston, TX, 5–8 May 2008).
15. H. K. Kotlar, International Patent Application WO/2005/124099.
16. H. K. Kotlar and P. Chen, International Patent Application WO/2005/124097.

7 Control of Naphthenate and Other Carboxylate Fouling

7.1 INTRODUCTION

Organic carboxylate salt scaling and/or naphthenate salt scaling is not such a widespread problem in the oil-production industry as inorganic or organic scaling (waxes and asphaltenes), but it has received considerable research attention in response to increased production in areas such as West Africa, where naphthenate deposits are particularly troublesome. These salts, formed when oil-soluble aliphatic carboxylic and naphthenic acids come in contact with metal cations in the produced water, cause tight emulsions, ragged interfaces, deposition and, ultimately, separation difficulties.[1]

The carboxylate salts (known as soaps) are usually sodium salts of long-chain, linear organic acids with one carboxylic acid group. The sodium carboxylate soaps accumulate at oil-water interfaces and cause emulsion problems in the form of thick emulsion pads in the separator. Low total dissolved solids produced waters are more susceptible to emulsion problems caused by carboxylate soaps. As the size of long-chain carboxylic acids increases, so does the amount of carboxylate deposit. Increases in brine pH also increased the amount of deposit.[2] Calcium salts of long-chain carboxylic acids are much more oil-soluble than the sodium salts and are therefore much less of a problem.

Naphthenic acids are carboxylic acids in which the alkyl chain is connected to one or more saturated cyclohexyl or cyclopentyl groups. Naphthenate soaps are usually salts of divalent anions, particularly calcium. They are more prone to forming deposits than the carboxylate soaps but can also cause emulsion problems.[3-5] Even fields with very low total acid number and fairly low calcium ion concentration can suffer from naphthenate deposition if damaging organic acids that lead to deposits are substantially present.[6] Although both carboxylate and naphthenate production problems can occur in the same field, it is more common that only one of the two problems dominates. For example, Malaysian fields mainly have carboxylate problems whereas in West Africa, naphthenates are the major problem.

It was traditionally believed that the "damaging" naphthenic acids had molecular weights in the range of 200–500. However, more recent analysis of naphthenate deposits from a wide range of fields has shown that the main naphthenic acid components are C80 tetra acids containing four to eight cyclic rings, sometimes referred to as ARN acids (Figure 7.1).[7-9,20] On contact with produced waters, the tetra acids

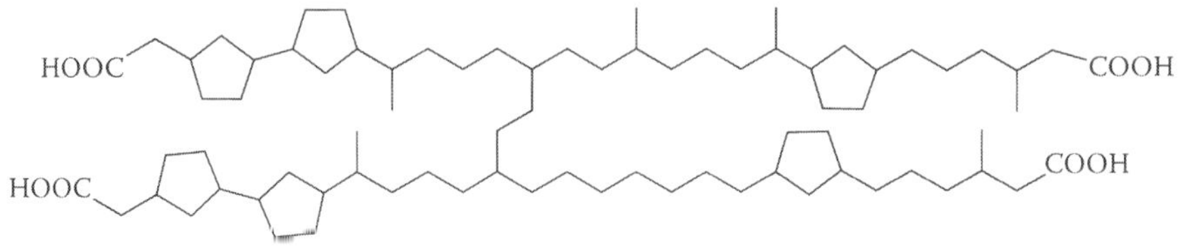

FIGURE 7.1 One of the naphthenic acids (ARN) responsible for naphthenate fouling.

form salts with divalent calcium ions, forming polymeric calcium naphthenate sticky solids, which harden on contact with air. The naphthenate solids foul pipelines and processing equipment, causing reduced production flow and, in worst-case scenarios, frequent unplanned shutdowns. However, many crudes that contain naphthenic acids do not cause an operational fouling problem.[10] As naphthenate ions have surface activity, they can exacerbate emulsion problems, although sodium carboxylate soaps are a worse source of this problem.[11–12]

7.2 NAPHTHENATE DEPOSITION CONTROL USING ACIDS

Naphthenic acids are weak acids that exist in equilibrium with naphthenate ions. The higher the pH, the more dissociated the acid will become and the more likely that soap formation will occur at the water-oil interface. Thus, the traditional way to avoid naphthenate deposition is to lower the pH by adding an acid with a lower pK_a value than naphthenic acids. Field experience has shown that lowering the pH to around 6.0 prevents the formation of naphthenate deposits. Further lowering of the pH has no additional benefit and causes worse corrosion problems.[13] Typical acids that have been used include:

- inorganic mineral acids such as phosphoric acid
- small organic acids such as acetic acid or glycolic acid
- surfactant acids such as dodecylbenzenesulfonic acid (DDBSA)

The addition of acid shifts the equilibrium away from naphthenate ions to naphthenic acids, which are less surface-active and less water-soluble and do not combine with metal ions to make salts. The most widely used acids to prevent naphthenate deposition appears to be acetic acid (CH_3COOH), followed by the stronger acid, phosphoric acid.[14–15] Acetic acid is fairly volatile and can cause 12-o'clock corrosion in the flowline. The use of glycolic acid ($HOCH_2COOH$) has been investigated and was found on one North Sea field to give less emulsion pad in the separator than acetic acid.[6] Mineral acids such as hydrochloric acid (HCl) have been used to temporarily remove naphthenate scaling problems. A mutual solvent such as monobutyl glycol ether ($C_4H_9OCH_2CH_2OH$) or isopropyl alcohol ($CH_3CH(OH)CH_3$) can be added to increase the solvency. Hydrocarbon solvents have also been used.[16] Phosphoric acid (H_3PO_4) has been used to prevent sodium carboxylate formation. DDBSA, often in combination with acetic acid, has been used for soaps control.[13]

7.3 LOW-DOSAGE NAPHTHENATE INHIBITORS

The amount of acid required to prevent naphthenate or carboxylate soap problems is based on the total water phase and its pH, not on the concentration of naphthenic or carboxylic acids. Thus, prevention of these problems by the injection of acids can be a fairly expensive procedure and dosages must be carefully controlled to avoid excessive corrosion problems. A more recent method to prevent naphthenate deposition is to add a low-dosage naphthenate inhibitor (LDNI). LDNIs are still in their infancy, but a few patents detailing chemical structures have appeared. For example, alcohols or alkyl ethoxylates reacted in a 2:1 ratio with phosphorus pentoxide (P_2O_5) give a mixture of monophosphate and diphosphate esters suitable as LDNIs (Figure 7.2).[17] Other phosphate esters of polyols have been claimed.[21] It is believed that these LDNIs exhibit surface-active properties that cause the inhibitors to align and concentrate in a layer at the oil-water interface and thereby prevent interactions between organic acids in the oil phase with cations or cation complexes in the water, that is, the LDNIs are therefore more interfacially active than naphthenic acid. In one field application, to prevent emulsion problems due to either naphthenates or LDNI, a demulsifier was added.[12] It is believed that the physical positioning and geometry of the naphthenate inhibitor blocks the growth of naphthenate salt crystals. It is preferred that the naphthenate inhibitors also avoid the formation of oil-in-water and water-in-oil emulsions. Dosages up to about 100 ppm can be used.

Phosphonate end-capped water-soluble polymers have been claimed as soap control agents.[18] An effective polymer in this class will scavenge the calcium from the crude oil and migrate into the water phase. Therefore, the efficacy of the soap control agent is directly proportional to the calcium ion concentration. The polymers are polyvinyl polymers and also contain sulfonate and/or carboxylate groups. They can also prevent inorganic scale deposition.

Quaternary ammonium or quaternary phosphonium compounds have also been claimed as LDNIs.[19] Examples are cocoalkylmethyl-bis(2-hydroxyethyl) ammonium chloride, cocoalkylmethyl [polyoxyethylene (15)] ammonium chloride, and tetrakis-hydroxymethyl phosphonium sulfate. The first two compounds are surfactants and would be expected to work by replacing naphthenate ions at the oil-water interface. The latter compound is also a biocide and sulfide scale dissolver. The same patent also claims linear compounds having at least two carboxylic acid or acrylic acid

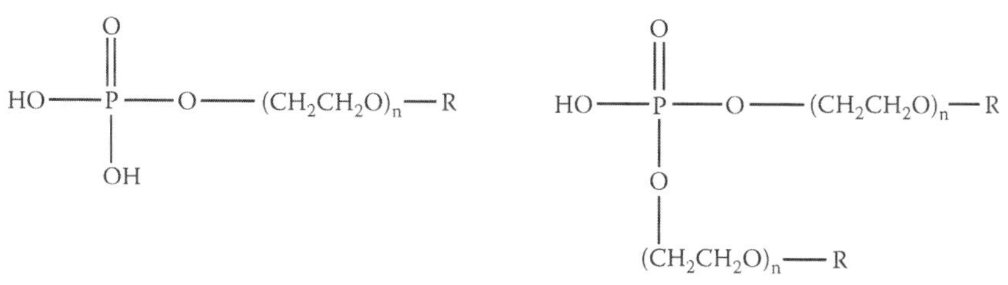

FIGURE 7.2 Monophosphate and diphosphate esters that together function as LDNIs: $n = 0–9$, R = alkyl.

functional moieties such as polyacrylic acids or polymaleates, useful also as inorganic scale inhibitors. In fact, any surfactant that can interact with a naphthenic acid to prevent a subsequent interaction with a metal ion to produce solids or emulsions could potentially be used.

REFERENCES

1. (*a*) R. A. Rodriguez and S. J. Ubbels, "Understanding Naphthenate Salt Issues in Oil Production," *World Oil* 228(8) (2007): 143–145. (*b*) C. Hurtevent and B. Brocart, *J. Disp. Sci. Techn.*, 2008, 29, 1496.
2. S. J. Dyer, G. M. Graham, and C. Arnott, "Naphthenate Scale Formation—Examination of Molecular Controls in Idealised Systems," SPE 80395 (paper presented at the International Symposium on Oilfield Scale, Aberdeen, UK, 29–30 January 2003).
3. A.-M. Dahl Hanneseth, M. Fossen, A. Silset, and J. Sjoblom, "Naphthenic Acid/Naphthenate Stabilized Emulsions and the Influence of Crude Oil Components" (paper presented at the 8th International Conference on Petroleum Phase Behavior and Fouling, Pau, France, 10–14 June 2007).
4. K. S. Sorbie, A. Shepherd, C. Smith, M. Turner, and R. A. Westacott, "Naphthenate Formation in Oil Production, General Theories and Field Observations," Proceedings of the Chemistry in the Oil Industry IX Symposium, Royal Society of Chemistry, Manchester, UK, 2005, 289.
5. T. Baugh, K. V. Grande, H. Mediaas, J. E. Vindstad, and N. O. Wolf, "The Discovery of High Molecular Weight Naphthenic Acids (ARN Acid) Responsible for Calcium Naphthenate Deposits," *Chemistry in the Oil Industry IX*, Royal Society of Chemistry, Manchester, UK, 2005, 275.
6. K. B. Melvin, C, Cummine, J. Youles, H. Williams, G. M. Graham, and S. Dyer, "Optimising Calcium Naphthenate Control in the Blake Field," SPE 114123 (paper presented at the SPE 9th International Conference on Oilfield Scale, Aberdeen, UK, 28–29 May 2008).
7. B. E. Smith, G. Fowler, J. Krane, B. Lutnaes, and S. J. Rowland, "Separation and Identification of High Molecular Weight Tetra Acids Responsible for Calcium Naphthenate Deposition" (paper presented at the 8th International Conference on Petroleum Phase Behavior and Fouling, Pau, France, 10–14 June 2007).
8. B. F. Lutnaes, O. Brandal, J. Sjoblom, and J. Krane, *Org. Biomol. Chem.*, 4 (2006): 616.
9. J. E. Vindstad, K. V. Grande, K. R. Hoevik, H. Kummernes, and H. Mediaas, "Applying Laboratory Techniques and Test Equipment for Efficient Management of Calcium Naphthenate Deposition at Oil Fields in Different Life Stages" (paper presented at the 8th International Conference on Petroleum Phase Behavior and Fouling, Pau, France, 10–14 June 2007).
10. B. Brocart, M. Bourrel, C. Hurtevent, J.-L. Volle, and B. Escoffier, "ARN-Type Naphthenic Acids in Crudes: Analytical Detection and Physical Properties," *Journal of Dispersion Science and Technology* 28 (2007): 331.
11. R. A. Rodriguez and S. J. Ubbels, "Understanding Naphthenate Salt Issues in Oil Production," *World Oil* 228 (2007): 143–145.
12. C. Hurtevent and S. Ubbels, "Preventing Naphthenate Stabilised Emulsions and Naphthenate Deposits on Fields Producing Acidic Crude Oils," SPE 100430 (paper presented at the SPE International Oilfield Scale Symposium, Aberdeen, UK, 31 May–1 June 2006).

13. M. S. Turner and P. C. Smith, "Controls on Soap Scale Formation, including Naphthenate Soaps—Drivers and Mitigation," SPE 94339 (paper presented at the SPE International Symposium on Oilfield Scale, Aberdeen, UK, 11–12 May 2005).
14. A. Goldszal, C. Hurtevent, and G. Rousseau, "Scale and Naphthenate Inhibition in Deep-Offshore Fields," SPE 74661 (International Symposium on Oilfield Scale, Aberdeen, UK, 30–31 January 2002).
15. S. J. Ubbels, "Preventing Naphthenate Stabilized Emulsions and Naphthenate Deposits during Crude Oil Processing," Proceedings from the 5th International Conference on Petroleum Phase Behaviour and Fouling, 13–17 June 2004.
16. J. E. Vindstad, A. S. Bye, K. V. Grande, B. M. Hustad, E. Hustvedt, and B. Nergård, "Fighting Naphthenate Deposition at the Heidrun Field," SPE 80375 (paper presented at the International Symposium on Oilfield Scale, Aberdeen, UK, 29–30 January 2003).
17. J. S. Ubbels, P. J. Venter, and V. M. Nace, World Patent Application WO2006025912, 2006.
18. C. R. Jones, International Patent Application WO/2005/085392.
19. C. Gallagher, J. D. Debord, S. Asomaning, J. Towner, and P. Hart, World Patent Application WO/2007/065107, 2007.
20. M. M. Mapolelo, L. A. Stanford, R. P. Rodgers, A. T. Yen, J. D. Debord, S. Asomaning, and A. G. Marshall, *Energy Fuels*, 2009, 23, 349.
21. M. Hellsten and I. Uneback, International Patent Application WO/2008/155333.

8 Corrosion Control during Production

8.1 INTRODUCTION

Internal and external corrosion of downhole tubing and equipment, subsea or surface pipelines, pressure vessels, and storage tanks is a major problem in the oil and gas industry.[1-3,14] Besides basic wastage of metal, either generally or locally, the consequences of electrochemical corrosion can also be embrittlement and cracking, all of which can lead to equipment failure. Corrosion of iron in steel requires the presence of water and aqueous species that can be reduced while the iron is oxidized. Oxygen, acid gases such as CO_2, H_2S, and natural organic acids in the produced fluids all contribute to corrosion. Corrosion is an electrochemical redox (reduction and oxidation) process whereby localized anodic and cathodic reactions are set up on the surface of the metal. The basic chemical corrosion processes are illustrated in Figure 8.1. Entrainment of hydrogen atoms into the metal is one cause of its embrittlement. Embrittlement and cracking of metals are unpredictable and give rise to conditions under which catastrophic failure may occur.

In an acid solution, the cathodic reaction is

$$2H^+ + 2e^- \rightarrow 2H$$

$$H + H \rightarrow H_2$$

The reaction of hydrogen atoms to form diatomic hydrogen is poisoned by sulfide species. Thus, the entry of hydrogen atoms into the metal matrix is accelerated by the presence of sulfides in the produced fluids (sour fluids) and may lead to sulfide stress cracking. These forms of corrosion generally occur later in service life and are not easily prevented by application of chemical inhibitors.

The removal of H^+ ions leaves OH^- (hydroxide) ions in solution. In neutral or basic solution, the cathodic reaction is

$$O_2 + 2H_2O + 4e^- \rightarrow 4OH^-$$

At the anode, iron is oxidized:

$$Fe \rightarrow Fe^{2+} + 2e^-$$

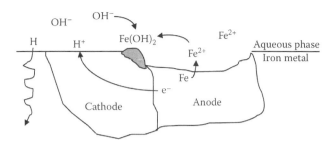

FIGURE 8.1 Corrosion cell reactions.

The generated hydroxide ions react with the iron(II) ions to form insoluble iron(II) hydroxide:

$$Fe^{2+} + 2OH^- \rightarrow Fe(OH)_2$$

In the presence of oxygen, iron(II) hydroxide can be oxidized to iron(III) hydroxide, $Fe(OH)_3$. There is little oxygen corrosion in production as the produced fluids are usually anaerobic. However, ingress of oxygen into well annuli, into the vapor space in tanks, and through pump packing becomes more commonplace in mature oilfields, which have less gas pressure, so traces of oxygen can be present. A laboratory study has shown that inhibition of corrosion in CO_2/O_2 systems was more difficult than in CO_2 alone but could be accomplished.[4] Oxygen corrosion is more prevalent in seawater injection wells where the level of dissolved oxygen in seawater is about 9 ppm. This is best treated by removing the oxygen from the water and/or using an oxygen scavenger. Besides oxygen corrosion, CO_2 corrosion ("sweet" corrosion) is very common in oil and gas production. H_2S corrosion ("sour" corrosion) is more severe than CO_2 corrosion. The key species that are chemically reduced in CO_2 corrosion are H^+, HCO_3^-, H_2CO_3, which are all present in the water phase.[5] After corrosion has taken place, this will lead to dissolved iron species as well as scales such as $FeCO_3$, which deposit on the steel. These can form a useful protective barrier reducing the apparent rate of general corrosion, however, localized corrosion can still take place under the scale layer, in so-called underdeposit corrosion.[6,7] For H_2S corrosion, the relevant reduced species are H_2S, H^+, and HS^-, which, after reaction with iron, can lead to iron sulfide scales such as FeS. Like $FeCO_3$, FeS can form a protective barrier hindering further corrosion.[8] CO_2 and H_2S corrosion rates generally increase with increasing temperature, pressure, and salinity (electrolyte concentration) of the produced fluids.

There are many types of corrosion that occur in the oilfield, including:

- general corrosion
- localized corrosion
- pitting and crevice corrosion
- galvanic corrosion
- erosion corrosion

Corrosion Control during Production

- microbially influenced corrosion (MIC)
- cracking corrosion

General corrosion is uniform wastage of metal along the whole tubing or flowline and the easiest to manage. Localized corrosion is more common and occurs at specific points. Galvanic corrosion is a process in which one metal corrodes preferentially when in electrical contact with a different type of metal and both metals in contact with an electrolyte. Pitting and crevice corrosion are similar types of extreme localized corrosion that leads to the creation of small holes in the metal. Pitting corrosion is particularly prevalent at high chloride ion concentrations. Erosion corrosion is also called flow-induced localized corrosion. Erosion corrosion is a complex materials degradation mechanism involving the combined effects of mechanical erosion and electrochemical corrosion. It is known to be induced by high shear stresses in strong flows. MIC is a very prevalent corrosion caused by chemical processes initiated by the metabolism of anaerobic microorganisms in the produced fluids.[9-12] The reduction of sulfate ions by sulfate-reducing bacteria (SRBs) under anaerobic conditions is of particular concern. Most MIC takes the form of pits that form underneath colonies of bacteria, also known as biofilms and which frequently develop within mineral and biodeposits. Biofilm creates a protective environment where conditions can become corrosive and corrosion is accelerated. Surface scale formation is one of the critical factors governing localized corrosion rates.[13]

8.2 METHODS OF CORROSION CONTROL

There are a number of ways of mitigating corrosion including:[14]

- using corrosion resistant alloys (CRAs)
- water removal (by pigging or dehydration)
- cathodic protection
- coatings
- corrosion inhibitors
- oxygen and H_2S scavengers
- biocides for preventing MIC (see Chapter 14)
- pH stabilization
- drag reduction

CRAs such as chrome steels are expensive but are sometimes used downhole where corrosion rates would be unacceptably high with normal carbon steels. In the North Sea and the Gulf of Mexico, tubular steel is usually 13Cr or 22Cr, with duplex steels being preferred for high-temperature wells. In other producing regions, carbon steel may be selected for downhole service, this decision being made by consideration of the relative costs of installing expensive CRAs against the operational costs of batch treatment with inhibitors. The presence of H_2S will also influence metallurgy selection. Regular and occasional downhole batch treatment by bullheading a corrosion inhibitor (tubing displacement) is often used in onshore wells to protect carbon steel. Emulsified corrosion inhibitors are claimed to perform better.[34]

Water removal by pigging can minimize localized corrosion where water collects in low points. Coatings including paints and plastics are sometimes used in the oil industry, especially for corrosion-sensitive locations. This is partly due to cost and partly due to the fact that coatings can be eroded away by abrasive sands/fines and other particles reexposing the steel.

Cathodic protection using impressed current devices and sacrificial anodes of aluminum or zinc (so-called galvanizing) is commonly used to protect external surfaces of flowlines or submerged structures such as rig legs. Sacrificial anodes are not used for internal corrosion control: they would be rapidly eroded and chemically attacked by the nature of the fluids within. However, deoiling tanks may have sacrificial anodes. Coatings would also suffer the same fate over time due to erosion.

Corrosion due to oxygen and H_2S can be reduced by the use of scavengers (see Chapters 15 and 16, respectively). This will not eliminate CO_2 corrosion, the concentration of which is usually much higher than that of oxygen and H_2S in produced fluids and thus expensive to scavenge. The concentration of H_2S, and other chemicals produced by bacteria that can cause corrosion, can be reduced by adding biocides (see Chapter 14).[15] Biocides are used both upstream and downstream of the separator. The largest use of biocides is to minimize MIC in water injection systems.

A method that has been used for corrosion control in wet gas pipelines is the use of pH stabilizers. pH stabilizers, such as metal hydroxides, metal carbonates, and bicarbonates or amines, can raise the pH of the produced aqueous fluids promoting formation of solid deposits such as $FeCO_3$, which forms a hard protective layer on the metal surface.[16] The pH stabilization technique is well-suited for use in combination with monoethylene glycol as hydrate inhibitor, because the pH stabilizer will remain in the regenerated glycol. This means that there is no need for continuous renewal of the pH stabilizer. The method has been used offshore Norway and the Netherlands. The pH stabilization technique has been used mostly for wet gas pipelines without H_2S in the gas but is now being taken into use also for pipelines in the Persian Gulf, where there are considerable amounts of H_2S in addition to CO_2.[6] An article describing efforts to develop corrosion inhibitors with glycol that prevent severe corrosion in wet gas systems has been published.[17]

Drag reduction, using drag-reducing agents (DRAs) can reduce the severity of flow-induced localized corrosion, and concomitantly, some film-forming corrosion inhibitors have drag-reducing properties (see Chapter 17 on DRAs).[18,19] Combinations of water-soluble high molecular weight drag-reducing polymers and surfactant-based film-forming corrosion inhibitors can give effective corrosion inhibition of iron and steel alloys in contact with oil-in-brine emulsions.[20]

8.3 CORROSION INHIBITORS

Corrosion inhibitors can be categorized as follows:[21,22]

- passivating (anodic)
- cathodic

Corrosion Control during Production

- vapor-phase or volatile
- film-forming

For protection of oil, condensate, and gas production lines, which are essentially anaerobic, the film-forming corrosion inhibitors (FFCIs) are mostly used, sometimes with synergists. These will be discussed in detail shortly, after a brief discussion of the other categories of corrosion inhibitors. Corrosion inhibitors suitable for acid-stimulation treatments are dealt with separately in Chapter 5 (Acid Stimulation).

Passivating corrosion inhibitors are not used in oil and gas production. They work best in low-salinity applications such as utility systems (freshwater or condensate water systems). A passivating corrosion inhibitor leads to the formation of a non-reactive thin surface film on the metal that stops access of corrosive substances to the metal, inhibiting further corrosion. All passivating inhibitors can accelerate corrosion if underdosed. Some passivating inhibitors require the presence of oxygen. These include:

- phosphate (PO_4^{3-}) and polyphosphates
- tungstate (WO_4^{2-})
- silicate (SiO_3^{2-})

Inhibitors that do not require the presence of oxygen include:

- chromate (CrO_4^{2-})
- nitrite (NO_2^-)
- molybdate (MoO_4^{2-})
- meta-, ortho-, and pyrovanadates ($NaVO_3$, Na_3VO_4, and $Na_4V_2O_7$, respectively)

Several of the passivating inhibitors are metal anionic species. Chromate is carcinogenic and, therefore, now not recommended for use. Phosphates or polyphosphates are nontoxic, but due to the limited solubility of calcium phosphate, it is difficult to maintain adequate concentrations of phosphates in many instances. Molybdate works in anaerobic conditions but works even better with oxygen present.[23,24] Nitrite is another good passivating or anodic inhibitor. If anodic inhibitors are used at too low a concentration, they can actually aggravate pitting corrosion, as they form a nonuniform layer with local anodes. Nitrite has been used to reduce corrosion in hot water bundles.[25] Vanadates in blends with 2,4 diamino-6-mercapto pyrimidine sulfate (DAMPS) have been shown to perform well as an H_2S corrosion inhibitor on carbon steel.[26]

Cathodic inhibitors are not used in production operations but have been used in drilling fluids.

An example of a cathodic inhibitor is zinc(II) ions in zinc oxide, which retards the corrosion by inhibiting the reduction of water to hydrogen gas.

Vapor-phase corrosion inhibitors (VpCIs) are organic compounds that have sufficient vapor pressure under ambient atmospheric conditions to essentially travel to the surface of the metal by gas diffusion and physically adsorbing onto the surface.[27] Examples are:

- dicyclohexylamine nitrite
- dicyclohexylamine carbonate
- diethylamine phosphate
- small volatile amines such as trimethylamine
- benzotriazole

In the presence of moisture, a VpCI molecule becomes polarized and different parts of the molecule will be attracted to the anode and the cathode of the metal. For example, cyclohexylamine nitrite will form the cyclohexylammonium cation and nitrite anion. The cation adsorbs to the metal and the hydrophobic part of the molecule forms a protective barrier to contaminants such as oxygen, water, chlorides, and other corrosion accelerators. With the protective barrier in place, the corrosion cell cannot form and corrosion is halted. The anion will also act as a corrosion inhibitor.

Small neutralizing volatile amine VpCIs are occasionally used in wet gas lines or gas coolers but are less practical as once-through gas-phase corrosion inhibitors since concentrations need to be relatively high. The objective of the use of the small amines is not to neutralize all the acid gas in the gas stream, only the amount that dissolves in any condensed water. Small amines are also used in closed-loop applications. Blends of light amines and imidazolines (discussed under Section 8.4 film-forming corrosion inhibitors) have been shown to be good for inhibition of vapor phase corrosion in gas pipelines.[28] Aminocarboxylate VpCIs have been used with film-forming corrosion inhibitors for oil and gas pipeline applications.[29]

8.4 FILM-FORMING CORROSION INHIBITORS

FFCIs are particularly useful for the prevention of chloride, CO_2, and H_2S corrosion. They can be deployed in continuous injection or batch treatment either downhole or at the wellhead. Encapsulated time-release pellets containing FFCIs have also been deployed downhole.[30,31] CI squeeze treatments in production and water supply wells have also been reported.[32,33] Emulsified CI compositions to prevent corrosion and hydrogen embrittlement in oil wells have been claimed.[34] Other examples are emulsified blends of thiophosphates and pyrophosphates.[35]

8.4.1 How Film-Forming Corrosion Inhibitors Work

The effectiveness of an FFCI is partly determined by the strength of its adsorption to the metal surface (or a ferrous scale surface such as siderite, iron carbonate) forming a protective layer that physically prevents corrosive chemicals such as water and chloride ions from penetrating to the metal surface.[36] FFCIs can be small molecules or polymers. However, many FFCIs are organic amphiphiles (surfactants) with a polar headgroup and a hydrophobic tail. The headgroups are designed to interact with iron atoms on the surface and the hydrophobic tails attract liquid hydrocarbons forming an oily film, which further prevents the corrosive aqueous phase from penetrating to the metal surface (Figure 8.2). If the tail is long, a protective double layer (bilayer) of surfactant FFCI with or without a cosurfactant or solvent can also be formed.[37] It has been shown that under some multiphase flow conditions, parts

Corrosion Control during Production

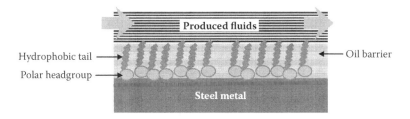

FIGURE 8.2 Effect of a film-forming corrosion inhibitor surfactant.

of the pipe wall are left unprotected by surfactant FFCIs, giving the possibility of localized corrosion.[38] There are also a number of corrosion inhibitors, both small molecules and polymers, which adsorb onto metal surfaces but do not have large hydrophobic tails, and therefore do not attract the liquid hydrocarbon phase to the metal surface. Thus, the "film" is probably made up of the inhibitor alone or as iron complexes, yet this still gives good corrosion inhibition. These will still be referred to as FFCIs. Examples are discussed later in this section. Computer modeling and quantitative structure-activity relationship (QSAR) analysis have been used to help design improved or more environment-friendly FFCIs.[39–41]

Some FFCIs are better than others at preventing corrosion underneath solid deposits such as siderite or weld corrosion.[42–44] Other FFCIs are particularly good at preventing erosion corrosion.[45] A novel multifunctional corrosion inhibitor formulation that is extremely effective at removing oily deposits ("schmoo") from the internal surface of pipeline has been reported.[46] The structures of the FFCIs in these categories are not described in the references given.

8.4.2 Testing Corrosion Inhibitors

There is a whole gamut of methods for corrosion inhibitor testing in the oil industry. These include the following:

- bubble or kettle test
- rotating cylinder electrode (RCE) test
- rotating disc electrode (RDE) test
- jet impingement test
- high-shear autoclave
- rotating cage test
- flow loop test
- wheel test
- static test

The objective of a corrosion test is to measure weight loss over time, observe surface changes, or measure current flow and interpret that as a corrosion current. In many of the above tests, metal coupons of the same composition as the pipe are placed in the apparatus. Corrosion rates are usually monitored using linear polarization resistance or electrochemical impedance spectroscopy.[47] From this, the rate of corrosion as the number of millimeters of metal loss per year can be calculated. Pitting corrosion can be observed using optical microscopy or scanning electron microscopy. In reality, the film-forming corrosion inhibitor will be exposed to turbulent fluid flow, which can lead to the inhibitor being stripped from the pipe walls if it does not adsorb well. Therefore, at some stage, a test with turbulent flow needs to be carried out to qualify an inhibitor. For low cost, initial chemical screening under low-shear conditions, the bubble test is often used. The RCE and RDE tests are fairly simple low-cost methods to test corrosion inhibitors under medium shear stress or turbulent regimes. More realistic but more expensive tests can be carried out in flow loops. The jet impingement test or high-shear autoclave are useful for very high shear conditions that lead to flow induced localized corrosion such as slug flow.[49] Flow loops can also be used.[50] Flow loop test procedures for investigating erosion corrosion have also been reported.[45,51] Techniques for measuring pitting and weld corrosion have been reported.[44] For multiphase well streams, the effect of the liquid hydrocarbon phase will also need to be addressed as the FFCI will partition between this phase and the aqueous phase. Preferential oil-wetting of the metal surface may also occur. The effects of solids in removing inhibitors from solution by adsorption may also need evaluating. For more information on FFCI test methods, the reader is referred to two useful, short reviews with many references comparing the laboratory tests methods for evaluating FFCIs.[52–54] Further, articles describing test procedures that can help users test and select inhibitors for CO_2/H_2S corrosion in a fast and cost-effective manner has also been published.[55] Finally, a review of analysis methods for FFCIs has also been published.[56]

Compatibility testing of FFCIs with other production chemicals often needs to be carried out. For example, many FFCIs adversely affect the performance of scale inhibitors and kinetic hydrate inhibitors (see Chapters 3 and 9). Other compatibility tests for FFCIs include thermal stability, foaming, solvent flashing, and compatibility with materials such as elastomeric seals. For example, in gas systems, it is frequently required that FFCIs possess low foam-creating properties. Further, an FFCI should not exacerbate the formation of emulsions, making oil-water separation more difficult.

8.4.3 Efforts to Develop More Environment-Friendly Film-Forming Corrosion Inhibitors

There has been a traditional preference for oil-soluble FFCIs in the oil and gas industry. For example, due to some oil-soluble FFCIs having better thermal stability, they perform better than water-soluble FFCIs in downhole continuous injection applications. Oil-soluble FFCIs will have a high bioaccumulation potential and tend to be toxic, but greener oil-soluble FFCIs have been developed.[57] There have also been developed better water-dispersible or water-soluble FFCIs. This is because many fields are in tail production and are producing more water than hydrocarbons. Efficiency can be improved by utilizing the majority fluid in the wells as the carrier for the inhibitor. Also, toxic aromatic carrier solvents used for oil-soluble FFCIs can be replaced by water or other more environment-friendly solvents for the

Corrosion Control during Production

water-soluble FFCIs such as alcohols and glycols. The solubility of FFCI surfactants can be changed by the hydrophilicity of the headgroup or by varying the length of the hydrophobic tail. However, if the tail is too short (usually less than 12 carbons), a good oily film of hydrocarbons with the FFCI will not form on the metal surface to prevent corrosion.[58] Conversely, FFCIs with very long hydrophobic tails (> C20) may be ripped off the walls at very high shear. The FFCI water solubility, r dispersibility can also be changed by using cosurfactants; although in some cases, this can lead to reduced film formation and film persistency. Cosurfactants may also exacerbate emulsion problems.

Many of the traditional FFCI surfactants discussed later, particularly nitrogenous surfactants, exhibit fairly high toxicity. Therefore, there has been a demand to develop less-toxic FFCIs and/or products, which readily biodegrade to smaller less toxic chemicals. One method to reduce toxicity is to reduce the length of the hydrophobic tail of the FFCI surfactant to about below eight or nine carbon atoms. This will also give a more water-soluble product with lower bioaccumulation potential. However, this reduction in chain length will greatly reduce the film-forming properties of the FFCI giving little corrosion protection. However, the efficacy of the FFCI can be increased by combination with a wetting agent, such as an ethoxylated alcohol having from about eight to ten carbon atoms. This has been demonstrated with ethoxyimidazolines.[59] For quaternary ammonium FFCIs or imidazolines, which could be protonated in acid media, a method to reduce their toxicity is to make the molecules zwitterionic, that is, by neutralizing the positive charge on the head group by adding anionic groups such as carboxylate groups. Monomeric or polymeric FFCIs, such as the biodegradable polyamino acids or various sulfur-nitrogen compounds, with little or no hydrophobic tails have also been developed, which are lower in toxicity than many surfactant FFCIs. These essentially hydrophilic polymers will not attract liquid hydrocarbons to help form a hydrophobic film at the metal surface, but they can cover the metal surface adequately by themselves to prevent corrosion. For example, polyaminoacids are now used as biodegradable combined corrosion and scale inhibitors. These and other categories of FFCIs are discussed below with their environmental properties, where known.

8.4.4 CLASSES OF FILM-FORMING CORROSION INHIBITORS

Regarding the structure of FFCIs, most contain heteroatoms in one or more headgroups, which bind via lone electron pairs to iron atoms on the metal surface. Typically, one finds nitrogen, phosphorus, sulfur, and oxygen atoms in the headgroup. The most common categories of surfactant FFCIs are:

- phosphate esters
- various nitrogenous compounds
- sulfur compounds often with other heteroatoms such as nitrogen

Biodegradable and low-toxicity polyaminoacids have also been used in environmentally sensitive areas. The various nitrogenous compounds include:

- amine salts of (poly)carboxylic acids
- quaternary ammonium salts and betaines (zwitterionics)
- amidoamines and imidazolines
- polyhydroxy and ethoxylated amines/amidoamines
- amides
- other heterocyclics

Amines such as fatty alkyldiamines and polyamines with hydrophobic tails have also been claimed as FFCIs.[60] Besides film-forming properties, amines will help neutralize corrosive carbonic acid (H_2CO_3) and hydrogen sulfide (H_2S) in the aqueous phase. Oxazolines, pyrrolinediones, and rosin amines have also been claimed as FFCIs in older patents but are not in general use today. Some FFCIs work synergistically together, for example, imidazolines and phosphate esters are often used together. Many other corrosion inhibitor formulations including two or more classes of FFCI are possible. Thus, one FFCI may be best at protecting the cathode on the ferrous metal surface, while another protects the anode, or a blend of two products may produce a better film. Potentially cheap and environment-friendly corrosion inhibitors based on natural products such as tobacco extract, molasses, and extracts of leaves and plants have been investigated but more work is needed to make them commercially competitive.[61,62] Small inorganic molecules such as passivating inhibitors can also be added to improve the corrosion protection or other synergists such as sodium thiosulfate or thiophosphate.[63,64] In certain applications, however, particularly in the presence of SRB, thiosulfate can cause pitting, crevice corrosion, and stress corrosion cracking of iron-base alloys. It can also act as an oxidizing agent and be reduced by thiosulfate-reducing bacteria further contributing to localized corrosion.[65]

There are many articles and conference papers discussing results on new FFCIs but rarely with the chemistry described. Therefore, most of the following references on FFCIs are taken from the patent literature.

8.4.4.1 Phosphate Esters

Phosphate esters, both monoesters and diesters, are good FFCIs and are often used in blends with other classes of FFCI. They are made by reacting alcohols or alkylphenols, or alkoxylated derivatives of such, with phosphating agents such as phosphorus pentoxide or orthophosphoric acid.[66] A mixture of both the monoester and diester is formed (Figure 8.3), which, having different hydrophilicities, will partition between the liquid hydrocarbon and water phases. Phosphate esters containing hydrophobic nonylphenol group have been shown to be considerably more effective FFCIs then linear or branched aliphatic phosphate esters.[67,68] Nonylphenol diesters seem to be more effective then the corresponding monoesters. Phosphate esters form fairly insoluble Fe(II) and Ca(II) salts, which may deposit on the pipe walls, hindering further corrosion.[69] Phosphate esters of poly-oxyalkylated thiols (e.g., octyl or dodecyl mercaptan reacted with various amounts of ethylene oxide) are especially useful as pitting FFCIs.[70] They have been claimed as FFCIs for ferrous metals, particularly in deep gas wells.[71] Poly(oxy-1,2-ethandiyl) tridecyl hydroxy phosphate has been claimed as a preferred FFCI in formulations with quaternary ammonium salt FFCIs such as didecyldimethyl quaternary ammonium chloride and thiocarbonyl

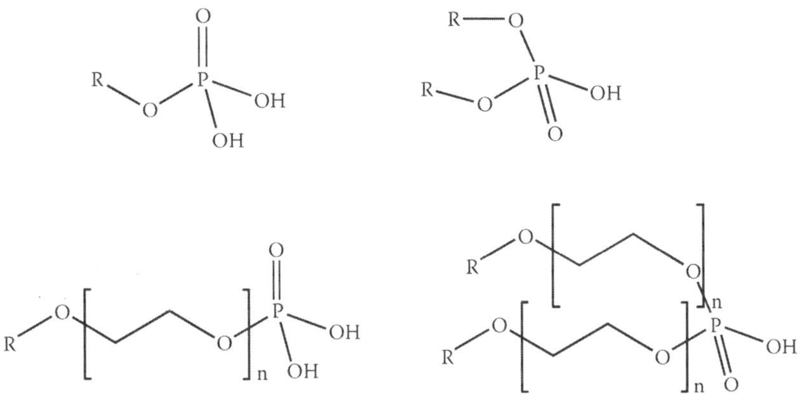

FIGURE 8.3 Structures of typical phosphate ester FFCIs.

compounds.[72] Amine salt reaction products of phosphate esters with various amines, such as acylated polyamines, morpholine, and ethoxylated fatty amines, have been claimed as inhibitors for general corrosion and cracking-type corrosion with improved environmental properties.[73,74]

8.4.4.2 Amine Salts of (Poly)Carboxylic Acids

Amine salts of fatty carboxylic acids have long been used in FFCI formulations. The amine is typically a trialkylamine, alkylpyridine, alkylquinoline, or imidazoline. Blends with mercaptocarboxylic acid amine salts have been claimed to give improved performance.[75] Oil-soluble dimer/trimer acid–based FFCIs are produced by the thermal condensation of functionalized C18 fatty acids (containing one or two double bonds, e.g., oleic and linoleic acids) to give varying amounts of C36 (dimerized) and C54 (trimerized) fatty acids. These dimer and/or trimer fatty acids are neutralized with an appropriate amine to produce a corrosion inhibitor.[76] A related patent claims the reaction products of maleic anhydride or fumaric acid with tall oil fatty acids, neutralized with an appropriate amine, to produce oil-soluble corrosion inhibitors.[77] Dimer/trimer mixtures are usually coformulated with materials such as fatty acid imidazolines and certain oils. C22 tricarboxylic acids neutralized with aminoethylethanolamine, an imidazoline, or an amidoamine have also been claimed as water-soluble FFCIs.[78] An FFCI with excellent film-forming and film-persistent characteristics may be produced by first reacting a polybasic acid or maleated fatty acids with a polyalcohol to form a partial ester, which is then reacted with imidazoline and/or fatty diamines to salt the ester.[79,80] The oil or water solubility of the product can be varied by the addition of cosurfactants. Amine salt FFCIs, specifically claimed to be environment-friendly, can be made by reacting an acid selected from the group consisting of a fatty acid anhydride and a 21-carbon dibasic acid with an alkylamine or imidazoline.[81] The product is formulated by dissolving the inhibitor in a fatty acid oil or ester, adding water-dispersing agents consisting of sulfonates and a long-chain ethoxylated alcohol and adjusting the viscosity with an alcohol such as isopropanol.

8.4.4.3 Quaternary Ammonium and Iminium Salts and Zwitterionics

Quaternary ammonium surfactants have long been known as FFCIs. They are rarely used alone but have been used in blends with other FFCIs. They are usually quite toxic, many of them being also useful as biocides, which can help prevent biofilm formation and thus underdeposit corrosion. Typical surfactants include benzalkonium chlorides, such as fatty alkyl benzyl dimethylammonium chlorides, alkyl pyridine quaternary surfactants, ethoxylated quaternary ammonium surfactants, and surfactants with two long alkyl chains such as didodecyldimethylammonium chloride (Figure 8.4). Didecyldimethyl quaternary ammonium chloride is claimed as a preferred quaternary surfactant FFCI for downhole applications either alone or blended with additional components such as phosphate esters and thiocarbonyl compounds.[72] Bis-quaternary ammonium surfactants, as well as amine oxides, based on derivatives of ethylene diamine have also been claimed.[82] All these quaternary surfactants can be made more biodegradable by placing weak links between the long alkyl chain and the quaternary nitrogen atom. The most common weak link is an ester group, which can be hydrolyzed. However, the resulting smaller, less surface-active quaternary ammonium compound produced degrades only very slowly. Quaternary surfactant anti-agglomerant low-dosage hydrate inhibitors (AAs) also function as FFCIs. Thus, when AAs are used on a continuous basis, it may be unnecessary for other specialized corrosion inhibitors (see Chapter 9).

Zwitterionics such as betaines can also function as FFCIs and are generally significantly less toxic than ordinary quaternary surfactants (zwitterionic imidazolines are discussed later). Betaines, which are amphoteric, also contain a quaternary ammonium center but the counter ion is now covalently bonded to the surfactant as a carboxylate group.[83,84] Long-chain alkyl propylenediamine FFCIs reacted via Michael addition with at least 1 mol acrylic acid also show low marine toxicity, lower than the original diamine.[85] They are probably zwitterionic in aqueous solution. In general, the toxicity of these products decreases with increasing acrylic acid substitution.

Zwitterionic water-soluble iminium compounds have also been claimed as FFCIs (Figure 8.5).[86] Preferred examples are made by reacting imines with acrylic acid.

Under some conditions, certain bis-quaternary ammonium surfactants have been claimed to perform better as FFCIs than mono-quaternary surfactants.[87,88] They can be made by reacting an epihalohydrin with 2 mol of a tertiary amine, such as dimethyldodecylamine, one of which can be protonated with acid. Quaternary or alkoxylated derivatives of hydrocarbon-soluble polyalkylenepolyamines show good CO_2 corrosion inhibition.[89] A biodegradable polymeric quaternary ammonium salt

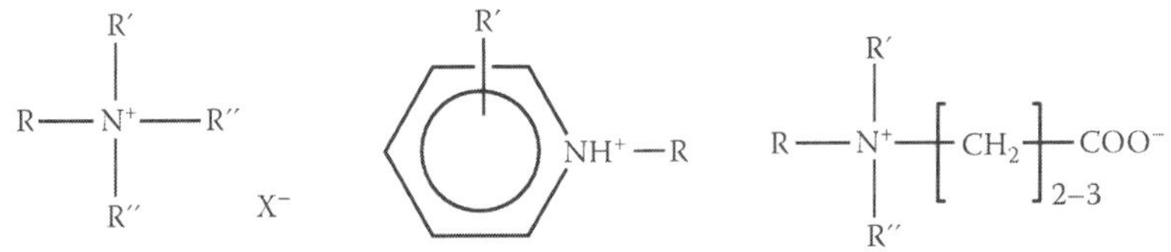

FIGURE 8.4 Quaternary ammonium, alkyl pyridine quaternary, and zwitterionic betaine FFCIs.

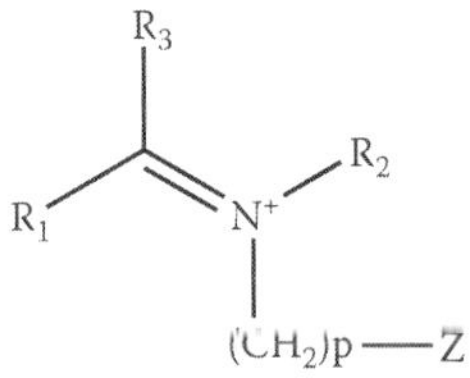

Corrosion Control during Production

FIGURE 8.5 Zwitterionic water-soluble iminium compounds where Z is carboxylate or various other anionic groups and at least one of R_1, R_2, or R_3 is a hydrophobic group.

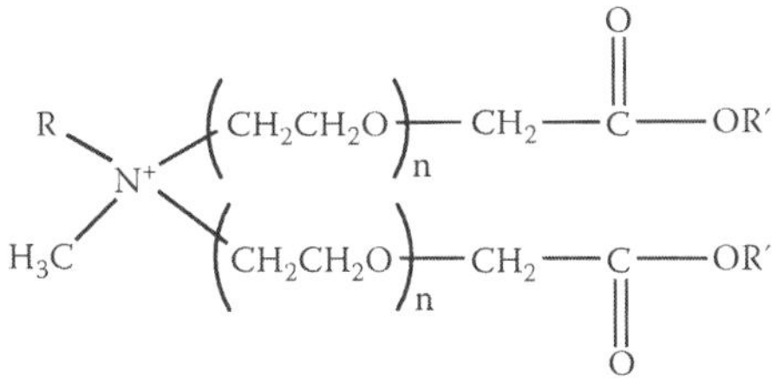

FIGURE 8.6 An example of doubly N-alkoxylated and carbonylated ammonium salt FFCIs.

biocide that also functions as an FFCI is made by reacting a polyepihalohydrin with a tertiary amine.[90] The polyepihalohydrin is prepared by a polymerization reaction of an epihalohydrin in the presence of a monomeric polyalcohol, such as glycerin. Typical amines are dodecyldimethylamine or imidazoline condensed with 4 mol of ethylene oxide or alkyl pyridines. Thiazine quaternary ammonium salts of polyepihalohydrin have also been claimed as FFCIs besides other uses.[91]

Quaternary surfactants with four alkyl or aryl groups attached to the nitrogen atom are poorly biodegradable. Therefore, their acute toxicity will persist for some time after discharge if the hydrophobic tail is not degraded. More biodegradable quaternary FFCIs can be made by introducing weak linkages in the hydrophobic tail(s) such as ester or amide groups. This is has been done for several new types of quaternary surfactant fabric conditioners but is less common for FFCIs. However, it has been found that doubly N-alkoxylated and carbonylated ammonium salts are effective FFCIs with good biodegradability.[92–92a] In one preferred example, a long-chain alkylamine or diamine is doubly alkoxylated, then reacted with chloroacetic acid, esterified, and then quaternized (Figure 8.6). Mono and bis-ester derivatives of pyridinium and quinolinium compounds have been claimed as environmentally friendly FFCIs.[92b] A biocide made by derivatizing metronidazole to make a quaternary surfactant also has good corrosion inhibition properties (Figure 8.7).[93]

8.4.4.4 Amidoamines and Imidazolines

Imidazolines are possibly the most common class of general corrosion FFCI used in the oil and gas industry and the most studied. Certain imidazoline-based FFCIs appear to perform well even in high-pressure, high-temperature (HPHT) conditions.[94] Although the basic imidazolines only offer poor to moderate performance at these conditions,[95] other HPHT FFCIs have been reported.[96] Imidazolines are

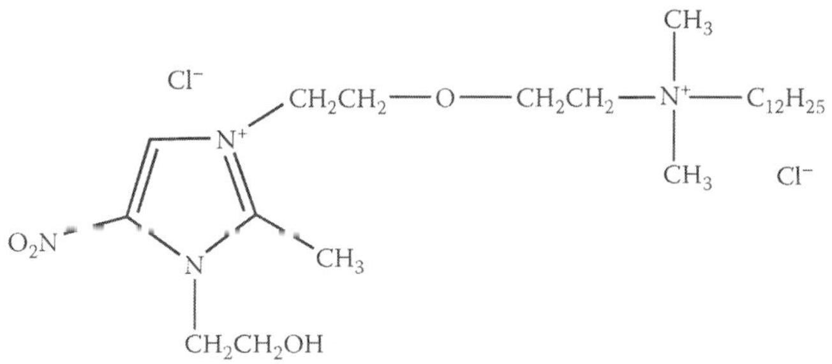

FIGURE 8.7 An FFCI based on the biocide metronidazole.

made by condensing a polyamine containing a 1,2-diaminoethane functionality with a carboxylic acid. This first forms an amide; a 2-alkylimidazoline is the main final product (Figure 8.8). N-substituted 2-alkylimidazoline products are formed if the 1,2-diaminoethane is substituted on one of the nitrogen atoms. For example, diethylenetriamine has been commonly used as the polyamine. With 2 mol of carboxylic acid, amidoimidazolines are formed, which are also well-known FFCIs. Fatty amines are usually used to make imidazolines, but ethercarboxylic acids can also be used.[97] Acyclic amidoamines are a side product of the reaction and are usually still present in imidazoline-based FFCIs. Bis-imidazolines can also be formed if the polyamine is big enough. A corrosion inhibitor comprising a dispersant, an imidazoline or bis-imidazoline, an amide, an alkyl pyridine, and a heavy aromatic solvent

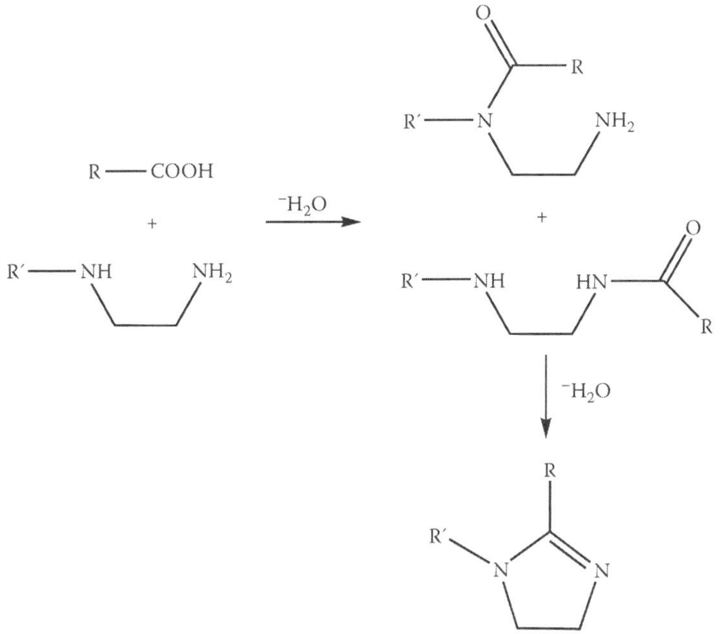

FIGURE 8.8 Formation of imidazolines.

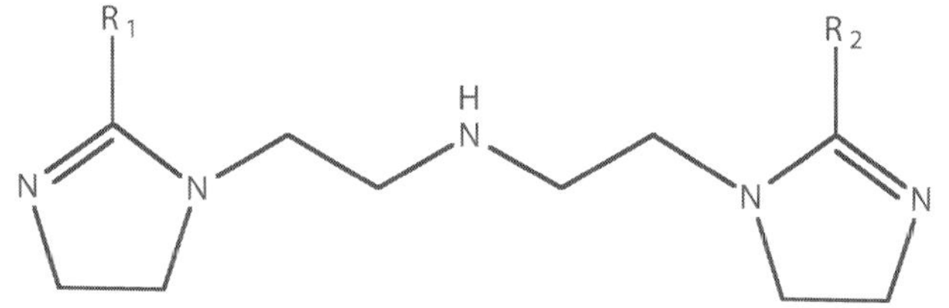

Corrosion Control during Production

FIGURE 8.9 Structure of bis-imidazolines.

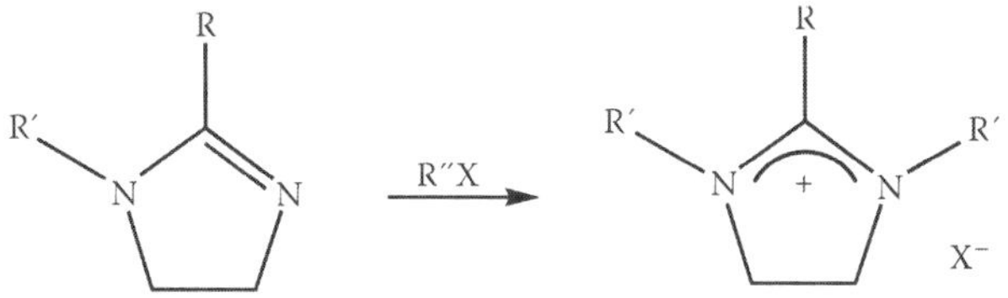

FIGURE 8.10 Protonation ($R'' = H$) or alkylation ($R = $ alkyl) of imidazolines.

has been patented as an FFCI formulation (Figure 8.9).[98] Protonation or alkylation of the imidazoline product leads to more hydrophilic imidazolinium salts, which also have corrosion inhibition properties (Figure 8.10). Protonation of imidazolines can also occur after injection due to acids in the produced water.

Tetrahydropyrimidines, six-ring analogues of imidazolines, and methylol derivatives thereof are also useful FFCIs.[100] Bis-amides made by reacting a diamine with dimer acids have also been claimed as FFCIs. An example is the reaction of the product of 2 mol of N-oleyl-1,3-propylenediamine and 1 mol dimer acid.[101]

A study of oleic imidazoline has been carried out using corrosion inhibitor testing, second harmonic generation at surfaces and molecular modeling techniques.[102] The results showed that the molecule is primarily bonding through the five-member nitrogen ring, which is lying planar to the metal surface, that the long hydrocarbon chain is playing an important role in the mechanism of inhibition, and that varying the chemistry of the pendant side chain does not affect the performance of the molecule to a major extent. In contrast, density functional theory and Monte Carlo simulations indicate that imidazolines favor a perpendicular adsorption to a metal surface, while their protonated (or alkylated) species adsorb in parallel positions over the metal surface.[103,104] Theory and electrochemical tests both showed that N-substituted 2-alkyl-imidazolines appear to perform better than unsubstituted 2-methylimidazolines. A QSAR has been developed to predict the performance of imidazoline FFCIs.[105] In another study using the RCE, a linear relationship was obtained between the minimum effective concentration and length of the hydrocarbon tail on the imidazoline.[106] With the hydrophobic chain length less than 12 carbons, no corrosion inhibition is observed. The pronounced effect of the hydrophobic group on corrosion inhibition of imidazolines was related to their ad-micelles' bilayer cohesive energies.

One study has shown that the chain length of the hydrophobic tail in a normal imidazoline FFCI that gives optimum performance (C18) gives poorest environmental properties.[107] Ethoxylated imidazolines are a common subclass of imidazoline

FFCI in which the N-nitrogen in the ring or side-chain amine is ethoxylated with varying amounts of ethylene oxide to provide more water-soluble products with lower bioaccumulation potential and toxicity.[108] The starting imidazoline is preferably made from reacting a fatty acid with 2,2-aminoethylamino ethanol or a diethylenetetramine. This has been demonstrated with ethoxyimidazolines.[59] As the number of carbon atoms in the fatty acid chain is reduced, the efficacy of the corrosion inhibitor is increased by combination with a wetting agent, preferably an ethoxylated alcohol having from ~8 to ~10 carbon atoms. Water-soluble alkoxyimidazoline FFCIs have been shown to perform synergistically with oligophosphate ester FFCIs (phosphate esters of ethoxylated polyols), especially at low dosages.[109] The alkoxyimidazoline (probably as the alkoxyimidazolinium ion due to protonation from the phosphate ester) protects the cathode of the electrochemical cell and the oligophosphate protects the anode—this dual mechanism increases confidence that the inhibitor blend will continue to give corrosion protection even in extreme situations where one of the components might have failed.

Another way to make an imidazoline FFCI more water-soluble and less toxic is to react a pendant alkyl amine group of an imidazoline intermediate with stoichiometric amounts of an organic carboxylic acid, such as, for example, acrylic acid ($CH_2=CH_2COOH$; Figure 8.11).[110,111] This results in an ampholytic imidazoline, which, it is believed, may hydrolyze in water to produce the amide amine. Quaternized amido imidazolines, quaternized imino imidazolines, and quaternized substituted diethylamino imidazolines, such as quaternized diacrylamino imidazolines, again made from amine-substituted imidazolines and acrylic acid, have also been claimed as FFCIs.[112,114] An interesting imidazoline phosphate ester/diester FFCI has been reported, which should exist as zwitterions in water.[99] Amines, amidoamines, and imidazoline surfactants derivatized with carboxylic groups (for example, by reaction with acrylic acid), blended together with mercaptocarboxylic acids with two to six carbon atoms ($HS(CH_2)_nCOOH$) have been claimed as synergistic FFCI compositions.[115] These formulations are claimed to have low toxicity. For example, thioglycolic acid (TGA; $HSCH_2COOH$), in combination with imidazolines, has been used for North Slope operations in Alaska.[116] Inhibitor compositions based on sulfhydryl acids (also called mercapto- or thiocarboxylic acids) and poly(ethyleneamino)imidazoline salts have also been claimed.[117] Amine imidazoline mixtures can also be reacted (rather than just blended at room temperature) with mercaptocarboxylic acids such as TGA to produce useful FFCIs.[118] It has also been shown that low-toxicity FFCIs can be made by reacting all reactive nitrogen atoms (those with hydrogen atoms) in imidazolines or amines and amidoamines, with acrylic acid and chloroacetic acid, and

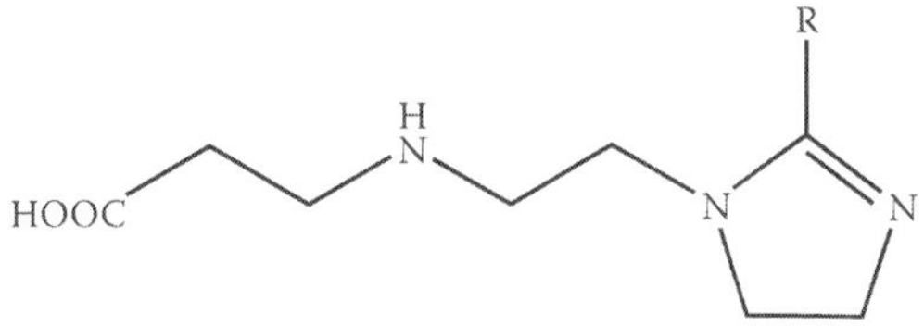

FIGURE 8.11 Example of an amineimidazoline reacted with one mole of acrylic acid.

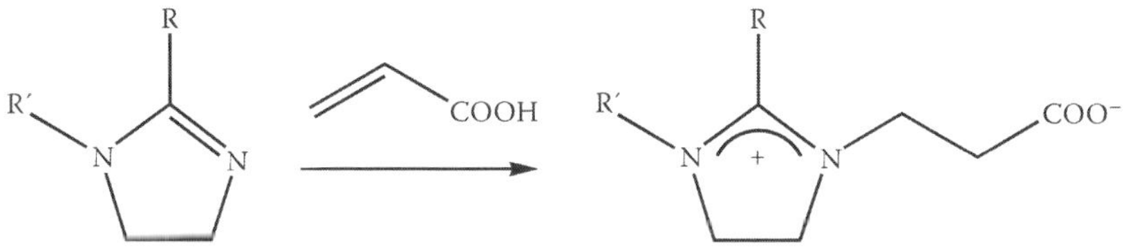

Corrosion Control during Production

FIGURE 8.12 Zwitterionic acrylated imidazoline FFCIs.

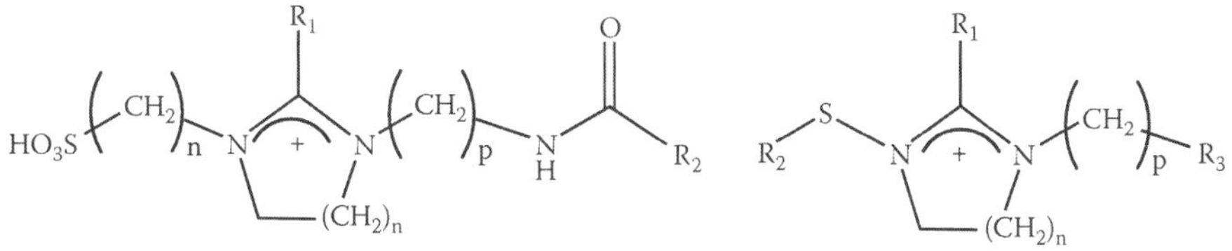

FIGURE 8.13 Examples of imidazoline derivatives containing sulfur.

then subsequent pH adjustment to 8–9 with a base.[119] Reaction of pendant amines in an imidazoline or amide with SO_2 gives products containing $-NSO_2$ groups that are also more water-soluble.[120]

Many imidazoline-based FFCIs have pendant groups on one of the nitrogen atoms in the ring structure (R' in Figure 8.8). These pendant groups often contain nitrogen atoms because polyamines such as diethylenetriamine or triethylenetetramine are used to make the imidazoline. It is possible that these pendant nitrogen atoms (or other heteroatoms with nonbonding electron lone pairs) can also interact with the metal surface enhancing the FFCI adsorption. However, it has been discovered that the presence of a pendant group to the imidazoline ring, which contains a heteroatom (nitrogen, oxygen, or sulfur) having a nonbonding pair of electrons available for interaction with a metal surface, is not required to achieve satisfactory corrosion inhibition.[121] In fact, it has been claimed that zwitterionic acrylated imidazolines that contain unsubstituted alkyl groups at the number 3 position, and thus, no hetero atoms or available nonbonding electrons, provide unexpectedly outstanding corrosion inhibition (Figure 8.12). An example is (N-propyl-2-heptadecenyl) imidazoline acrylate, which exists as the imidazolinium species.

Derivatives of imidazolinium ions containing sulfur have also been claimed as FFCIs.[122] Examples are given in Figure 8.13.

8.4.4.5 Amides

Amide derivatives of long-chain amines have been proposed as environmentally acceptable FFCIs in oil-production applications.[123] Unfortunately, such materials can be difficult to formulate and adversely affect the oil-water separation process. As mentioned earlier, amidoamines are present as by-products from the reaction of polyethyleneamines with carboxylic acids; imidazolines are the main product. Polymethylenepolyaminedipropionamides have been claimed as low-marine toxicity CO_2 corrosion inhibitors.[124] They are made by condensation of acrylamide with

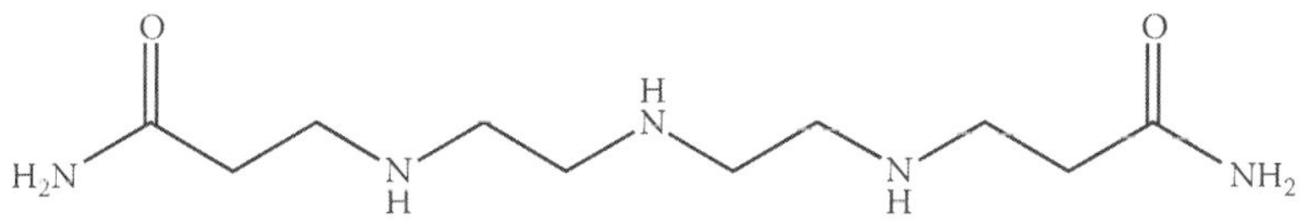

FIGURE 8.14 An example of a polymethylenepolyaminedipropionamide FFCI.

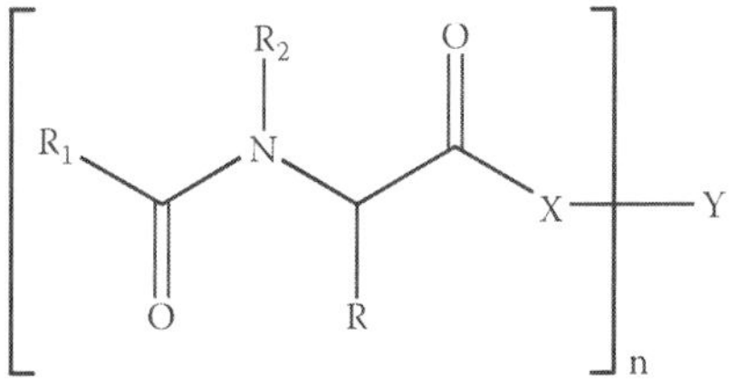

FIGURE 8.15 Acylated derivatives of amino acids appended to a suitable backbone Y via a link group X.

a polyamine such as diethylenetriamine, triethylenetetraamine, or bis(propylamino) ethylenediamine and contain no major hydrophobic groups. An example is the diamide given in Figure 8.14. Blends with mercaptoacids, such as TGA, work synergistically to improve the performance.[125]

A class of amide FFCI claimed to have high biodegradability are those based on acylated derivatives of amino acids appended to a suitable backbone (Figure 8.15).[126] For example, N-decanoyl-l-aspartic acid can be reacted (via the anhydride) with polyols, polyamines, and hydroxyamines to form useful FFCIs.

8.4.4.6 Polyhydroxy and Ethoxylated Amines/Amides

Ethoxylation of fatty amines or diamines with ethylene oxide gives ethoxylated amines, which are useful FFCIs (Figure 8.16). A biodegradable link, such as an amide group, between the hydrophobic tail and the ethoxylated nitrogen atom can make the product more environmentally attractive.[127] For example, tallow-$CONH(CH_2)_3NHCH_2CH_2OH$ is claimed to perform better as an FFCI than tallow-$NH(CH_2)_3NH_2$ ethoxylated with 3 mol of ethylene oxide.

Ethoxylation makes the amine molecule more water-soluble as well as giving extra sites (oxygen atoms) for adsorption to the metal surface. Another way to introduce

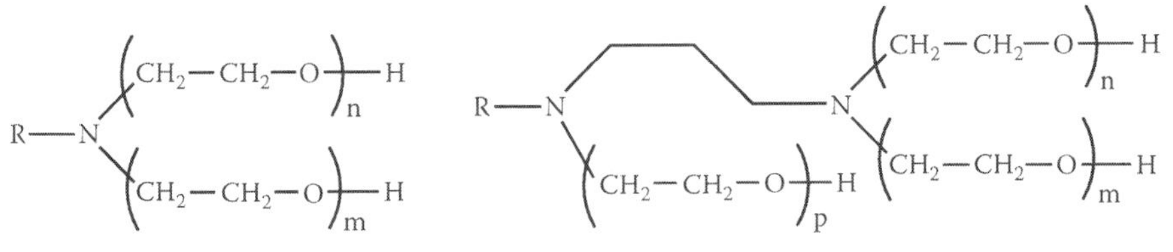

FIGURE 8.16 Ethoxylated fatty amine and diamine FFCIs. R can be an alkyl group or R'C(=O).

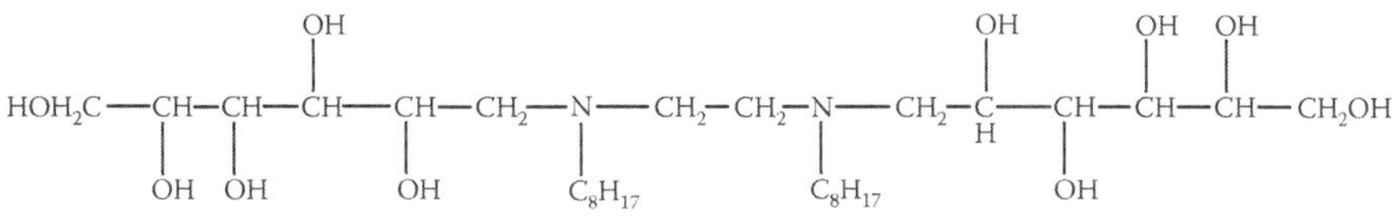

Corrosion Control during Production

FIGURE 8.17 N,N'-dioctyl-N,N'-bis(1-deoxyglucityl)ethylenediamine.

water solubility with oxygen atom functionality is to incorporate hydroxyl groups into the amine. Examples claimed to be good FFCIs are deoxyglucityl derivatives of alkylamines (Figure 8.17).[82]

8.4.4.7 Other Nitrogen Heterocyclics

Methyl substituted nitrogen-containing aromatic heterocyclic compounds such as 2,5-dimethylpyrazine, 2,3,5,6-tetramethylpyrazine, or 2,4,6-trimethylpyridine reacted with an aldehyde, such as 1-dodecanal, or ketone, such as 5,7-dimethyl-3,5,9-decatrien-2-one, to produce useful oil- and gas-well corrosion inhibitors.[128] The same nitrogen-containing aromatic heterocyclic compounds reacted with dicarboxylic acid monoanhydrides such as 2-dodecen-1-yl succinic anhydride have also been claimed for the same purpose.[129]

Downhole FFCIs can be made by reacting mixed nitrogen-containing mainly heterocyclic compounds (made from an unsaturated aldehyde and a polyamine such as acrolein and ethylenediamine) with a carboxylic acid, organic halide, or an epoxide.[129]

8.4.4.8 Sulfur Compounds

Thiosulfate ions and mercaptocarboxylic acids have been mentioned earlier as synergists for nitrogenous FFCIs. In fact, a number of sulfur compounds are particularly good at preventing cracking corrosion. FFCIs containing water-soluble mercaptocarboxylic acids (e.g., TGA, which is relatively low in toxicity) have been used successfully in high-shear stress applications; however, used alone, they are only partially effective at inhibiting corrosion in a CO_2-saturated environment (Figure 8.18) The reason that mercapto compounds perform as corrosion inhibitors has been shown to be due to oxidation of the mercapto group (–SH) to disulfide groups (–S–S–), which form complexes with iron ions at the metal surface.[65] Thus, the disulfide 3,3'-dithiodipropionic acid (DTDPA) has been shown to perform equally as well as TGA against general corrosion despite the fact that DTDPA is an oxidizing agent and TGA is a reducing agent (Figure 8.18). In fact, DTDPA is expected to work better than TGA against localized corrosion. Further, mercaptoalcohol (MA) was

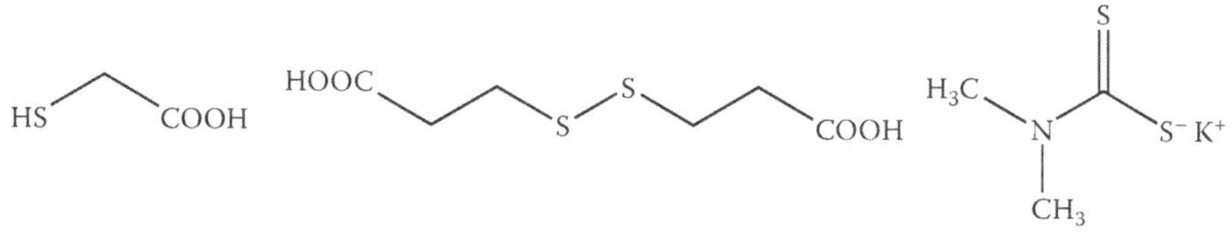

FIGURE 8.18 Thioglycolic acid, 3,3'-dithiodipropionic acid and potassium dimethyl dithiocarbamate.

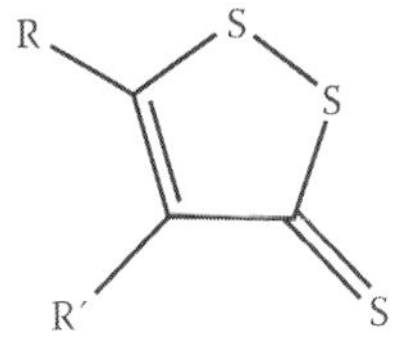

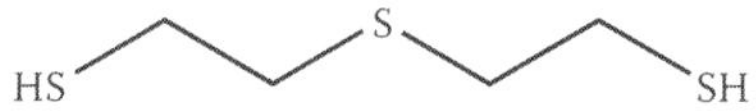

FIGURE 8.19 The structure of trithiones (left) and 2-mercaptoethyl sulfide.

a better inhibitor for both general and localized/pitting corrosion than TGA or DTDPA. MA is predominantly a cathodic inhibitor, while TGA is an anodic inhibitor. Thiocarbonyls such as potassium dimethyl dithiocarbamate are also synergists for FFCIs (Figure 8.18).

There are few patents claiming sulfur compounds without other heteroatoms as FFCIs. However, trithiones, with three sulfur atoms, and a suitable dispersant have been claimed as good inhibitors of CO_2 corrosion in sweet wells (Figure 8.19).[130] Trithiones, such as a quaternary of 4-neopentyl-5-*t*-butyl-1,2-dithiole-3-thione, have been particularly used in an environment where metal failure through stress cracking is a concern.[64,116] Trithiones have also been claimed in FFCI formulations for limited oxygen systems blended with thiophosphates, quaternaries, polyphosphate esters, and cyclic amidines such as fatty carboxylic acid salts of an imidazoline.[131] The thiophosphates are made by reacting oxyalkylated fatty alcohols with P_2S_5. Other claimed sulfur-only FFCIs include mercaptan-based products with one or more –SH groups and optionally a sulfido group, such as 2-mercaptoethyl sulfide (Figure 8.19).[132]

Thiophosphorus compounds have also been claimed to prevent organic and other naphthenic acid corrosion, although this is usually a high-temperature downstream problem.[133] Other claimed inhibitors of naphthenic acid corrosion include phosphorous acid and certain sulfur- and phosphorus-free aromatic compounds substituted with nitrogen, containing functional groups at the 5- or 3,5-position.[134,135]

A number of sulfur-nitrogen–based corrosion inhibitors for ferrous metals have been developed. Interestingly, many of them do not have a hydrophobic tail and therefore will not attract liquid hydrocarbon to the metal surface to form an oily film. Examples are the natural amino acids cysteine and cystine and their decarboxylated analogues cysteamine and cystamine, which are good synergists for polyaminoacids FFCIs.[136] Sulfur-nitrogen compounds, such as benzothiazoles, are more common for corrosion inhibition of other metals such as copper. However, ether carboxylic acids based on alkoxylated mercaptobenzothiazoles are also claimed to inhibit ferrous metal corrosion (Figure 8.20).[137] Another class of sulfur-nitrogen compounds claimed as FFCIs is based on the reaction product of a compound containing a carbonyl group (e.g., monoaldehydes), an amine (e.g., alkyl monoamines), and a thiocyanate (e.g., ammonium thiocyanate).[138]

Thiocarboxylic acids have very unpleasant odors and alternatives have been sought. 2,5-Dihydrothiazoles have been claimed as volatile corrosion inhibitors in gas wells (Figure 8.21).[139] However, they have limited water solubility. Thiazolidines have been claimed as more water-soluble FFCIs (Figure 8.21).[116] The synthesis

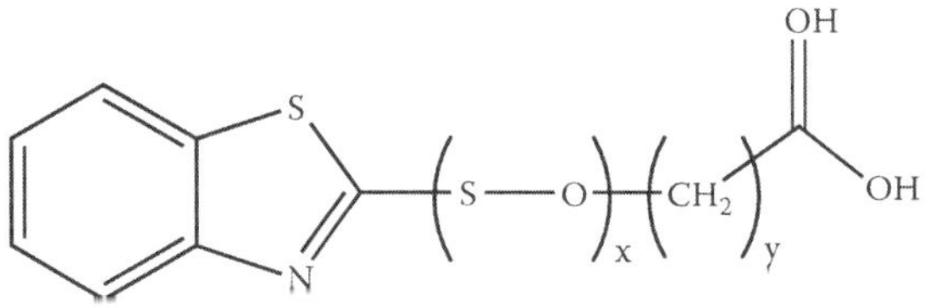

FIGURE 8.20 Ether carboxylic acids based on alkoxylated mercaptobenzothiazoles. The acid form or an amine salt of the acid can be used.

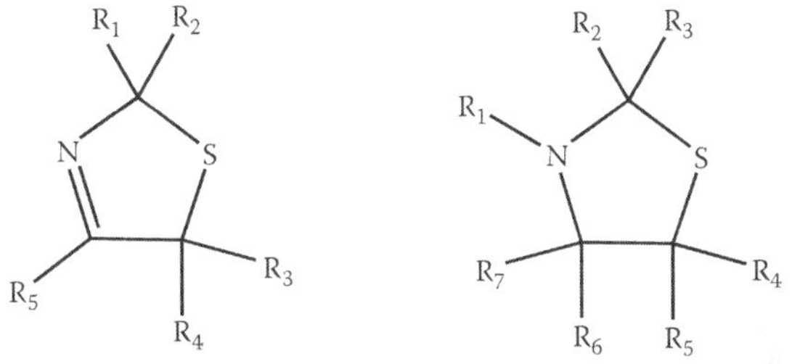

FIGURE 8.21 The structure of 2,5-dihydrothiazoles and thiazolidines.

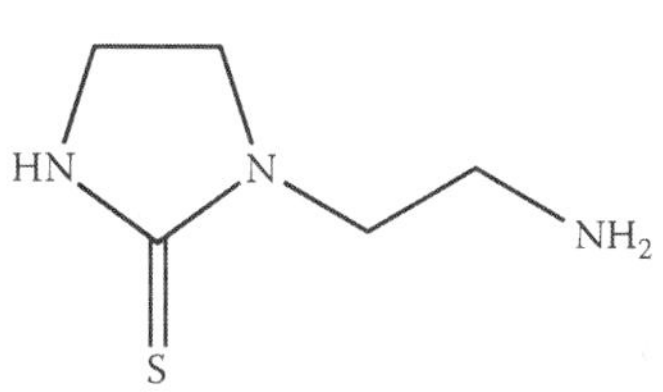

FIGURE 8.22 1-(2-aminoethyl)-2-imidazolidinethione.

method involves reacting a dihydrothiazole such as 2,5-dihydro-5,5-dimethyl-(1-methylethyl)thiazole, with a mixture comprising formic acid and an aldehyde. Preferred thiazolidine products have no large hydrophobic groups and are therefore not typical surfactants.

Small nonsurfactant imidazolidinethiones have been claimed as low toxicity, readily biodegradable FFCIs.[140] A specific example is 1-(2-aminoethyl)-2-imidazolidinethione made by reacting thiourea and diethylenetriamine, although other polyalkylene polyamines could be used (Figure 8.22).

Low molecular weight polyfunctional polymers prepared by polymerizing vinyl monomers, such as alkenoic acids and esters thereof, such as N-hydroxyethylacrylate, in the presence of mercaptan chain transfer agents such as dodecylmercaptan ($C_{12}H_{25}SH$) are claimed to be particularly good for downhole CO_2 corrosion inhibition.[141]

A class of FFCI claimed to have increased biodegradability and reduced toxicity is salts of amidomethionine derivatives and amines (Figure 8.23).[142] Examples of

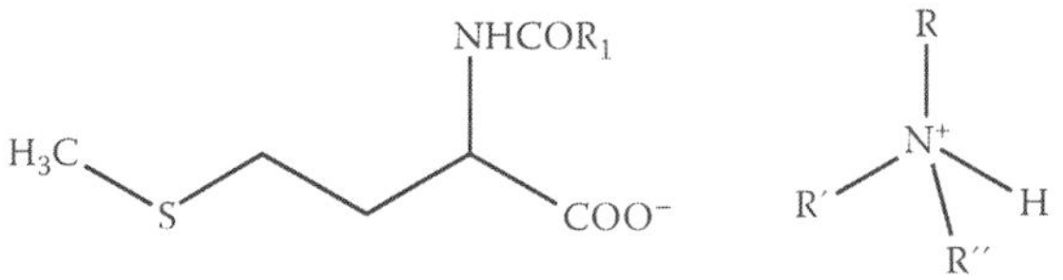

FIGURE 8.23 Salts of amidomethionine derivatives and amines.

amidomethionines are cocoyl- or octanoylmethionine. Examples of suitable amines include morpholine, triethanolamine, and dibutylamine.

8.4.4.9 Polyaminoacids and Other Polymeric Water-Soluble Corrosion Inhibitors

Phosphates and organic phosphonates have long been used as scale inhibitors and are also regularly used for corrosion protection in water treatment systems.[143–146] (see also Chapter 3 on scale control). An example is 2-hydroxy-2-phosphonoacetic acid $((HO)_2POCH(OH)COOH)$. However, by themselves, they are not usually effective enough for corrosion protection in the harsh environment of oil and gas production. Organic carboxylates and polycarboxylates also give some protection such as certain substituted carboxymethoxysuccinic acid compounds, aminohydroxysuccinic acids, or oligomers and polymers of tartaric acid (also known as polyepoxysuccinic acids).[147–149] Many of these organic polycarboxylates will also function as scale inhibitors.

Later, it was been found that polyaminoacids with pendant carboxylate groups function very well as scale inhibitors, but they also gave fairly good CO_2 corrosion protection although not as good as the best FFCIs at the time.[150,151] The best known and cheapest examples of polyaminoacids are salts of polyaspartic acid, although glutamic acid can be used (Figure 8.24).[152] Used alone, polyaspartates showed best potential in low-chloride, low-pH conditions, which are fairly rare in petroleum production.[107] The performance of polyaspartates can be improved by adding amino thiol or amino disulfide compounds.[153] Particularly effective inhibitor compositions are the natural amino acids cysteine and cystine and their decarboxylated analogues cysteamine and cystamine in combination with polyaspartic acid. These are very environment-friendly compositions having low toxicity and good biodegradability.

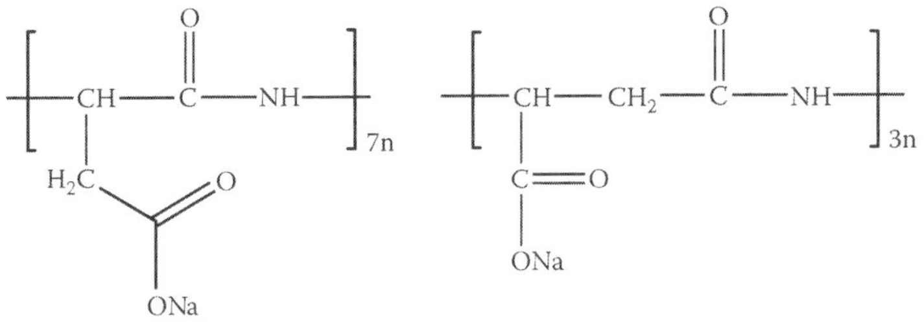

FIGURE 8.24 Sodium polyaspartate.

Another corrosion inhibitor patent claims the use of a substantially water-soluble polymer of an acidic amino acid and at least one water-soluble salt of molybdenum (molybdate) or zinc(II).[154] Aspartate-containing polymers have also been shown to work synergistically with alkylpolyglucosides as a new class of green corrosion inhibitor.[155,156] A corrosion- and scale-inhibitor package based on polyaspartate with added synergists has been developed and used in the North Sea. A study of 30 laboratory-synthesized polypeptides concluded that none of them were better than commercial polyaminoacids of the polyaspartate type.[157] Further, an ordered polymer gave better corrosion inhibition than a random polymer and the performance dropped considerably below a polymer molecular weight of 1,000. The authors also state that they have better, commercial, low-toxicity, nonpeptidic FFCIs.

Polyaspartates are made by hydrolyzing water-insoluble polysuccinimide. Environment-friendly amide derivatives of polysuccinimide, as well as amides of lactobionic acid, with C12–18 tails have been shown to perform well as sour gas corrosion inhibitors.[158] Polysuccinimide can also be reacted with hydroxyamines (e.g., ethanolamine) or hydroxylamine to give polymers with pendant hydroxyethyl or hydroxamic groups (R–C(=O)NHOH), respectively. Some of these polymers are claimed to be biodegradable scale-and corrosion inhibitors, although the homopolymers of 2-hydroxyethylaspartamide or 2-hydroxyethylglutamide are not biodegradable.[159,160] Other less biodegradable water-soluble polymers with pendant hydroxamic acid groups also function as corrosion inhibitors. An example is the reaction product of polyacrylamide with hydroxylamine.[161]

REFERENCES

1. (*a*) A. W. Peabody, *Control of Pipeline Corrosion*, 2nd ed., ed. R. L. Bianchetti, Houston, TX: NACE International, 2001. (*b*) B. Craig, *Oilfield Metallurgy and Corrosion*, 3rd ed., Houston, TX: MetCorr (NACE), 2004.
2. H. H. Uhlig and R. Winston Revie, *Corrosion and Corrosion Control*, 3rd ed., New York: Wiley-Interscience, 1985.
3. (*a*) G. V. Chilingar, R. Mourhatch, and G. Al-Qahtani, *The Fundamentals of Corrosion and Scaling: A Handbook for Petroleum and Environmental Engineers*, Houston, TX: Gulf Publishing Company, 2008. (*b*) M. Davies and P. J. B. Scott, *Oilfield Water Technology*, Houston, TX: National Association of Corrosion Engineers (NACE), 2006.
4. R. L. Martin, "Corrosion Consequences of Oxygen Entry into Sweet Oilfield Fluids," SPE 71470 (paper presented at the SPE Annual Technical Conference and Exhibition, New Orleans, LA, 30 September–3 October 2001).
5. E. Dayalan, F. D. de Moraes, J. R. Shadley, S. A. Shirazi, and E. F. Rybicki, "CO_2 Corrosion Prediction in Pipeflow under $FeCO_3$ Scale-Forming Conditions," Paper 51 (paper presented at the NACE CORROSION Conference, 1998).
6. R. Nyborg, "Controlling Internal Corrosion in Oil and Gas Pipelines. Business Briefing: Exploration and Production," *Oil and Gas Review* 2 (2005): 71.
7. J. A Billman, "Antibiofoulants: A Practical Methodology for Control of Corrosion Caused by Sulfate-Reducing Bacteria," *Materials Performance*, 36 (1997): 43.
8. W. Sun and S. Nesic, "A Mechanistic Model of H_2S Corrosion of Mild Steel," Paper 07655 (paper presented at the NACE CORROSION Conference, 2007).

9. J.-L. Crolet, "Microbial Corrosion in the Oil Industry: A Corrosionist's View," in *Petroleum Microbiology*, eds. B. Ollivier and M. Magot, Washington, DC, ASM Press, 2005, 143.
10. D. Pope, *Microbiologically Influenced Corrosion in Pipelines*, Houston, TX: Gulf Publishing Co., 2000.
11. S. W. Borenstein, *Microbiologically Influenced Corrosion Handbook*, New York: Woodhead, 1994.
12. (*a*) R. Javaherdashti, *Microbiologically Influenced Corrosion: An Engineering Insight*, London: Springer, 2008. (*b*) B. J. Little and J. S. Lee, *Microbiologically Influenced Corrosion*, Hoboken, NJ: Wiley-Interscience, 2007.
13. W. Sun and S. Nesic, "Basics Revisited: Kinetics of Iron Carbonate Scale Precipitation in CO_2 Corrosion," Paper 06365 (paper presented at the NACE CORROSION Conference, 2006).
14. H. G. Byars, *Corrosion Control in Petroleum Production*, TCP 5, 2nd ed., Houston, TX: Forbes Custom Publication, December 1999.
15. P. R. Rincon, J. P. McKee, C. E. Tarazon, and L. A. Guevara, "Biocide Stimulation in Oilwells for Downhole Corrosion Control and Increasing Production," SPE 87562 (paper presented at the SPE International Symposium on Oilfield Corrosion, Aberdeen, UK, 28 May 2004).
16. A. Dugstad and M. Seiersten, "pH-stabilization, a Reliable Method for Corrosion Control of Wet Gas Pipelines," SPE 87560 (paper presented at the SPE International Symposium on Oilfield Corrosion, Aberdeen, UK, 28 May 2004).
17. S. Ramachandran, S. Mancuso, K. A. Bartrip, and P. Hammonds, "Inhibition of Acid Gas Corrosion in Pipelines Using Glycol for Hydrate Inhibition," *Materials Performance* 45 (2006): 44–47.
18. G. Schmitt, "Drag Reduction by Corrosion Inhibitors—A Neglected Option for Mitigation of Flow Induced Localized Corrosion," *Materials and Corrosion* 52(5) (2001): 329.
19. G. Schmitt, M. Bakalli, and M. Hörstemeier, "Contribution of Drag Reduction to the Pexrformance of Corrosion Inhibitors in One- and Two-Phase Flow," Paper 615 (paper presented at the NACE CORROSION Conference, 2007).
20. J. D. Johnson, S.-L. Fu, M. J. Bluth, and R. A. Marble, U.S. Patent 5939362, 1999.
21. J. Palmer, W. Hedges, and J. Dawson, eds., *Working Party Report on the Use of Corrosion Inhibitors in Oil and Gas Production*, EFC39 (European Federation of Corrosion), Maney, 2004.
22. (*a*) V. S. Sastri, *Corrosion Inhibitors: Principles and Applications*, Chichester, UK: Wiley, 1998. (*b*) I. L. Rosenfeld, *Corrosion Inhibitors*, Moscow: M. Khimia Publisher, 1977.
23. A. M. S. El Din and L. Wang, "Mechanism of Corrosion Inhibition by Sodium Molybdate," *Desalination* 107 (1996): 29.
24. M. Saremi, C. Dehghanian, and M. Mohammadi Sabet, "The Effect of Molybdate Concentration and Hydrodynamic Effect on Mild Steel Corrosion Inhibition in Simulated Water Cooling," *Corrosion Science* 48 (2006): 1404.
25. E. Sletfjerding, A. Gladsø, S. Elsborg, and H. Oskarsson, "Boosting the Heating Capacity of Oil-Production Bundles Using Drag-Reducing Surfactants," SPE 80238 (paper presented at the International Symposium on Oilfield Chemistry, Houston, TX, 5–7 February 2003).
26. T. A. Ramanarayanan and H. L. Vedage, U.S. Patent 5279651, 1994.
27. B. Boyle, "A Look at Developments in Vapor Phase Corrosion Inhibitors," *Metal Finishing* 102(5) (2004): 37.
28. R. L. Martin, Paper 337 (paper presented at the NACE CORROSION Conference, 1997).

29. M. Kharshan and A. Furman, "Incorporating Vapor Corrosion Inhibitors (VCIs) in Oil and Gas Pipeline Additive Formulations," Paper 236 (paper presented at the NACE CORROSION Conference, 1998).
30. S. E. Campbell, U.S. Patent 7135440, 2006.
31. S. J. Weghorn, C. W. Reese, and B. Oliver, "Field Evaluation of an Encapsulated Time Release Corrosion Inhibitor," Paper 07321 (paper presented at the NACE CORROSION Conference, 2007).
32. S. Kokal, K. Raju, and A. Biedermann, "Cost Effective Design of Corrosion Inhibitor Squeeze Treatments for Water Supply Wells," SPE 53143 (paper presented at the Middle East Oil Show and Conference, Bahrain, 20–23 February 1999).
33. H. A. Nasr-El-Din, H. R. Rosser, and M. S. Al-Jawfi, "Formation Damage Resulting from Biocide/Corrosion Inhibitor Squeeze Treatments," SPE 58803 (paper presented at the SPE International Symposium on Formation Damage Control, Lafayette, LA, 23–24 February 2000).
34. E. C. French, W. F. Fahey, and J. G. Harte, U.S. Patent 5027901, 1991.
35. R. L. Martin, J. P. Mullen, P. E. Brown, and T. G. Braga, U.S. Patent 5753596, 1998.
36. (a) S. Ramachandran and K. Bartrip, "Molecular Modeling of Binary Corrosion Inhibitors," Paper 3624 (paper presented at the NACE CORROSION Conference, 2003). (b) A. Swift, A. J. Paul, and J. C. Vickerman, "Investigation of the Surface Activity of Corrosion Inhibitors by XPS and ToF SIMS," *Surface Interface Analysis* 20(1) (1993): 27.
37. S. S. Shah, T. G. Braga, B. A. Oude Alink, and J. Mathew, U.S. Patent 5456767, 1995.
38. H. Wang, H. Wang, H. Shi, C. Kang, and P. W. Jepson, "Why Corrosion Inhibitors Do Not Perform Well in Some Multiphase Conditions: A Mechanistic Study," Paper 2276 (paper presented at the NACE CORROSION Conference, 2002).
39. W. P. Singh, J. Ahmed, G. H. Lin, Y. Kang, and J. O'M Bockris, "About a Chemical Computational Approach to the Design of Green Inhibitors," Paper 33 (paper presented at the NACE CORROSION Conference, 1995).
40. W. P. Singh, G. Lin, J. O'M. Bockris, and Y. Kang, "Designing Green Corrosion Inhibitors Using Chemical Computational Methods," Paper 208 (paper presented at the NACE CORROSION Conference, 1998).
41. W. H. Durnie, "Modeling the Functional Behavior of Corrosion Inhibitors," Paper 4401, (paper presented at the NACE CORROSION Conference, 2004).
42. W. H. Durnie, M. A. Gough, and J. A. M. DeReus, Paper 5290 (paper presented at the NACE CORROSION Conference, 2005).
43. J. A. M. DeReus, E. L. J. A. Hendriksen, M. E. Wilms, Y. N. Al-Habsi, W. H. Durnie, and M. A. Gough, "Test Methodologies and Field Verification of Corrosion Inhibitors to Address under Deposit Corrosion in Oil and Gas Production Systems," Paper 5288 (paper presented at the NACE CORROSION Conference, 2005).
44. A. E. Jenkins, W. Y. Mok, C. G. Gamble, and G. E. Dicken, "Development of Green Corrosion Inhibitors for Preventing under Deposit and Weld Corrosion," SPE 87558 (paper presented at the SPE International Symposium on Oilfield Corrosion, Aberdeen, UK, 28 May 2004).
45. S. Ramachandran, Y. S. Ahn, K. A. Bartrip, V. Jovancicevic, and J. Bassett, "Further Advances in the Development of Erosion Corrosion Inhibitors," Paper 5292 (paper presented at the NACE CORROSION Conference, 2005).
46. D. I. Horsup, T. S. Dunstan, and J. H. Clint, "A Breakthrough Corrosion Inhibitor Technology for Heavily Fouled Systems," Paper 7690 (paper presented at the NACE CORROSION Conference, 2007).
47. M. Stern and A. L. Geary, "Electrochemical Polarization. A Theoretical Analysis of Shape of Polarization Curves," *Journal of Electrochemical Society* 104 (1957): 56.

48. T. Hong, Y. H. Sun, and W. P. Jepson, "Study on Corrosion Inhibitor in Large Pipelines under Multiphase Flow Using EIS," *Corrosion Science* 44 (2002): 101.
49. A. E. Jenkins, W. Y. Mok, C. G. Gamble, and S. R. Keenan, "Development of Green Corrosion Inhibitors for High Shear Applications," Paper 4370 (paper presented at the NACE CORROSION Conference, 2004).
50. D. Abayarantha, A. Naraghi, and N. Grahmann, "Inhibitor Evaluations Using Various Corrosion Measurement Techniques in Laboratory Flow Loops," Paper 21 (paper presented at the NACE CORROSION Conference, 2000).
51. M. Tandon, K. P. Roberts, J. R. Shadley, S. Ramachandran, E. F. Rybicki, and V. Jovancicevic, "Flow Loop Studies of Inhibition of ErosionCorrosion in CO_2 Environments With Sand," Paper 6597 (paper presented at the NACE CORROSION Conference, 2006).
52. S. Papavinasam, R. W. Revie, M. Attard, A. Demoz, and K. Michaelian, "Comparison of Laboratory Test Methodologies to Evaluate Corrosion Inhibitors for Oil and Gas Pipelines," *Corrosion* 59(10) (2003): 897.
53. S. Papavinsam, R. W. Revie, and M. Bartos, "Testing Methods and Standards for Oil Field Corrosion Inhibitors," Paper 4424 (paper presented at the NACE CORROSION Conference, 2004).
54. S. Papavinasam, R. W. Revie, T. Panneerselvam, and M. Bartos, "Standards for Laboratory Evaluation of Oil Field Corrosion Inhibitors," *Materials Performance* 46 (2007): 46.
55. (*a*) S. D. Kapusta, "Corrosion Inhibitor Testing and Selection for Exploration and Production: A User's Perspective," *Materials Performance* 38 (1999): 56. (*b*) S. Stewart, V. Jovancicevic, C. M. Menedez, and J. Maloney, New Corrosion Inhibitor Evaluation Approach for Highly Sour Conditions, Paper 09360, NACE, Corrosion 2009 Conference and Exposition, Atlanta, GA, March 22–26, 2009.
56. A. J. Son, "Developments in the Laboratory Evaluation of Corrosion Inhibitors: A Review," Paper 7618 (paper presented at the NACE COROSION, 2007).
57. C. W. Bowman, W. Y. Mok, S. R. Keenan, C. G. Gamble, and S. Jarrett, "Environmental Constraints of Oil Soluble Corrosion Inhibitors: Challenges and Opportunities," Proceedings of Chemistry in the Oil Industry IX, Royal Society of Chemistry, 2005, 95.
58. V. Jovancicevic, S. Ramachandran, and P. Prince, "Inhibition of CO_2 Corrosion of Mild Steel by Imidazolines and Their Precursors," Paper 18 (paper presented at the NACE CORROSION Conference, 1998).
59. T. G. Braga, R. L. Martin, J. A. McMahon, B. A. Oude Alink, and B. T. Outlaw, U.S. Patent 6338819, 2002.
60. A. H. Schroeder, T. A. Ching, S. Suzuki, and K. Katsumoto, International Patent Application WO/1988/005039.
61. S. Taj, A. Sidekkha, S. Papavinsam, and E. W. Revie, "Some Natural Products as Green Corrosion Inhibitors," Paper 7630 (paper presented at the NACE CORROSION Conference, 2007).
62. R. L. Martin, "Unusual Oilfield Corrosion Inhibitors, SPE 80219 (paper presented at the International Symposium on Oilfield Chemistry," Houston, TX, 5–7 February 2003).
63. N. J. Phillips, J. P. Renwick, J. W. Palmer, and A. J. Swift, "The Synergistic Effect of Sodium Thiosulfate on Corrosion Inhibition" (paper presented at the 7th International Symposium on Oil Field Chemicals, Geilo, 1996).
64. R. L. Martin, U.S. Patent 3959177, 1976.
65. V. Jovancicevic, Y. S. Ahn, J. Dougherty, and B. Alink, "CO_2 Corrosion Inhibition by Sulfur-Containing Organic Compounds," Paper 7 (paper presented at the NACE CORROSION Conference, 2000).
66. (*a*) A. Naraghi, U.S. Patent 5611991, 1997. (*b*) A. Naraghi and N. Grahmann, U.S. Patent 5611992, 1997.

67. Baker Petrolite, *Materials Performance* (Suppl.) 1999.
68. B. Alin, B. Outlaw, V. Jovancicevic, S. Ramachandran, and S. Campbell, "Mechanism of CO_2 Corrosion Inhibition by Phosphate Esters," Paper 37 (paper presented at the NACE CORROSION Conference, 1999).
69. H. Yu, J. H. Wu, H. R. Wang, J. T. Wang, and G. S. Huang, "Corrosion Inhibition of Mild Steel by Polyhydric Alcohol Phosphate Ester (PAPE) in Natural Sea Water," *Corrosion Engineering Science and Technology* 41 (2006): 259.
70. T. J. Bellos, U.S. Patent 4311662, 1982.
71. B. T. Outlaw, B. A. Oude Alink, J. A. Kelley, and C. S. Claywell, U.S. Patent 4511480, 1985.
72. R. L. Martin, G. F. Brock, and J. B. Dobbs, U.S. Patent 6866797, 2005.
73. R. L. Martin, U.S. Patent 4722805, 1988.
74. R. L. Martin, European Patent EP567212, 1993.
75. A. Naraghi and P. Prince, International Patent Application WO/1997/008264.
76. A. B. Gainer, U.S. Patent 4197091, 1980.
77. D. E. Knox and E. R. Fischer, U.S. Patent 4927669, 1990.
78. E. R. Fischer and P. G. Boyd, U.S. Patent 5759485, 1998.
79. J. A. Alford, P. G. Boyd, and E. R. Fischer, U.S. Patent 5174913, 1992.
80. E. R. Fischer, J. A. Alford, and P. G. Boyd, U.S. Patent 5292480, 1994.
81. B. A. Miksic, A. Furman, and M. Kharshan, U.S. Patent 6800594, 2004.
82. R. J. Goddard and M. E. Ford, U.S. Patent Application 0180794, 2006.
83. D. Leinweber and M. Feustel, International Patent Application WO/2006/040013.
84. A. Chalmers, I. G. Winning, D. McNaughtan, and S. McNeil, "Laboratory Development of a Corrosion Inhibitor for a North Sea Main Oil Line Offering Enhanced Environmental Properties and Weld Corrosion Protection," Paper 6487 (paper presented at the NACE CORROSION Conference, 2006).
85. P. J. Clewlow, J. A. Haslegrave, N. Carruthers, D. S. Sullivan II, and B. Bourland, U.S. Patent 5427999, 1995.
86. G. R. Meyer, U.S. Patent 6171521, 2001.
87. K. M. Henry and K. D. Hicks, International Patent Application WO/2006/019585.
88. K. M. Henry, R. Meyer, K. D. Hicks, and D. I. Horsup, "The Design and Synthesis of Improved Corrosion Inhibitors," Paper 5282 (paper presented at the NACE CORROSION Conference, 2005).
89. A. W. Ho, International Patent Application WO/1993/007307.
90. A. Naraghi and N. Obeyesekere, International Patent Application WO/2006/034101.
91. P. M. Quinlan, U.S. Patent 4371497, 1983.
92. (*a*) U. Dahlmann and M. Feustel, International Patent Application WO/2003/008668. (*b*) L. Tiwari, International Patent Application WO/2008/157234.
93. J. Y. Huang, L. S. Zheng, C. Y. Fu, U. E. Qu, and J. G. Liu, "The Inhibition Effects of a New Heterocyclic Bisquaternary Ammonium Salt in Simulated Oilfield Water," *Anti-Corrosion Methods and Materials* 51 (2004): 272.
94. S. Ramachandran, Y. S. Ahm, M. Greaves, V. Jovancicevic, and J. Bassett, "Development of High Temperature, High Pressure Corrosion Inhibitors," Paper 6377 (paper presented at the NACE CORROSION Conference, 2006).
95. H. Chen, T. Hong, and W. P. Jepson, "High Temperature Corrosion Inhibition Performance of Imidazoline and Amide," Paper 35 (paper presented at the NACE CORROSION Conference, 2000).
96. N. Obeyesekre, A. Naraghi, L. Chen, S. Zhou, and S. Wang, "Novel Corrosion Inhibitors for High Temperature Applications," Paper 5636 (paper presented at the NACE CORROSION Conference, 2005).
97. M. Feustel and P. Klug, U.S. Patent 6372918, 2002.
98. S. Kanwar and P. Eaton, International Patent Application WO/1997/007176.

99. L. Xiao, W. Qiao, H. Guo, and J. Qu, "Synthesis of an Imidazoline Phosphate Surfactant and Its Application on Corrosion Inhibition," *Tenside Surfactants Detergents* 5 (2008): 244.
100. (*a*) B. A. Oude Alink, U.S. Patent 4212843, 1980. (*b*) B. A. Oude Alink and B. T. Outlaw, U.S. Patent 4343930, 1982.
101. J. Levy, U.S. Patent 4344861, 1982.
102. A. Edwards, C. Osborne, D. Klenerman, M. Joseph, O. Ostovar, and M. Doyle, "Mechanistic Studies of the Corrosion Inhibitor Oleic Imidazoline," *Corrosion Science* 36(2) (1994): 315.
103. D. Turcio-Ortega, T. Pandiyan, J. Cruz, and E. Garcia-Ochoa, "Interaction of Imidazoline Compunds with Fe_n ($n = 1-4$ Atoms) as a Model for Corrosion Inhibition: DFT and Electrochemical Studies," *Journal of Physical Chemistry C* 111 (2007): 9853.
104. Y. Duda, R. Govea-Rueda, M. Galicia, H. I. Beltrán, and L. S. Zamudio-Rivera, "Corrosion Inhibitors: Design, Performance, and Simulations," *Journal of Physical Chemistry B* 109(47) (2005): 22674.
105. W. H. Durnie and M. A. Gough, "Characterization, Isolation and Performance Characteristics of Imidazolines: Part II. Development of Structure-Activity Relationships," Paper 03336 (paper presented at the NACE CORROSION Conference, 2003).
106. V. Jovancicevic, S. Ramachandran, and P. Prince, "Inhibition of Carbon Dioxide Corrosion of Mild Steel by Imidazolines and Their Precursors," *Corrosion* 55(5) (1999): 450.
107. R. L. Martin, B. A. Alink, J. A. McMahon, and R. Weare, "Further Advances in the Development of Environmentally Acceptable Corrosion Inhibitors," Paper 98 (paper presented at the NACE CORROSION Conference, 1999).
108. R. L. Martin, J. A. McMahon, and B. A. Oude Alink, U.S. Patent 5393464, 1995.
109. W. M. McGregor, "Novel Synergistic Water Soluble Corrosion Inhibitors," SPE 87570 (paper presented at the SPE International Symposium on Oilfield Corrosion, Aberdeen, UK, 28 May 2004).
110. N. E. Byrne and J. D. Johnson, U.S. Patent 5322640, 1994.
111. P. J. Clewlow, J. A. Haselgrave, N. Carruthers, and T. M. O'Brien, European Patent Application EP526251, 1994.
112. G. R. Meyer, U.S. Patent 6303079, 2001.
113. G. R. Meyer, U.S. Patent 6448411, 2002.
114. G. R. Meyer, U.S. Patent 6599445, 2003.
115. J. D. Watson and J. G. Garcia Jr., U.K. Patent Application GB2319530, 1997.
116. B. A. M. O. Alink, and B. T. Outlaw, U.S. Patent 6419857, 2002.
117. T. E. Pou and S. Fouquay, International Patent Application WO/1998/041673.
118. G. R. Meyer, U.S. Patent, 6696572, 2004.
119. A. Naraghi, H. Montgomerie, and N. U. Obeyesekere, European Patent EP1043423, 2000.
120. A. Naraghi, U.S. Patent 6063334, 2000.
121. G. R. Meyer, International Patent Application WO/2004/092447.
122. T. Gu, Y. Hu, Y. Tang, Z. Liu, L. Huang, Z. Yang, Y. Huang, J. Wang, H. Chang, H. Yu, G. Li, and J. Cao, International Patent Application WO/2007/112620.
123. D. Darling and R. Rakshpal, "Green Chemistry Applied to Scale Inhibitors," *Materials Performance* 37 (1998): 42.
124. T. E. Pou and S. Fouquay, U.S. Patent 6365100, 2002.
125. T. E. Pou and S. Fouquay, International Patent Application WO/1999/039025.
126. R. M. Thompson, International Patent Application WO/1999/059958.
127. K. Overkempe, E. Parr, W. John, and J. C. Speelman, International Patent Application WO/2003/054251.
128. D. S. Treybig, U.S. Patent 4676834, 1987.
129. (*a*) D. S. Treybig and J. L. Potter, U.S. Patent 4725373, 1988. (*b*) R. G. Martinez, D. S. Treybig, and T. W. Glass, U.S. Patent 4762627, 1988.

Corrosion Control during Production   223

130. R. H. Hausler, B. A. Alink, M. E. Johns, and D. W. Stegmann, European Patent Application EP0275651, 1988.
131. R. L. Martin and E. W. Purdy, U.S. Patent 4339349, 1982.
132. M. D. Greaves, C. M. Menendez, and Q. Meng, International Patent Application WO/2008/091429.
133. M. J. Zetlmeisl, U.S. Patent 5863415, 1999.
134. G. Sartori, D. C. Dalrymple, S. C. Blum, L. M. Monette, M. S. Yeganeh, and A. Vogel, U.S. Patent 6706669, 2004.
135. M. S. Yeganeh, S. M. Dougal, G. Sartori, D. C. Dalrymple, C. Zhang, S. C. Blum, and L. M. Monette, U.S. Patent 6593278, 2003.
136. J. C. Fan, L.-D. G. Fan, and J. Mazo, International Patent Application WO/2000/075399.
137. U. Dahlmann, M. Feustel, and R. Kupfer, International Patent Application WO/2002/092583.
138. P. R. Petersen, L. G. Coker, and D. S. Sullivan III, U.S. Patent 4938925, 1990.
139. B. A. M. O. Alink, R. L. Martin, J. A. Dougherty, and B. T. Outlaw, U.S. Patent 5197545, 1993.
140. P. Prince, International Patent Application WO/1998/051902.
141. Y. Wu and R. A. Gray, U.S. Patent 5135999, 1992.
142. D. Leinweber and M. Feustel, International Patent Application WO/2007/087960.
143. V. Jovancicevic and D. Hartwick, "Recent Developments in Environmentally-Safe Corrosion Inhibitors," Paper 226 (paper presented at the NACE CORROSION Conference, 1996).
144. L. W. Jones, U.S. Patent 4554090, 1985.
145. G. Woodward, G. P. Otter, K. P. Davis, and R. E. Talbot, Patent 6814885, 2004.
146. K. P. Davis, G. P. Otter, and G. Woodward, International Patent Application WO/2001/057050.
147. C. G. Carter and V. Jovancicevic, U.S. Patent 5135681, 1992.
148. C. G. Carter, V. Jovancicevic, J. A. Hartman, and R. P. Kreh, U.S. Patent 5183590, 1992.
149. C. G. Carter, L.-D. G. Fan, J. C. Fan, R. P. Kreh, and V. Jovancicevic, U.S. Patent 5344590, 1994.
150. A. J. McMahon and D. Harrop, Paper 32 (paper presented at the NACE CORROSION Conference, Orlando, 1995).
151. D. I. Bain, G. Fan, J. Fan, and R. J. Ross, "Scale and Corrosion Inhibition by Thermal Polyaspartates," Paper 120 (paper presented at the NACE CORROSION Conference, 1999).
152. W. J. Benton and L. P. Koskan, U.S. Patent 5607623, 1997.
153. J. C. Fan, L.-D. G. Fan, and J. Mazo, International Patent Application WO/2000/075399.
154. J. C. Fan, L.-D. G. Fan, U.S. Patent 6277302, 2001.
155. H. A. Craddock, S. Caird, H. Wilkinson, and M. Guzmann, "A New Class of "Green" Corrosion Inhibitors: Development and Application," SPE 104241 (paper presented at the SPE International Oilfield Corrosion Symposium, Aberdeen, UK, 30 May 2006).
156. M. Guzmann, U. Ossmer, and H. Craddock, International Patent Application WO/2007/063069.
157. N. Obeyesekre, A. Naraghi, and J. S. McMurray, "Synthesis and Evaluation of Biopolymers as Low Toxicity Corrosion Inhibitors for North Sea Oil Fields," Paper 1049 (paper presented at the NACE CORROSION Conference, 2001).
158. G. Schmitt and A. O. Saleh, "Evaluation of Environmentally Friendly Corrosion Inhibitors for Sour Service," Paper 335 (paper presented at the NACE CORROSION Conference, 2000).
159. J. Tang, R. T. Cunningham, and B. Yang, U.S. Patent 5750070, 1998.
160. J. Tang, S.-L. Fu, and D. H. Emmons, U.S. Patent 6022401, 2000.
161. D. W. Fong and B. S. Khambatta, U.S. Patent 5308498, 1994.

9 Gas Hydrate Control

9.1 INTRODUCTION

Gas hydrates are ice-like clathrate solids that are formed from water and small hydrocarbons at elevated pressures and at lower temperatures (Figure 9.1).[1–2] The temperature below which hydrates can form increases with increasing pressure and can be as high as 25–30°C (77–86°F). Typical pressure-temperature conditions for formation of gas hydrates are shown in Figure 9.1. Gas hydrates are most commonly encountered in subsea or cold climate wet gas or multiphase (oil/water/gas) pipelines, where they can block the flow of fluids, but they can also be formed during drilling, completion, and workover operations as well as in gas-processing facilities, gas injection lines, and aqueous chemical injection in gas lift lines if the pressure-temperature conditions are right. Many multiphase production lines are designed to operate without fear of hydrate formation but problems may occur if a shut-in occurs and the fluids cool, untreated, to within the hydrate-forming envelope. Further, if a subsea well is shut-in, a hydrate plug can also form below the wellhead unless the necessary precautions are taken. The prevention of gas hydrate plugging of flow lines is considered one of the main production issues to deal within deepwater field developments.[3]

In gas hydrates, the water molecules form an open structure containing cages held together by hydrogen bonding. These cages are occupied by small molecules, such as small hydrocarbons, which stabilize the clathrate structure through van der Waals interactions. There are three structures for gas hydrates that can form, each of which exhibit different pressure-temperature equilibrium curves: structure I, structure II, and structure H. Structure I can be formed when the natural gas is very rich in methane and contains almost no C3–4 hydrocarbon components. Structure II is the most common gas hydrate structure encountered in the field since it is stable whenever a natural gas mixture contains some propane or butanes besides methane. Structure H is very rarely encountered in the oil industry. It is stabilized by methane in small cages and fairly large hydrocarbons such as methylcyclopentane or benzene in large cages.

There are a number of methods for preventing gas hydrate formation and deposition.[4] They include the following:

- Keep the pressure low and outside the hydrate stable zone
- Keep the temperature above the hydrate equilibrium temperature at the system pressure by passive heat retention or active heating
- Separate out the water (dehydration)
- Modify the gas phase with another gas

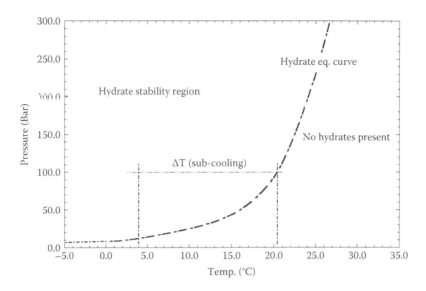

FIGURE 9.1 Pressure-temperature graph for a typical natural gas hydrate.

- Conversion of water to transportable hydrate particles without the use of chemicals
- Chemical treatment

Keeping the pressure low on a continuous basis is rarely done since production rates would be uneconomically low in many cases, although depressurization could be performed during shut-in. However, depressurization cannot be done in very deep water due to the hydrostatic pressure of the water keeping the fluids in the hydrate-forming region. Several methods are used to raise the temperature of a pipeline so as to avoid gas hydrate formation. The simplest method is to insulate the pipe. Burying the pipeline can help to some degree, as well as putting insulation materials or a vacuum around the pipe, although this latter method is expensive and will not work in extended shutdown situations. Another method is to wrap the pipeline wall with an electric resistance, heat-tracing cable, or a heat-tracing tube containing a circulating hot fluid, which elevates the temperature of the pipeline wall. Another method, which is used on several North Sea fields, is direct electric heating of a pipeline using an alternating current.[5] This is expensive to use on a continual basis so it is used only in extended shutdown situations. On another North Sea field a bundle pipeline system is used where hot water is injected from the platform side into the annulus of the bundle keeping the multiphase fluids outside the hydrate-forming region.[6]

Subsea separation of oil and water has been carried out on one field in the North Sea, but the technology is still in its infancy.[7] Another method, which at first seems counterintuitive, is to inject a gas such as N_2 or CO_2 to raise the pressure threshold for hydrate formation.[8] A technique that could represent a major breakthrough in deepwater hydrate control, although not a panacea for all oilfield applications, is to convert produced water to transportable hydrate particles without the use of chemicals. This technique is often called "cold flow"; "CONWHYP" or "Hydraflow" are alternative names to describe this technique.[9-14] The basis of the technology is the use of a recirculation stream containing hydrate particles. These particles are fed into a fast-cooled (shock-chilled) water-containing well stream in which the water is dispersed or emulsified in the liquid hydrocarbons by mixers. The hydrate particles seed further hydrate growth in the water droplets from the inside out and grow quickly and in a controlled manner. The hydrate particles are dry, nondepositing, and nonagglomerating and will eliminate free water from the rest of the transport system. It would appear to work best at high subcoolings. A cold-flow method that does not require recycling of hydrates has also been claimed.[15] One possible problem with cold flow is whether free water (before it reaches the mixers) might agglomerate during an unplanned shut-in. The technique does not work for wet gas fields or very high water-cut fields where flow of hydrate particles is impossible. In addition, possible increased inorganic scale-deposition problems, due to removal of free water as gas hydrate and subsequent higher concentrations of ions in solution, need to be addressed. An onshore field trial of cold-flow technology has been considered by one American oil company. For very high water cuts, it has been suggested to inject additional water or brine to allow complete gas conversion to hydrates and the hydrate slurry to be transportable in excess water, with or without liquid hydrocarbons present.[16] A special type of hydrate anti-agglomerant chemical might be needed (see Section 9.2.3). The same principle of shock chilling has also been proposed to avoid wax deposition (see Chapter 10).

9.2 CHEMICAL PREVENTION OF HYDRATE PLUGGING

Chemical treatment to prevent hydrate plugging can be accomplished with three different classes of chemical, all of which are now used in the field:

- thermodynamic hydrate inhibitors (THIs)
- kinetic hydrate inhibitors (KHIs)
- anti-agglomerants (AAs)

The last two categories are known collectively as low-dosage hydrate inhibitors (LDHIs), reflecting the much lower dosage requirements compared with THIs. One oil company has used the acronym THI, meaning threshold hydrate inhibitors, as another expression for KHIs. In this book, THI refers specifically to thermodynamic hydrate inhibitors.

9.2.1 THERMODYNAMIC HYDRATE INHIBITORS

THIs are sometimes called "hydrate antifreeze." They are by far the most common chemical class used to prevent hydrate formation.[1] They work by changing the bulk thermodynamic properties of the fluid system, thereby shifting the equilibrium conditions for gas hydrate formation to lower temperatures or higher pressures. Thus, they can be used to prevent hydrate formation and also "melt" existing hydrate deposits. THIs are added at very high concentrations, in some cases, up to one to two barrels of THI per barrel of water.

The most commonly used classes of THIs are alcohols, glycols, and salts. Methanol (CH_3OH) and monoethylene glycol (MEG, $HOCH_2CH_2OH$) are widely used to protect against hydrate formation in production, workover, and process operations and for melting hydrate plugs. Diethylene glycol (DEG) and triethylene glycol (TEG) are also sometimes used to prevent hydrate formation although they are less powerful. TEG is mainly used for gas drying, that is, adsorbing water in gas-flow lines or processing facilities. In South America, ethanol is commonly used as a THI since it is low cost and available in large quantities from sugar fermentation. However, above ca. 5.6 mol%, ethanol can form a binary hydrate with methane and can result in significantly less hydrate inhibition than would be expected from ice-melting depression.[17] Although methanol and glycols are relatively low-cost chemicals, because they are dosed at such high concentrations, it is often economically worthwhile to recover the THI and use it again. MEG regeneration facilities are widely used today, but methanol regeneration is much less common. Methanol and glycols are rarely used on a continuous basis in oil fields due to the high volumes of water required to be treated, but glycols in particular are used on a continuous basis in condensate and gas fields, MEG being the most common.[18] It is important to get the dosage of THI correct, since underinhibition (using a dosage that does not fully protect against hydrate formation) can, at certain, doses increase the plugging potential.[19-20]

Besides alcohols and glycols, the only other common chemical class used to prevent hydrate formation is salts such as sodium chloride, calcium chloride, and potassium formate. These salts are commonly used in drilling fluids to suppress hydrate formation, sometimes in combination with glycols. Halide salts are less suited for injection into production lines due to the high concentrations needed, produced water incompatibility (for calcium salts), and increased corrosion potential. However, the use of the low corrosive salt, potassium formate, has been proposed.[21] Other chemicals that have been investigated as THIs include water-soluble solvents such as dimethylformamide and N-methyl pyrrolidone, ethanolamines, and other alcohols besides ethanol, which form azeotropes with water, for example, iso-propanol. All these organic chemicals are more expensive and less powerful thermodynamic inhibitors than methanol and MEG and are not currently used in the field.

The performance of THIs is usually expressed as the temperature change in the hydrate equilibrium curve (suppression) at a given pressure and inhibitor concentration. For example, a 20 wt.% solution of methanol in water lowers the hydrate equilibrium temperature by approximately 10°C (18°F), whereas 32 wt.% MEG is needed for the same temperature suppression. A rough guide to calculate the amount of organic THI needed to avoid hydrate formation was first formulated by Hammerschmidt in the following equation:[22]

$$\Delta T = \frac{Ks}{M(100-s)}$$

where ΔT is the hydrate suppression (°F), K is a constant (in this case Hammerschmidt's Fahrenheit-based constant, 2,335) depending on the nature of the inhibitor and on the heat of formation of the hydrate, s is concentration of the THI in the water phase (wt.%), and M is the molecular weight of the inhibitor. Clearly, the molecular weight of the inhibitor is crucial to the performance. Generally, the lower the molecular weight, the higher the performance. That is why methanol is a better THI than MEG, which again is better than DEG, and so forth. For the same reason, methanol should be more powerful at melting hydrate plugs than the glycols although THI density affects the performance (see also Section 9.3.1 on hydrate plug removal). However, the above Hammerschmidt equation takes no account of the distribution of the THI between the gas or liquid phases or the pressure in the system and gives poor results at high inhibitor concentrations. It can give particularly bad results for methanol since methanol partitions significantly both the gas and liquid hydrocarbon phases, much more so than the glycols. Thus, in multiphase transportation systems where the GOR (gas/oil ratio) is high and water-cut low, glycols are often preferred over methanol since MEG partitions almost exclusively into the bulk water phase. Regeneration of MEG is also simpler. In practice, the engineer will usually inject the THI at a somewhat higher dosage than required to be absolutely sure there will not be any hydrate formation.

More accurate equations to calculate the subcooling provided by thermodynamic inhibitors have been developed.[23-25] The simplest way today is to use a PVT model, which takes into account the distribution of the inhibitor between the phases. Commercial software of PVT models with added modules designed for hydrate calculations are available from several companies, research institutes, and universities. However, even these models can sometimes give inadequate results especially at high pressures, high inhibitor concentrations, or when the water also contains significant concentrations of salts, although efforts to minimize these problems have been made. A rough guide to compare the performance of a number of THIs is given in Table 9.1 below using data generated by two commercial software packages. The table illustrates values of the subcooling provided at various THI concentrations in the water phase, that is, losses to the gas or liquid hydrocarbon phase are not taken into account. Available values for HCOOK (potassium formate) are taken from the literature.[21]

The best method, if done properly, to determine the equilibrium point for hydrate formation, with or without added THIs (but not LDHIs), is to do it experimentally. This is done by dissociating preformed hydrate and determining

TABLE 9.1
Calculated Subcoolings (°C) for Various Thermodynamic Inhibitors in the Water Phase

Concentration (wt.%)	MeOH	EtOH	MEG	DEG	TEG	NaCl	HCOOK
5	2.0	1.4	1.05	0.63	0.46	1.96	
10	4.2	3.0	2.25	1.4	1.05	4.3	2.5
20	9.3	6.6	5.2	3.3	2.7	10.7	7.1
30	15.3	10.7	9.0	5.9	5.0	15.0	12.9
35	18.6	13.0	11.35	7.5	6.5	—	
40	22.2	15.4	14.0	9.3	8.2	—	

the hydrate formation temperature (it is not done by cooling and measuring the onset temperature for hydrate formation as this temperature will vary due to hydrate formation being a stochastic process). Thus, once some hydrate has been formed, the system is heated. Heating must be very slow (< 0.2°C/h is best) as you approach the expected equilibrium temperature and good mixing of all phases is essential. The point at which the last particle of hydrate dissociates and the plot of the pressure curve returns to the plot while cooling is taken as the equilibrium point for hydrate formation.[131] A stepwise heating method has been shown to give even more reliable results than using constant heating, in less measuring time.[132]

9.2.1.1 Operational Issues with THIs

There are a number of operational issues regarding the use of THIs. The problem of THI phase distribution (i.e., loss of THI to the gas or liquid hydrocarbon phase), and thus, determining the dosage rate has been highlighted already. Other potential problems are mentioned briefly below:

- High inhibitor volumes requiring high transport costs, large storage tanks, injection lines, and so forth.
- Toxicity and flammability of methanol.
- The cost of building and maintaining methanol or glycol inhibitor regeneration facilities. Fouling, salting out, and corrosion in regeneration facilities are common.
- Downstream pollution of the hydrocarbons (gas or liquids) can lead to refinery problems. Pollution is known to occur particularly for methanol, but also for MEG, and can reduce the value of the hydrocarbons if their concentration is above the required limits.
- THIs increase the potential for scale and naphthenate formation. NaCl may deposit at high salinities, or the potential for carbonate or sulfate scales may be reached, or increased above the original MIC (minimum inhibitor concentration) of the scale inhibitor. Methanol and ethanol are more detrimental than glycols.[26-27]
- Some methanol will partition into the oil phase and can act as a wax (paraffin) precipitant by lowering the wax appearance temperature.
- Pumping the inhibitor to the required location in the event of an unplanned shut-in can be tricky. The pump and injection line capacity may be too small to treat the amount of water expected. This could be during deepwater drilling or subsea multiphase transportation.
- The viscosity of the inhibitor, especially the higher glycols, is relevant in very cold environments (gas processing) and narrow, long, cold injection lines.
- Freezing of the inhibitor in gas processing systems.
- Methanol and glycols have high biological and chemical oxygen demand, which can limit their discharge allowance.

For subsea multiphase transportation over long distances, the high thermodynamic inhibitor volumes and expensive storage, injection, and regeneration facilities that would be needed motivated the search for other methods to prevent gas hydrate formation. This led to the development of LDHIs in the nineties. An extensive review up to 2005 has been published on this subject.[28] The two classes of LDHIs, KHIs and AAs, are discussed below.

9.2.2 Kinetic Hydrate Inhibitors

9.2.2.1 Introduction to KHIs

The key components in all known KHI formulations are water-soluble polymers often with other smaller organic molecules added as performance enhancers (synergists). Generally, the polymers show little partitioning to liquid hydrocarbon phases, yet these phases often affect KHI performance. KHIs delay gas hydrate nucleation and, usually, also crystal growth for a period dependent on the subcooling and to some extent the pressure in the system.[29–31] In subsea multiphase transportation, this enables produced fluids to be transported to the process facilities before gas hydrate formation and deposition occurs in the line. Thus, any need for long shut-ins will be critical in determining the field applicability of a KHI. Generally, field applications of most of the commercial KHIs are limited to a maximum of 9–10°C (16–18°F) subcooling in the production line because the required delay time before hydrate formation is most often in the regions of days. Higher subcoolings (driving forces) would give shorter delay times before hydrate formation occurred. Thus, KHIs are not applicable for most deepwater fields where the subcooling and pressure are both high. Of course, if the residence time of the produced fluids is small, it may be possible to use KHIs at higher subcoolings than 9–10°C (16–18°F).

KHIs have been used commercially in the field since about 1995. They are added at low concentrations, less than 1 wt.% of the water phase and often around 0.3–0.5 wt.%. This can be contrasted with the 20–60 wt.% needed for THIs such as methanol or glycols. In one field application, CAPEX savings of U.S.$40 million were realized by choosing KHI technology instead of methanol injection and regeneration.[32]

Many water-soluble polymers have been shown to work as KHIs. There are two key structural features in a KHI polymer. First, the polymer needs functional groups that can hydrogen-bond to water molecules or gas hydrate particle surfaces. These are usually amide groups. The second key feature is a hydrophobic group adjacent to or bonded directly to each of the amide groups.[33] An example of such a polymer, which was also the first KHI to be discovered, is polyvinylpyrrolidone (PVP; Figure 9.2).[34–35] The performance of KHIs is sometimes quoted as the subcooling at which a multiphase fluid can be transported without hydrate formation for a period (hold time) of 48 h at a given pressure. For PVP, without any synergists, that subcooling is only 3–4°C (5.4–7.2°F) at 70–90 bar. The mechanism by which PVP and other improved KHIs work is the subject of debate. Two major mechanisms have been proposed. The first suggests that KHI polymers perturb the water structure to such a degree that gas hydrate particles cannot grow to the critical nuclear size, at which point growth becomes spontaneous.[28] Molecular modeling studies indicate that this happens for some KHIs including PVP.[36] However, a neutron diffraction

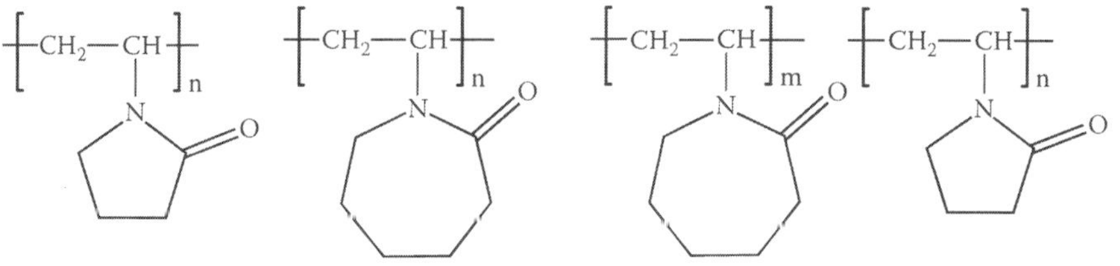

FIGURE 9.2 Poly-N-vinyl lactam polymers/polyvinylpyrrolidone, polyvinylcaprolactam, and vinylpyrrolidone/vinylcaprolactam copolymer.

study showed that PVP does not affect the water structure in propane-water systems before and during gas hydrate formation.[37] The second mechanism suggests that KHI polymers adsorb onto the surfaces of growing hydrate particles limiting their growth, and possibly deforming the hydrate cavities.[39] This can occur before or after the particles reach the critical nuclear size so they can act as nucleation and crystal growth inhibitors. KHI polymer classes may adsorb in different in ways onto hydrate surfaces. In general, it is assumed that the hydrophobic groups on the polymer mimic small hydrocarbon guest molecules and interact with open cavities on the hydrate surface. The amide groups anchor the polymer on the surface through hydrogen bonding. There is clear evidence for the second mechanism for some KHI polymers from studies on the growth inhibition of tetrahydrofuran (THF) hydrate[38] and small-angle neutron scattering as well as from molecular modeling studies.[39–41] For some polymers, both mechanisms may be operating. Interestingly, another molecular modeling study showed that methane hydrate initially nucleates into a phase consistent with none of the common bulk crystal structures but containing structural units of all of them.[41] Another possible mechanism that has been proposed is that the highest performing KHIs adsorb most strongly to surfaces (e.g., pipe walls or produced particles such as silica) that would otherwise act as heteronucleation sites for gas hydrate formation.[42]

The performance of polymeric KHIs has been shown to be considerably reduced in systems containing high concentrations of H_2S.[43] High concentrations of CO_2 also have a significant negative effect on the performance of KHIs. The reason is unclear but may be related to the relatively high solubility of these gases in water, compared with small hydrocarbons, and the fact that they are also clathrate hydrate formers.

Most KHI field applications are based on polymers from one of two classes. They are:

- vinyl lactam polymers and copolymers
- hyperbranched polyesteramides

Blends of these two polymer classes can be used synergistically. KHIs can also be used in conjunction with THIs for added subcooling protection.[44–46] Two other classes of commercial KHI include a biodegradable polyester pyroglutamate polymer, useful for low subcooling applications, and a polyisopropylmethacrylamide claimed to work at higher subcoolings than the two common classes of KHI mentioned above. All these KHIs are discussed below.

9.2.2.2 Vinyl Lactam KHI Polymers

PVP belongs to the class of vinyl lactam polymers but its performance is low, and its commercial use has been superseded by vinyl caprolactam polymers, which are useful for field applications up to about 9–10°C (16–18°F).[47–52] It should be mentioned that butylated PVP, a commercial polymer, and vinyl pyrrolidone copolymers with small alkylacrylates perform better as KHIs than PVP but not as good as the best vinyl caprolactam polymers.[53–54] The simplest of the vinyl caprolactam polymers is the homopolymer polyvinylcaprolactam (PVCap) although copolymers such as vinyl pyrrolidone/vinyl caprolactam (VP/VCap) are also commercially used (Figure 9.2). N-methyl-N-vinyl acetamide/vinyl caprolactam copolymer (VIMA/VCap) has been shown to outperform PVCap and was for a time used commercially but is no longer available due to the cost of the VIMA monomer (Figure 9.3).[55] Vinyl caprolactam/dimethylaminoethylmethacrylate (DMAEMA) copolymers are also high-performing KHI polymers (Figure 9.3).[56] In fact, DMAEMA-based copolymers with methacrylamide or N,N-dimethylacrylamide and without any vinyl lactam component have been claimed as KHIs.[57] Vinyl caprolactam/vinyl pyridine copolymers have been claimed as KHIs with corrosion inhibitor (CI) properties.[58] Copolymers of vinyl caprolactam with short polyethoxylated methacrylates have been claimed as better KHIs than PVCap. Vinyl lactam polymers generally have poor biodegradability. However, a commercial product that is a graft copolymer of polyethyleneglycol and VCap monomer, optionally with other vinyl monomers such as vinyl acetate, has been developed, which shows greater than 20% biodegradation over 60 days in the OECD306 seawater test.[60] Very recently a new graft polymer has been developed by the same company with 53% biodegradation in 28 days and 98% in 60 days.

The ideal molecular weight (Mw) for a KHI polymer (or oligomer) for optimum performance is around 1,500–3,000. At molecular weights lower than 1,000, the performance drops drastically, and at increasing molecular weights above 3,000–4,000, the performance drops slowly but does not disappear. A low molecular weight polymer has the advantage of maintaining low viscosity of the injected formulation. Thus, samples of commercial PVCap, the highest performing, currently commercial vinyl lactam polymer, have molecular weights of about 2,000–4,000. However, there is a claim and evidence that a bimodal distribution of molecular weights gives increased performance.[61] The bimodal distribution can

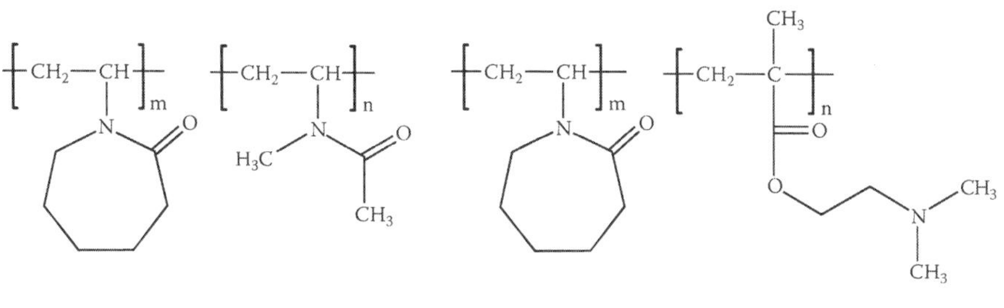

FIGURE 9.3 N-methyl-N-vinyl acetamide/vinyl caprolactam copolymer and vinyl caprolactam/dimethylaminoethylmethacrylate copolymer.

be made from a single polymerization or by mixing two polymers with unequal molecular weight distributions. Ultralow molecular weight PVCap, when dosed at 0.5 wt.% based on the water phase, has been shown to delay gas hydrate formation in a natural gas-water system at 13°C (23.4°F) subcooling for over 48 h at 70 bar.[62] By "natural gas," it is meant a normal raw gas mixture that leads to structure II hydrates. For structure I hydrate, which is predicted to occur in a minority of fields, the subcooling performance is lower for PVCap. Other polymers (with unpublished structures) have been reported that give better performance on structure I hydrate.[63] This ultralow molecular weight PVCap polymer was tested in a wheel loop in various fluid systems. Depending on the fluid type (gas only or with condensate or crude oil) the performance varied from 11 to 19°C (from 20 to 34°F) subcooling at constant pressure for a successful test that lasted 2–3 days.[64] The variation in results underlines the importance of testing the KHI performance with the correct produced fluids at operating conditions. Besides multiphase production, VCap-based KHIs have also been claimed for use in fracturing fluids.[65]

The polymerization process for making vinyl caprolactam polymers will affect its performance and cloud point water. It has been shown that PVCap made by polymerization of VCap in butyl glycol (2-butoxyethanol) gave better KHI performance than PVCap made in isopropanol after the effect of the 2-butoxyethanol as synergist had been taken into account.[66–67] High–cloud point polymers are also needed for high-saline environments. PVCap has a cloud point of 30–37°C (86–99°F) in distilled water (depending on the molecular weight and polymerization procedure) and will be insoluble in high-saline brines due to the salting-out effect. Thus, there is a possibility of KHI deposition ("gunking") on the pipe walls near the injection point if the cloud point is lower than the temperature of the produced fluids, although one service company unofficially claims that such deposits do not build up beyond a thin layer. High–cloud point copolymers of VCap with more hydrophilic monomers, such as those in Figures 9.2–9.3 have been made commercially available. The most common high–cloud point copolymer on the market today is low molecular weight 1:1 vinyl pyrrolidone/vinyl caprolactam copolymer (1:1 VP/VCap), which has a performance a little lower than PVCap of the same molecular weight. However, a KHI will still perform well as a cloudy solution above its cloud point but at a temperature lower than the deposition point.

Various synergists have been proposed to boost the performance of vinyl caprolactam (VCap) polymers. Besides 2-butoxyethanol discussed above, 2-isobutoxyethanol has been proposed as a better synergist.[68] Small quaternary salts such as tetrabutylammonium bromide (TBAB) have been used as VCap polymer synergists in the field for some years.[69] More biodegradable alternatives to TBAB with ester linkages have been claimed.[70] Small, quaternary ammonium zwitterionic molecules such as tributylammoniumpropylsulfonate, have also been shown to work as synergists.[71] The key feature in all these quaternary molecules is a quaternary nitrogen (or phosphorus) atom with three or more butyl or pentyl groups. Their structures are related to the quaternary ammonium AA surfactants (see Section 9.2.3 on AAs). Amine oxides such as tributylamine oxide are also good synergists for VCap polymers but have not been used commercially.[72] Certain substituted amines, alkylenediamines or polyamines, and derivatives of these have been patented as synergists for KHIs such as VCap-based polymers.[73] Small polyetheramines have been used in the field as synergists for VCap polymers such as PVCap.[74,75] Polyetheramines have also been used as AAs in gas wells (see later section on AAs). In addition, various classes of cationic or nonionic surfactant, or a sugar, have been claimed as synergists for VCap-based polymers.[76] Surfactant examples are N-alkyl pyrrolidones, poly(ethylene glycol), or polyethylene oxide propylene oxide and examples of sugars are sorbitol, mannitol, fructose, and/or sucrose. Polyethyleneoxide is also a synergist for PVCap.[77]

9.2.2.3 Hyperbranched Polyesteramide KHIs

Besides vinyl lactam polymers, the other major class of KHI polymer that is in commercial use is the hyperbranched polyesteramides.[78] Like the VCap polymers, they are also useful for field applications up to about 10°C (18°F) subcooling. Some of these polymers are claimed to have higher biodegradability than PVCap, even over 20% by the 28-day OECD306 test. Polymers with high–cloud points, useful in systems with up to 12% salt water, have also been developed: their reduced performance is counteracted by the hydrate suppression obtained from the salt. Polyesteramides are made by condensing a cyclic acid anhydride with diisopropanolamine in a ratio of $n{:}n+1$ where n is an integer (by varying n, one can vary the molecular weight of the polymer). This gives a polymer with hydroxyl groups at the tips. The hyperbranching is caused by the dialkanolamine, which has three reactive groups. By adding a third molecule to the reaction mixture, such as an imine, the tips of the polymer can be modified to become more hydrophilic. The preferred choice for the cyclic acid anhydride appears to be cis-1,2-cyclohexanedicarboxylic anhydride (or hexahydrophthalic anhydride) and the imine could be 3,3-iminobis(N,N-dimethylpropylamine) or 1-methylpiperazine (Figure 9.4). The feature that singles out the polyesteramides from all other KHI polymers is their hyperbranching. All other KHI polymers are linear snake-like molecules, most of them with a polyvinyl backbone. To use an analogy for KHI bonding to hydrate surfaces, the polyesteramides are like a hand clutching a hydrate ball, whereas other KHIs are like a single long finger attached to the ball. The molecular weight of the polymer does not need to be very high either. For example, polymers with molecular weights of 1,500–2,000 give good KHI performance. This is roughly the same figure found for polyvinyl-based KHIs such as PVCap. In the case of the polyesteramide shown in Figure 4, the hydrophobic group that can interact with open hydrate cages (and/or perturb the water structure) is the cyclohexyl ring, although the methyl group of the diisopropanolamine may make a contribution. The amide, and to a lesser extent ester, groups take part in hydrogen bonding to water molecules. Hyperbranched polyesteramides are claimed to perform better on structure I hydrates than the VCap polymers.[28]

Various synergists have been claimed to enhance the performance of hyperbranched polyesteramides.[79–80] One example is the reaction product of polyethyleneimine with formaldehyde and caprolactam. This gives polymers with pendant caprolactam rings just as one finds in PVCap. Another KHI synergist is made by reacting N-methyl butylamine with formaldehyde and polyacrylamide. Other synergists mentioned are nonpolymeric surfactants with caprolactam or alkylamide head groups. A small amount of a quaternary AA can also be formulated with the

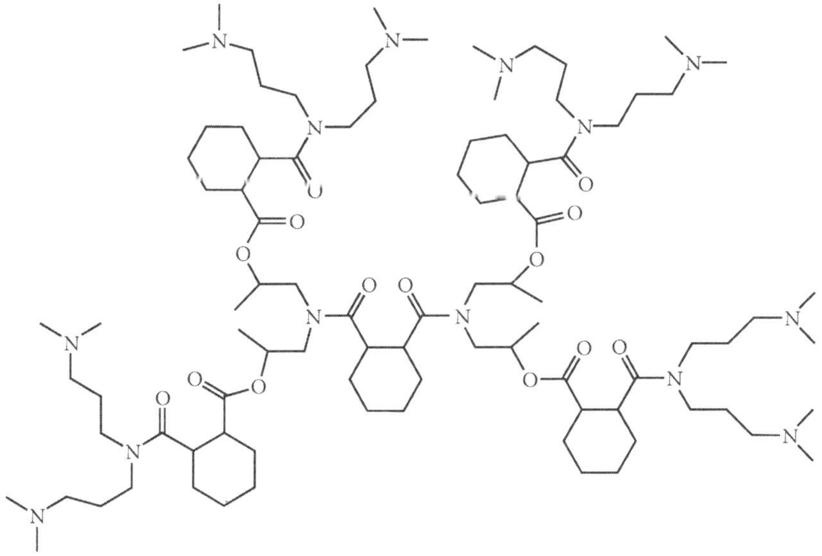

FIGURE 9.4 Example of a hyperbranched poly(ester amide) structure formed from cis-1,2-cyclohexanedicarboxylic anhydride, diisopropanolamine, and 3,3-iminobis(N,N-dimethylpropylamine).

KHI blend to improve the performance. It is likely that vinyl caprolactam polymers are synergists for hyperbanched polyesteramides, due to their different structures, although no examples are given in the literature.

9.2.2.4 Compatibility of KHIs

Studies have shown that the performance of commercial KHIs is detrimentally affected by some surfactant film-forming CIs. However, compatible CIs have now been developed.[81–84] Conversely, KHIs can also negatively impact the performance of CIs. These effects may be due to preferential adsorption of KHIs on the walls of the pipe or due to CI-KHI surfactant-polymer interactions. Also, many classes of surfactants, both nonionic, cationic, and amphoteric can accelerate hydrate formation especially if they have good foaming or emulsifying properties giving better contact of the gas (or gas dissolved in oil) with the water phase.[85] KHI-compatible CIs probably avoid these properties. Scale inhibitors in general are not detrimental to the performance of neutral KHIs, and can in fact enhance the performance in some cases.[86] KHIs have been combined with wax inhibitors in a single injection package.[87]

9.2.2.5 Pyroglutamate KHI Polymers

Polyesters or polyethyleneimines with pendant pyroglutamate groups have been developed as high–cloud point, biodegradable KHIs. One of these polymers is already in use on a low subcooling, small field application in Europe.[88–90] An example is given in Figure 9.5. The pendant groups resemble those found in PVP, although the point of attachment to the polymer backbone is different. Therefore, the subcooling performance is not expected to be very high, but it is claimed to be somewhat

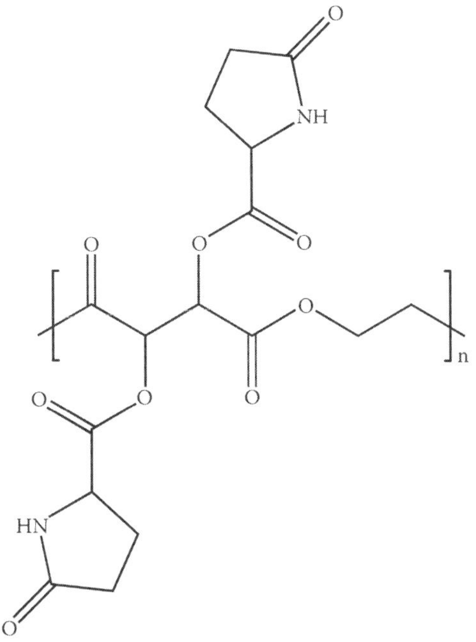

Gas Hydrate Control

FIGURE 9.5 One class of pyroglutamate polyester kinetic hydrate inhibitors.

better than PVP. Polymers in this class, for example, with fatty acid ester groups, appear to show some performance as hydrate AAs. The polyesters are made by condensing substituted dicarboxylic acids with diols or polyols followed by reaction with pyroglutamic acid. Some of these materials also show AA properties.[90]

9.2.2.6 Poly(di)alkyl(meth)acrylamide KHIs

There are many other categories of KHI polymers that have not been commercialized due to cost and availability. Possibly, the most studied examples of these are polyalkyl(meth)acrylamides and polydialkyl(meth)acrylamides, one example of which is now commercially available (see below; Figure 9.6).[91–94] In particular, it has been shown that polymers of the methacrylamide monomers give better performance

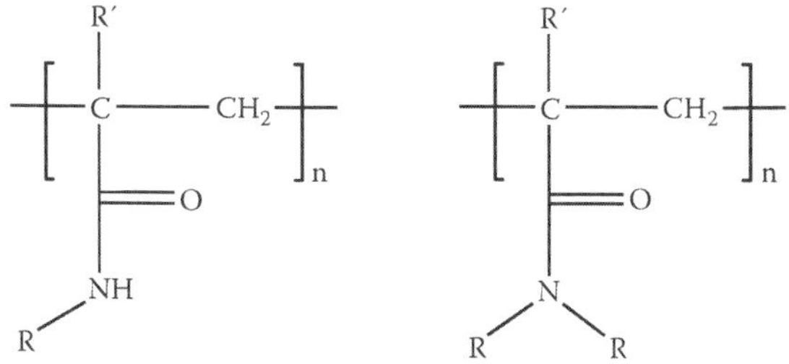

FIGURE 9.6 Polyalkyl(meth)acrylamides and polydialkyl(meth)acrylamides (R = alkyl or RNR = cycloimino, R′ = H or CH_3).

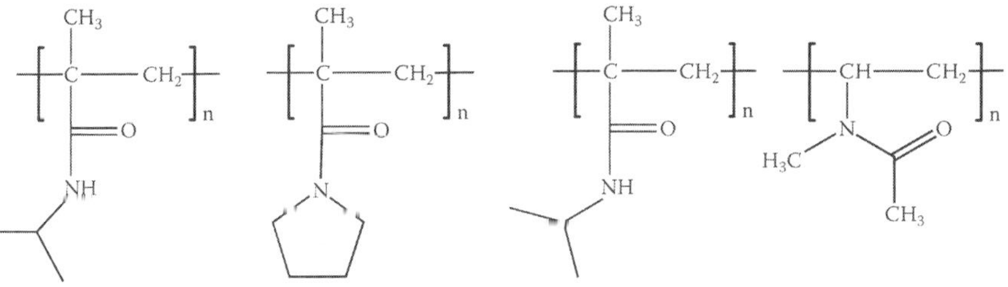

FIGURE 9.7 Polyisopropylmethacrylamide, polymethacryloylpyrrolidine, and N-vinyl-N-methyl acetamide/isopropylmethacrylamide 1:1 copolymer.

than the acrylamide monomers.[95] Thus, placing the extra methyl group on the polyvinyl backbone makes the polymer less flexible giving less movement of the pendant alkylamide groups in water (lower entropy) giving improved kinetic inhibition. This idea was also shown to work for other KHI polymer classes. A paper showing that poly(isopropylmethacrylamide) chains have smaller conformational changes in aqueous solution than poly(isopropylacrylamide) has been published.[96]

The best methacrylamide polymer appears to be made from isopropylmethacrylamide (IPMA), followed by methacryloylpyrrolidine. Similar to observations with PVCap, polyIPMA made from IPMA polymerized in 2-butoxyethanol performed better polyIPMA made in isopropanol.[97] In fact, a 1:1 IPMA copolymer with VIMA gave superior performance to IPMA homopolymer at unoptimized molecular weights (Figure 9.7).[98] This same copolymer effect with VIMA was discussed earlier for PVCap versus 1:1 VIMA/VCap copolymer, the latter being the better KHI. However, at very low molecular weights, polyIPMA (or oligoIPMA) is just as effective as the copolymer with VIMA. OligoIPMA is now commercially available, but no field applications have yet been carried out. Unpublished laboratory tests by the vending service company indicate that its performance is superior to commercial PVCap technology. It has a cloud point of about 35°C (95°F) in fresh water.

9.2.2.7 Other Classes of KHIs

Other classes of noncommercial polymers that have been shown to perform well as KHIs are shown in Figures 9.8–9.11. They include alkylamide derivatives of

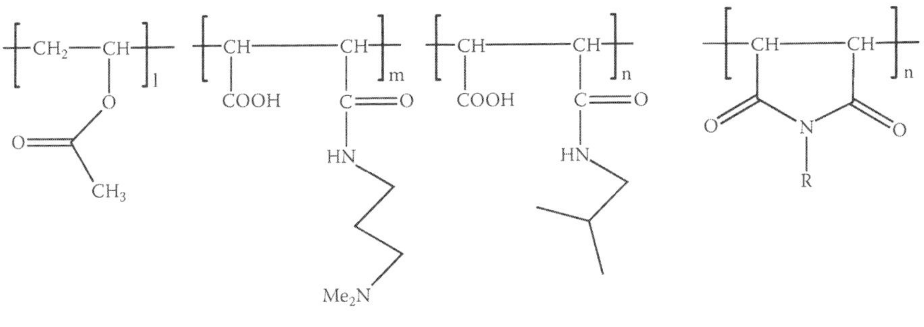

FIGURE 9.8 Alkylamide derivatives of maleic copolymers and polymaleimides (usually copolymerized with hydrophilic monomers to improve water solubility).

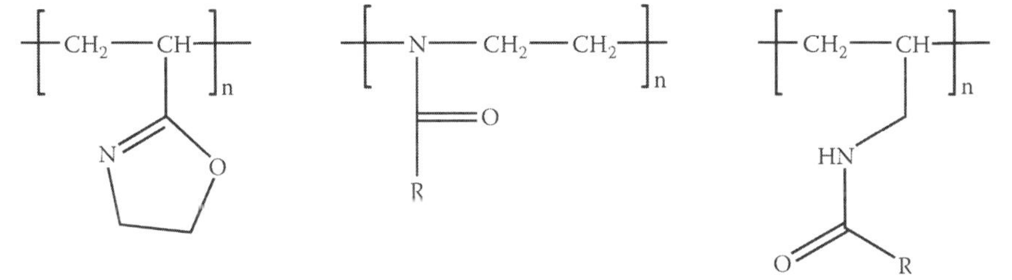

FIGURE 9.9 Polyvinyloxazolines, polyalkyloxazolines (or polyacylethyleneimines), and polyallylamides.

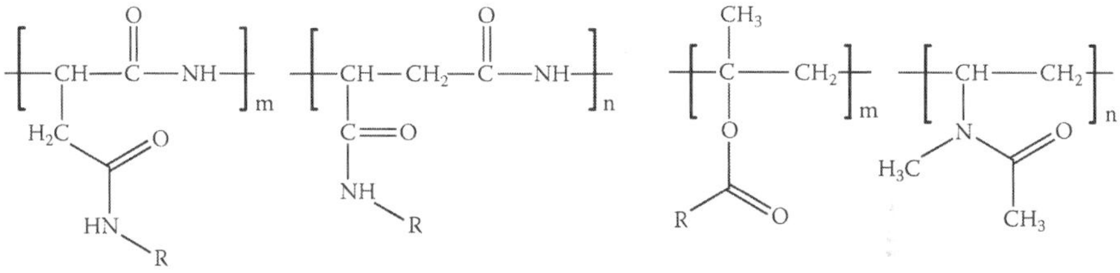

FIGURE 9.10 Polyaspartamides and vinyl alkanoate/N-methyl-N-vinyl acetamide copolymers.

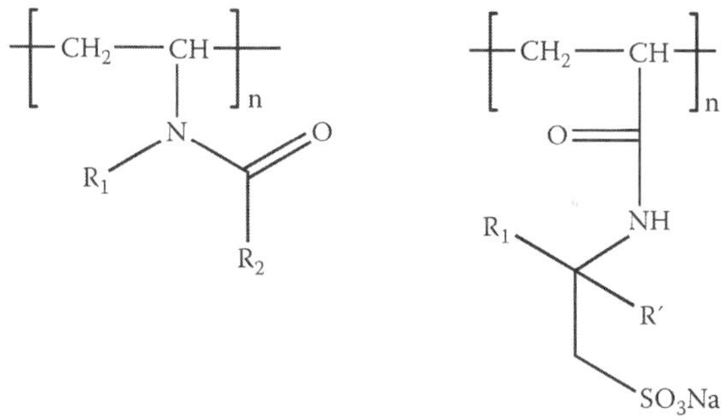

FIGURE 9.11 Polyvinylalkanamides and modified AMPS polymers (R = C1–6 and R' = H or CH$_3$).

maleic copolymers,[99] polymaleimides,[100] polyalkyloxazolines (or polyacylethyleneimines),[101] polyvinyloxazolines, polyallylamides,[102] polyaspartamides,[103] starch derivatives,[104] vinyl alkanoate/VIMA copolymers,[91] polyvinylalkanamides,[105] and modified acrylamidopropylsulfonic acid (AMPS) polymers.[106] Polyaspartamides also showed good biodegradability but performed somewhat worse than a commercial 1/1 VP/VCap copolymer on structure II gas hydrate. Note that polyvinyloxazolines do not contain amide groups, as do other classes of KHI polymers. Also note that the amide groups in vinyl alkanoate/VIMA copolymers are not in the same monomer as the hydrophobic groups, which are pendant to the alkanoate functionality. The modified AMPS polymers gave best performance when the hydrophobic R group was five carbons. This suggests that the dominant inhibition mechanism for these polymers is perturbing the water structure, preventing hydrate nucleation, rather than attaching to hydrate particle surfaces as the R group is not an ideal size for such interactions.

Since the early nineties, it has been known that antifreeze proteins (AFPs) not only inhibit ice growth but also inhibit hydrate formation. AFPs, active fragments of these AFPs, and mimetics thereof have been claimed as KHIs.[107] The AFPs or active fragments, originally isolated from fish, insects, plants, fungi, or bacteria, can presumably be made in an enzymatic process. Such a process would allow for large-scale manufacture, as isolation of large quantities of AFPs from these sources is prohibitively expensive. Interestingly, certain AFPs have the ability to eliminate the memory effect from melted hydrates.[108] The memory effect is the ability of water formed from melting hydrates to more easily form hydrates when cooled, than water that has not been treated this way. It is generally believed that the hydrate must not be melted more than a few degrees centigrade above the hydrate equilibrium temperature for the memory effect to be retained: at increased temperatures, the effect would be lost. A hypothesis for this memory effect is that water formed from melted hydrates retains some of the hydrate structure at the molecular level, such as partial cages or polyhedral clusters. However, a neutron-scattering study on methane hydrate melted to < 1 K above the equilibrium temperature concluded that there is no significant difference between the structure of water before the hydrate formation and the structure of water after the hydrate decomposition.[188] A second theory for the memory effect has been proposed. In a molecular-dynamics computer simulation study of methane hydrate decomposition, it was concluded that the effect relates more to the persistence of a high concentration, and retarded diffusion, of methane in the melt than it does to the persistence of metastable hydrate precursors.[189–190] A similar conclusion was drawn in an unpublished NMR spectroscopic study of THF hydrate and methane hydrate.[191]

A commercial terpolymer originally designed for cosmetic use has been shown to have good performance as a KHI but no reports of its field use have been published (Figure 9.12).[109] Polyalkoxylated amines, for example, propoxylated derivatives of triethanolamine, have also been shown to have KHI activity although lower than most commercial KHIs (Figure 9.13). They may possibly make good synergists for other KHIs.[110]

Non-polymeric imidazolium-based ionic liquids have been shown to perform as thermodynamic and kinetic hydrate inhibitors, albeit at fairly extreme conditions.[111]

9.2.2.8 Performance Testing of KHIs

There are several methods of testing the performance of KHIs. If an operator is interested in qualifying a KHI for field use, the minimum hold time at the worst-case subcooling field conditions is usually determined[43,112] (the hold time is the duration between the moment when the system enters the hydrate stability domain and the onset of hydrate formation). This will give a conservative value of the dosage required to eliminate hydrate problems since the flowline is not at the maximum subcooling and pressure all the time. As mentioned earlier, the tests should be run at the system pressure since several studies have indicated that hold times vary with pressure even at the same subcooling.[29–31] Shut-in tests with no flow must also be

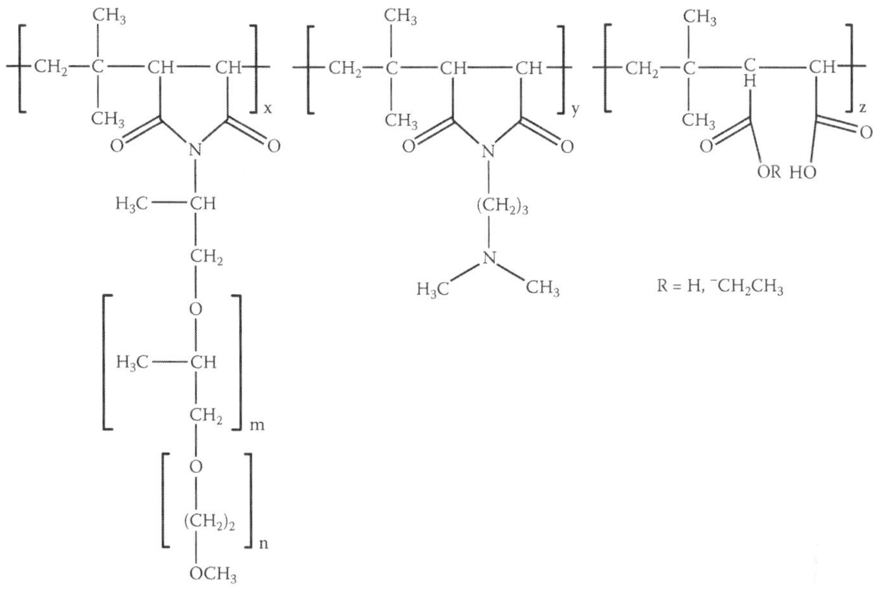

FIGURE 9.12 A maleic-based terpolymer kinetic hydrate inhibitor.

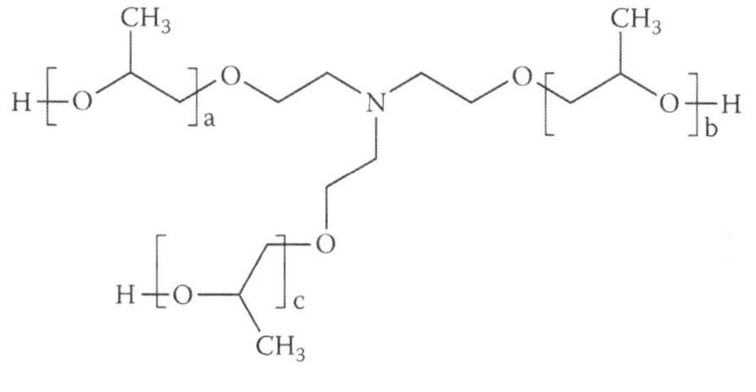

FIGURE 9.13 The structure of a preferred polyalkoxylated amine where $a + b + c = 14.9$.

included for field verification under planned or unplanned shut-in conditions, as well as tests in the presence of other production chemicals, particularly CIs: many film-forming CIs are known to drastically reduce the KHI performance. The tests can be carried out in autoclaves, rocking cells, pipe wheels, or loops with the actual field fluids and gas composition. Tests in large loops and pipe wheels are generally the last and best step before field trials or field implementation of the KHI. This is because test results from small, "clean" laboratory equipment are most influenced by the stochastic nature of hydrate formation. Thus, a test needs to be repeated a number of times to judge the performance. It has been suggested that theoretically, an infinite number of identical KHI tests in the laboratory should be run, and that the worst result should be taken as the performance expected in the field at the same PT and fluid conditions.[112] In practice, operators will usually not risk operating the production line right at the performance limit of a KHI, but at perhaps 1–2°C less subcooling (or higher dosage) than was qualified in the laboratory or pilot tests. In fact, a hydrate plug formed in the presence of a KHI is harder to dissociate than a plug formed without KHI present.[186]

If it is only of interest to rank the performance of KHIs, other methods also become available. For example, one can cool the inhibited system without agitation/flow to a preset temperature in the thermodynamic hydrate-forming region and then begin agitation. The induction time to hydrate formation can then be measured for each KHI.[113] Alternatively, constant cooling into the hydrate region with agitation/flow can be carried out and the onset of hydrate formation measured: the lower the onset temperature, the better the KHI.[114] The cooling can also be ramped, such that the temperature is held constant for a number of hours before cooling again to a higher subcooling, and so forth. This makes it easier to detect the onset of hydrate formation because the pressure is held constant at each ramped temperature. A new method has been developed that claims to give less scattering in the induction time (or hold time) data due to the stochastic nature of hydrate formation.[115] This method entails forming hydrates, then melting the hydrates to just above the equilibrium temperature for a short time so as to preserve hydrate "precursor" structures in the water phase (the so-called memory effect) and then cooling to a preset temperature in the hydrate-forming region. The holdtime is claimed to be more repeatable with this method because hydrate formation is mainly triggered by heterogeneous germination. However, different KHI polymers will affect the rate and temperature of hydrate dissociation to different degrees, so you do not have a constant amount of precursors in all experiments on cooling, which may affect the ranking of KHIs adversely. Injection of the KHI after dissociation of the hydrates could overcome this. More work is needed on this new method. Another method for testing KHIs uses differential scanning calorimetry and relies on making stable water-in-oil emulsions.[193] The length of the delay times to first detected hydrate formation are greater with an emulsion than without and appear more scattered.

Induction time and crystal growth inhibition experiments on THF hydrate (structure II) at atmospheric pressure can give information on the performance of some KHIs (and crystal growth–modifying AAs), but can be misleading for other KHIs because THF is very water-soluble and the inhibition mechanism may be different from a real gas system.[116] For example, low molecular weight PVP gave best results in real gas hydrate systems whereas high molecular weight PVP was best at inhibiting the growth of THF hydrate crystals.[117] In addition, some polymers are not soluble in THF-water mixtures but are in water alone.

KHI polymers are relatively expensive and used at a higher continuous dosage compared with other production chemicals. However, their use can give substantial capital and operational cost savings compared with alternate hydrate control strategies. KHI/THI blends can also be used. A method to recover and reuse a KHI polymer has been claimed although no practical field applications are known.[118]

9.2.3 Anti-Agglomerants

AAs are a class of LDHI, the best of which can prevent hydrate plugging at higher subcoolings than KHIs. There are two subclasses of AAs:

Gas Hydrate Control 243

- production or pipeline AAs
- gas-well AAs

All AAs allow hydrates to form but they prevent them from agglomerating and, subsequently, accumulating into large masses. A pipeline AA enables the hydrates to form as a transportable nonsticky slurry of hydrate particles dispersed in the liquid hydrocarbon phase. Gas-well AAs disperse hydrate particles in an excess of water. Products in both categories are in commercial use.

Pipeline AAs cannot be used on gas fields because they require a liquid hydrocarbon phase. In general, the water cut for pipeline AAs should be below approximately 50%, otherwise the hydrate slurry gets too viscous to transport. A method to overcome this problem by adding water to the fluid mixture in an amount sufficient to lower the gas/water ratio sufficiently to achieve a pumpable hydrate slurry has been disclosed.[119] AAs can perform at higher subcoolings than the KHIs and are thus applicable for severe and/or deepwater applications.

9.2.3.1 Emulsion Pipeline AAs

There are two mechanisms by which the pipeline AA effect can be accomplished. In one mechanism, a surfactant is injected that forms a special kind of water-in-oil emulsion. This emulsion confines hydrates to form within the water droplets, and the hydrates never agglomerate. The end product is a slurry of hydrate particles in a hydrocarbon phase. Dosage levels are around 0.8–1.0 wt.% based on the water phase. Many classes of surfactant have been claimed as emulsion AAs including diethanolamides, dioctylsulfosuccinates, sorbitans, ethoxylated polyols, ethoxylated fatty acids, and ethoxylated amines. The best examples of emulsion AAs appear to be polymeric surfactants such as those based on polyalkyleneglycol derivatives of polyalkenyl succinic anhydrides, or carboxylic acid hydroxycarbylamide, substituted or nonsubstituted, and a carboxylic acid monoethanolamide or diethanolamide containing three to six carbon atoms.[120–122] Compositions comprising at least one ester, associated with a nonionic cosurfactant of the polymerized (dimer and/or trimer) carboxylic acid type have also been found to perform well as emulsion AAs.[123] The emulsion AA additive can be recycled.[124] Pilot and field trials have been carried out on one product with mixed success but no full field applications have been carried out.[125–126] Work on this class of AA has now been disbanded by the original authors. There are a couple of drawbacks to the emulsion AA technology. First, the water phase has to be thoroughly emulsified before entering hydrate-forming conditions; otherwise, hydrate agglomeration and deposition are likely. Can this be guaranteed in the field? Second, hydrates form from condensed water on the upper walls of the pipe during laminar flow or during shut down. The condensed water problem and also slurry transportation difficulties may be overcome when the flow is not stratified (higher liquid volumes and flow rates).

9.2.3.2 Hydrate-Philic Pipeline AAs

The other mechanism by which pipeline AAs work was discovered in the early nineties.[28,127] In this mechanism, the surfactant molecule has a head group that is "hydrate-philic" (seeks hydrate crystal surfaces) and a tail that is hydrophobic

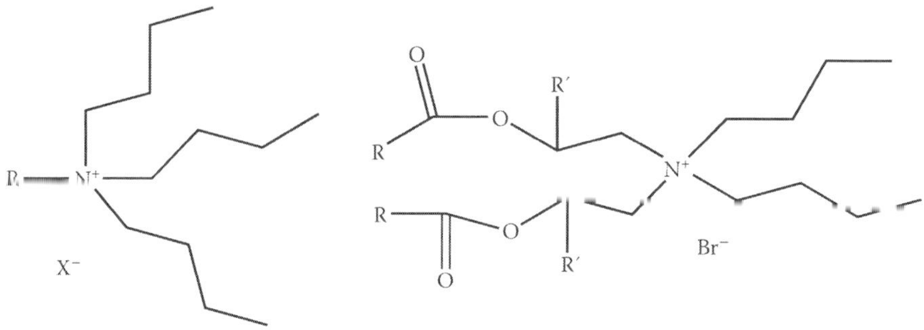

FIGURE 9.14 General structures of the butylated single- and twin-tail quaternary AAs. R = long alkyl chain with optionally a spacer group, R' = H or CH_3, X = counter ion.

(oil-loving). When several of these surfactant molecules are attached to the hydrate crystal surface, further growth is disrupted and the crystal becomes hydrophobic. It is then easily dispersed in the liquid hydrocarbon phase. Many classes of surfactants that are not hydrate-philic actually accelerate hydrate formation and have been studied for use in natural gas storage and transportation as gas hydrate. Several classes of surfactants have been shown to have hydrate-philic AA properties, such as caprolactam and alkylamide surfactants, but only one class has become commercialized and is in use today. These are quaternary ammonium surfactants with head groups that in most cases contain two or more n-butyl, n-pentyl, or iso-pentyl groups (Figure 9.14).[128] These were discovered in the early nineties, and there are now about five quaternary AAs commercially available from various sources. The quaternary head group could also be phosphonium-based, but this would make the AAs more expensive. In general, butyl groups rather than pentyl groups are used in commercial pipeline AAs, due to the ease of manufacture and costs. Like KHIs, the performance of quaternary AAs are also dependent on the composition of the hydrocarbon fluid. The performance of AAs has been shown to decrease as the naphthenic acid content of a crude oil increases.[187] A possible explanation, assuming the AA studied was a cationic surfactant, is that naphthenate anions are ion pairing with the AA, increasing its oil solubility and moving it away from the oil-water interface where it needs to be to function. A high phenol content in a crude oil can also be detrimental to the AA performance. Quaternary pipeline AAs show fairly good corrosion inhibition, and, since they are injected at several thousand parts per million, there is sometimes no need for an extra specialized CI.

Regarding the mechanism of quaternary AAs, there is evidence that the butyl or pentyl groups penetrate open cavities and get embedded in the hydrate crystal surface preventing the surfactants from detaching.[28] This may be why they work better than other surfactants that can interact with hydrate surfaces but not get embedded. The quaternary surfactant AAs generally perform better at increasing salinity.[129] Polymeric surfactants (for example, PVCap or hyperbranched polyesteramides with long, hydrophobic tails attached) do not appear to work well as "hydrate-philic" pipeline AAs.

All classes of quaternary surfactant AAs currently on the market have a single hydrophobic tail. A twin-tail quaternary surfactant AA has been in commercial use but will now no longer be manufactured (Figure 9.14). The twin tail AA has ester linkages giving it fairly good biodegradability.[130] It partitions mostly in the oil phase, whereas the single-tail AAs partition mostly in the aqueous phase. The best single-tail AAs are claimed to work to very high subcoolings (possibly without limit if the dosage is high enough), whereas the twin-tail AAs appear to work up to about 14–15°C (25–27°F) subcooling at a dosage of only 0.25–0.3 wt.% (active surfactant). The single-tail AAs require a higher dosage and do not perform as well in fresh water; performance is usually better as the salinity increases, given the same subcooling. They can also adversely affect the discharged water quality. Further, the performance of AAs under shut-in/cooldown conditions seems to depend strongly on the rate at which the system is cooled.[112] Field applications of the single-tail quaternary AA have been reported since 2000.[133–136] They have also been found to work in highly sour systems.[137]

The first version of the twin-tail quaternary AA had no ester spacer groups and had poor biodegradability. The second version of the twin-tail quaternary AA was made from N-butyl diethanolamine and had ethylene spacer groups between the ester groups and the quaternary nitrogen atom.[138] It was found that addition of a small alkyl branch (methyl or ethyl) to the spacer group improved the performance of the AA.[139] Specifically, the performance of shut-in/start-up loop test with preformed hydrates slurries were improved. A longer spacer group such as propylene gave worse performance. The twin-tail quaternary AA also contained a certain amount of unquaternized amine in the finished product. The amine actually improves the performance of the AA as well as helping break emulsions. A field application of the twin-tail quaternary AA has been reported.[140] The shelf life of this AA is only about 1 year due to degradation of the ester groups. Manufacture of this twin-tail quaternary AA has been discontinued, due to additional costs in having the chemical approved under the European Community's new REACH environmental regulations. The operator plans to switch to the use of KHI as the maximum subcooling on this condensate field has dropped and is now within the performance range of this class of LDHI.

The original single-tail quaternary surfactant AAs are poorly biodegradable and fairly toxic. However, the toxicity is claimed to be reduced by addition of anionic, nonionic, or amphoteric surfactants or polymers (anionic surfactants appear most preferable).[141–143] Examples are sodium dodecyl sulfate and ammonium alkyl ether sulfate. Interestingly, this also improves the performance of the AA or allows a lower active dosage to be used. Another company has claimed the use of a single-tail quaternary AA with an anionic surfactant counterion as an ion-pair amphiphile blend. This makes it more oil-soluble.[144] Examples include benzyl dimethyl cocoamine/tall oil fatty acid ion pair. A field application has been reported.[145] The improvement in performance of the quaternary AA by adding an anionic surfactant may be due to the anionic surfactant following the quaternary surfactant down to the hydrate crystal surface as an ion pair, giving extra coverage of the crystal surface and enhancing the hydrophobicity of the crystals.

Another oil-soluble quaternary AA, claimed to be useful for high water cuts and subcoolings up to 16.7°C (30°F), has been reported by a service company.[146–147]

The same company has an international patent application on AAs claiming a wide range of quaternary ammonium and phosphonium surfactants, some with degradable ester or amide linkages, although some betaine and amine oxide surfactants are also mentioned.[148a] A further more specific patent claims certain amido quaternary surfactants as AAs.[148b] An example is N,N-dibutyl-cocoamidopropyl carbomethoxy betaine. The commercialized AA is further claimed to be less toxic than other commercial AAs, to perform at lower dosages than the original quaternary AAs, to give better-produced oil/water quality, and to not destabilize asphaltenes. In laboratory tests, it has been proven to prevent hydrate blockages even at water cuts up to 80%. It is generally believed that AA applications are limited to below about 50% water cut; otherwise, the slurry is too viscous to transport. However, AAs could theoretically work at higher water cuts if the salinity is very high. In such a situation, not all the water may be converted to hydrates. This is because the unconverted water will have a higher concentration of salts and may be sufficiently thermodynamically inhibited that further hydrates will not form. This highly saline water, together with the liquid hydrocarbon, can act as the transport medium for the hydrate particles.

The performance of the original single-tail quaternary AAs is claimed to be improved by blending it with an amine salt and optionally a solvent.[149] The amine salt contained preferably alkyl or hydroxyalkyl groups with one to three carbon atoms or an ammonium salt could be used. Other quaternary surfactant AAs such as dialkoxylated quaternary ammonium compounds, quaternized N,N'-dibutylaminoalkyl ether carboxylates, and quaternized alkylaminoalkyl diesters have been patented.[150–152]

Another patent application describes a method to make hydroxy quaternary surfactant AAs from halohydrins (Figure 9.15). It states that reaction of hindered amines such as tributylamine with alkyl halides is difficult, often giving poor yields of single-tail quaternary AAs.[153] By reacting a halohydrin, such as epichlorohydrin, with a long-chain alcohol, the 1,2-hydroxyhalide reaction product will react more easily with tributylamine to form a high yield of quaternary AA because the reaction proceeds via a protonated epoxide.

Quaternary surfactant AAs based on oxazolidinium compounds have also been patented.[154] Particularly useful compounds are shown in Figure 9.16. The

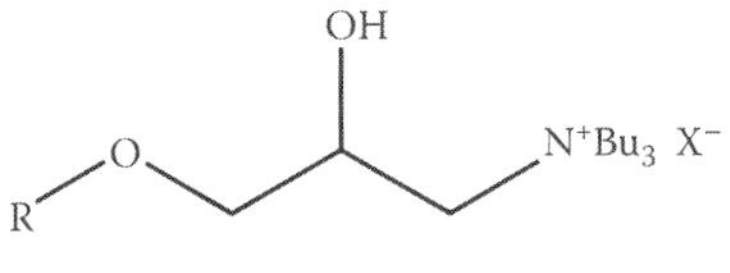

FIGURE 9.15 Hydroxy quaternary AAs made from halohydrins.

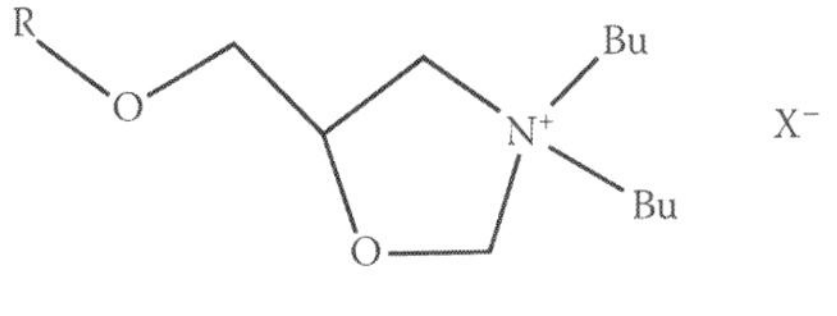

FIGURE 9.16 Examples of oxazolidinium AA compounds where R is preferably C12–14.

oxazolidinium compounds are formed by the reaction of a halohydrin or an epoxide with a secondary amine and an aldehyde or a ketone. The oxazolidinium compounds are formed directly and do not require the reaction of a preformed oxazolidine with an alkylating agent.

Since quaternary ammonium surfactant AAs are generally fairly toxic and only partially biodegradable, attempts to find greener, nonquaternary AAs have been the subject of research, although the subcooling performance appears to be compromised. Examples are caprolactam and alkylamide surfactants with head groups identical to the pendant groups found in VCap- and alkylacrylamide-based KHIs.[155–156] Zwitterionic quaternary AAs have also been investigated.[157] Also, certain alkoxylated and/or acylated nonquaternary nitrogen-containing compounds have been claimed.[158] Further, biodegradable biosurfactants, such as rhamnolipid biosurfactants have been shown to exhibit AA behavior with a model oil and THF hydrate.[159–160] The performance is improved by adding alcohol cosurfactants. None of these classes of AA are at present commercially available.

9.2.3.3 Performance Testing of Pipeline AAs

AAs can be tested for performance in rocking cells, autoclaves, pipewheels, and loops. Rocking cells are a good way of screening AA candidates at a variety of test conditions and are often used by the service companies as the first step in qualifying a product.[142] A metal ball is rocked back and forth in a cell containing hydrate-forming fluids with an observation window (e.g., sapphire) and sensors at the end of the cell detect when hydrate deposits prevent the ball from touching the ends. Visual observations are paramount in such tests. Stirred autoclaves with observation windows and torque measurements have also been used.[155] Certain pumps in a pipe loop may break up hydrate particles contributing to the AA effect. If one is only interested in studying the build-up of hydrate deposits on the pipe loop walls, this can be done by warming up the fluids after each circulation to melt any transportable hydrates and recooling in the section of pipe after the pump.[139] Wheel pipes avoid the use of pumps altogether.[161–162] Changes in the torque on the wheel indicate hydrate deposition as well as visual observations in one section of the wheel. One usually needs to test the performance of an AA in three situations at varying water cut:

1. Cooling into the hydrate region under flowing conditions
2. Restart after a shut-in period with a preformed hydrate slurry
3. Cooling into the hydrate region during shut-in, then a restart after a shut-in period

9.2.3.4 Natural Surfactants and Nonplugging Oils

Some oils have been shown to give no hydrate plugs in laboratory tests at varying water cuts[28,163–165] It appears they contain natural surfactants that prevent hydrate agglomeration and deposition, although it is uncertain which mechanism is operating, the emulsion AA mechanism or the "hydrate-philic" mechanism. In addition, no research group has managed to determine the exact structure of the active components in these nonplugging oils although the naphthenic acids, maltenes, and asphaltenes are suspected for the AA behavior. A large percentage of these components do not necessarily

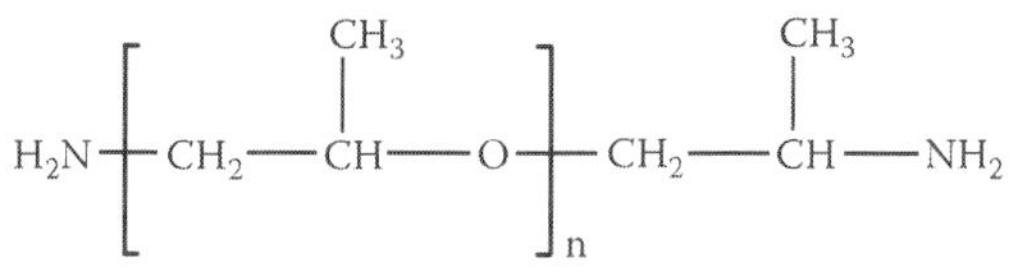

$$H_2N {-\!\!\!\Big[} CH_2 - \underset{\underset{\displaystyle CH_3}{|}}{CH} - O {\Big]\!\!\!-}_n CH_2 - \underset{\underset{\displaystyle CH_3}{|}}{CH} - NH_2$$

FIGURE 9.17 Structure of a typical polyetherdiamine AA.

prevent the oil from forming hydrate plugs in a multiphase system. Further, the non-plugging property of a crude oil appears to be a whole oil property and not those of a "magic" component. In theory, it should be possible to carry out multiphase transportation with a nonplugging oil, although no fields have yet been reported to have been developed with this hydrate plug prevention strategy.[166] However, one operator benefits from knowledge of their nonplugging oils in their ongoing hydrate prevention strategy.[167] In addition, one North Sea operator no longer uses thermodynamic hydrate (THI) on a subsea multiphase pipeline as the line was found not to plug with hydrates, even after shut-in/start-ups, when THI was inadvertently stopped or underdosed.

9.2.3.5 Gas-Well AAs

Polyetherpolyamines, such as polyetherdiamines (particularly of the oxypropylene type), have been shown to disperse hydrates in an excess of water (Figure 9.17).[28,168–170] Molecular weights of these polymers are low, usually under 500. Quaternized derivatives gave better performance in such tests than PVCap, or several hydrate-philic quaternary pipeline AAs.[171]

Polyetheramines have been used downhole for several years to prevent hydrate blockages in over 100 gas wells, mostly onshore, but also offshore.[172] A strong synergistic effect was discovered at a certain ratio of methanol and polyetheramine.[173] Use of the polyetheramines in completion and fracturing fluids has also been patented.[174] As mentioned earlier in this chapter, polyetheramines are good synergists for VCap-based KHIs. A pipeline field application of a polyetheramine AA has also been reported.[170] In this case, the AA is injected and, if hydrate buildup in the line is "detected," a THI is injected until the line is clear. Injection of the AA can then be continued. The rate of THI injection would also be increased in a shut-in to give thermodynamic protection.[184] Polyetheramines and their derivatives are also good CIs, particularly against H_2S corrosion (see Chapter 8 on corrosion control).

Related to polyetheramine chemistry, certain polypropoxylated polyamines have been shown to display AA properties in oil/brine/gas systems as long as there is good agitation of the fluids.[192]

9.3 GAS HYDRATE PLUG REMOVAL

Under normal production operations, hydrate plugs can form both in the top of a cold climate or subsea well during shut-in or in the topside flowline.[175] There are various ways of removing a hydrate plug including:

- depressurization
- extended reach coiled tubing or tractors

Gas Hydrate Control 249

- heated wireline
- hot oiling
- chemical treatment

Recent advances in multiphase-flow modeling capabilities allow operators to conduct detailed assessments of the impact of blowdown with a hydrate plug present at any point within the wellbore or pipeline.[18] Industry has also been developing new tools that can be deployed at short notice to depressurize the pipeline or wellbore. Depressurization on both sides of the plug is recommended to avoid hydrate plugs breaking loose suddenly and becoming dangerous projectiles. In one deepwater North Sea field, where the seabed temperature is below 0°C (32°F), experiments showed that hydrate plugs will always convert into ice plugs during depressurization at subzero temperatures. Currently, there are no technologies that can guarantee the removal of an ice plug that is located (for example) 20 km downstream of the pipeline inlet. Various extended-reach coiled tubing capabilities for ice or hydrate plug removal are under development, which will be especially useful for long, deepwater tie-backs. Hot oiling to the tubing-casing annulus has been shown to melt dry tree well hydrate plugs.[176]

There are two methods of removing a hydrate plug by chemical treatment:

- use of thermodynamic inhibitors (THIs)
- heat-generating chemicals

The first method is commonly used in the field, but the second method has only rarely been used.

9.3.1 Use of Thermodynamic Hydrate Inhibitors

As the earlier section on THIs mentioned, these chemicals are not only useful to inhibit gas hydrate formation but can melt hydrate (or ice) plugs that have formed. Methanol and MEG are the preferred THIs for hydrate plug melting. However, they do not always work as planned in remediation operations.[177] This is because the hydrate plug properties are critical in determining the melting efficiency of THIs.[178–179] The plug properties, the density and viscosity of the THI, and the contact area between the THI and the plug are the most important factors in determining the melting efficiency.[180] For very porous plugs, the low viscosity of methanol seems to be beneficial during melting. Otherwise, denser MEG can be more efficient for other types of plugs. MEG is denser than oil, but methanol is not. Hence, MEG will penetrate oil above the hydrate plug, and not just the plug itself. Inhibitor mixtures of methanol/MEG and $CaCl_2$/methanol did not result in higher melting efficiencies. Potassium formate could be a potential low-corrosive hydrate plug meter. Aqueous solutions can be up to 75 wt.% and are very dense.

9.3.2 Heat-Generating Chemicals

Heat-generating chemicals have been used to remove both hydrate and wax deposits. The first system, developed originally for wax removal, uses a blend of an ammonium salt and sodium nitrite (to make ammonium nitrite) and an acid catalyst (e.g., acetic acid) or precatalyst, which generates acid in situ. If a precatalyst such as an acid anhydride is used, heat is generated after a built-in time delay.[181–182] This chemical system has been called the "SGN process." The acid catalyzes decomposition of the ammonium nitrite to yield sodium chloride, nitrogen, water, and heat. It has been used to remove hydrate plugs in subsea Christmas trees.[183] The second method relies on a simple in-situ acid-base reaction (e.g., hydrochloric acid plus sodium hydroxide) to generate heat.[184] Due to the rapid reaction kinetics, the acid and base must be mixed where required, such as on top of a hydrate plug in the top of a well. The use of heat-generating chemicals does entail a certain risk if not designed and deployed correctly. As a worst-case scenario, loss of control of the exothermic reaction could lead to unexpected stress to pipeline materials and potential failure.

REFERENCES

1. E. D. Sloan Jr. and C. A. Koh, *Clathrate Hydrates of Natural Gases*, 3rd ed., Boca Raton, FL: CRC Press, Taylor & Francis Group, 2008.
2. Y. F. Makogon, *Hydrates of Hydrocarbons*, Tulsa, OK: PennWell Publishing Company, 1997.
3. (a) J. Carroll, *Natural Gas Hydrates—A Guide for Engineers*, Boston, MA: Gulf Professional Publishing, Elsevier Science, 2003. (b) E. D. Sloan, "A Changing Hydrate Paradigm—From Apprehension to Avoidance to Risk Management," *Fluid Phase Equilibria* 228–229 (2005): 67.
4. S. Mokhatab, R. J. Wilkens, and K. J. Leontaritis, "A Review of Strategies for Solving Gas-Hydrate Problems in Subsea Pipelines," *Energy Sources, Part A: Recovery, Utilization and Environmental Effects* 29 (2007): 39.
5. O. Urdahl, A. H. Boernes, K. J. Kinnari, and R. Holme, "Operational Experience by Applying Direct Electrical Heating for Hydrate Prevention," SPE 85015, *SPE Production & Facilities* 19(3) (2004): 161.
6. R. G. Harris and J. Clapham, *Offshore* 59(2) 1999.
7. *Offshore* 66(9) (2006).
8. K. J. Kinnari, C. Labes-Carrier, K. Lunde, and L. Aaberge, International Patent Application WO/2006/027609.
9. R. Larsen, A. Lund, V. Andersson, and K. W. Hjarbo, "Conversion of Water to Hydrate Particles," SPE 71550 (paper presented at the SPE Annual Technical Conference and Exhibition, New Orleans, LA, 30 September–3 October 2001).
10. A. Lund, D. Lysne, R. Larsen, and K. W. Hjarbo, International Patent Application WO/2000/025062.
11. L. D. Talley, D. J. Turner, and D. K. Priedeman, International Patent Application WO/2007/095399.
12. L. Talley, "Hydrate Inhibition Via Cold Flow: No Chemicals or Insulation," Proceedings of the 6th International Conference on Gas Hydrates Vancouver, British Columbia, Canada, 6–10 July 2008.
13. D. Merino-Garcia and S. Correra, "Cold Flow: A Review of a Technology to Avoid Wax Deposition," *Petroleum Science Technology* 26 (2008): 446.
14. H. Haghighi, R. Azarinezhad, A. Chapoy, R. Anderson, and B. Tohidi, "Hydraflow: Avoiding Gas Hydrate Problems," SPE 107335 (paper presented at the SPE Europec/EAGE Annual Conference and Exhibition, London, UK, 11–14 June 2007).

15. F. J. P. C. M. G. Verhelst, A. Twerda, J. P. M. Smeulers, M. C. Peters, M. Adrianus, S. P. C. Belfroid, and W. Schiferli, International Patent Application WO/2008/023979.
16. R. Azarinezhad, A. Chapoy, F. Ahmadloo, R. Anderson, and B. Tohidi, "Hydraflow: A Novel Approach in Addressing Flow Assurance Problems," Proceedings of the 6th International Conference on Gas Hydrates Vancouver, British Columbia, Canada, 6–10 July 2008.
17. R. Anderson, A. Chapoy, J. Tanchawanich, H. Haghighi, J. Lachwa-Langa, and B. Tohidi, "Binary Ethanol-Methane Clathrate Hydrate Formation in the System CH_4-C_2H_5OH-H_2O: Experimental Data and Thermodynamic Modelling," Proceedings of the 6th International Conference on Gas Hydrates Vancouver, British Columbia, Canada, 6–10 July 2008.
18. U. C. Klomp and A. P. Mehta, "An Industry Perspective on the State-of-the Art of Hydrates Management," Proceedings of the 5th International Conference on Gas Hydrates, Trondheim, Norway, 12–16 June 2006.
19. M. H. Yousif, "Effect of Underinhibition with Methanol and Ethylene Glycol on the Hydrate-Control Process," SPE 50972, *SPE Production & Facilities* 13(3) (1998): 184.
20. P. Hemmingsen and X. Li, "Hydrate Plugging Potential in Underinhibited Systems," Proceedings of the 6th International Conference on Gas Hydrates Vancouver, British Columbia, Canada, 6–10 July 2008.
21. F. H. Fadnes, T. Jakobsen, M. Bylov, A. Holst, and J. D. Downs, "Studies on the Prevention of Gas Hydrates Formation in Pipelines Using Potassium Formate as a Thermodynamic Inhibitor," SPE 50688 (paper presented at the SPE European Petroleum Conference, The Hague, 20–22 October 1998).
22. E. G. Hammerschmidt, "Formation of Gas Hydrates in Natural Gas Transmission Lines," *Industrial & Engineering Chemistry* 26 (1934): 851.
23. K. K. Østergaard, R. Masoudi, B. Tohidi, A. Danesh, and A. C. Todd, *Journal of Petroleum Science and Engineering* 48 (2005): 70.
24. M. D. Jager, A. L. Ballard, and E. D. Sloan, "Comparison between Experimental Data and Aqueous-Phase Fugacity Model for Hydrate Prediction," *Fluid Phase Equilibria* 232 (2005): 25.
25. A. L. Ballard and E. D. Sloan, "The Next Generation of Hydrate Prediction: Part III. Gibbs Energy Minimization Formalism," *Fluid Phase Equilibria* 218 (2004): 15.
26. R. Masoudi, B. Tohidi, A. Danesh, A. C. Todd, and J. Yang, "Measurement and Prediction of Salt Solubility in the Presence of Hydrate Organic Inhibitors," SPE 87468, *SPE Production & Operations* 21(2) (2006): 182.
27. M. B. Tomson, A. T. Kan, and G. Fu, "Inhibition of Barite Scale in the Presence of Hydrate Inhibitors," SPE 87437, *SPE Journal* 10(3) (2005): 256.
28. M. A. Kelland, "History of the Development of Low Dosage Hydrate Inhibitors," *Energy & Fuels* 20 (2006): 825.
29. M. Arjmandi, B. Tohidi, A. Danesh, and A. C. Todd, "Is Subcooling the Right Driving Force for Testing Low-Dosage Hydrate Inhibitors?," *Chemical Engineering Science* 60 (2005): 1313–1321.
30. J.-P. Peytavy, P. Glénat, and P. Bourg, "Kinetic Hydrate Inhibitors—Sensitivity Towards Pressure and Corrosion Inhibitors," IPTC 11233 (paper presented at the International Petroleum Technology Conference, Dubai, UAE, 4–6 December 2007).
31. M. A. Kelland, J.-E. Iversen, K. Moenig, and K. Lekvam, "Feasibility Study for the Use of Kinetic Hydrate Inhibitors in Deep-Water Drilling Fluids," *Energy & Fuels* 22 (2008): 2405.
32. N. J. Phillips and M. Grainger, "Development and Application of Kinetic Hydrate Inhibitors in the North Sea," SPE 40030 (paper presented at the Proceedings of the Annual Gas Technology Symposium, Calgary, Alberta, Canada, 15–18 March 1998).

33. M. Varma-Nair, C. A. Costello, K. S. Colle, and H. E. King, "Thermal Analysis of Polymer-Water Interactions and Their Relation to Gas Hydrate Inhibition," *Journal of Applied Polymer Science* 103(4) (2007): 2642.
34. E. D. Sloan, U.S. Patent 5420370, May 30, 1995.
35. J. Long, J. Lederhos, A. Sum, R. L. Christiansen, and E. D. Sloan, Proceedings of the 73rd Annual GPA Convention, New Orleans, LA, 7–9 March 1994.
36. R. Hawtin and P. M. Rodger, "Polydiversity in Oligomeric Low Dosage Gas Hydrate Inhibitors," *Journal of Materials Chemistry* 16 (2006): 1934.
37. N. Aldiwan, Y. Lui, A. Soper, H. Thompson, J. Creek, R. Westacott, E. D. Sloan, and C. Koh, "Neutron Diffraction and EPSR Simulations of the Hydrate Structure around Propane Molecules before and during Gas Hydrate Formation," Proceedings of the 6th International Conference on Gas Hydrates Vancouver, British Columbia, Canada, 6–10 July 2008.
38. (*a*) T. Y. Makogon, R. Larsen, C. A. Knight, and E. D. Sloan Jr., "Melt Growth of Tetrahydrofuran Clathrate Hydrae and Its Inhibition," *Journal of Crystal Growth* 179 (1997): 258. (*b*) H. E. King, J. L. Hutter, M. Y. Lin, and T. Sun, *J. Chem. Phys.*, 2000, 112, 2523.
39. (*a*) B. J. Anderson, J. W. Tester, G. P. Borghi, and B. L. Trout, "Properties of Inhibitors of Methane Hydrate Formation via Moecular Dynamics," *Journal of the American Chemical Society* 127 (2005): 17852. (*b*) A. Cruz-Torres, A. Romero-Martinez, and A. Galano, "Computational Study on the Antifreeze Glycoproteins as Inhibitors of Clathrate-Hrydrate Formation," *Chemphyschem* 9(11) (2008): 1630.
40. T. J. Carver, M. G. B. Drew and P. M. Rodger, "Inhibition of Crystal Growth in Methane Hydrate," *Journal of the Chemical Society, Faraday Transactions* 91 (1995): 3449.
41. (*a*) D. A. Gomez Gualdron and P. B. Balbuena, "Classical Molecular Dynamics of Clathrate-Methane-Water-Kinetic Inhibitor Composite Systems," *Journal of Physical Chemistry* 111 (2007): 15554. (*b*) R. W. Hawtin, D. Quigley, and P. M. Rodger, "Gas Hydrate Nucleation and Cage Formation at a Water/Methane Interface," *Physical Chemistry Chemical Physics* 10 (2008): 4853.
42. H. Zeng, H. Lu, E. Huva, V. K. Walker, and J. A. Ripmeester, "Differences in Nucleator Adsorption May Explain Distinct Inhibition Activities of Two Gas Hydrate Kinetic Inhibitors" *Chemical Engineering Science* 63 (2008): 4026.
43. U. C. Klomp and A. P. Mehta, "Validation of Kinetic Inhibitors for Sour Gas Fields," IPTC-11374 (paper presented at the International Petroleum Technology Conference, Dubai, UAE, 4–6 December 2007).
44. B. Huang, Y. Wang, S. Zhang, and Y. Ao, "Kinetic Model of Fixed Bed Reactor with Immobilized Microorganisms for Removing Low Concentration SO," *Journal of Natural Gas Chemistry* 16 (2007): 81.
45. S. Szymczak, K. Sanders, M. Pakulski, and T. Higgins, "Chemical Compromise: A Thermodynamic and Low-Dose Hydrate-Inhibitor Solution for Hydrate Control in the Gulf of Mexico," SPE 96418, *SPE Project Facilities & Construction* 1(4) (2006): 1.
46. L. W. Clark and J. Anderson, "Low Dosage Hydrate Inhibitors (LDHI): Further Advances and Developments in Flow Assurance Technology and Applications Concerning Oil and Gas Production Systems," IPTC 11538 (paper presented at the International Petroleum Technology Conference, Dubai, UAE, 4–6 December 2007).
47. E. D. Sloan, U.S. Patent 5432292, 1995.
48. E. D. Sloan, R. L. Christiansen, J. Lederhos, V. Panchalingam, Y. Du, A. K. W. Sum, and J. Ping, U.S. Patent 5639925, 1997.
49. C. B. Argo, R. A. Blaine, C. G. Osborne, and I. C. Priestly, "Commercial Deployment of Low Dosage Hydrate Inhibitors in a Southern North Sea 69 Kilometer Wet-Gas Subsea Pipeline," SPE 37255 (paper presented at the SPE International Symposium on Oilfield Chemistry, Houston, TX, February 1997).

50. L. D. Talley and G. F. Mitchell, "Application of Kinetic Hydrate Inhibitor in Black-Oil Flowlines," SPE 56770 (paper presented at the SPE Annual Technical Conference and Exhibition, September 1999).
51. S. B. Fu, L. M. Cenegy, and C. Neff, "A Summary of Successful Field Applications of A Kinetic Hydrate Inhibitor," SPE 65022 (paper presented at the SPE International Symposium on Oilfield Chemistry, Houston, TX, 13–16 February 2001).
52. (*a*) K. Bakeev, R. Myers, J.-C. Chuang, T. Winkler, and A. Krauss, U.S. Patent 6242518, 2001. (*b*) M. Angel, S. Stein, and K. Neubecker, International Patent Application WO/2001/066602.
53. T. Namba, Y. Fujii, T. Saeki, and H. Kobayashi, International Patent Application WO96/37684, 1996.
54. M. Angel, K. Neubecker, and S. Stein, International Patent Application WO2004/042190, 2004.
55. L. D. Talley and M. Edwards, "First Low Dosage Hydrate Inhibitor is Field Proven in Deepwater," *Pipeline and Gas Journal* 226 (1999): 44.
56. V. Thieu, K. Bakeev, and J. S. Shih, U.S. Patent 6359047, 2002.
57. D. J. Freeman, D. J. Irvine, J. Kitching, and C. S. Rogers, International Patent Application WO/2006/051265.
58. K. Bakeev, J.-C. Chuang, T. Winkler, M. A. Drzewinski, and D. E. Graham, U.S. Patent 6281274, 2001.
59. U. Dahlmann, M. Feustel, and C. Kayser, U.S. Patent 7297823, 2007.
60. (*a*) A. Maximilian, K. Neubecker, and A. Sanner, U.S. Patent 6867262, 2005. (*b*) R. Widmaier, L. Wegmann, A. Mauri, K. Mathauer, W. Jahnel, L. H. Taboada, K. Neubecker, and A. Khvorost, U.S. Patent Application 20080255326.
61. K. Colle, L. D. Talley, and J. M. Longo, World Patent Application WO 2005/005567, 2005.
62. B. Fu, S. Neff, A. Mathur, and K. Bakeev, "Application of Low-Dosage Hydrate Inhibitors in Deepwater Operations," SPE 78823, *SPE Production and Facilities* 17 (2002): 133.
63. M. Jurek, M. Alexandre, and S. Bell, "New Approaches to Low Dose Gas Hydrate Treatment," *Chemicals in the Oil Industry*, Royal Society of Chemistry, Manchester, 5–7 November 2007.
64. A. Rasch, A. Mikalsen, T. Austvik, L. H. Gjertsen, and X. Li, "Evaluation of a Kinetic Inhibitor with Focus on the Pressure and Fluid Dependency," Proceedings of the 4th International Conference on Gas Hydrates, Yokohama, Japan, 19–23 May 2002.
65. M. Pakulski and J. C. Dawson, U.S. Patent 7067459, 2006.
66. B. Fu, "The Development of Advanced Kinetic Hydrate Inhibitors," Proceedings of Chemistry in the Oil Industry VII, Royal Society of Chemistry, Manchester 13–14 November 2002, 264.
67. J. M. Cohen, P. F. Wolf, and W. D. Young, U.S. Patent Application 5723524, 1998.
68. P. Klug and F. Holtrup, European Patent Application EP 0933415A2, 1999.
69. S. Duncum, A. R. Edwards, and C. G. Osborne, International Patent Application WO96/04462, 1996.
70. D. Leinweber and M. Feustel, International Patent Application WO/2006/092208, 2006.
71. M. Arjmandi, S. Ren, and B. Tohidi, "Anti-agglomeration and Synergism Effect of Quaternary Ammonium Zwitterions," Proceedings of the 5th International Conference on Gas Hydrates, Trondheim, Norway, 12–16 June 2005.
72. P. Klug, M. Feustel, and V. Frenz, International Patent Application WO98/03615, 1998.
73. I. K. Meier, R. J. Goddard, and M. E. Ford, U.S. Patent Application 7452848, 2008.
74. D. Hurd and M. Pakulski, "Uncovering a Dual Nature of Polyether Amines Hydrate Inhibitors," Proceedings of the 5th International Conference on Gas Hydrates, Trondheim, Norway, 12–16 June 2006.

75. K. Bakeev, R. Myers, and D. E. Graham, U.S. Patent 6180699, 2001.
76. M. J. Jurek, International Patent Application WO/2007/143489.
77. J. D. Lee and P. Englezos, "Enhancement of the Performance of Gas Hydrate Kinetic Inhibitors with Polyethylene Oxide," *Chemical Engineering Science* 60 (2005): 5323.
78. U. C. Klomp, WO Patent Application 01/77270, 2001.
79. G. T. Rivers and D. L. Crosby, International Patent Application WO/2004/022909.
80. G. T. Rivers and D. L. Crosby, WO Patent Application WO/2004/022910.
81. J. Anderson and L. W. Clark, "Development of Effective Combined Kinetic Hydrate Inhibitor/Corrosion Inhibitor (KHI/CI) Products," Proceedings of the 5th International Conference on Gas Hydrates, 12–16 June 2006, Trondheim, Norway.
82. B. Fu, "Development of Non-Interfering Corrosion Inhibitors for Sour Gas Pipelines with Co-injection of Kinetic Hydrate Inhibitors," Paper 07666 (paper presented at the NACE CORROSION Conference, Nashville, TN, 11–15 March 2007).
83. A. W. R. MacDonald, SPE, M. Petrie, J. J. Wylde, SPE, A. J. Chalmers, and M. Arjmandi, "Clariant Oil Services, Field Application of Combined Kinetic Hydrate and Corrosion Inhibitors in the Southern North Sea: Case Studies," SPE 99388 (paper presented at the SPE Gas Technology Symposium, Calgary, Alberta, Canada, 15–17 May 2006).
84. J. Moloney, W. Mok, and C. Gamble, "Corrosion and Hydrate Control in Wet Sour Gas Transmission Systems," SPE 115074 (paper presented at the SPE Asia Pacific Oil and Gas Conference and Exhibition, Perth, Australia, 20–22 October 2008).
85. M. Pakulski, "Accelerating Effect of Surfactants on Gas Hydrates Formation," SPE 106166 (paper presented at the International Symposium on Oilfield Chemistry, Houston, TX, 28 February–2 March 2007).
86. R. Masoudi and B. Tohidi, "Experimental Investigation on the Effect of Commercial Oilfield Scale Inhibitors on the Performance of Low Dosage Hydrate Inhibitors (LDHI)," Proceedings of the 5th International Conference on Gas Hydrates, Trondheim, Norway, 12–16 June 2006.
87. T. A. Swanson, M. Petrie, and T. R. Sifferman, "The Successful Use of Both Kinetic Hydrate and Paraffin Inhibitors Together in a Deepwater Pipeline with a High Water Cut in the Gulf of Mexico," SPE 93158, Proceedings of the SPE International Symposium on Oilfield Chemistry, Houston, TX, 2–4 February 2005.
88. D. Leinweber and M. Feustel, International Patent Application WO/2006/084613.
89. D. Leinweber and M. Feustel, International Patent Application WO/2007/054226.
90. (*a*) M. Arjmandi, D. Leinweber, and K. Allan, "Development of A New Class of Green Kinetic Hydrate Inhibitors" (paper presented at the 19th International Oilfield Chemical Symposium, Geilo, Norway, March 2008). (*b*) D. Leinweber and M. Feustel, U.S. Patent Application 20080214865. (*c*) D. Leinweber, A. R. Roesch, and M. Feustel, U.S. Patent Application 20090054268.
91. K. S. Colle, C. A. Costello, L. D. Talley, J. M. Longo, R. H. Oelfke, and E. Berluche, International Patent Application WO96/08672, 1996.
92. T. Namba, Y. Fujii, T. Saeki, and H. Kobayashi, International Patent Application WO96/38492, 1996.
93. M. A. Kelland, T. M. Svartaas, and J. Ovsthus, "A New Class of Kinetic Hydrate Inhibitor," Proceedings of the 3rd Natural Gas Hydrate Conference, Salt Lake City, July 1999.
94. M. A. Kelland, P. M. Rodger, and T. Namba, International Patent Application WO98/53007, 1998.
95. K. S. Colle, C. A. Costello, L. D. Talley, R. H. Oelfke, and E. Berluche, International Patent Application WO96/41786, 1996.
96. Y. Tang, Y. Ding, and G. Zhang, "Role of Methyl in Phase Transition of Poly(*N*-Isopropylmethcrylamide)," *Journal of Physical Chemistry B* 112(29) (2008): 8447.
97. V. Thieu, K. Bakeev, and J. S. Shih, U.S. Patent 6451891, 2002.

98. L. D. Talley and R. H. Oelfke, International Patent Application WO97/07320, 1997.
 99. M. A. Kelland and P. Klug, International Patent Application WO98/23843, 1998.
100. K. S. Colle, C. A. Costello, and L. D. Talley, Canadian Patent Application 96/2178371, 1996.
101. K. S. Colle, L. D. Talley, R. H. Oelfke, and E. Berluche, International Patent Application WO96/08673, 1996.
102. K. S. Colle, C. A. Costello, L. D. Talley, R. H. Oelfke, and E. Berluche, International Patent Application WO96/41834, 1996.
103. (a) M. A. Kelland, L. Del Villano, and R. Kommedal, "A Class of Kinetic Hydrate Inhibitor with Good Biodegradability," *Energy & Fuels* 22 (2008): 3143. (b) M. A. Kelland, "Additives for Inhibiting Gas Hydrate Formation," International Patent Application WO/2008/023989.
104. K. Kannan, A. D. Punase, Low-Dosage, High Efficiency, and Environment-Friendly Inhibitors: A New Horizons in Gas Hydrates Mitigation in Production Systems, SPE 120905, SPE International Symposium on Oilfield Chemistry, The Woodlands, TX, April 20–22, 2009.
105. K. S. Colle, L. D. Talley, R. H. Oelfke, and E. Berluche, International Patent Application 96/41784, 1996.
106. D. G. Peiffer, C. A. Costello, L. D. Talley, and P. J. Wright, International Patent Application WO99/64718, 1999.
107. (a) V. Walker, J. A. Ripmeester, and H. Zeng, International Patent Application WO/03/087532, 2003. (b) H. Zeng, A. Brown, B. Wathen, J. A. Ripmeester, and V. K. Walker, "Antifreeze Proteins: Adsorption to Ice, Silica and Gas Hydrates," Proceedings of the 5th International Conference on Gas Hydrates, Trondheim, Norway, 2005, 13–16 June.
108. V. Walker, H. Zeng, R. Gordienko, M. Kuiper, E. Huva, Z. Wu, D. Miao, and J. Ripmeester, "The Mysteries of the Memory Effect and Its Elimination with Antifreeze Proteins," Proceedings of the 6th International Conference on Gas Hydrates Vancouver, British Columbia, Canada, 6–10 July 2008.
109. M. A. Kelland, unpublished results.
110. (a) C. R. Burgazli, World Patent Application WO2004/111161, 2004. (b) C. R. Burgazli, R. C. Navarrete, and S. L. Mead, "New Dual Purpose Chemistry for Gas Hydrate And Corrosion Inhibition," Paper 2003-070 (paper presented at the Petroleum Society's Canadian International Petroleum Conference, Calgary, Alberta, Canada, 10–12 June 2003) (Also *Journal of Canadian Petroleum Technology* 44 (2005): 47).
111. C. Xiao and H. Adidharma, *Chem. Eng. Sci.*, 2009, 64, 1522.
112. U. Klomp, "The World of LDHI: From Conception to Development to Implementation," Proceedings of the 6th International Conference on Gas Hydrates Vancouver, British Columbia, Canada, 6–10 July 2008.
113. M. Arjmandi, S.-R. Ren, J. Yang, and B. Tohidi, "Anti-Agglomeration and Synergism Effect of Quaternary Ammonium Zwitterions," Proceedings of the 4th International Conference on Natural Gas Hydrates, Yokohama, Japan, 19–23 May 2002.
114. M. A. Kelland, T. M. Svartaas, and L. A. Dybvik, "Control of Hydrate Formation by Surfactants and Polymers," SPE 28506, Proceedings of the SPE 69th Annual Technical Conference and Exhibition, New Orleans, LA, October 1994.
115. (a) C. Duchateau, J.-L. Peytavy, P. Glenat, T.-E. Pou, M. Hidalgo, and C. Dicharry, "Laboratory Evaluation Of Kinetic Hydrate Inhibitors: A New Procedure for Improving the Reproducibility of Measurements," Proceedings of the 6th International Conference on Gas Hydrates Vancouver, British Columbia, Canada, 6–10 July 2008. (b) C. Duchateau, J-L. Peytavy, P. Glenat, T.-E. Pou, M. Hidalgo, and C. Dicharry, *Energy Fuels*, 2009, 23(2), pp. 962–966.
116. L. D. Talley, G. F. Mitchell, and R. H. Oelfke, *Annals of the New York Academy of Sciences* 912 (2000): 314.

117. M. J. Anselme, M. J. Reijnhout, and U. C. Klomp, International Patent Application WO93/25798, 1993.
118. L. D. Talley and K. Colle, International Patent Application WO/2006/110192, 2006.
119. L. D. Talley, International Patent Application WO/2007/111789.
120. A. Sugier, J. P. Durand, U.S. Patent 5244878, 1993.
121. J. P. Durand, A. S. Delion, P. Gateau, and M. Velly, European Patent Application 740048, 1995.
122. M. Velly, M. Hillion, A. Sinquin, and J. P. Durand, European Patent Application EP 905350.
123. A. Sinquin, C. Dalmazzone, A. Audibert, and V. Pauchard, U.S. Patent Application US2006015102.
124. A. Rojey, M. Thomas, A.-S. Delion, and J.-P. Durand, U.S. Patent 5816280, 1998.
125. T. Palermo, A. Sinquin, H. Dhulesia, and J. M. Fourest, Proceedings of Multiphase, BHR Group, 1997, 1333.
126. T. Palermo, C. B. Argo, S. P. Goodwin, and A. Henderson, "Flow Loop Tests on a Novel Hydrate Inhibitor to be Deployed in the North Sea ETAP Field," Proceedings of the 3rd International Conference on Natural Gas Hydrates, *Annals of the New York Academy of Sciences*, 2000.
127. U. C. Klomp, V. C. Kruka, and R. Reijnhart, "Low Dosage Hydrate Inhibitors and How They Work," Proceedings of Symposium *Controlling Hydrates, Waxes and Asphaltenes*, IBC Conference, Aberdeen, October 1997.
128. U. C. Klomp, V. C. Kruka, and R. Reijnhart, WO Patent Application 95/17579, 1995.
129. R. Azarinezhad, A. Chapoy, R. Anderson, and B. Tohidi, OTC 19485, "HYDRAFLOW: A Multiphase Cold Flow Technology for Offshore Flow Assurance Challenges" (paper presented at the Offshore Technology Conference, Houston, TX, 5–8 May 2008).
130. U. C. Klomp and R. Reijnhart, International Patent Application WO96/34177, 1996.
131. L. H. Gjertsen and F. H. Fadnes, "Measurements and Predictions of Hydrate Equilibrium Conditions, Gas Hydrates: Challenges for the Future," *Annals of the New York Academy of Sciences* 912 (2000): 722–734.
132. B. Tohidi, R. W. Burgass, A. Danesh, K. K. Ostergaard, and A. C. Todd, "Improving the Accuracy of Gas Hydrate Dissociation Point Measurements," *Annals of the New York Academy of Sciences* 912 (2000).
133. L. M. Frostman, C. G. Gallagher, S. Ramachandran, and K. Weispfennig, "Ensuring Systems Compatibility for Deepwater Chemicals," SPE 65006 (paper presented at the SPE International Symposium on Oilfield Chemistry, Houston, TX, 13–16 February 2001).
134. L. M. Frostman, "Anti-Agglomerant Hydrate Inhibitors for Prevention of Hydrate Plugs in Deepwater Systems," SPE 63122 (paper presented at the SPE Annual Technical Conference and Exhibition, Dallas, TX, 1–4 October 2000).
135. L. M. Frostman and J. L. Przybylinski, "Successful Applications of Anti-agglomerant Hydrate Inhibitors," SPE 65007 (paper presented at the SPE International Symposium on Oilfield Chemistry, Houston, TX, 13–16 February 2001).
136. A. P. Mehta, P. B. Herbert, E. R. Cadena, and J. P. Weatherman, "Successful Applications of Anti-Agglomerant Hydrate Inhibitors," OTC 14057, Proceedings of the Offshore Technology Conference, Houston, TX, 6–9 May 2002.
137. V. Thieu and L. M. Frostman, "Use of Low-Dosage Hydrate Inhibitors in Sour Systems," SPE 93450 (paper presented at the SPE International Symposium on Oilfield Chemistry, Houston, TX, 2–4 February 2005).
138. A. Buijs, G. van Gurp, T. Nauta, R. Smakman, and A. M. Wit-Van Grootheest, U.S. Patent 6379294, 2002.
139. U. C. Klomp, International Patent Application WO99/13197, 1999.

140. U. C. Klomp, M. Le Clerq, and S. Van Kins, "The First Use of Hydrate Anti-Agglomerant for a Fresh Water Producing Gas/Condensate Field," Proceedings of the 2nd Petromin Deepwater Conference, Shangri-La, Kuala Lumpur, Malaysia, 18–20 May 2004.
141. G. T. Rivers, L. M. Frostman, J. L. Pryzbyliski, and J.-A. McMahon, U.S. Patent 6620330, 2003.
142. D. L. Crosby, G. T. Rivers, and L. M. Frostman, International Patent Application WO/2005/116399.
143. B. Fu, C. Houston, and T. Spratt, "New Generation LDHI with an Improved Environmental Profile," Proceedings of the 5th International Conference on Gas Hydrates, Trondheim, Norway, 13–16 June 2005.
144. P. A. Spratt, International Patent Application WO/2006/052455.
145. L. Cowie, W. Shero, N. Singleton, N. Byrne, and L. Kauffman, *Deepwater Technology*, Gulf Publishing Co., 2003.
146. (*a*) R. Alapati and A. Davies, "Oil-Soluble LDHIs Represents New Breed of Hydrate Inhibitor," *Journal of Petroleum Technology* (2007): 28. (*b*) R. Alapati, J. Lee, and D. Beard, Flow Assurance Chemistry Found Effective at High Watercuts, *World Oil*, November 2008, 39.
147. R. Alapati, J. Lee, and D. Beard, "Two Field Studies Demonstrate That New AA LDHI Chemistry is Effective at High Water Cuts without Impacting Oil/Water Quality," OTC 19505 (paper presented at the Offshore Technology Conference, Houston, TX, 5–8 May 2008).
148. (*a*) V. Panchalingam, M. G. Rudel, and S. H. Bodnar, International Patent Application WO/2005/042675. (*b*) V. Panchalingam, M. G. Rudel, and S. H. Bodnar, U.S. Patent 7,381,689, 2008.
149. J. L. Przybylinski and G. T. Rivers, U.S. Patent 6596911B2, 2003.
150. U. Dahlmann and M. Feustel, M. U.S. Patent 7183240, 2007.
151. U. Dahlmann and M. Feustel, M. U.S. Patent 7214814, 2007.
152. U. Dahlmann and M. Feustel, M. U.S. Patent, 7323609, 2008.
153. G. T. Rivers, International Patent Application WO/2008/008697.
154. G. T. Rivers, J. Tian, and J. A. Hackerott, International Patent Application WO/2008/063794.
155. M. A. Kelland, T. M. Svartaas, J. Ovsthus, T. Tomita, and K. Mizuta, "Studies on Some Alkylamide Surfactant Gas Hydrate Anti-Agglomerants," *Chemical Engineering Science* 61 (2006): 4290.
156. Z. Huo, E. Freer, M. Lamar, B. Sannigrahi, D. M. Knauss, E. D. Sloan, and E. D., "Hydrate Plug Prevention by Anti-Agglomeration," *Chemical Engineering Science* 56 (2001): 4979.
157. M. A. Kelland, T. M. Svartaas, J. Ovsthus, T. Tomita, and J. Chosa, "Studies on Some Zwitterionic Surfactant Gas Hydrate Anti-Agglomerants," *Chemical Engineering Science* 61 (2006): 4048.
158. M. Hellsten and H. Oskarsson, International Patent Application WO/2007/107502.
159. (*a*) A. Firoozabadi and J. D. York, "Comparing Effectiveness of Rhamnolipid Biosurfactant with a Quaternary Ammonium Salt Surfactant for Hydrate Anti-Agglomeration," *Journal of Physical Chemistry B* 112 (2008): 845. (*b*) J. D. York and A. Firoozabadi, "Alcohol Cosurfactants in Hydrate Antiagglomeration," *Journal of Physical Chemistry B* 112 (2008): 10455.
160. A. Firoozabadi, J. D. York, "Use of Biosurfactants in Hydrate Antiagglomeration," SPE 116214 (paper presented at the SPE Annual Technical Conference and Exhibition, Denver, CO, 22–24 September 2008).
161. O. Urdahl, A. Lund, P. Mork, and T. Nilsen, "Inhibition of Gas Hydrate Formation by Means of Chemical Additives: I. Development of an Experimental Set-up for Characterization of Gas Hydrae Inhibitor Efficiency with Respect to Flow Properties and Deposition," *Chemical Engineering Science* 50(5) (1995): 863.

162. D. Lippmann, D. Kessel, and I. Rahimian, "Gas Hydrate Nucleation and Growth Kinetics in Multiphase Tranport Systems," Proceedings of the 5th International Offshore and Polar Engineering Conference, The Hague, The Netherlands, 11–16 June 1995.
163. A. Sinquin, X. Bredzinsky, and V. Beunat, "Kinetic of Hydrates Formation: Influence of Crude Oils," SPE 71543, Proceedings of the SPE Annual Technical Conference and Exhibition, New Orleans, LA, 30 September–3 October 2001.
164. P. V. Hemmingsen, X. Li, J.-L. Peytavy, and J. Sjoblom, "Hydrate Plugging Potential of Original and Modified Crude Oils," *Journal of Dispersion Science and Technology* 28 (2007): 371.
165. K. Erstad, S. Hoeiland, T. Barth, and P. Fotland, "Isolation and Molecular Identification of Hydrate Surface Active Components in Petroleum Acid Fractions," Proceedings of the 6th International Conference on Gas Hydrates (ICGH 2008), Vancouver, British Columbia, Canada, 6–10 July 2008.
166. T. Palermo, A. Mussumeci, and E. Leporcher, "Could Hydrate Plugging Be Avoided Because of Surfactant Properties of the Crude and Appropriate Flow Conditions," OTC016681 (paper presented at the Offshore Technology Conference Houston, TX, 2004).
167. (*a*) R. M. T. Camargo, M. A. L. Goncalves, J. R. T. Montesami, C. A. B. R. Cardoso, and K. Minami, "A Perspective View of Flow Assurance in Deepwater Fields in Brazil," OTC 16687 (paper presented at the Offshore Technology Conference, Houston, TX, 2004). (*b*) K. Kinnari and P. Fotland, Statoil, personal communication.
168. M. Pakulski, U.S. Patent 6331508, 2001.
169. M. Pakulski, U.S. Patent 5741758, 1998.
170. M. Pakulski, "Twelve Years of Laboratory and Field Experience for Polyether Polyamine Gas Hydrate Inhibitors," Proceedings of the 6th International Conference on Gas Hydrates, Vancouver, British Columbia, Canada, 6–10 July 2008.
171. M. Pakulski, U.S. Patent European Patent Application 6025302, 2000.
172. D. Lovell and M. Pakulski, "Hydrate Inhibition in Gas Wells Treated With Two Low Dosage Hydrate Inhibitors," SPE 75668 (paper presented at the SPE Gas Technology Symposium, Alberta, Canada, 2002).
173. D. Budd, D. Hurd, M. Pakulski, and T. D. Schaffer, "Enhanced Hydrate Inhibition in Alberta Gas Field," SPE 90422 (paper presented at the SPE Annual Technical Conference and Exhibition, Houston, TX, 26–29 September 2004).
174. M. Pakulski and J. C. Dawson, U.S. Patent 6756345, 2004.
175. S. R. Davies, J. A. Boxall, C. Koh, and E. D. Sloan, P. V. Hemmingsen, K. J. Kinnari, and Z.-G. Xu, "Predicting Hydrate Plug Formation in a Subsea Tieback," SPE 115763 (paper presented at the SPE Annual Technical Conference and Exhibition, Denver, CO, 21–24 September 2008).
176. A. F. Harun, T. E. Krawietz, and M. Erdogmus, "When Flow Assurance Fails: Melting Hydrate Plugs in Dry-tree Wells," *World Oil* 228 (2007): 51.
177. A. F. Harun, T. E. Krawietz, and M. Erdogmus, "Hydrate Remediation in Deepwater Gulf of Mexico Dry-Tree Wells: Lessons Learned," *SPE Production & Operations* 22(4) (2007): 472.
178. T. Austvik, X. Li, and L. H. Gjertsen, "Hydrate Plug Properties—Formation and Removal of Plugs," Proceedings of Gas Hydrates: Challenges for the Future, *Annals of the New York Academy of Sciences* 912 (2000): 294.
179. X. Li, L. H. Gjertsen, and T. Austvik, "Thermodynamic Inhibitors for Hydrate Plug Melting," Proceedings of Gas Hydrates: Challenges for the Future, *Annals of the New York Academy of Sciences* 912 (2000): 822.
180. X. Li, L. H. Gjertsen, and T. Austvik, "Melting Hydrate Plugs by Thermodynamic Inhibitors—Plug Properties and Melting Efficiencies," Proceedings of the 4th International Conference on Gas Hydrates, Yokohama, Japan, 19–23 May 2002.

181. C. N. Khalil, European Patent Application EP0909873.
182. C. N. Khalil, N. D. O. Rocha, and L. C. F. Leite, U.S. Patent 6035933, 2000.
183. L. C. C. Marques, C. A. Pedroso, and L. F. Neumann, "A New Technique to Solve Gas Hydrate Problems in Subsea Christmas Trees," SPE 77572, *SPE Production & Facilities* 19(4) (2004): 253.
184. J. Chatterji and J. E. Griffith, U.S. Patent Application 5713416, 1998.
185. R. Von Flatern, "Pulling the Plug from Deep Pipe," *Offshore Engineer* September (2006): 71.
186. A. C. Gulbrandsen and T. M. Svartaas, "Influence of Formation Temperature and Inhibitor Concentration on the Dissociation Temperature for Hydrates Formed with Poly Vinyl Caprolactam," Proceedings of the 6th International Conference on Gas Hydrates, Vancouver, British Columbia, Canada, 6–10 July 2008.
187. S. Gao, "Investigation of Interactions between Gas Hydrates and Several Other Flow Assurance Elements," *Energy & Fuels* 22(5) (2008): 3150.
188. P. Buchanan, A. K. Soper, H. Thompson, R. E. Westacott, J. L. Creek, G. Hobson, and C. A. Koh, "Search for Memory Effects in Methane Hydrate: Structure of Water Before Hydrate Formation and after Hydrate Decomposition," *Journal of Chemical Physics* 123 (2005): 164507.
189. P. M. Rodger, "Methane Hydrate: Melting and Memory," *Annals of the New York Academy of Sciences* 912 (2000): 474.
190. L. Ding, C. Geng, Y. Zhao, X. He, and H. Wen, "Molecular Dynamics Simulation for Surface Melting and Self-Preservation Effect of Methane Hydrate," *Science in China Series B: Chemistry* 51(7) (2008): 651.
191. W. G. Chapman, S. Gao, M. Yarrison, K. Song, and W. House, "Equilibrium and Dynamics of Gas Hydrates," Paper 7, Proceedings of the 10th International Conference on PPEPPD, Snowbird, UT, Engineering Conferences International, NY, 16–21 May 2004.
192. M. A. Kelland, T. M. Svartaas, and L. Dybvik Andersen, *J. Petr. Eng.*, 2009, 64, 1.
193. J. W. Lachance, C. Koh, and E. D. Sloan, *Chem. Eng. Sci.*, 2009, 64, 180.

10 Wax (Paraffin) Control

10.1 INTRODUCTION

Wax (or paraffin) deposition can be both a downhole and topside problem, blocking the flow of hydrocarbons fluids as they are cooled.[1-3] Waxes are solids made up of long-chain (> C18), normal or branched alkane compounds that are naturally present in crude oils and some condensates.[4] Some cyclic alkanes and aromatic hydrocarbons may also be present. It has been established conclusively that normal alkanes (n-paraffins) are predominantly responsible for pipeline wax deposition. Waxes in crudes are usually harder to control than those in condensates as they are of longer-chain alkanes. When the molecular size is 16–25 carbon atoms, soft mushy waxes are observed. Hard crystalline waxes have 25–50 or more carbons in the chain. The melting point of the paraffins increases as the size of the molecule increases. Generally, the higher the melting point, the more difficult it is to keep the paraffin from forming deposits.

In the reservoir, at high temperature and pressures, any waxes within the oil will be in solution. As the crude oil temperature drops, wax will begin to precipitate from the crude oil, usually as needles and plates. In addition, as the pressure drops during production, loss of low molecular weight hydrocarbons (light ends) to the gas phase reduces the solubility of the waxes in the oil. The wax appearance temperature (WAT) or cloud point is the temperature at which the first wax crystals begin to precipitate from the crude oil. The WAT can be as high as 50°C (122°F) for some oils and depends on the pressure, oil composition (in particular, the concentration of light ends), and bubble point. Wax precipitation/deposition is normally a problem at a higher temperature than gas hydrate formation.[5] The oil that reaches the sales tank will often contain paraffinic solids. This oil, because of the loss of light ends and the lower-temperature environment, has lost much of its ability to hold the waxes in solution. Typical problems caused by wax deposition include:

- reduction or plugging of pipework, blocking flow—this can be downhole if the well temperature is low, or topside in cold climate or subsea transportation
- increased fluid viscosity leading to increased pumping pressure
- restartability issues caused by wax gel strength
- reduced operating efficiency and process upsets with interruptions to production or shutdowns
- costly and technically challenging removal, especially in deepwater pipelines
- safety hazard due to deposits interfering with the operation of valves and instruments
- disposal problems associated with accumulated wax

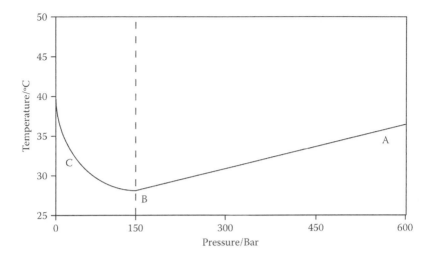

FIGURE 10.1 A typical phase diagram for wax precipitation.

Figure 10.1 illustrates a typical phase diagram for wax precipitation. Position A is the reservoir pressure with the oil undersaturated. As the fluids are produced, the pressure falls and the light ends expand in greater proportion to the dissolved waxes increasing their solubility and yielding a lower WAT. At the bubble point (B), the volume ratio of light ends to heavy ends is at its greatest so the WAT is at its lowest. Reducing the pressure still further (C) results in liberation of dissolved gases and light ends to the gas phase. This reduces the solubility of the wax, and so, the WAT increases. This effect can change the WAT by as much as 15°C (27°F) compared with stock tank oil at 1 bar.

Wax precipitation has three effects:

- Gradual pipe restrictions due to deposition at the wall.
  - The resulting reduced flow is due to a combination of reduced pipe diameter and an increase in pipe wall roughness due to the wax deposit. Complete blockage is rare.
- Increased fluid viscosity
  - This leads to subsequent pressure loss in the line. In worst cases, gelling of the fluid can totally stop production.
- Formation of wax gel
- This can occur if a pipeline is shut in, the fluids cool below the pour point, and a gel results at zero shear rate. If the yield stress of the gel is higher than the maximum pipeline pressure, then the line cannot be restarted.

10.1.1 Wax Deposition

Wax deposition is considered to occur by two primary mechanisms:

Wax (Paraffin) Control

1. If the pipe wall is colder than the WAT, wax can form and deposit at the wall. This can occur even if the bulk fluid is above the WAT. This is called the "molecular diffusion mechanism."
2. Already precipitated wax close to the pipe wall will move to a region of lower velocity at the pipe wall and deposit. This is known as shear dispersion.

Thus, wax deposition by mechanism 1 can occur below and above the WAT. Mechanism 2 only operates below the WAT. The rate of wax deposition depends largely on mechanism 1 above, and the hydrodynamics since the rate of wax stripping from the walls is considered crucial. Although up to eight different mechanisms have been proposed in the literature to contribute to the transport of either solid or liquid waxes toward the cold wall, the only one with a significant flux toward the wall is the so-called molecular diffusion.[6] That is to say, liquid waxes are driven toward the cold wall by a concentration gradient. However, a reliable universal model for wax deposition is still to be agreed upon. For example, a laboratory study using multiphase fluids concluded that there has to be a different wax deposition mechanism operating for some fluids than those based on conventional diffusion theory.[6] Another pilot loop study demonstrated that the wax deposit could result from flow patterns caused by rheological factors.[6] Computer models for determining the rate and amount of wax deposition are commercially available. All wax models are based on empirical correlations and require comprehensive analytical data for the aliphatic and aromatic components of crude up to about C50.

For some oils, wax deposition can begin in the lower part of the well if it is cool enough. Such wells can be intermittently shut in and "dewaxed" via use of chemicals, hot oil, or water, wireline cutters, or through-flow line tools. Major wax deposition problems can occur in subsea flow lines and risers, especially in cold or deep water where the wall temperature is very low, unless it is well insulated or heated. Wax deposition and the WAT are both affected by the amount and type of asphaltenes in the crude or condensate. In general, a significant reduction in wax deposition is observed for crudes containing a high proportion of asphaltic components. The idea that the WAT is a function of asphaltene surface area has been proposed and supported by experiment.[7]

10.1.2 Increased Viscosity and Wax Gelling

The second wax-related problem is increased viscosity and even gelling of the oil due to high amounts of wax precipitation in the oil. Very waxy crudes are usually more prone to this problem. On cooling, the waxes separate out as platelike crystals, which interact together to form a three-dimensional network in which liquid oil becomes trapped, resulting in increased oil viscosity, decreased oil flowability, and pressure loss in the pipeline.

If a crude oil production pipeline is shut-in, the fluids may cool further and a gel may result at zero shear rate. If the yield stress of the gel is higher than the maximum pipeline pressure, then the line cannot be restarted.[8a] Experimental results on 24 crudes showed that approximately 2 wt.% precipitated wax is sufficient to cause gelling of a virgin waxy crude.[8b] The experimentally measured "pour point" is a rough measure of the temperature where the crude oil begins to gel. It is often 10–30°C (18–54°F) below the WAT and may not be reached during continuous production but only during shut-in. The presence of asphaltenes has been shown to lower the pour point of a crude, whereas naphthenic acids increase the pour point.[9] Other factors such as flow regime and water will affect the viscosity and gelling point.

10.2 WAX CONTROL STRATEGIES

Mixing of waxy crude oils, particularly heavy crudes, with a diluent is sometimes applied to avoid wax problems in pipelines. The diluent can be a gas condensate, natural gas liquids, or light crude with a lower WAT or pour point. As a result of this mixing, the wax content of the waxy crude is diluted, which reduces its WAT or pour point to lower temperatures.

There are several other ways to control wax gelling and the buildup of wax deposition downhole and in flowlines:

- insulation
- mechanical removal
  - pigging of transportation lines
  - wireline cutters downhole
- heating
  - downhole
  - flowline
- wax dissolvers
- wax (paraffin) inhibitors, pour-point depressants (PPDs), and dispersants
- magnets
- shock chilling (cold flow)
- ultrasonics
- microbial treatment

Only chemical wax control (removal and prevention) will be discussed in detail in this chapter.

Pigging of pipelines to prevent the buildup of wax deposits is very common and is done at regular intervals.[10] A wax deposition simulator can be used to determine the pigging frequency and pressure drop across the line. Pigging is often carried out in conjunction with continuous wax inhibitor treatments, which reduces the rate of wax deposition and softens the deposited waxes. Cutting out wax mechanically downhole is a relatively simple procedure and requires extensive well shutdown time. Moreover, it is extremely inefficient inasmuch as substantial amounts of wax often remain in the well.[11]

Insulating a pipeline can help keep the transported fluids above the wax deposition temperature. Vacuum insulation has been carried out on a number of subsea fields.[12] Although expensive, pipeline heating can be a viable option for subsea multiphase transportation particularly if it can simultaneously alleviate more serious gas hydrate problems. Heating can be carried out using electricity (inductive heating) or hot water, either in bundles or pipe-in-pipe systems.[13–15]

Shock chilling or cold flow has been discussed as a method to prevent wax deposition in oil transportation, particularly in deep water where the temperature is very low and wax deposition severe.[16–17] The idea is to cool some of the waxy oil as quickly as possible, thereby precipitating the wax in the bulk fluid as much as possible rather than on the pipe walls. The first-formed seed particles of wax will then provide sites for further growth of wax crystals. Eventually, all the wax will be formed as a dispersion of transportable wax particles with little or no deposition on the pipe walls. This nonchemical method is still at the research stage. The same method is currently being researched for preventing gas hydrate blockages (see Chapter 9 on gas hydrate control). Methods whereby wax is caused to deposit on scale particles, induced by electric or electromagnetic fields, rather than on the pipe walls have also been claimed.[18]

Permanent magnets and electromagnets (or magnetic fluid conditioning, MFC) have been used to prevent wax deposition, mostly downhole. The technology is much used in Asia, less in the West. By 1995, as many as 14,440 magnets had been installed in China with a claimed good success rate.[19] Multiple magnets are claimed to work better.[20] Pulsed electric or magnetic field have also been shown to reducing the viscosity of crudes.[21] However, as with scale control, the laboratory and field success rate has been variable. Negative field results may result because the operator uses the technology at field conditions beyond the performance of the product. Some laboratory studies have shown a positive effect of magnets in reducing wax deposition and viscosity.[22–23] The effect is reported to be greater in condensate systems rather than crude oil.[24] The presence of water and its salinity may also impact magnetic wax treatment although effects of magnets on wax deposition have been seen in purely hydrocarbon systems.[25] A university project for DeepStar, a deepwater R&D program sponsored by a consortium of oil companies, produced results showing that MFC does indeed lower the WAT.[26] The decrease was small (ca. 2°F), although the MFC system was not optimized.

Various ways of placing a heating element downhole to melt and remove wax have been reported. A less expensive process using an inductance heating element and avoiding the variable resistance heating method has been patented.[27] Polar water-wet glass lining tubing or epoxy phenolic coating compound have been used to prevent waxes adhering to surfaces downhole.[28]

Microbial treatment has been shown to remove wax from oil wells.[29] Oil production increased, and 16 cycles of thermal-washing treatment and 44 cycles of additives were eliminated from the wells during 4 months of testing. Considerable economic profit was achieved. Microbial remediation of wax is a fairly expensive technique on a large scale. However, the performance is described as very good for wells treated in China.[28] The metabolic by-products such as organic acids and biosurfactants also have the ability to disperse wax and other organic deposits and increase their solubility in produced fluids.[30] The use of enzymes has also been studied for use in subsea oil transportation.[30] For example, specific enzymes for C20–30 alkanes were identified. The initial studies showed that the enzymes could lower the pour point of a crude by some few degrees celsius. More dramatic was the decrease in viscosity at seabed temperatures.

Ultrasonic treatments downhole has been shown in laboratory tests to remove wax deposits and has been proposed as a new downhole treatment. The method is suggested to be particularly effective at treating long sections of pay where chemical methods may be too expensive.[31]

10.3 CHEMICAL WAX REMOVAL

There are several methods of using chemicals to remove wax deposits, which can be summarized as follows:

- hot-oiling
- wax solvents
- thermochemical packages

10.3.1 HOT-OILING AND RELATED TECHNIQUES

Hot-oiling, to melt and dissolve wax deposits, has been carried out as long as there has been oil production. In the hot-oiling method, produced crude is heated to a temperature well above the melting point of the wax and is then circulated down through the annulus of the well and returned to a hot-oil heating system via the production tubing. The purpose here is for the hot oil to melt and/or dissolve the wax so that it can be removed from the well in liquid form. This is an expensive method since the crude must be put through a heater treater along with a demulsifier to facilitate the removal of solids and water therefrom. Therefore, it is not usually used to clear subsea flowlines. During the hot-oiling process, a wax dispersant, which is often a petroleum sulfonate, is added to the crude as it is being heated. The wax dispersant assists in dispersing the melted wax in the hot-oil phase. Formation damage due to hot-oiling has been reviewed.[32] The technique of hot-oiling can be dangerous, particularly with wells producing a crude having a low flash point. A way of avoiding excessive temperatures has been claimed. Thus a mixture of water, an alkyl aralkyl polyoxyalkylene phosphate ester surfactant, a mutual solvent composed of a blend of an alcohol selected from the group consisting of aliphatic alcohols, glycols, polyglycols, and glycol esters and an aromatic hydrocarbon (e.g., toluene or xylene) can dissolve wax when the said mixture is heated to a temperature of 15–20°C (59–68°F) higher than the melting point of the wax to be removed.[33] Steam or hot water has been used to melt waxes downhole. This may cause corrosion and emulsion problems.

10.3.2 WAX SOLVENTS

The practice of using wax solvents downhole and in pipelines to remove wax deposits is common. Many toxic solvents have been used in the past such as benzene, chlorinated solvents, and combinations thereof, together with other hydrocarbons. Carbon disulfide, which has a low flash point, is still used occasionally onshore in the United States. The most popular solvents used today include substituted aromatics, such as toluene and xylenes (or distillates containing a high aromatic content), and blends with gas oil. One patent claims a solvent composition comprising a substantially pure aromatic hydrocarbon such as toluene or xylenes (lower flash point) and an aliphatic and/or alicyclic hydrocarbon such as petroleum naphtha. The wax solvent composition may contain a surfactant.[34] A laboratory study suggested that the use of hot solvents (rather than hot oiling) such as hot xylene has the greatest potential benefit of removing wax downhole.[35] Aromatic solvents will also remove asphaltene deposits. Wax deposition was successfully removed in a subsea flowline using a xylene slug swept by produced gas from the well.[36] Testing indicated that the problem resulted from the interaction between glycol and produced condensate. Overtreatment with glycol and low-produced water volumes contributed to the precipitation of paraffin and eventual plugging of the flowline.

Aromatics and most alkyl-substituted aromatics are now classified as marine pollutants. Greener and nonhazardous wax solvents based on limonene (a terpene) have been claimed.[37] Limonene with an alkyl glycol ether and various other polar chemicals have also been claimed.[38] Limonene has good biodegradability, shows some toxicity toward fish, and has a flash point of 45–50°C (113–122°F). A successful wax solvent job was carried out on a large-scale subsea flowline in the North Sea: the wax dissolver was a proprietary blend of terpenoid extracts with a good environmental rating.[39] However, another study concluded that no solvents were capable of fully dissolving wax deposits, formed from a condensate, at seabed temperatures (ca. 4–6°C [39–43°F]); heating was necessary.[40]

A chemical dispersant-solvent package, combined with the right mechanical application, has been used to remove a wax blockage in a deep-water pipeline, in which earlier remedial treatments had failed and compounded the problem.[41]

10.3.3 Thermochemical Packages

Heat can be generated in situ by a chemical reaction, which can be used to melt wax deposits. The use of this remediation technique does entail a certain risk, if not designed and deployed correctly. As a worst-case scenario, loss of control of the exothermic reaction could lead to unexpected stress to pipeline materials and potential failure.

Heat can be generated thermochemically by a number of reactions such as the reaction of an acid and base. For example, acetic acid, generated by mixing water with acetic acid anhydride, can be metered separately from an aqueous solution of sodium hydroxide into the downhole tubular or pipeline.[42] Such acid-base combinations can be difficult to carry out in practice as the neutralization reaction cannot be controlled/delayed and it may also cause corrosion. A preferred neutralization reaction that may be less corrosive is that of a sulfonic acid such as dodecylbenzene sulfonic acid with an amine such as isopropylamine.[43]

A more promising thermochemical process is the acid-catalyzed decomposition of ammonium nitrite generated in situ.[44] This reaction is the basis for a number of thermochemical packages used both downhole and subsea to remove wax deposits. For example, such a system has been reported to stimulate oil wells on and offshore United States.[35,45] Previous attempts to stimulate these wells with wax solvents and

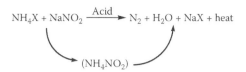

FIGURE 10.2 Decomposition of a thermochemical wax removal treatment. X = Cl or NO_3.

acid systems were unsuccessful. The exothermic decomposition of ammonium nitrite reaction employs a mixture of sodium nitrite and ammonium nitrate or chloride in aqueous solution: the reaction products being nitrogen, water, and sodium nitrate or chloride (Figure 10.2). For downhole applications, the reaction rate is controlled to generate predetermined amounts of heat at a previously established well depth. The reaction occurs as soon as the forming salts are mixed, in the presence of HCl as catalyst, the control of the reaction being done by buffering the pH of the solution in the range from 5.0 to 8.0. The reaction is faster at a lower pH. Control is maintained such that the reaction begins gradually and progresses slowly as the solution is displaced throughout the production string at constant rate. Some distance above the perforations, the reaction rate increases and produces huge amounts of heat, the temperature reaching a thermal maximum, heat being lost to the environment, with consequent reduction in the temperature of the spent solution. This process is limited to noncalciferous formations, since that kind of formation could react with the acid catalyst. It is also more expensive than hot-oiling of solvents and may not be economical as a means of stimulation in low-producing wells.

A closely related thermochemical system, which has been successfully used to remove wax deposits downhole or in deepwater pipelines in Brazil, is based on a redox decomposition of high- and low-valence nitrogen compounds releasing nitrogen. This system has been called Nitrogen Generating System, or SGN after the original Portuguese.[46] High concentrations of ammonium chloride (or sulfate) and sodium nitrite are emulsified as a water-in-oil emulsion using a surfactant. When a catalytic amount of an acid is added to the mixture, a redox decomposition occurs, releasing nitrogen gas and forming water and sodium chloride. The reason for the decomposition is that ammonium nitrite, formed in situ, is not stable at low pH. Early versions of the SGN method used an organic acid such as acetic acid as catalyst.[47] However, the decomposition was not controllable enough. Better versions use an anhydride, which reacts slowly with water in the emulsion to generate acid in situ.[48] The anhydride, dissolved in a suitable solvent, can be added to the emulsion as it is being pumped. Claimed anhydrides include a polyanhydride such as poly(adipic) anhydride or preferably a copolyanhydride such as poly(adipic-co-sebacic) anhydride solubilized in a polar organic solvent such as chloroform. The emulsion is pumped into the flowline and the acid generated in situ. The heat generated from the SGN reaction is sufficient to melt the wax. The wax dissolves in the organic solvent part of the emulsion. The release of nitrogen gas provides agitation to the liquid system speeding up the dissolution process. A research project has been carried out using coated capsules of acid catalyst (such as citric acid), which has a delayed release of acid.[49] For downhole applications, an emulsion is injected down the annulus containing the nitrogen-based SGN redox system, an organic nonpolar solvent, an emulsion activator (sodium persulfate), and a viscoelastic polymer.[50] As the temperature of the emulsion rises, the activator decomposes, the pH drops to ~4, and a redox reaction takes place generating heat and a high-quality stable foam. The penetration of the foam into the porous media is enough to ensure sufficient wax removal and production enhancement.

A solid material controlled-release thermochemical system uses a polymeric material capable of being hydrated mixed with a catalyst.[51] On rehydration, the catalyst is released, which generates H⁺ ions in solution, which catalyze the decomposition of the ammonium nitrite–generating chemicals. Considerable heat can also be generated by the decomposition of hydrogen peroxide by transition metal ions such as iron(III) ions (Fenton's reaction). The peroxide is broken down into a hydroxide ion and a hydroxyl free radical. The hydroxyl free radical is the primary oxidizing species and can be used to oxidize and break apart organic molecules. No field applications have been reported. Another thermochemical wax removal method, for which there are no field reports, is to generate heat to melt wax from sudden exothermic crystallization of supersaturated solutions, such as sodium acetate.[52] Yet another method to generate heat downhole is to react in situ an imidazolidone derivative, such as 2-imidazolidone, with an acid such as sulfuric acid. The reaction generates heat, shock, and CO_2.[53] This system has been claimed mainly as a gentle, fracturing stimulation technique.

10.4 CHEMICAL WAX PREVENTION

10.4.1 TEST METHODS

This section discusses the laboratory methods for determining the need for and the performance of wax inhibitors and PPDs.

To determine the potential for wax deposition of a crude or condensate, the WAT or cloud point is first measured. There are a variety of ways to do this, which have been reviewed.[54–55] Techniques such as viscosity measurement and filter plugging can only be used under very favorable circumstances.[56] The same is true of differential scanning calorimetry (DSC).[57] Further, conventional American Society for Testing and Materials (ASTM) procedures for cloud point determination are not applicable to dark crude oils and also do not account for potential subcooling of the wax. In one study, these methods failed to identify the potential for paraffin deposition in a Gulf of Mexico pipeline.[58]

A reliable method to determine the WAT consists of determining the temperature at which wax deposits begin to form on a cooled surface exposed to warm, flowing oil.[59–61] Other good techniques include:

- cross-polarization microscopy [62–64]
- laser and collimator solids detection system [65–66]
- Fourier transform infrared spectroscopy[67]
- high-pressure, cold-stage microscope technique[68]
- sonic testing device[69]

A flow cell with videomicroscope observations regarding agglomeration and deposition has also been constructed. Strong wax aggregates could be observed in uninhibited crude, while weakly bonded clusters were observed in inhibited crude. Notably, lower cloud points were observed using this technique than more traditional methods.[70]

The rate of deposition of wax is usually determined using the cold-finger technique,[71] the Chilton-Colburn analogy, and a single model tuning parameter.[69,72–74] A multi–cold-finger apparatus has been described.[75] Alternatively, a coaxial shearing cell wax deposition apparatus can be used.[76] A more elaborate apparatus for measuring wax deposition rates is a flow loop or tube-blocking rig. Capillary tube-blocking rigs measure the change in pressure across a microbore capillary due to the buildup of wax on the internal walls of the capillary. The capillary or loop is either jacketed or placed in a cooling bath.[77–82] Large-diameter pilot loops have also been used.[83]

The pour point of a waxy crude is traditionally carried out using the simple ASTM D-97 (or IP 15) Pour Point Test, although one should also measure viscosity. The repeatability of the test can be poor but allows a ranking of PPD efficiency. The dosage of PPD needed in a laboratory test will not be representative and will almost always be higher than that necessary for field use. ASTM pour-point measurements involve periodically tilting a tube containing a thermometer and treated or untreated oil samples over a progressively declining temperature range and observing for movement. The shear imparted by this technique is extremely low and can represent very small changes in viscosity. Therefore, viscosity measurements avoid the ambiguity introduced by this method.[69] The pour point is claimed to be best measured with a rheological mechanical spectrometer.[65] The strength of waxy gel can be determined through yield stress tests using viscometry.[84]

The viscosity of the wax inhibitor or PPD formulated in a solvent also needs investigation especially for cold and/or deepwater subsea injection through umbilicals and capillaries. High-pressure viscometry has been found to be beneficial for product qualification.[85] Instability in the polymer solution phase behavior of paraffin inhibitor formulations at high pressures at cold temperatures may be the reason behind some industry umbilical line failures. The instability is dependent on the inhibitor chemistry and its concentration. Most wax inhibitors are not inherently highly soluble at cold temperatures even in good solvents. However, better inhibitor solubility, even at reduced temperatures, can be achieved using a combination of a weak to moderate wax solvent and a second strong wax solvent.[86] Exemplary weak to moderate wax solvents include benzene, toluene, xylene, ethyl benzene, propyl benzene, trimethyl benzene, and mixtures thereof. Exemplary strong wax solvents include cyclopentane, cyclohexane, carbon disulfide, decalin, and mixtures thereof.

10.4.2 Wax Inhibitors and Pour-Point Depressants

Chemicals are needed to prevent both wax deposition and wax gelling. Therefore, the chemicals must affect the WAT and the wax pour point, respectively. Chemicals can also modify wax crystals so they do not agglomerate and deposit. Chemicals that affect the WAT are usually referred to as wax inhibitors or wax crystal modifiers. Chemicals that affect the pour point are referred to as PPDs or flow improvers. Since both classes of chemicals must somehow interact with the wax crystallization process, there is a good deal of overlap in the chemistry and mechanisms of the two classes. Thus, most wax inhibitors also function as PPDs. Wax dispersants act in a different manner. Wax inhibitors and PPDs must be deployed to pipelines before the bulk temperature drops below the WAT.

The concentration of wax inhibitor or PPD needed depends on the severity of the wax problem, but 100–2,000 ppm covers most application needs. Many applications with wax inhibitors do not totally prevent wax deposition in the whole flowline but can reduce the frequency for wax-removal treatments such as pigging or hot oiling. PPDs are often used to keep the wax from gelling during shut-in when the fluids might cool below the pour point. One Asian pipeline used PPD to the cost of U.S.$12 million per annum, until further investigation revealed that the cooling of the crude oil was found to be less pronounced than estimated earlier and could be transported safely without the use of PPD.[87]

There are various ways of designing a chemical that will interfere with the wax nucleation and crystallization process. In short, part of the molecule must interact or co-crystallize with the wax to modify or interfere with its crystallization. Another part of the molecule can prevent further wax growth by covering sites where new wax molecules would attach. This prevents the formation of structured wax lattices at the pipe wall. Clearly, there are at least two mechanisms as one study with cold-finger deposition showed.[88] One class of chemical greatly reduced wax deposition on a cold finger, leaving a transparent solution. The second class of chemical gave an opaque solution and was obviously acting on the wax in the bulk solution. Several studies suggest that effective wax inhibitors create weaker deposits, which are more susceptible to removal by shear forces in the flow field.[73] At the microscopic scale, wax inhibitors have been shown to have a dramatic effect on wax crystal morphology: rather than the usual platelike growth exhibited by pure wax, highly branched microcrystalline meshes are observed.[89]

The main classes of wax inhibitors and PPDs can be summarized as follows:

- ethylene polymers and copolymers
- comb polymers
- miscellaneous branched polymers with long alkyl groups

As we shall see, formulating these products in aromatic solvents can increase their performance. High concentrations of PPDs or wax inhibitor polymers can be difficult to handle in cold climates due to their high solidification temperature. Oil-in-water emulsification can be used to prevent them gelling or allow more than one production chemical to be deployed in the same line.[90–92] A mix of a weak to moderate wax solvent, such as toluene or xylene, and a strong wax solvent, such as cyclopentane or cyclohexane, can also avoid gelling of the wax inhibitor in cold climates.[93]

The comb polymers are most effective as wax inhibitors but can work synergistically with ethylene copolymers, such as ethylene/vinyl acetate (EVA) copolymer, and surfactants.[94] As the name suggests, comb polymers resemble a comb in that they have a polyvinyl backbone with many long-chain side groups (Figure 10.3). Actually,

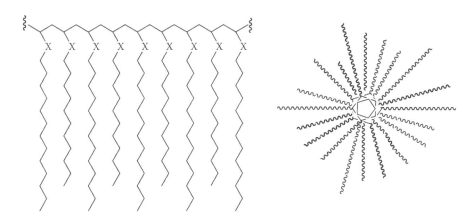

FIGURE 10.3 Traditionally depicted structure of a comb polymer (left). X is a spacer group. The structure looking down the helical backbone (right).

a better description is a "hair curler" because in an atactic polymer, the side chains point away from the backbone in many directions. As a general guideline, wax inhibitors or wax crystal modifiers can prevent wax deposition up to 10–15°C (18–27°F) below the WAT at typical doses of 50–200 ppm. Often, the pour-point depression is greater than the WAT depression, up to 30°C (54°F) has been observed given a very high PPD dose.[95] Besides being able to co-crystallize with waxes through van der Waals interactions, comb polymers place a steric hindrance on the wax crystal that interferes with proper alignment of new incoming wax molecules and growth terminates. This prevents the wax crystals from adhering together and sometimes prevents sticking to pipe walls. The pour point is also normally reduced. The growth inhibition by a comb polymer has been modeled using poly(octadecyl acrylate) on octacosane.[96] The model predicts that poly(octadecyl acrylate) will prevent wax growth at low supersaturations, where growth occurs largely at step defects, and slows growth at higher supersaturations, where island nucleation is important.[97] The optimum length of the side chains in comb polymers depends on the length of alkanes in the waxes.[98] Thus, in general, higher molecular weight waxes are best inhibited with long side chain comb polymers.[99] In essence, a good match between the lengths of the wax alkanes and side chains is needed. However, for very long waxes (C30+) there is no cost-effective, synthetic way to introduce equally long alkyl chains into a comb polymer. Laboratory studies show that some of the best commercial comb polymers, and also EVAs, cannot cope with such long alkane waxes.[100] The frequency of the long side chains in comb polymers is also critical. One study showed that 60% C18 side chains in a polyacrylate ester gives optimum performance on one type of wax.[101] The rest of the side chains were methyl groups since the C18 esters are made by base-catalyzed transesterification of polyacrylate methylester. A wax deposition study on various crude oils using wax characterization by NMR spectroscopy demonstrated that the stickier waxes have branched carbon chains while the nonsticky waxes are mainly straight-chain paraffins.[102] Thus, branching in the alkyl side chains of polymeric wax inhibitors may be beneficial.

10.4.3 Ethylene Polymers and Copolymers

Amorphous high molecular weight polyethylene has been used as a wax crystal modifier or PPD in the past, although better PPDs are made by copolymerizing ethylene with larger monomers.[103] The larger monomers act as branches disrupting the normal alkanes for crystallizing as wax. Examples are:

- ethylene/small alkene copolymers
- ethylene/vinyl acetate (EVA) copolymers
- ethylene/acrylontrile copolymers

Poly(ethylene butene) and poly(ethylene-*b*-propylene) have been investigated as wax-deposition inhibitors.[104] The use of 1,000 ppm of poly(ethylene butene), which was shown to reduce the yield stress of a gelled crude tenfold, actually increased the initial deposition rate.[105]

By far, the most common and well known of the ethylene copolymers is EVA. EVA copolymers are preferably random copolymers (Figure 10.4) and of low molecular weight.[106] The key parameter that determines the effectiveness of EVA copolymer is the percentage of vinyl acetate in the copolymer. Pure polyethylene would be expected to co-crystallize with structurally similar wax and have little impact on the crystallization process. Increasing vinyl acetate content lowers the crystallinity and aids solubility due to higher polarity. Some vinyl acetate can even be partially hydrolyzed.[107] The side chains in the vinyl acetate groups disrupt the wax crystallization process lowering the WAT or pour point. However, a high level of vinyl acetate decreases co-crystallization with waxes and has a negative impact on the performance. Generally about 25–30% vinyl acetate content gives optimum performance for EVA copolymers.[108] Interestingly, EVA has been shown to be a wax-nucleating agent as well as a growth inhibitor. These two mechanisms were identified from the analysis of the cloud point, the chemical composition of the crystals, and the observations of the crystal habit.[109] The presence of asphaltenes in the crude also affects the performance of EVA.[110]

EVA copolymers are not as effective PPDs as comb polymers. In one study, EVA copolymers dosed at 200 ppm gave a reduction in yield stress of three orders of magnitude for the C36 wax, whereas the reduction is one order of magnitude for C32 and only threefold for the C28 wax. This decrease in efficiency with decreasing wax carbon number indicates that the EVA materials would not provide an adequate reduction in yield stress to ensure against gelation in pipeline transport.[111] An experimental and

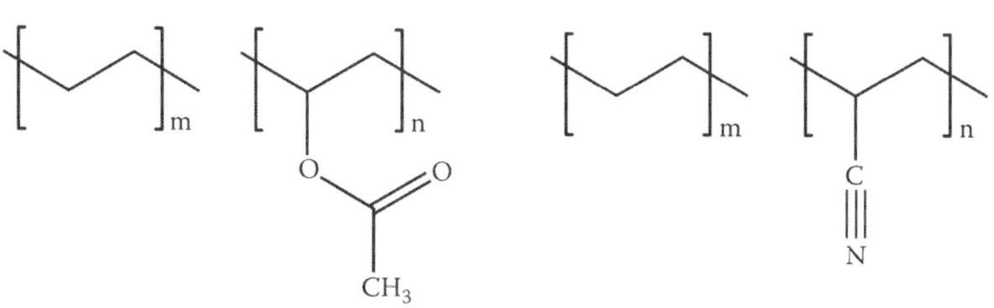

FIGURE 10.4 Ethylene/vinyl acetate (left) and ethylene/acrylonitrile copolymers (right).

molecular simulation study indicated that side chains introduced by propylene were a benefit to the affinity between the EVA-type molecules and alkanes in the wax plane, comparing with those branches introduced by butylenes.[112] Partially hydrolyzed EVA, containing hydroxyl groups, has been claimed as a superior wax inhibitor.[113]

Another cheap, polar monomer that can be copolymerized with ethylene to make wax inhibitors or PPDs is acrylonitrile (Figure 10.4). Ten percent to 20% acrylonitrile is preferred.[114]

10.4.4 Comb Polymers

Comb polymers are generally regarded as the most effective class of wax inhibitors. However, in severe wax cases with high WAT, even these inhibitors may not solve the deposition problem over extended time periods. This is because the bulk of the wax alkanes are considerably longer than the alkyl chains in the comb polymer.[115] Hence, periodic pigging or some other wax-removal technique may be needed.[60] Comb polymers are usually made from one of two classes of monomers, (meth) acrylic acid, or maleic anhydride or both.

10.4.4.1 (Meth)acrylate Ester Polymers

There are many reports of the use of acrylate or methacrylate ester polymers as wax inhibitors or PPDs (Figure 10.5). The ester groups are made using long-chain alcohols and should be at least 16 carbons in length. The side chains are spaced differently in methacrylate polymers compared with acrylate polymers, the former giving a better effect on pour point and deposition.[107] Polymers with longer alkyl side chains than 18 carbons would perform best as wax inhibitors in most cases, as the alkanes in most waxes are considerably longer. However, the cost of alcohols longer than 18 carbons (stearyl alcohol) is high and thus the wax inhibitor becomes more expensive. In one laboratory study, the optimum chain length of the alkyl ester chain was 20–24 carbon atoms and the optimum molecular weight was 30–40,000 Da for waxes with predominantly 20–29 carbon atoms.[116] Further, it was suggested that the melting point of the PPD should match the melting point of the waxes for limiting the growth of the crystals. Another study showed that polymethacrylate wax inhibitors (with alkyl side chains with a maximum of 18 carbon atoms) tested at a dosage of 100 ppm suppressed the WAT of lower molecular weight paraffin (C–24) solutions,

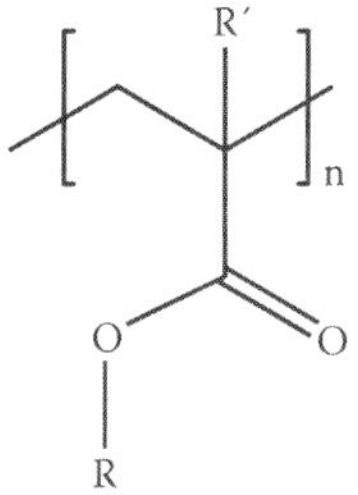

FIGURE 10.5 Acrylate ester ($R' = H$) and methacrylate ester ($R' = CH_3$) polymers as wax inhibitors or PPDs. R is preferably a long linear alkyl chain.

but had little or no effect for higher molecular weight paraffin (C–36) solutions.[66] As mentioned earlier, not all the (meth)acrylate groups need to be long chains for optimum performance. About 60% C18 side chains gave maximum WAT depression in a polyacrylate ester, the remaining side groups being methyl esters.[101]

It is important for the service company to have a range of comb polymers available for wax treatment because the length and proportions of the wax alkane chains varies from crude to crude and this will affect their performance. Comb polymers with a specific "U" distribution of alkyl chain lengths have been proposed to give a more universal performance.[117]

An improved class of flow improver is stearyl acrylate copolymerized with a small percentage of hydroxyethyl acrylate and subsequently esterifying the hydroxyl groups with stearic acid chloride. In this way, a longer side chain can be introduced into the molecule, avoiding the use of costly C20+ alcohols such as behenyl alcohol.[118] Grafting alkyl (meth)acrylate chains onto polyvinyl backbones, such as that of EVA, can also be used.[119]

There have been claims for a number of improvements on the basic polyalkylmethacrylate esters by using copolymers. For example, copolymers of (meth)acrylic acid ester of C16+ alcohols with a small percentage of hydrophilic (meth)acrylic acid, vinyl pyridine, or N-vinyl pyrrolidone have been claimed as improved PPDs and flow improvers.[117,120–121] Stearyl acrylate/allyl polyglycol copolymers can be used as flow improvers, preferably blended with synergists such as polyisobutylene and alkylphenol formaldehyde resins.[122] The presence of polar units confers a dispersing character to the polymer, which permits avoidance of deposition of wax on the pipe walls. Terpolymers containing (meth)acrylate esters with a specific mix of alkyl chain lengths and a third vinyl comonomer such as 2- or 4-vinyl pyridine, styrene, vinyl acetate, or vinyl benzoate have also been claimed as wax inhibitors.[123]

A synergistic blend of two polyalkyl(meth)acrylate has been claimed as a PPD.[124] The two polymers are a polyalkyl(meth)acrylate having on onset of crystallization at temperatures above 15°C (59°F) and a polyalkyl(meth)acrylate having an onset of crystallization or segregation at temperatures of 15°C (27°F) or lower, with the proviso that there is a temperature difference of at least 5–10°C (9–18°F) between the onset of crystallizations of the two polymers. A later study of polyalkyl(meth)acrylates led to a new invention whereby two such polymers with different long chain lengths was shown to perform better than polymers of just one chain length.[76] The polyesters are preferably made by base-catalyzed transesterification of polymethyl(meth)acrylate with alkanols and the preferred molecular weight is 20,000–30,000 Da. A mix of 7% C18, 58% C20, 30% C22, and 6% C24 alkanols is commercially available. The use of varying carbon chain lengths in the side chains is probably related to the varying length of alkanes in wax crystals. These polymers can also be used in synergistic combinations with long chain polyethyleneimine derivatives and oil-soluble film-forming surfactants. Branching in polyalkyl(meth)acrylates containing nitrogen functional groups is claimed to result in improved performance as a PPD.[125] The branching is supplied by adding a very minor amount of a divinyl monomer such as divinylbenzene or butylene-1,4-diacrylate such that cross-linking is minimized.

Addition of xylene has been shown to improve the performance of commercial polymeric wax inhibitors.[60] For example, poly(behenylacrylate) gave a lower pour point when mixed with xylene or other aromatic solvents used to remove asphaltenes.[126] A trichloroethylene-xylene (TEX) binary system, without any polymeric wax inhibitor, showed a unique ability to effect substantial pour-point depression and improve transport properties for a wide range of waxy crudes in cold spot tests.[127] This led to improved inhibition of paraffin deposition by the TEX additive in comparison with the performance of some tested, commercial antiparaffin chemical products.

10.4.4.2 Maleic Copolymers

Next to using (meth)acrylic acid monomers to make comb polymers, the other most important and equally as cheap monomer is maleic anhydride. Maleic anhydride does not easily polymerize by itself but does so readily in the presence of other vinyl comonomers. To contain long alkyl side chains, the anhydride can be derivatized to monoester, diester, thioester, or imide groups using long-chain alcohols, alkyl mercaptans, and alkylamines. In many polymers of this class, one begins with copolymers of maleic anhydride and long-chain α-olefins such as 1-octadecene. This places a long chain on every fourth carbon atom in the backbone. The anhydride groups can then be derivatized to increase the density of long side chains. Copolymers of maleic anhydride have a regular alternating monomer structure, ABABAB, and so forth. This can give these copolymers better properties as wax inhibitors over random copolymers. Thus, copolymers of (meth)acrylic acid esters with C16+ alcohols and maleic anhydride have been claimed as improved wax inhibitors over polyalkyl(meth) acrylates (Figure 10.6).[128–129] The performance can be further improved by adding EVA copolymer as synergist and surfactant-based wax-settling additives.[130]

A study on a maleic anhydride copolymer, probably with side chains of up to 18 carbons atoms, gave poor performance when the wax was made up of C24+ alkanes.[115] Thus, the wax inhibitor cannot cope with very long-chain alkanes unless it has chains of similar length. A very revealing study on mostly derivatives of α-olefin/maleic anhydride polymers has been reported.[131] Using branched alcohols to make maleic esters gave worse results than using normal alcohols. Monoesters of maleic polymers clearly performed better than diesters in lowering the cloud

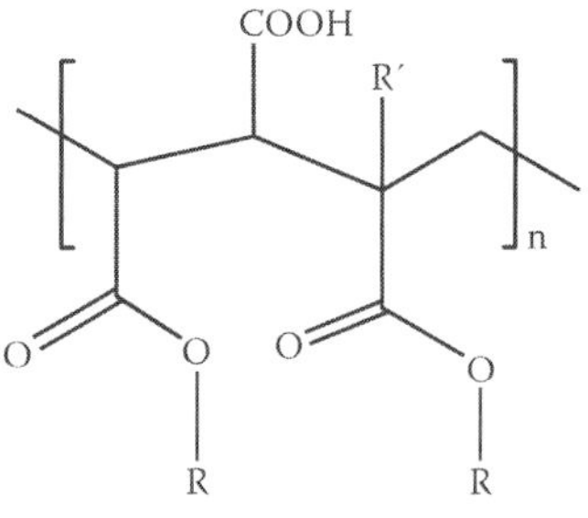

FIGURE 10.6 Monoester of maleic anhydride/(meth)acrylate ester copolymers. R = long alkyl chain, R' = H or CH_3.

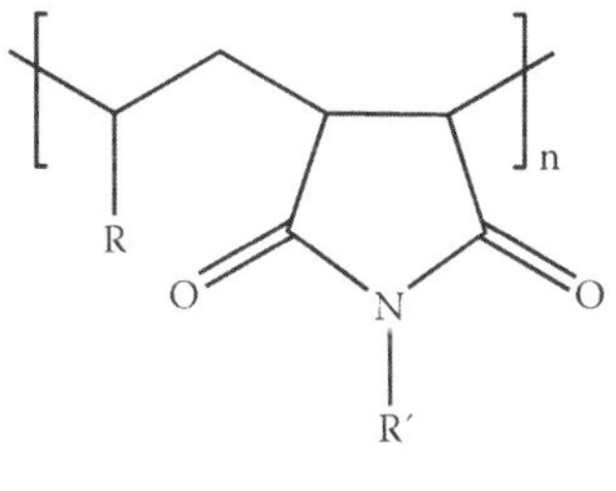

FIGURE 10.7 Alkyl maleimide/α-olefin copolymers. R and R' are preferably long alkyl chains.

point, whereas the two classes of esters gave similar pour-point depressions. Further, monoesters with side chains of 20 or fewer carbon atoms were better PPDs than esters with longer side chains. This was deduced assuming the esters form a barrier to promote growth of wax crystals and therefore do not need to be so bulky, as they do not need deep penetration into the wax crystals. High molecular weight polymers performed best as PPDs. Thus, a high molecular weight octadecene/maleic anhydride copolymer derivatized into a monoester with short-length alcohols would be an optimum wax inhibitor and PPD.

Maleic anhydride/α-olefin copolymers can be derivatized with long alkylamines to increase the frequency of long side chains. For example, maleic anhydride/α-olefin copolymers subsequently reacted with C18 alkylamines to make a maleimide have been claimed as PPDs (Figure 10.7).[132] Poly(maleic anhydride octadecene) modified with octadecyl amine was shown to be a good wax deposition inhibitor.[105,133] Another claimed class of PPDs is copolymers of meth(acrylic) acid esters with C16+ alcohols and nitrogen-containing alkenes including alkyl maleimides.[134]

Another way to introduce long alkyl chains into maleic copolymer is to use alkyl vinyl ethers. Thus, octadecylvinyl ether/maleic anhydride copolymer and derivatives have been claimed as PPDs.[132] Aliphatic glycol ether solvents are claimed as synergists for these types of comb polymers as well as vinyl pyrrolidone/eicosene copolymers.[135] Poly(maleic anhydride/ethyl vinyl ether) modified with docosanyl amine is also reported to give good wax inhibition.[105]

Polyolefin polymers can be grafted with unsaturated monomers such as maleic anhydride. This is another way of introducing long alkyl groups into the side chains if the maleic anhydride is derivatized, for example, maleimides can be used (Figure 10.8). Maleic derivatives can also be grafted onto EVA copolymer.[136]

Other unsaturated carboxylic acids monomers besides maleic acid can be used to make wax inhibitors. For example, long-chain alkyl fumarate/vinyl acetate copolymer were shown to be good as flow improver for high waxy Indian crude oils. Fumaric acid is the *trans* form of maleic acid.[137] Novel carbon monoxide/dialkyl fumarate copolymers have been claimed as wax crystal modifiers and flow improvers.[138] Alkyl fumarate copolymers are claimed to work synergistically with the reaction product of an alkanolamine, such as triethanolamine with a long-chain acylating agent such as a C8–20 alkyl succinic anhydride.[139]

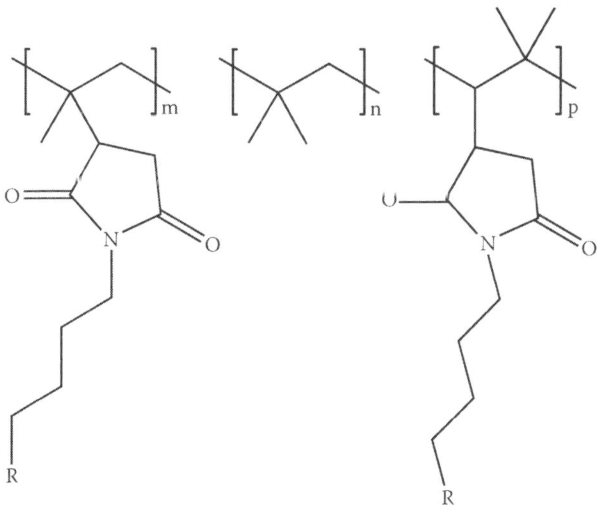

FIGURE 10.8 Grafted polyisobutylene-alkyl maleimide polymers. R is preferably > C12.

10.4.5 MISCELLANEOUS POLYMERS

Alkyl phenol-formaldehyde resins have been claimed as PPDs and flow improvers (Figure 10.9). By themselves, they are not as effective as comb polymers described earlier but they make useful synergists.[140] To make the resins, phenol is first reacted with a long chain α-olefin and then condensed with aldehyde. This class with shorter alkyl groups is also useful as asphaltene dispersants and as demulsifiers when polyalkoxylated.

The reaction of long-chain phosphoric ester surfactants with sodium aluminate is reported to give a high molecular weight material that lowers the WAT of waxy crude oils. A mixture of esters was used with maximum alkyl chain lengths of only 12 carbons atoms. Esters with longer alkyl chains were not reported but might have given even better results.[141]

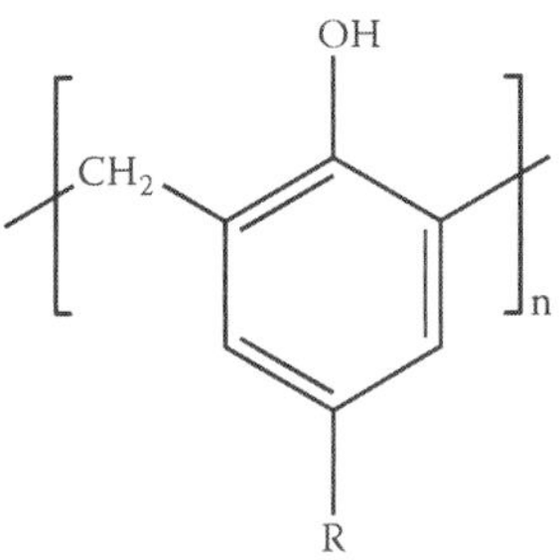

FIGURE 10.9 Alkyl phenol-formaldehyde resins.

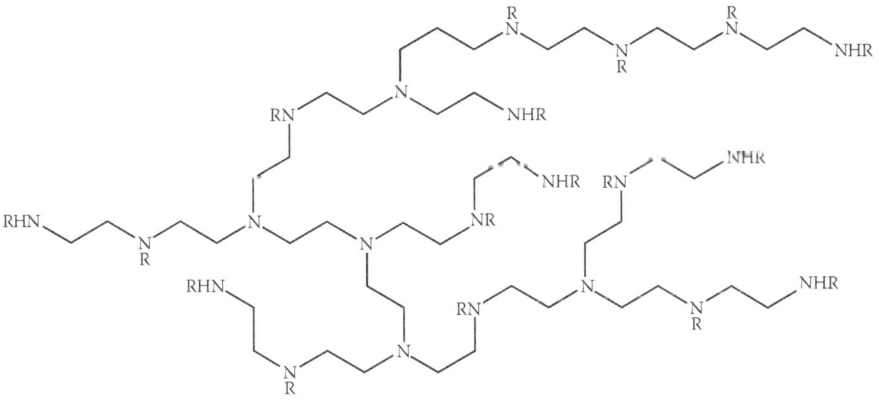

FIGURE 10.10 An example of a derivatized polyethyleneimine wax inhibitor, R = R′CH(OH)CH$_2$ where R′ is an alkyl group containing 10–22 carbon atoms. The ratio of tertiary/secondary/primary amines is approximately 1:2:1.

Branched polymers can also be used as wax inhibitors. For example, the reaction of branched polyethyleneimine (PEI) with 1,2-epoxyoctadecane gives a polymer with many pendant C18 groups, which performs well as a wax deposition inhibitor (Figure 10.10). The molecular weight of the PEI was 1,800 Da.[142] Dendrimeric hyperbranched polyesteramides, preferably with long pendant alkyl chains, have also been claimed as wax inhibitors.[143] One of many examples is a condensation product of succinic acid and diisopropanolamine in which the hydroxyl end groups were nearly all esterified with a fatty acid such as stearic or behenic acid. The molecular weight (Mw) was 3–4,000 Da. Nonvinyl, branched backbones can also be found in polyalkanolamines. Thus, hexatriethanolamine oleate esters have been shown to be good PPDs.[144]

10.4.6 Wax Dispersants

Wax dispersants are surfactants that adsorb onto pipe surfaces reducing the adhesion of waxes to the surface. This could be by changing the wettability of the surface to water-wet or by creating a weak layer on which wax crystals grow and are later sheared off by turbulent fluid flow. Some wax dispersants probably function by adsorbing and water-wetting the surface of the pipe. Wax dispersants also will adsorb to growing wax crystals thereby reducing the tendency for them to stick together. A good dispersant formulation will also function to penetrate accumulated deposits of wax, adsorbing on individual particles and enabling them to move freely into the surrounding oil. The overall effect is less accumulation on the pipe wall. Typical dosages are 50–300 ppm. Wax dispersants may be dosed continuously for inhibitory effect or by higher dosage batch treatment to achieve remedial benefits. Dispersants are frequently blended with polymeric wax inhibitors to enhance their performance. Used on their own, wax dispersants have had limited success in the field. However,

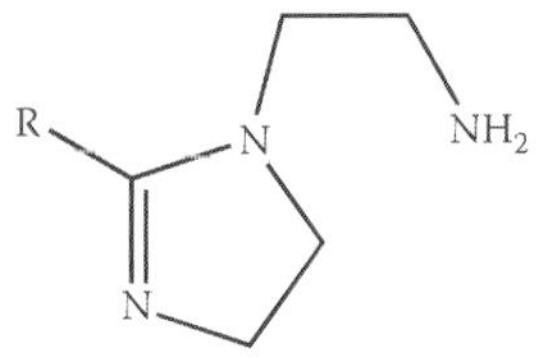

FIGURE 10.11 2-Aminoethyl-2-alkyl-imidazoline surfactants.

the successful evaluation and field trial of wax dispersants in a New Mexico field, where EVA and other wax crystal modifiers had failed, has been reported.[145] The dispersants used were water- and oil-soluble surfactants. Typical low-cost surfactants used as wax dispersants are alkyl sulfonates, alkyl aryl sulfonates, fatty amine ethoxylates, and other alkoxylated products.

There are a few other reports on the sole use of surfactant dispersants to prevent wax deposition or gelling. One paper showed that some surfactants were able to lower the pour point of crude oils by breaking up the three-dimensional network of wax crystals.[146] Another paper describes the use of a synergistic blend of film-forming ionic surfactants as wax dispersants and wax antisticking agents.[147] The surfactants have an ability to absorb onto bare surfaces of tubings and equipment, making them oleophobic. Details of the structures of the surfactants were not reported. However, good film-forming surfactants are also good corrosion inhibitors, such as imidazolines. Surfactants are not only useful as antisticking agents. Long-chain imidazolines, including their dimeric and trimeric forms, have been claimed to reduce pour points as much as 30°C (54°F; Figure 10.11).[148] The use of surfactants such as imidazolines rather than polymers overcomes the problem of gelling of the PPD in cold climates.

Polyethyleneimine-based wax inhibitors mentioned in the previous section are claimed to work synergistically in combination with polyalkyl(meth)acrylate polymers and optionally an oil-soluble, film-forming surfactant or corrosion inhibitor.[76] Examples given of oil-soluble film-forming surfactants are N-tallyl-1,3-propylene diamine or oil-soluble ethoxylated versions thereof, imidazolines such as N-2-aminoethyl-2-oleyl-imidazoline and phosphate esters. This has been borne out in a study where an oleic imidazoline corrosion inhibitor improved the performance of two wax inhibitors tested.[71] A molecular modeling project has shown that the presence of a protective corrosion inhibitor film, such as that formed from oleic imidazoline, does generate an ordered layer of long-chain alkane molecules with a structure akin to that found in the bulk alkane wax crystals. While such an alkane layer will increase the efficiency of corrosion inhibition, it is suggested that the layer may act as a nucleation site for wax deposition.[149] If wax does nucleate on the surfactant layer, the wax is not sticking directly to the surface. Over time, the surfactant and wax nuclei may be sheared off the surface together as the bonding between the surfactant and the surface is weak, thus, keeping the surface free from wax. It has also been reported that the high molecular weight fractions from crudes adsorb onto metal surfaces preventing wax deposition.[150–151]

It has been reported that adding surfactants to promote emulsification can reduce the tendency for wax deposition, obviously, when water is present.[152] It is properties that induce tighter emulsions, such as lower interfacial tension and greater shear rate, that lead to this effect. Furthermore, the wax that did deposit from an emulsion was softer (lower average molecular weight) than wax that deposits in the absence of any chemical. Another observation was that the wax that deposited in the presence of some commercial polymer-based wax inhibitors could be even harder (higher average molecular weight) than the deposit formed in the absence of any chemical additive.

10.4.7 POLAR CRUDE FRACTIONS AS FLOW IMPROVERS

An interesting and potentially low-cost method of obtaining a flow improver is to take out polar extracts from crude and distillate oils using a supercritical gas. The best examples of such a gas are carbon dioxide, ethylene, propylene, or C1–3-fluoroalkanes.[153] The extracts contain asphaltenes, resins, and aromatics. These natural flow improvers can be combined with EVA copolymer or comb polymers for synergistic effect. Besides these more polar extracts, it has been shown that high concentrations of cyclo/branched alkanes (> 50 wt.%) in a crude enhanced the activity of a wax crystal modifier. This result may be due to a structural effect, that is, loose packing of crystals from the steric effect of naphthenic and branched structures.[154]

It has been discovered that heavy fractions of paraffinic lubes produced over dewaxing catalysts are effective as wax crystal modifiers. The heavy fraction of a paraffinic lube suitable for use as a wax crystal modifier is derived from a Fischer-Tropsch product that has been catalytically dewaxed.[155] Olefin waxes that have been selectively and partially oxidized can advantageously lower the pour point of a hydrocarbon composition and/or decrease the average wax particle size of waxy precipitates in a hydrocarbon composition.[156]

10.4.8 DEPLOYMENT TECHNIQUES FOR WAX INHIBITORS AND PPDS

By far, the most common deployment method for wax inhibitors and PPDs is injection at the wellhead. However, if wax is a problem downhole and a capillary string is available, these chemicals can also be injected downhole. A technique used in China is to add cylindrical solid wax inhibitor placed in a tubular vessel or attached to the bottom of an oil pump into the well.[159] The inhibitor gradually dissolves over time as a slow-release system. Theoretically, squeeze applications do not appear to be economic since wax inhibitors are, of necessity, very similar in polarity to naturally occurring crudes. Thus, they have little affinity to adsorb to the reservoir rock and therefore tend to return rapidly. This creates a short effective treatment life. However, there has been a report of successful squeeze treatments, some lasting in excess of 6 months.[69,157] Details of the chemistry and retention mechanism are not given. Two important aspects to control are wettability and design for a minimum inhibitor concentration in the produced fluids.

Due to the polymeric nature of most wax inhibitors and PPDs, solutions of these chemicals can be very viscous, making them difficult to handle. Therefore, deployed compositions are often low-concentration mixtures or solutions of the active ingredient in a solvent, often an aromatic hydrocarbon. One way to improve the concentration of the active ingredients and keep the viscosity low is to use an emulsion. One method describes the use of an aqueous external dispersion comprising a wax dispersant and an organic crystal modifier composition dispersed through a continuous water phase.[158] The dispersant comprises a nonionic surfactant, such as an ethoxylated aliphatic alcohol, and is present in the dispersion in an amount sufficient to impart at least meta-stability to the dispersion.

REFERENCES

1. S. Misra, S. Baruah, and K. Singh, "Paraffin Problems in Crude Oil Production and Transportation: A Review," *SPE Production and Facilities* 10 (1995): 50.
2. E. D. Burger, T. K. Perkins, and J. H. Striegler, "Studies of Wax Deposition in the Trans Alaska Pipeline," *SPE Journal of Petroleum Technology* 33 (1981): 1075.
3. J. R. Becker, *Crude Oil: Waxes Emulsions and Asphaltenes*, Tulsa, OK: PennWell Publishing, 1997, Chap. 13.
4. W. D. McCain, Jr., *The Properties of Petroleum Fluids*, Tulsa, OK: PennWell Publishing Company, 1990.
5. N. F. Carnahan, "Paraffin Deposition in Petroleum Production," *Journal of Petroleum Technology* 41 (1989): 1024.
6. (*a*) D. Merino-Garcia and S. Correra, "Kinetics of Waxy Gel Formation from Batch Experiments" (paper presented at the 7th International Conference on Petroleum Phase Behavior and Fouling, 25–29 June 2006). (*b*) A. Bruno, C. Sarica, H. Chen, and M. Volk, "Paraffin Deposition during the Flow of Water-in-Oil and Oil-in-Water Dispersions in Pipes," SPE 114747 (paper presented at the SPE Annual Technical Conference and Exhibition, Denver, CO, 21–24 September 2008). (*c*) A. Benallal, P. Maurel, J. F. Agassant, M. Darbouret, G. Avril, and Eric Peuriere, "Wax Deposition in Pipelines: Flow-Loop Experiments and Investigations on a Novel Approach," SPE 115293 (paper presented at the SPE Annual Technical Conference and Exhibition, Denver, CO, 21–24 September 2008).
7. P. Kriz and S. I. Andersen, "Effect of Asphaltenes on Crude Oil Wax Crystallization," *Energy & Fuels* 19 (2005): 948.
8. (*a*) J. J. Magda, H. El-Gendy, K. Oh, M. D. Deo, A. Montesi, and R. Venkatesan, *Energy Fuels*, 2009, 23, 1311. (*b*) H. Li, J. Zhang, and D. Yan, "Correlations Between the Pour Point/Gel Point and the Amount of Precipitated Wax for Waxy Crudes," *Petroleum Science and Technology* 23 (2005): 1313.
9. G. E. Oliveira, C. R. E. Mansur, E. F. Lucas, G. González, and W. F. de Souza, "The effect of Asphaltenes, Naphthenic Acids, and Polymeric Inhibitors on the Pour Point of Paraffin Solutions," *Journal of Dispersion Science and Technology* 28 (2007): 349.
10. Q. Wang, C. Sarica, and T. Chen, "An Experimental Study on Mechanics of Wax Removal in Pipeline," SPE 71544 (paper presented at the SPE Annual Technical Conference and Exhibition, New Orleans, LA, 30 September–3 October 2001).
11. D. R. Galloway, U.S. Patent 5168929, 1992.
12. S. Feeney, "Project Case Histories and Future Applications of Vacuum Insulated Tubing," (paper presented at the AIChE Spring National Meeting, Houston, TX, 14–18 March 1999).

13. R. C. Sarmento, G. A. S. Ribbe, and L. F. A. Azevedo, "Wax Blockage Removal by Inductive Heating of Subsea Pipelines," *Heat Transfer Engineering* 25 (2004): 2.
14. K. A. Esakuhl, G. Fung, G. Harrison, and R. Perego, "Active Heating for Flow Assurance Control in Deepwater Flowlines," OTC 15188 (paper presented at the Offshore Technology Conference, Houston, TX, 5–8 May, 2003).
15. L. D. Brown, J. Clapham, C. Belmear, R. Harris, A. Loudon, S. Maxwell, and J. Stout, "Design of Britannia's Subsea Heated Bundle for a 25 Year Service Life," OTC 11017 (paper presented at Offshore Technology Conference, Houston, TX, 2–5 May 1999).
16. A. K. Mehrotra, "'Cold Flow' Deposition Experiments with Wax-Solvent Mixtures under Laminar Flow in a Flow-Loop with Heat Transfer" (paper presented at the 9th International Conference on Petroleum Phase Behavior and Fouling, Victoria, BC, June 15–19, 2008).
17. (a) D. Merino-Garcia and S. Correra, "Cold Flow: A Review of a Technology to Avoid Wax Deposition," *Petroleum Science and Technology* 26 (2008): 446. (b) C. B. Argo, P. Bollavaram, T. Y. Makogon, N. Oza, M. Wolden, R. Larsen, A. Lund, K. W. Hjarbo, International Patent Application WO200405918.
18. (a) M. Juenke and L. Rzeznik, U.S. Patent Application 20080067129. (b) D. Stefanini, International Patent Application WO/2008/017849.
19. B. Wang and L. Dong, "Paraffin Characteristics of Waxy Crude Oils in China and the Methods of Paraffin Removal and Inhibition," SPE 29954 (paper presented at the SPE International Meeting on Petroleum Engineering, Beijing, China, 14–17 November 1995).
20. J. W. McDonald, K. J. Humphreys, R .D. Humphreys, K. R. Kopecky, and G. W. Adams, U.S. Patent 5804067, 1998.
21. R. Tao and X. Xu, "Reducing the Viscosity of Crude Oil by Pulsed Electric or Magnetic Field," *Energy & Fuels* 20(5) (2006): 2046.
22. N. P. Tung, N. Van Vuong, B. Q. K. Long, N. Q. Vinh, P. V. Hung, V. T. Hue, and L. D. Hoe, "Studying the Mechanism of Magnetic Field Influence on Paraffin Crude Oil Viscosity and Wax Deposition Reductions," SPE 68749 (paper presented at the SPE Asia Pacific Oil and Gas Conference and Exhibition, Jakarta, Indonesia, 17–19 April 2001).
23. N. Rocha, C. González, L. C. do C. Marques, and D. S. Vaitsman, "A Preliminary Study on the Magnetic Treatment of Fluids," *Petroleum Science and Technology* 18 (2000): 33.
24. W. A. Cañas-Marin, J. D. Ortiz-Arango, U. E. Guerrero-Aconcha, and C. Lira-Galeana, *AIChE Journal* 52(8) (2006): 2887.
25. L. C. C. Marques, N. O. Rocha, A. L. C. Machado, G. B. M. Neves, L. C. Vieira, and C. H. Dittz, "Study of Paraffin Crystallization Process under the Influence of Magnetic Fields and Chemicals," SPE 38990 (paper presented at the SPE Latin American and Caribbean Petroleum Engineering Conference, Rio de Janeiro, Brazil, 30 August–3 September, 1997).
26. "Magnetic Fluid Conditioners' Success Depends on Factors," *Offshore* 58(9) 1998.
27. J. W. McManus, E. Winckler, and J. Backus, U.S. Patent 4538682, 1985.
28. B. Wang and L. Dong, "Paraffin Characteristics of Waxy Crude Oils in China and the Methods of Paraffin Removal and Inhibition," SPE 29954 (paper presented at the SPE International Meeting on Petroleum Engineering, Beijing, China, 14–17 November 1995).
29. (a) Z. He, B. Mei, W. Wang, J. Sheng, S. Zhu, L. Wang, and T. F. Yen, *Petroleum Science and Technology* 21 (2003): 201. (b) K. Duncan, L. Gieg, and I. Davidova, "Paraffin Control in Oil Wells Using Anaerobic Microorganisms" (paper presented at the 14th Annual International Petroleum Environmental Conference, Houston, 8 November 2007). (c) B. Soni and B. Lal, U.S. Patent Application 20090025931.

30. (a) I. Lazar, A. Voicu, C. Nicolescu, D. Mucenica, S. Dobrota, I. G. Petrisor, M. Stefanescu, and L. Sandulescu, "The Use of Naturally Occurring Selectively Isolated Bacteria for Inhibiting Paraffin Deposition," *Journal of Petroleum Science and Engineering* 22 (1999): 161. (b) H. K. Kotlar, A. Wentzel, M. Throne-Holst, S. Zotchev, and T. Ellingensen, "Wax Control by Biocatalytic Degradation in High Paraffinic Crude Oils," SPE 106420 (paper presented at the SPE International Symposium on Oilfield Chemistry, Houston, TX, 28 February–2 March 2007).

31. (a) P. M. Roberts, A. Venkitaraman, and M. M. Sharma, "Ultrasonic Removal of Organic Deposits and Polymer-Induced Formation Damage," *SPE Drilling & Completion* 15(1) (2000): 19. (b) B. Wang and C. Wang, International Patent Application WO/2009/000177.

32. K. M. Barker, "Formation Damage Related to Hot Oiling," SPE 16230, *SPE Production Engineering* 4(6) (1989): 371.

33. W. B. Walton, U.S. Patent 4813482, 1989.

34. C. L. Thierheimer, U.S. Patent 4925497, 1990.

35. T. J. Straub, S. W. Autry, and G. E. King, "An Investigation into Practical Removal of Downhole Paraffin by Thermal Methods and Chemical Solvents," SPE 18889 (paper presented at the SPE Production Operations Symposium, Oklahoma City, OK, 13–14 March 1989).

36. J. C. Bailey and S. J. Allenson, "Paraffin Cleanout in a Single Subsea Flowline Environment: Glycol to Blame?," OTC 19566 (paper presented at the Offshore Technology Conference, Houston, TX, 5–8 May 2008).

37. C. J. Bushman, International Patent Application WO/2008/024488.

38. J. A. Blunk, U.S. Patent 6176243, 2001.

39. H. A. Craddock, K. Mutch, K. Sowerby, S. McGregor, J. Cook, and C. Strachan, "A Case Study in the Removal of Deposited Wax from a Major Subsea Flowline System in the Gannet Field," SPE 105048 (paper presented at the SPE International Symposium on Oilfield Chemistry, Houston, TX, 28 February–2 March 2007).

40. R. A. Molland, P. Fotland, P. Clark, A. Valle, and B. H. Ovreas, "Handling Wax Deposition in a 116 km Sub-Sea, Single Phase Condensate Pipeline" (paper presented at the 19th Oilfield Chemical Symposium, Geilo, Norway, 9–12 March 2008).

41. S. Szymczak, G. L. Poole, G. Brock, and G. Casey, "Successful Pipeline Cleanout: Lessons Learned from Cleaning Paraffin Blockage from a Deepwater Pipeline," SPE 115658 (paper presented at the Annual Technical Conference and Exhibition, Denver, CO, 22–24 September 2008).

42. J. S. Als, U.S. Patent 6984614, 2006.

43. P. R. Hart and M. J. Brown, U.S. Patent 5484488, 1996.

44. E. A. Richardson and R. F. Scheuerman, U.S. Patent 4178993, 1979.

45. J. P. Ashton, L. J. Kirspel, H. T. Nguyen, and D. J. Credeur, "In-Situ Heat System Stimulates Paraffinic-Crude Producers in Gulf of Mexico," SPE 15660, *SPE Production Engineering* 4(3) (1989): 157.

46. N. O. Rocha and C. N. Khalil, "Controlling Wax Deposition in Offshore Production Systems—Petrobras Experience" (paper presented at the IBC Global Conferences Ltd: Focus on Hydrates, Waxes and Asphaltenes, Aberdeen, 1999).

47. C. N. Khalil, U.S. Patent 5639313, 1997.

48. C. R. De Souza and C. N. Khalil, U.S. Patent 6003528, 1999.

49. D. A. Nguyen, "Fused Chemical Reactions to Remediate Paraffin Plugging in Sub-Sea Pipelines," Ph.D. thesis, University of Michigan, 2004.

50. N. O. Rocha, C. N. Khalil, L. C. F. Leite, and R. M. Bastos, "A Thermochemical Process for Wax Damage Removal," SPE 80266 (paper presented at the SPE International Symposium on Oilfield Chemistry, Houston, TX, 5–7 February 2003 2003).

51. D. Sarkar, S. T. Arrington, R. J. Powell, I. D. Robb, and B. L. Todd, International Patent Application WO/2008/032067.
52. D. G. Clarke, U.S. Patent 6348102, 2002.
53. J. J. Habeeb and R. L. Espino, U.S. Patent 6354381, 2002.
54. T. G. Monger-McClure, J. E. Tackett, and L. S. Merrill, "Comparisons of Cloud Point Measurement and Paraffin Prediction Methods," SPE 54519, *SPE Production & Facilities* 14(1) (1999): 4.
55. J. A. P. Coutinho and J.-L. Daridon, "The Limitations of the Cloud Point Measurement Techniques and the Influence of the Oil Composition on Its Detection," *Petroleum Science and Technology* 23 (2005): 1113.
56. S. U. Amadi, A. Y. Dandekar, G. A. Chukwu, S. Khataniar, S. L. Patil, W. F. Haslebacher, and J. Chaddock, "Energy Sources, Part A, Recovery," *Utilization and Environmnetal Effects* 27 (2005): 831.
57. J. M. Letoffe, P. Claudy, M. Garcin, and J. L. Volle, "Crude Oils: Characterization of Waxes Precipitated on Cooling by DSC and Thermomicroscopy," *Fuel* 74 (1995): 810.
58. D. D.Erickson, V. G. Niesen, and T. S. Brown, "Thermodynamic Measurement and Prediction of Paraffin Precipitation in Crude Oil," SPE 26604 (paper presented at the SPE Annual Technical Conference and Exhibition, Houston, TX, 3–6 October 1993).
59. V. R. Kruka, E. R. Cadena, and T. E. Long, "Cloud-Point Determination for Crude Oils," *Journal of Petroleum Technology* 47(8) (1995): 681.
60. O. O. Bello, S. O. Fasesan, C. Teodoriu, and K. M. Reinicke, "An Evaluation of the Performance of Selected Wax Inhibitors on Paraffin Deposition of Nigerian Crude Oils,"*Petroleum Science and Technology* 24 (2006): 195.
61. C. E. Ijeomah, A. Y. Dandekar, G. A. Chukwu, S. Khataniar, S. L. Patil, and A. L. Baldwin, "Measurement of Wax Appearance Temperature under Simulated Pipeline (Dynamic) Conditions,"*Energy & Fuels* 22 (2008): 2437.
62. M. C. Garcia, L. Carbognani, A. Urbina, and M. Orea, "Correlation Between Oil Composition and Paraffin Inhibitors Activity," SPE 49200 (paper presented at the SPE Annual Technical Conference and Exhibition, New Orleans, LA, 27–30 September 1998).
63. A. Hammami and M. A. Raines, "Paraffin Deposition from Crude Oils: Comparison of Laboratory Results With Field Data," *SPE Journal* 4(1) (1999): 9.
64. K. Karan, J. Ratulowski, and P. German, "Measurement of Waxy Crude Properties Using Novel Laboratory Techniques," SPE 62945 (paper presented at the SPE Annual Technical Conference and Exhibition, Dallas, TX, 1–4 October 2000).
65. K. A. Ferworn, A. Hammami, and H. Ellis, "Control of Wax Deposition: An Experimental Investigation of Crystal Morphology and an Evaluation of Various Chemical Solvents," SPE 37240 (paper presented at the SPE International Symposium on Oilfield Chemistry, Houston, TX, 18–21 February 1997).
66. K. S. Wang, C. H. Wu, J. L. Creek, P. J. Shuler, and Y. C. Tang, "Evaluation of Effects of Selected Wax Inhibitors on Wax Appearance and Disappearance Temperatures," *Petroleum Science and Technology* 21 (2003): 359.
67. R. M. Roehner and F. V. Hanson, "Determination of Wax Precipitation Temperature and Amount of Precipitated Solid Wax Versus Temperature for Crude Oils Using FT-IR Spectroscopy," *Energy & Fuels* 15(3) (2001): 756.
68. T. S. Brown, V. G. Niesen, and D. D. Erickson, "The Effects of Light Ends and High Pressure on Paraffin Formation," SPE 28505 (paper presented at the SPE Annual Technical Conference and Exhibition, New Orleans, LA, 25–28 September 1994).
69. J. R. Becker, "Paraffin-Crystal Modifier Studies in Field and Laboratory," SPE 70030 (paper presented at the SPE Permian Basin Oil and Gas Recovery Conference, Midland, TX, 15–16 May 2001).

70. N. F. Magri, B. Kalpakci, and L. Nuebling, "Evaluation of Paraffin Crystal Modifiers by Dynamic Videomicroscopy," SPE 37241 (paper presented at the SPE International Symposium on Oilfield Chemistry, Houston, TX, 18–21 February 1997).
71. K. S. Wang, C. H. Wu, J. L. Creek, P. J. Shuler, and Y. C. Tang, "Evaluation of Effects of Selected Wax Inhibitors on Paraffin Deposition," *Petroleum Science and Technology* 21 (2003). 369.
72. K. Weispfennig, "Advancements in Paraffin Testing Methodology," SPE 64997 (paper presented at the SPE International Symposium on Oilfield Chemistry, Houston, TX, 13–16 February 2001).
73. D. W. Jennings and K. Weispfennig, "Effects of Shear on the Performance of Paraffic Inhibitors: Coldfinger Investigation with Gulf of Mexico Crude Oils," *Energy & Fuels* 20(6) (2006): 2457.
74. L. Dong, H. Xie, and F. Zhang, "Chemical Control Techniques for the Paraffin and Asphaltene Deposition," SPE 65380 (paper presented at the SPE International Symposium on Oilfield Chemistry, Houston, TX, 13–16 February 2001).
75. S. Vaage, "A Novel Approach on Wax Inhibitor Qualification" (paper presented at the 19th Oilfield Chemical Symposium, Geilo, Norway, 9–12 March 2008).
76. S. N. Duncum, K. James, and C. G. Osborne, International Patent Application WO98021446, 1998.
77. N. Fong and A. K. Mehrotra, "Deposition under Turbulent Flow of Wax-Solvent Mixtures in a Bench-Scale Flow-Loop Apparatus with Heat Transfer," *Energy & Fuels* 21(3) (2007): 1263.
78. I. Gjermundsen and M. Duenas Diez, "Wax Deposition; A Comparison between Measurements and Predictions from a Commercial Model" (paper presented at the 7th International Conference on Petroleum Phase Behavior and Fouling, 25–29 June 2006).
79. N. V. Bhat and A. K. Mehrotra, "Modeling the Effects of Heat Transfer and Shear on Composition and Growth of Deposit-Layer from 'Waxy' Mixtures in a Pipeline under Laminar Flow" (paper presented at the 7th International Conference on Petroleum Phase Behavior and Fouling, 25–29 June 2006).
80. B. F. Towler and S. Rebbapragada, "Mitigation of Paraffin Wax Deposition in Cretaceous Crude Oils of Wyoming," *Journal of Petroleum Science and Engineering* 45 (2004): 11.
81. R. Brockmann, "HPHT Flow Loop for Testing Wax and Scale Solid Formation" (paper presented at the 7th Int. Oil Field Chemicals Symposium, Geilo, 17–20 March 1996).
82. R. Brockmann, "Experimental Study of Pressure and Gas Content Effects on Wax Appearance Temperatures" (paper presented at the IBC Conference: Controlling Hydrates, Waxes and Asphaltenes, Aberdeen, UK, 16–17 September 1996).
83. J. L. Creek, H. J. Lund, J. P. Brill, and M. Volk, "Wax Deposition in Single Phase Flow," *Fluid Phase Equilibria* 158 (1999): 801.
84. J. F. Tinsley, R. K. Prud'homme, and X. Guo, "Effect of Polymer Additives upon Waxy Deposits" (paper presented at the 7th International Conference on Petroleum Phase Behavior and Fouling, 25–29 June 2006).
85. D. W. Jennings and M. E. Newberry, "Application of Paraffin Inhibitor Treatment Programs in Offshore Developments," OTC 19154 (paper presented at the Offshore Technology Conference, 5–8 May 2008, Houston, TX).
86. D. W. Jennings, U.S. Patent Application 20070213231.
87. J. Prasad, V. Sharma, P. C. Philip, S. R. Nimoria, and P. K. Verma, "Flow Assurance in 30″ Subsea Pipeline without the usage of PPD," OTC 19166 (paper presented at the Offshore Technology Conference, Houston, TX, 5–8 May 2008).
88. M. Greenaway, "Analytical Solutions for Wax Deposition," *Chemistry in the Oil Industry VIII*, Royal Society of Chemistry, 2003.

89. J. L. Hutter, S. Hudson, C. Smith, A. Tetervak, and J. Zhang, "Banded Crystallization of Tricosane in the Presence of Kinetic Inhibitors During Directional Solidification," *Journal of Crystal Growth* 273 (2004): 292.
90. J. S. Manka, J. S. Magyar, and R. P. Smith, "A Novel Method to Winterize Traditional Pour Point Depressants," SPE 56571 (paper presented at the SPE Annual Technical Conference and Exhibition, Houston, TX, 3–6 October 1999).
91. M.-F. Delamotte, D. Faure, and D. Tembou N'Zudie, U.S. Patent Application 20070062101, 2007.
92. (*a*) A. Capelle, U.S. Patent, 4110283, 1978. (*b*) M. Guzmann, Y. Liu, R. Konrad, and D. Franz, International Patent Application WO/2008/125588.
93. D. W. Jennings, U.S. Patent Application 20040058827.
94. M. J. Wisotsky, U.S. Patent 4153423, 1979.
95. K. S. Pedersen and H. P. Rønningsen, "Influence of Wax Inhibitors on Wax Appearance Temperature, Pour Point, and Viscosity of Waxy Crude Oils," *Energy & Fuels* 17(2) (2003): 321.
96. D. M. Duffy and P. M. Rodger, "Wax Inhibition with Poly(Octadecyl Acrylate)," *Physical Chemistry Chemical Physics* 4 (2002): 328.
97. D. M. Duffy and P. M. Rodger, "Modeling the Activity of Wax Inhibitors: A Case Study of Poly(Octadecyl Acrylate)," *Journal of Physical Chemistry B* 106(43) (2002): 11210.
98. D. M. Duffy, C. Moon, and P. M. Rodger, "Computer-Assisted Design of Oil Additives," *Molecular Physics* 102 (2004): 203.
99. A. Hennessy, A. Neville, and K. J. Roberts, "In Situ SAXS/WAXS and Turbidity Studies of the Structure and Composition of Multihomologous *n*-Alkane Waxes Crystallized in the Absence and Presence of Flow Improving Additive Species," *Crystal Growth & Design* 4(5) (2004): 1069.
100. S. Y. Cho and H. S. Fogler, "Efforts on Solving the Problem of Paraffin Deposit: I. Using Oil-Soluble Inhibitors," *Journal of Industrial and Engineering Chemistry* 5 (1999): 123.
101. D. M. Duffy, C. Moon, J. L. Irwin, A. F. Di Salvo, P. C. Taylor, M. Arjmandi, A. Danesh, S. R. Ren, A. Todd, B. Tohidi, M. T. Storr, L. Jussaume, J.-P. Montfort, and P. M. Rodger, "Chemistry in the Oil Industry" (paper presented at the Symposium VIII, Manchester, England, 2003).
102. S. M. Bucaram, "An Improved Paraffin Inhibitor," *Journal of Petroleum Technology* 19(2) (1967): 150.
103. J. F. Tinsley, R. K. Prud'homme, and X. Guo, "Effect of Polymer Additives Upon Waxy Deposits" (paper presented at the 7th International Conference on Petroleum Phase Behavior and Fouling, 25–29 June 2006).
104. J. F. Tinsley, R. K. Prud'homme, X. Guo, D. H. Adamson, S. Callahan, S. Amin, S. Shao, R. M. Kriegel, and R. Saini, "Novel Laboratory Cell for Fundamental Studies of the Effect of Polymer Additives on Wax Deposition from Model Crude Oils," *Energy & Fuels* 21(3) (2007): 1301.
105. C. J. Dorer Jr. and K. Hayashi, U.S. Patent 4623684, 1986.
106. (*a*) O. E. Lindeman and S. J. Allenson, "Theoretical Modeling of Tertiary Structure of Paraffin Inhibitors," SPE 93090 (paper presented at the SPE International Symposium on Oilfield Chemistry, The Woodlands, TX, 2–4 February 2005). (*b*) J. B. Taraneh, G. Rahmatollah, A. Hassan, and D. Alireza, Fuel Processing Technology, 2008, 89, 973.
107. L. A. McDougall, A. Rossi, and M. J. Wisotsky, U.S. Patent 3693720, 1972.
108. (*a*) A. L. C. Machado, E. F. Lucas, and G. Gonzalez, "Poly(Ethylene-*co*-Vinyl Acetate) (EVA) as Wax Inhibitor of a Brazilian Crude Oil: Oil Viscosity, Pour Point and Phase Behavior of Organic Solutions," *Journal of Petroleum Science and Engineering* 32 (2001): 159. (*b*) E. Marie, Y. Chevalier, F. Eydoux, L. Germanaud, and P. Flores, "Control of *n*-Alkanes Crystallization by Ethylene-Vinyl Acetate Copolymers," *Journal of Colloid Interface Science* 290 (2005): 406.

109. A. L. C. Machado and E. F. Lucas, "Influence of Ethylene-*co*-Vinyl Acetate Copolymers on the Flow Properties of Wax Synthetic Systems," *Journal of Applied Polymer Science* 85 (2002): 1337.
110. H. S. Ashbaugh, X. Guo, D. Schwahn, R. K. Prud'homme, D. Richter, and L. J. Fetters, "Interaction of Paraffin Wax Gels with Ethylene/Vinyl Acetate Co-polymers," *Energy & Fuels* 19 (2005): 138.
111. C. Wu, J.-L. Zhang, W. Li, and N. Wu, "Molecular Dynamics Simulation Guiding the Improvement of EVA-Type Pour Point Depressant," *Fuel* 84 (2005): 2039.
112. L. A. McDougall, A. Rossi, and M. J. Wisotski, U.S. Patent 3693720, 1972.
113. K. L. Mtz, R. A. Latham, and R. J. Statz, European Patent EP 345008, 1989.
114. M. Del Carmen García, "Crude Oil Wax Crystallization: The Effect of Heavy *n*-Paraffins and Flocculated Asphaltenes,"*Energy & Fuels* 14(5) (2000): 1043.
115. H. K. Singhal, G. C. Sahai, G. S. Pundeer, and L. Chandra, "Designing and Selecting Wax Crystal Modifier for Optimum Field Performance Based on Crude Oil Composition," SPE 22784 (paper presented at the SPE Annual Technical Conference and Exhibition, Dallas, TX, 6–9 October 1991).
116. J.-F. Brunelli and S. Fouquay U.S. Patent 6218490, 2001.
117. H. Wirtz, S.-P. Von Halasz, M. Feustel, and J. Balzer, U.S. Patent 5349019, 1994.
118. G. Meunier, R. Brouard, B. Damin, and D. Lopez, U.S. Patent 4608411, 1986.
119. W. Ritter, C. Meyer, W. Zoellner, C.-P. Herold, and S. Tapavicza, U.S. Patent 5039432, 1991.
120. P. Gateau, A. Barbey, and J. F. Brunelli, U.S. Patent 6750305.
121. (*a*) M. Feustel, M. Krull, and H.-G. Oschmann, U.S. Patent 6821933, 2004. (*b*) O. E. Shmakova-Lindeman, International Patent Application WO/2005/098200.
122. O. E. Shmakova-Lindeman, U.S. Patent Application 20050215437, 2005.
123. M. Mueller and H. Gruenig, U.S. Patent 5281329, 1994.
124. R. Eckert and B. Vos, B., Canadian Patent 1231659, 1988.
125. D. Chanda, A. Sarmah, A. Borthakur, K. V. Rao, B. Subrahmanyam, and H. C. Das, "Combined Effect of Asphaltenes and Flow Improvers on the Rheological Behavior of Indian Waxy Crude Oil," *Fuel* 77 (1998): 1163.
126. V. A. Adewusi, "An Improved Inhibition of Paraffin Deposition from Waxy Crudes," *Petroleum Science and Technology* 16 (1998): 953.
127. E. Barthell, A. Capelle, M. Chmelir, and K. Dahmen, U.S. Patent 4663491, 1987.
128. H. P. M. Tomassen, C. van de Kamp, M. J. Reynhout, and J. Lin, U.S. Patent 5721201, 1998.
129. J. A. Day, M. J. Reynhout, and H. P. M. Tomassen, U.S. Patent 5585337, 1996.
130. A. J. Son, R. B. Graugnard, and B. J. Chai, "The Effect of Structure on Performance of Maleic Anhydride Copolymers as Flow Improvers of Paraffinic Crude Oil," SPE 25186 (paper presented at the SPE International Symposium on Oilfield Chemistry, New Orleans, LA, 2–5 March 1993).
131. H. T. Le, U.S. Patent 4992080, 1991.
132. X. H. Guo, M. Herrera-Alonso, J. F. Tinsley, and R. K. Prudhomme, *Preprint Papers— American Chemical Society, Division of Petroleum Chemistry* 50 (2005): 318.
133. B. Wahle, C.-P. Herold, W. Zoellner, L. Schieferstein, and D. Oberkobusch, U.S. Patent 5006621, 1991.
134. A. M. Robinson, J. R. Stromberg, M. J. Jurek, and K. Bakeev, U.S. Patent Application 20020166995, 2002.
135. J. Balzer, M. Feustel, M. Krull, and W. Reimann, U.S. Patent 5439981, 1995.
136. A. Borthakur, D. Chanda, S. R. Dutta Choudhury, K. V. Rao, and B. Subrahmanyam, "Alkyl Fumarate–Vinyl Acetate Copolymer as Flow Improver for High Waxy Indian Crude Oils," *Energy & Fuels* 10(3) (1996): 844.
137. A. O. Patil, S. Zushma, E. Berluche, and M. Varma-Nair, U.S. Patent 6444784, 2002.

138. J. S. Manka, K. L. Ziegler, and D. R. Nelson, U.S. Patent 6017370, 2000.
139. D. J. Martella and J. J. Jaruzelski, European Patent EP0311452, 1989.
140. D. O. Gentili, C. N. Khalil, N. O. Rocha, and E. F. Lucas, "Evaluation of Polymeric Phosphoric Ester-Based Additives as Wax Deposition Inhibitors," SPE 94821 (paper presented at the SPE Latin American and Caribbean Petroleum Engineering Conference, Rio de Janeiro, Brazil, 20–23 June 2005).
141. S. N. Duncum, K. James, and C. G. Osborne, U.S. Patent 6140276, 2000.
142. P. F. Van Bergen, M. A. Van Dijk, and A. J. Zeeman, International Patent Application WO/2006/056578.
143. A. A. Hafiz and T. T. Khidr, "Hexa-Trithanolamine Oleate Esters as Pour Point Depressant for Waxy Crude Oils," *Journal of Petroleum Science and Engineering* 56 (2007): 296.
144. G. G. McClaflin and D. L. Whitfil, "Control of Paraffin Deposition in Production Operations," SPE 12204 (paper presented at the Annual Technical Conference and Exhibition, San Francisco, October 1983).
145. L. K. Verma and Mukhdeo, "Effects of Surfactants on Pour Point of Crude Oil" (paper presented at the 1st Natl. Conv. of Chem. Eng., Calcutta, 21–23 February 1986).
146. D. Groffe, P. Groffe, S. Takhar, S. I. Andersen, E. H. Stenby, N. Lindeloff, and M. Lundgren, "A Wax Inhibition Solution to Problematic Fields: A Chemical Remediation Process," *Petroleum Science and Technology* 19 (2001): 205.
147. R. L. Martin, H. L. Becker, and D. Galvan, U.S. Patent Application 20070051033, 2007.
148. M. A. San-Miguel and P. M. Rodger, "The Effect of Corrosion Inhibitor Films on Deposition of Wax to Metal Oxide Surfaces," *Journal of Molecular Structure: THEOCHEM* 506 (2000): 263.
149. J. J. Hanke, "An Experimental Study on the Nature of the Film Forming Characteristics of Crude Oil Fractions on Steel Surfaces and Their Influence on Paraffin Deposition," Ph.D. thesis, Pet. Eng., University of Texas, 1967.
150. B. F. Birdwell, "Effects of Various Additives on Crystal Habit and Other Properties of Petroleum Wax Solutions," Ph.D. thesis, University of Texas, 1964.
151. S. Ahn, K. S. Wang, P. J. Shuler, J. L. Creek, and Y. Tang, "Paraffin Crystal and Deposition Control by Emulsification," SPE 93357 (paper presented at the SPE International Symposium on Oilfield Chemistry, The Woodlands, TX, 2–4 February 2005).
152. M. Feustel, H.-G. Oschmann, and U. Kentschke, U.S. Patent 6803492, 2004.
153. M. Del Carmen García, L. Carbognani, M. Orea, and A. Urbina, *Journal of Petroleum Science and Engineering* 25 (2000): 99.
154. A. R. Bishop, A. L. Ansell, W. B. Genetti, M. A. Daage, D. F. Ryan, E. B. Sirota, J. W. Johnson, and P. Brant, U.S. Patent Application 20060219597.
155. E. J. Baralt and H. Yang, U.S. Patent Application 20070095723, 2007.
156. J. B. Dobbs, "A Unique Method of Paraffin Control in Production Operations," SPE 55647 (paper presented at the SPE Rocky Mountain Regional Meeting, Gillette, WY, 15–18 May 1999).
157. D. J. Poelker, T. J. Baker, and J. W. Germer, U.S. Patent 5858927, 1999.
158. S. Gao, Y. Jiang, and A. Rodriguez, "Characterization of Oilfield Waxes Using Nuclear Magnetic Resonance Spectroscopy" (paper presented at the AICHE Annual Meeting, Philadelphia, PA, 16–21 November 2008).
159. B. Wang, C. Yu, and D. Gao, Chinese Patent CN1487048, 2004.

11 Demulsifiers

11.1 INTRODUCTION

Emulsions are colloidal dispersions, droplets of one phase dispersed in a second phase.[1-3] Crude oil is almost always produced as a water-in-oil emulsion, that is, water droplets stabilized in a continuous crude oil phase (Figure 11.1). Free-produced water may also be present depending on the water cut. The water and dissolved salts in the emulsion must be separated out before the oil is acceptable for further transportation or treatment at a refinery. This process is called "demulsification" or "dehydration." In refineries, the process of removing salty washwater from crude is called desalting. The sales specification for crude oil gives a maximum value for both water and solids in the form of a bottom solids and water value. Typically, the desired maximum water content will be in the range of 0.2–0.5%, and the acceptable maximum salt content will be 10–25 lb/1,000 bbl, although the refinery may set tighter specifications for water and salt than this.

The water separated from a water-in-oil emulsion usually contains dispersed oil as an oil-in-water emulsion (Figure 11.1). This also needs treatment (deoiling), usually with flocculants, to get the residual oil below a regulated level (often about 10–30 ppm depending on the region) before the water can be approved for discharge (see also Chapter 13).

Emulsions are formed due to turbulence in the production tubing and pipeline and especially when passing through chokes such as at the wellhead. The water-in-oil emulsions are stabilized by solid particles, resins (natural surfactants) and/or asphaltene molecules present in the crude oil.[4-7] The resin fractions of crude oil are a large family of polar molecules containing S, O, or N atoms. Soaps, formed by the interaction of oil-soluble carboxylic acids or naphthenic acids with cations in the water phase also precipitate and stabilize emulsions.[8] One study showed that contrary to well-established beliefs, compounds constituting the majority of polar components of the material collected from water-in-crude oil emulsion droplets are not asphaltenes but rather smaller, polar molecules some of which do not have an aromatic ring.[7] However, the aqueous sampling technique may have excluded asphaltenes. Film-forming corrosion inhibitors are surfactants and they also stabilize emulsions.[9] Particularly difficult emulsions can be formed as a result of acid stimulation and the back-produced aqueous fluids will thus require treatment with demulsifiers and flocculants.[10]

Solids are very instrumental in stabilizing emulsions, also by forming Pickering emulsions.[11-13] These includes fines, colloidal scale particles, corrosion products, and even precipitated wax. One study showed that solid content, not asphaltene content or any other crude oil parameter investigated, is by far the best single predictor for

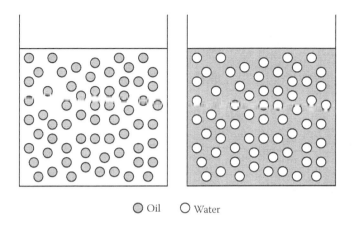

FIGURE 11.1 Oil-in-water (i.e., water continuous) (left) and water-in-oil emulsions (right). The droplets sizes are not to scale.

gauging emulsion stability.[14] Heating the emulsion at the process facilities solubilizes any wax solids and destabilizes the emulsion to some extent.[15] Heating is usually good for demulsification, but if carboxylate- or naphthenate-stabilized emulsions are present, then heating can make the emulsion worse. This is because heating causes the pH to rise (less CO_2 in the water phase) making the surfactants more polar. Besides heating the emulsion to destabilize it, the most important method of resolving the emulsion is to add a demulsifier. The concentration of the added demulsifier is usually in the range of 5–500 ppm based on the water phase. Sufficient mixing/dispersing is required to get the demulsifier to the oil-water interface, and time must be allowed for the coalesced droplets to phase separate. Any emulsion that is not broken is referred to as "slop" or "rag." The application of electric current (electrostatic coalescence) is often the final stage of emulsion treatment. It should be noted that overtreatment of many demulsifiers can restabilize an emulsion. Simple sulfonated surfactants, such as α-olefin sulfonates, are an exception. Therefore, there is an optimum dosage for a demulsifier.

A prerequisite for good demulsification is that the oil-water emulsion should be as free of gas as possible. If the crude contains significant gas, then the formation of gas bubbles will cause unnecessary agitation restricting the ability of the chemical demulsifier to produce a clean interface. Gas may be beneficially removed upstream in a dedicated gas separator.

Once the emulsion is broken and the bulk of the water has been separated out, the oil is ready for further transportation to the refinery. There is still a little water remaining in the oil. The oil should be as free from water as possible, usually less than 1%. The salt content in the residual water in the oil is also critical, as this can upset catalytic refining processes and cause corrosion and heat exchanger fouling. Therefore, fresh water is usually added to the oil to dilute the salt concentration in the residual water and the salty water separated out by the use of demulsifiers (desalters) and gravity separation. In this way, most of the salt is removed and the oil is then ready for refining. This process is known as "desalting."

The water separated out at the processing facilities contains residual oil in the form of an oil-in-water emulsion or reverse emulsion and solids. The oil and solids can be separated out mechanically or using chemical flocculants, also known as water clarifiers, deoilers, or reverse emulsion breakers. This is the subject of the next chapter.

11.2 METHODS OF DEMULSIFICATION

Gravity separation of the aqueous phase, normally at elevated temperatures, in the presence of chemical demulsifiers is the most widely used technology for demulsification of water-in-crude oil emulsions.[16-17] Other techniques such as electrostatic demulsification, hydrocyclones and centrifugation may also be deployed. Other methods such as thermal flash methods and demulsification by sonication have also been proposed.[18-19] Thermal treatment of an emulsion can help break it in two ways. First, the oil becomes less viscous on heating such that it is easier for water droplets to sink through it. Second, emulsions are usually less stable at higher temperatures. If they reach the so-called phase-inversion temperature, they will go from being water-in-oil emulsions to oil-in-water emulsions.

A microwave technology has been claimed for treatment of hard-to-treat emulsions, especially the rag layer.[20] During the separation of water and solids from crude oil or bitumen, a rag layer (or slop oil) containing a substantial amount of organic material is often formed between the water-in-oil emulsion layer and the bulk water layer. The presence of natural and added surfactants strongly influences the formation of this layer, which is difficult to treat.

11.3 WATER-IN-OIL DEMULSIFIERS

11.3.1 Theory and Practice

In offshore production, demulsifiers, particularly, are usually injected at the processing facilities before the separator. However, there may be additional benefits in injecting a demulsifier further upstream at the wellhead, or even downhole if a capillary string is available. Emulsions that are only slowly resolved will need adequate time to separate. The fluids are also hotter further upstream, which is beneficial and will also result in faster emulsion resolution. In addition, the viscosity of the multiphase fluid will be reduced if the emulsion is resolved, leading to lower drag forces and higher transportation rates.[21]

The important interfacial properties governing water-in-oil emulsion stability are shear viscosity, dynamic tension, and dilational elasticity.[22] Interfacial tension alone is not an emulsion-stabilizing factor. There have been many efforts to correlate demulsifier effectiveness with some of the physical properties governing emulsion stability.[23-24] However, our understanding in this area is still limited. Therefore, choosing a demulsifier is still largely a "black art" in which often, a range of demulsifiers and blends is taken to the process facilities and tested by trial and error. It should be underlined that testing with live emulsion at the processing facilities must be carried out to optimize the final selection of demulsifier.

There are various parameters that can be changed regarding the environment of an emulsion for it to be destabilized including viscosity, density, water-cut, emulsion age, and control of the emulsifying agents.[25] Oils with a high viscosity hold up more and larger water droplets. Raising the temperature, adding a diluent, or certain chemicals can lower the viscosity of an emulsion. Lowering the viscosity increases both the rate at which droplets settle and the mobility of water droplets, the speed at which they coalesce and separate out. Besides lowering the viscosity, heating an emulsion decreases the density of the oil at a greater rate than that of the water and thus allows more rapid settling of the water. Highly saline produced waters with higher densities should exhibit faster water droplet settling. Heavy oils are less easy to dehydrate, as their density is closer to that of water. The stability of an emulsion varies with water cut. Generally, low water cuts are harder to resolve, as the water droplets are further apart. Older emulsions tend to be more stable. Breaking an emulsion as soon as possible after its formation will be beneficial. For example, injection of a demulsifier at the wellhead may give better or faster emulsion resolution than injection further away at the processing facilities. Emulsifying agents can be natural chemicals present in the hydrocarbon phase such as asphaltenes, resins, and naphthenic acids, or they can be production chemicals added to the well stream. Film-forming corrosion inhibitors are typically surfactants, which can often stabilize emulsions. Hence, careful selection of the corrosion inhibitor may help reduce the use of demulsifiers further downstream.

The factors that control demulsification are interfacial rheology, rate of mass transfer or demulsifier molecules to the interface (to suppress interfacial tension gradients), and steric effects associated with colloids or macromolecules adsorbed at the oil-water interface.[26] A reduction in the characteristic relaxation time and the interfacial viscosity and elasticity have been directly correlated with the half-life of emulsion droplets and the effectiveness of a demulsifier. Thus, good demulsifiers exhibit short relaxation times and significantly lower the dilational viscoelasticity of the oil-water interfacial layers.[27]

There are three main processes by which demulsifiers "unlock" emulsions:[28]

1. Flocculation: Water droplets are brought together like a cluster of fish eggs.
2. Coalescence: the emulsifying film that once stabilized the water droplets in the emulsion is ruptured and the water droplets grow large enough to settle out as a separate phase. Larger droplets have less surface tension so anything that can be done to increase droplet size will help in the separation process.
3. Solids wetting: Solids that stabilize the emulsion are dispersed in the hydrocarbon phase or water-wetted and removed with the water.

A large range of combinations of flocculating demulsifiers, coalescing demulsifiers, and wetting agents together with a solvent are used in the industry. A synergistic blend of an asphaltene stabilizer (dispersant) and a primary demulsifier have been shown to perform well in breaking emulsions stabilized by asphaltenes.[29]

11.3.2 TEST METHODS AND PARAMETERS FOR DEMULSIFIER SELECTION

Bottle testing is the best way to evaluate demulsifier products but it demands a lot of time in laboratory tests and field trials, and it can sometimes give inaccurate results. There have been a number of ideas to create an easier, more accurate, and faster method to evaluate demulsifiers, providing the customer the correct products that meet their needs.

Many of the larger demulsifier suppliers present relative solubility number (RSN) values to the customer to establish a parameter for demulsifier performance selection. RSN numbers are most often used to categorize demulsifiers for second-stage desalting, for example, where you may need to know that the demulsifier is still present in the second stage and not all removed in the wash water at the primary stage separator. RSN analysis is associated with the hydrophilic and hydrophobic characteristics of a product. It is also known as water number. RSN was first established some time ago,[30] using a dioxane/benzene 96:4 blend, which was later adapted for a safer solvent system using toluene and ethylene glycol dimethyl ether (EGDE) 2.6:97.4.[31] RSN is to some extent related to hydrophilic-lipophilic balance (HLB), but the RSN determination method, interpretation, and use are somewhat different. RSN determination method is as follows: 1.0 g of the respective products (demulsifiers formulated or not) is weighed directly into a beaker already containing 30 ml of the standard solution: 2.6 % toluene and 97.4 % EGDE previously prepared. The product is dissolved using a magnetic stirring bar. Water is added like a conventional titration and the end point is when the solution achieves a visually persistent cloudy aspect. The RSN result is the water volume (in milliliters) required to achieve the cloud point. Thus, as the hydrophilicity of a product increases, so does the RSN. Very hydrophilic demulsifier products are not usually recommended because they will migrate to the aqueous phase during separation, requiring additional water treatment.

It is very important to point out that just considering RSN is not appropriate, it is also important to consider the chemical architecture of the demulsifier molecule for an appropriate product selection. The RSN is useful for product selection within a demulsifier chemical family. Comparing RSN between different chemical families may not result in a good approach in most cases of performance evaluation. Another important chemical characteristic is the water/octanol partition coefficient that is more related to water quality and treatment. It is also related to RSN and HLB.

Nearly all demulsifier testing uses the simple bottle test, which is simply a visual assessment of water drop and interface quality but (in the hands of an experienced technician) is capable of providing product blends that will work in the processing plant. Dose optimization will be finalized in the field. A range of demulsifiers from different structural classes is normally tested. The usual bottle test method is to add a dose of the demulsifier to the emulsion in bottles or measuring cylinders and observe the speed and amount of separation of oil and water. Many technicians now regularly use digital cameras to demonstrate effectiveness. In one study, 10 bottle-test performance parameters (four describing water drop, three describing oil dryness, and three describing the oil-water interface) were analyzed using several statistical methods: analysis of variance, multivariate correlations, cluster analysis, and principal component analysis.[32] Good demulsifiers are able to drop water rapidly, provide relatively clean interfaces, and produce dry, saleable oil.

There are a number of more advanced analytical techniques that can be used to characterize an emulsion or the effect of a demulsifier. Interfacial tension measurements showed that the kinetics of demulsifier adsorption at the water-crude oil interface (or, alternatively, the Gibbs elasticity modulus) is correlated to the phase separation rate: the higher the kinetics, the faster the separation rate.[33] Another method is to measure the current flow between electrodes in the emulsion sample.[34] The demulsifier producing the most rapid rate of change of current flow is identified as the most effective demulsifier in breaking the water-in-oil emulsion. The use of the dielectric constant as a criterion for screening, ranking, and selection of demulsifiers for emulsion resolution has also been investigated and was shown to be effective in screening and ranking of demulsifiers.[35] A critical electric field technique has been developed to determine demulsifier performance. It has also played a significant role in chemical demulsifier research.[36-37] Other methods for improving the evaluating of demulsifier performance have been proposed.[38]

11.3.3 CLASSES OF WATER-IN-OIL DEMULSIFIER

It is difficult to categorize classes of water-in-oil demulsifiers, as over the years, chemical suppliers have developed an ever-increasing range of products in attempts to destabilize emulsions. However, many water-in-oil demulsifiers are polymeric nonionic chemicals, many with complex comb or branched structures, with molecular weights of about 2,000–50,000. However, anionic and cationic polymers can be used depending on the emulsion-stabilizing chemicals in the feedstream or as wetting agents. The most common classes of water-in-oil demulsifier can be summarized in the following list:

- polyalkoxylate block copolymers and ester derivatives
- alkylphenol-aldehyde resin alkoxylates
- polyalkoxylates of polyols or glycidyl ethers
- polyamine polyalkoxylates and related cationic polymers (mainly for oil-in-water resolution)
- polyurethanes (carbamates) and polyalkoxylate derivatives
- hyperbranched polymers
- vinyl polymers
- polysilicones (also as demulsifier boosters)

Examples from all these categories are discussed later. A number of potentially biodegradable demulsifiers, only some of which belong to the above categories, are also discussed below. Blends of a demulsifier and a polyalkylene glycol ether have been claimed to give better emulsion breaking.[39] The latter chemical presumably acts a wetting agent on any solids present.

Most classes of water-in-oil demulsifiers are oil-soluble and are deployed as solutions in hydrocarbon solvents. The solvent for the demulsifier can have a significant

Demulsifiers

impact on emulsion resolution, and therefore, the same demulsifier in more than one solvent should be tested for performance.[15] Mixed aromatic/low-alcohol solvents are commonly used. Since oil-in-water emulsions are complex and often stabilized by more than one mechanism, many commercial demulsifiers are mixtures of two or more classes of chemical working in synergy. Oil-soluble demulsifiers can be dispersed in an aqueous solvent by adding a water-soluble surfactant. In this way, one can avoid the use of toxic and/or flammable/combustible organic solvents.[40]

Most of the listed demulsifiers above are neutral or slightly basic molecules. However, some emulsions are best treated with acidic demulsifiers usually made by derivatizing one of the above classes with a phosphorus oxide or oxyacid. Emulsions must not be overtreated with these demulsifiers since injection of an acidic chemical induces corrosion and the phosphoric acid portion of the demulsifier reacts with calcium in the brine to form hydroxyapatite and other scale solids.[41] Overtreatment can also restabilize an emulsion. As mentioned earlier, when mineral solids are present, heavy water-in-oil emulsions are difficult to treat. One study showed that adsorption of a demulsifier onto solids changes the wettability of the solids and promotes adhesion of the oil droplets to the solids, thus reducing the effective density difference between the oil and the water and hindering oil-water separation.[42]

As can be seen from the above list of demulsifiers, many classes of demulsifier contain polyalkoxylate chains. Polyalkoxylates can be made by ring-opening ethylene oxide (EO), propylene oxide (PO), butylene oxide (BO), or tetrahydrofuran (THF) using a base such as an amine or alcohol (Figure 11.2). The beauty of using this chemistry is that one can choose a wide variety of substrates with which to tether the polyalkoxylate chains, and one can easily vary the HLB of the molecule as well as the molecular weight.[43] Polyoxyethylene chains are very hydrophilic, polypropylene oxide chains mildly hydrophobic, and polyoxybutylene chains very hydrophobic. Thus, one can make a range of products using the same alcohol or amine base with varying side chains of EO, PO, and BO molecules with varying partition coefficients and interfacial activity. EO and PO are by far the most common alkylene oxides used due to their lower cost. Two studies have shown that a partition coefficient of 1.0 (equal partitioning between the oil and water phases) gave optimum demulsifier performance for a range of polyalkoxylate demulsifiers.[44-45]

Many of the polyalkoxylate classes of demulsifiers can be derivatized to produce molecules of higher molecular weight and varying HLB. For example, they can be cross-linked with multifunctional reagents such as diisocyanates, dicarboxylic acids, bisglycidyl ethers, dimethylolphenol, and trimethylolphenol. Increasing the branching at constant molecular weight has been shown to improve the demulsifier performance.[44]

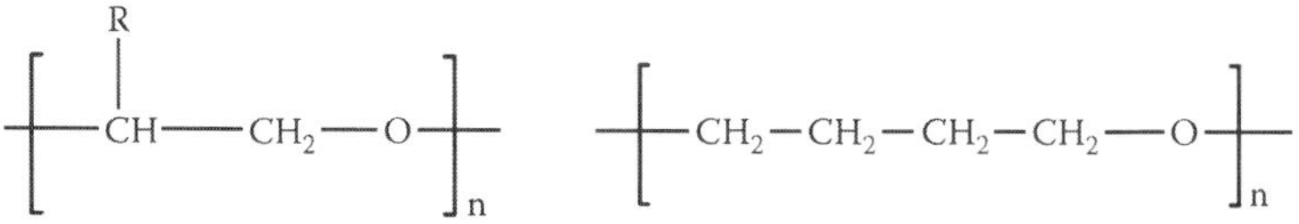

FIGURE 11.2 Structures of polyalkylene oxides (left) and polytetrahydrofuran (right). R = H (EO), R = Me (PO), R = Et (BO).

The most important classes of demulsifiers are discussed in more detail below. At the end, a section is given on patents describing efforts to make more biodegradable demulsifiers, a demand that has arisen due to environmental policies in some areas such as the North Sea basin.

11.3.3.1 Polyalkoxylate Block Copolymers and Ester Derivatives

PO can be base-catalyzed, ring-open polymerized to give polypropyleneglycols usually with a maximum molecular weight of about 4,000 Da. The hydroxyl groups on the ends of these polymers can be ethoxylated with EO to form EO/PO/EO block copolymers, which are linear demulsifiers. These copolymers are fairly poor demulsifiers by themselves.[46] The more hydrophilic EO/PO/EO block copolymers have been shown to perform better than the more hydrophobic PO/EO/PO block copolymers.[47] Higher molecular weight polypropylene glycol or polybutylene glycol (6,000 < M_w < 26,000 Da) formed by a special catalytic process can be ethoxylated to give demulsifiers with improved performance.[48] THF can be ring-open polymerized with alkylene oxides to give polyalkylene glycol block copolymers.[49] All these block polymers can be reacted with dicarboxylic acids such as maleic acid, fumaric acid, adipic acid, and aminocarboxylic acids or pyromellitic dianhydride to produce polyalkoxylate esters with even higher molecular weights and often improved performance.[44,50] Polyalkoxylate esters of maleic anhydride–oleic acid adducts have also been shown to be good demulsifiers, in some tests, better than nonyl phenol formaldehyde resin alkoxylates.[51-52]

Esters of polyalkoxylate block copolymers can be made anionic or cationic. For example, polyalkylene glycols can also be transesterified with an anionic diacid monomer or diester such as dimethyl 5-sulfoisophthalate to give anionic functionality.[53] Polycondensation of an EO/PO block copolymer, an oxalkylated fatty amine, and a dicarboxylic acid also gives demulsifiers.[54] A linear terpolymer structure results. The nitrogen atoms can be quaternized to give a cationic polymer. It is possible to create a highly branched polyesteramine by incorporation of a polyfunctional EO/PO polymer.

11.3.3.2 Alkylphenol-Aldehyde Resin Alkoxylates

Possibly the most common class of demulsifier is the alkylphenol-aldehyde resin alkoxylates (Figure 11.3). This class has been around for many decades because of their consistent high performance and ease of manufacture. Choosing different-length alkyl groups on the phenol ring can vary the hydrophobic tails. A selected alkylphenol is condensed with an aldehyde, usually formaldehyde, to form a polymer resin, which is then alkoxylated with varying amounts of EO and PO. In this way, a range of resin alkoxylates can be made. The alkylphenols are mainly synthesized isoalkylphenols substituted at the *ortho* and *para* positions, or the natural product cardanol can be used, which contains an unsaturated alkenyl group at the *meta* position.

Molecular weights for alkylphenol-aldehyde resin alkoxylates are preferably in the range of 5,000–50,000 Da. Below a molecular weight of 4,000 Da, the performance is poor.[55] The resin alkoxylates have been claimed to need a polydispersity ($Q = M_w/M_n$) of at least 1.7, preferably 1.7–5.0, for optimum demulsifier performance.[56] Early resin products produced by direct condensation reaction resulted in a too high

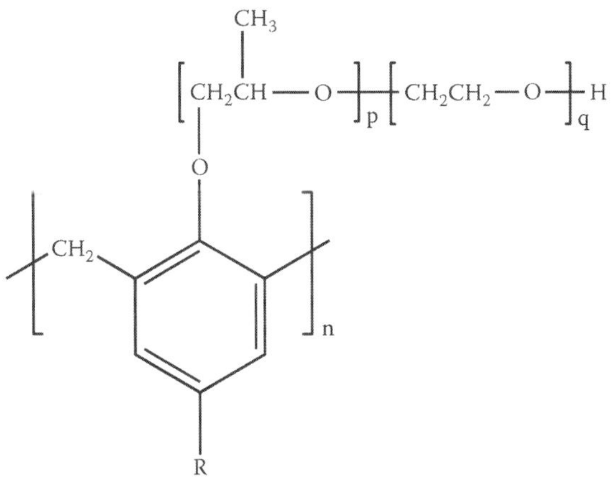

FIGURE 11.3 Alkylphenol-aldehyde resin alkoxylates.

molecular weight material due to undesirably high cross-linking in the product and did not give optimum demulsifier performance. A synthetic procedure to avoid this has been patented.[57] Such alkylphenol-aldehyde resins are usually linear. However, cyclic tetramers can be formed in high yield by using an alkylphenol and paraformaldehyde (not formaldehyde) in a nonpolar solvent such as xylene. These can then be alkoxylated to give the final products.[58]

There has been considerable discussion as to the environmental impact of alkylphenol-aldehyde resin alkoxylate demulsifiers. Alkylphenols such as nonylphenol, which are also formed by degradation of ethoxylated alkylphenols, are known to be endocrine disrupters in marine species.[59] These small molecules are believed to be present in trace quantities as unreacted monomers in the finished demulsifier or they may be formed as degradation products after discharge of the water. Alkylphenol-aldehyde resin alkoxylates are almost entirely oil-soluble and have low human toxicity. Recombinant yeast assay tests showed no link between the chemistry and potential endocrine disruption in the marine environment.[60] However, biodegraded products were not tested.

Since about the mid-nineties, research has been conducted to try to find alternative demulsifiers to the alkylphenol-aldehyde resin alkoxylates, which are not made from alkylphenols and therefore have no estrogenic activity. For example, it has been found that certain alkylphenol-free aromatic aldehyde resins, which have a functional group capable of alkoxylation, but no alkyl radical on the aromatic ring, exhibit excellent action as oil-in-water emulsion breakers and are not suspected of having a hormonelike action, although no environmental data were presented.[61–62] Typical substituents on the aromatic ring precursor can be –NHR, –COOR, –OR, or –CONHR (R = H or alkyl). Specific examples are resorcinol, hydroquinone, ethyl salicylate, *p-N,N*-dibutylaminophenol, butyl *p*-hydroxybenzoate, resorcinol octadecyl ether, and *p*-methoxyphenol.

There are many variations of alkylphenol-aldehyde resin alkoxylates that have been patented. Instead of formaldehyde, glyoxylic acid (HOOCCHO) can be used to give more hydrophilic alkylphenol aldehyde resin alkoxylates with carboxylic acid groups.[63] In contrast, increased hydrophobicity can be achieved by using an aldehyde such as benzaldehyde.[64] Dialdehydes such as glyoxal can be used to give more complex structures.[65] In another example, the hydroxyl groups in the resin alkoxylates can be esterified with vinyl monomers, such as maleic anhydride or acrylic acid, and then polymerized to form more complex demulsifiers with even higher molecular weights.[66–67] Reaction of diisocyanates with alkylphenol-aldehyde resin alkoxylates also gives a range of higher molecular weight cross-linked demulsifiers.[68] Siloxane cross-linked demulsifiers have been claimed, prepared by reacting alkylphenol-formaldehyde resin alkoxylates or polyalkylene glycols with one or more silicon-based cross-linkers.[69] An example of a cross-linker is tetraethoxy silane, $(EtO)_4Si$. Alkylphenol-aldehyde resin alkoxylates can be reacted with phosphorus pentoxide, phosphorous oxychloride, or phosphoric acid to produce acidic phosphate ester demulsifiers.[70–71] Such demulsifers are claimed to give improved performance over currently used demulsifers (at the time of patenting) by providing more rapid water separation as well as lower basic sediments and water in the shipping crude. Bis-phenols such as 2,2-bis-(4-hydroxyphenyl)-propane can be used to make more complex alkylphenol-aldehyde resin alkoxylates.[72] Diamines, such as ethylenediamine (EDA), can be included in the resin condensation process to make alkoxylated alkylphenol-formaldehyde-diamine polymers.[73] Alkoxylate resins can be reacted with ethylene carbonate to produce a range of demulsifiers.[74]

11.3.3.3 Polyalkoxylates of Polyols or Glycidyl Ethers

Branching in polyalkoxylates improves the demulsifier efficiency. Thus, compared with straight-chain EO/PO/EO block copolymers formed from glycols, alkoxylation of polyols with more than two OH groups will give a branched structure. Typical polyols are glycerol, pentaerythritol, and trimethylolpropane. Polyalkoxylates of diglycidyl ethers are well-known demulsifiers. A typical diglycidyl ether is the diglycidyl ether of bis-phenol, known as bis-phenol A (Figure 11.4). Several variations on this theme have been patented.[75–76] For example, reaction of an alkoxylated polyol with the glycidyl ether of an alkylphenol such as cardanol has been claimed to give useful demulsifiers.[77] The polyol may optionally be cross-linked before reaction with the aromatic hydrocarbon and the cross-linking agent may be a diepoxide.

Polyalkoxylates of polyols can be partially cross-linked with a vinyl monomer such as acrylic acid to increase the molecular weight and improve demulsifier efficiency. For example, an EO/PO/EO block copolymer and oxyalkylated copolymer of

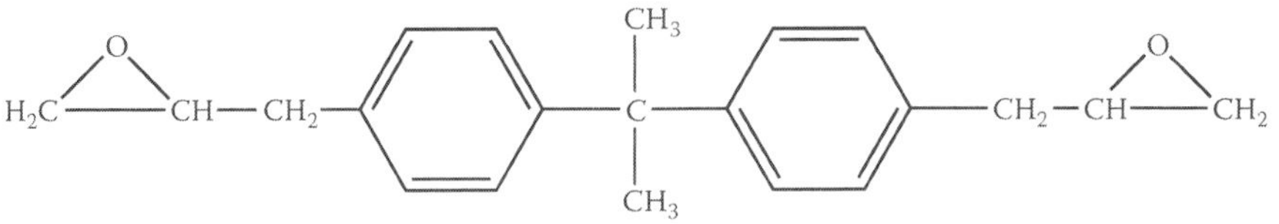

FIGURE 11.4 Bis-phenol A, a diglycidyl ether.

2-amino-2-hydroxymethyl 1,3-propanediol can be partially cross-linked with acrylic acid.[78] In another invention, epoxidized fatty acid esters can be opened with alcohols or carboxylic acids to form polyols and the resulting hydroxyl groups reacted with alkylene oxides, EO and PO.[79]

Diepoxyglycidyl or diglycidyl compounds can be used to make a number of other demulsifiers. Reaction of these compounds with amines and, optionally, a second amine-containing group, which includes a tertiary amine group, and subsequent alkylation (quaternization) leads to a range of quaternized aliphatic polyhydroxyetheramines cationic demulsifiers.[80]

11.3.3.4 Polyamine Polyalkoxylates and Related Cationic Polymers

There are a number of small polyalkyleneamines commercially available, which can be derivatized with various amounts of EO and PO to produce branched demulsifiers. Examples of polyethyleneamines are EDA, diethylenetriamine (DETA), triethylenetetraamine (TETA), and tetraethylenepentamine (TEPA) (Figure 11.5). Commercial TETA and higher polyethyleneamines also contain minor amounts of ring compounds such as aminoethylpiperidines. One study on DETA-based demulsifiers showed that roughly equal amounts of EO and PO in the side chains gave optimum performance.[81]

Polyalkyleneamines with higher molecular weights, with at least 50 recurring ethylene imine or propylene imine units, can be reacted with various amounts of EO and PO to produce demulsifiers.[82] These polyamine polyalkoxylates are claimed to perform even better when blended with isoalkylphenol formaldehyde resin alkoxylates.[83] Polyamine alkoxylates must have a polydispersity ($Q = M_w/M_n$) of at least 1.7, preferably 1.7–5.0, for optimum demulsifier performance.[56]

Products based on ring-opening reactions of epoxidized fatty acid esters with amines, diamines, or polyamines, after subsequent alkoxylation, have been shown to have an excellent breaking effect even at a very low concentration.[84] A typical product is made by reacting an amine, such as coconut amine or TEPA, with soybean oil epoxide and then alkoxylating the intermediate with various amounts of EO and PO.

Polyamines, such as DETA, can be first reacted with a dicarboxylic acid, such as adipic acid, and then alkoxylated to produce demulsifiers containing amide groups. The products can be optionally quaternized with dimethyl sulfate.[85] Quaternized cationic polymeric demulsifiers can also be manufactured from the reaction product of a polyoxyalkylene glycol, epichlorohydrin, a polyol containing at least two hydroxyl groups (one or more of which can be optionally alkoxylated), and a end-capped polyoxyalkylene diamine or triamine.[86] Smaller, quaternized fatty amine ethoxylates have also been claimed as water-in-oil demulsifiers.[87]

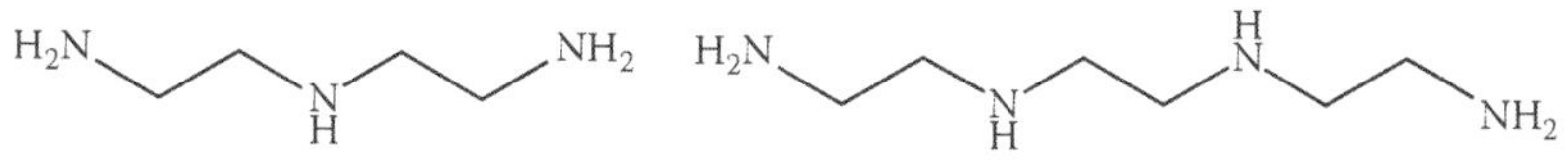

FIGURE 11.5 DETA and TETA. Each proton on the nitrogen atoms can be alkoxylated.

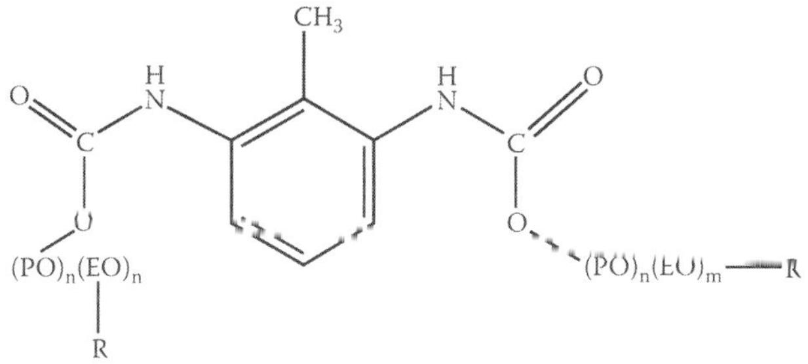

FIGURE 11.6 The structure of a polyurethane polyalkoxylate (dicarbamate) formed from 2,6-toluene diisocyanate. The 2,4 isomer can also be used.

11.3.3.5 Polyurethanes (Carbamates) and Polyalkoxylate Derivatives

Polyurethane alkoxylates are a well-known class of demulsifier.[88-89] They contain carbamate functional groups and are made by condensing polyisocyanates, such as toluene diisocyanate, and polyglycols or polyalkoxylates with terminal hydroxyl groups (Figure 11.6). If polyglycols are used, both hydroxyl end groups can react with a diisocyanate to produce a high molecular weight polyurethane demulsifier. The ratio of EO and PO in the polyglycol can be varied to obtain a range of products with different surface activities. Another modification is to use substantially EO-containing polyurethanes and attach hydrophobic groups along the hydrophilic backbone.[90]

Alkylphenol-aldehyde resin alkoxylates and polyurethanes acted synergistically when added simultaneously, rendering water separation rates significantly higher than those observed when used individually.[91] Polyurethanes aided sedimentation of water at moderate concentrations (ca. 200 ppm) by "bridging" nearby droplets, but they retarded coalescence when added at significantly higher concentrations, even when alkylphenol-aldehyde resin alkoxylates were present.

11.3.3.6 Hyperbranched Polymers

A wide variety of nondendrimeric, highly functional, hyperbranched polymers have been claimed as demulsifiers including hyperbranched polycarbonates, hyperbranched polyesters, hyperbranched polyethers, hyperbranched polyurethanes, hyperbranched polyurea polyurethanes, hyperbranched polyureas, hyperbranched polyamides, hyperbranched polyether amines, and hyperbranched polyesteramides.[92] An example of a hyperbranched polymer is made by reacting adipic acid, glycerol, and glycerol monostearate and subsequently reacting the hydroxyl groups with an alkylisocyanate.

Polyethyleneimine (PEI) is a commercial, low-cost hyperbranched polymer. Its molecular weight (M_w) can vary greatly from about 1,000–1,000,000 Da. PEI can be alkoxylated with EO, PO, and/or BO to form demulsifiers. This class performs synergistically with oxyethylated isoalkylphenol formaldehyde resins.[56]

Another class of hyperbranched polymers can be made by reacting an acrylate ester, such as methyl methacrylate, with ammonia and EDA.[93] The products, termed polyamidoamines, can be quaternized with, for example, epichlorohydrin or 2-hydroxy-3-chloropropyl trimethyl ammonium chloride to form polyamidoamines having pendent quaternary ammonium moieties.[94-95] These products function as demulsifiers for breaking or resolving emulsions of the water-in-oil type as well as oil-in-water type emulsions.

11.3.3.7 Vinyl Polymers

A wide variety of vinyl polymers have been claimed as demulsifiers. In general, they contain both hydrophilic and hydrophobic parts. This can be carried out by polymerizing vinyl monomers such as (meth)acrylic acid or maleic anhydride, hydroxyethyl(meth)acrylate, or (meth)allyl alcohol, which can be subsequently alkoxylated with EO, PO, and BO under basic conditions (Figure 11.7). Alternatively, (meth)acrylic acid or maleic anhydride can be reacted with one of the many classes of polyalkoxylates described earlier in this chapter, and the new ester monomers polymerized or grafted onto existing polymers.[96-100] Another patent claims that polymerizing a mixture of methacrylate esters, some of which are hydrophilic by ethoxylation and some are hydrophobic based on alcohols, gives good demulsifiers.[101] If maleic anhydride is reacted with polyalkoxylates such as EO/PO/EO block copolymers and polymerized with acrylic acid, the remaining carboxylic acid groups can be esterified with a second polyalkoxylate, such as the alkylphenol formaldehyde resin alkoxylates.[102]

Hydrophobic groups can also be introduced by using monomers such as vinyl carboxylate esters or acrylate esters.[103] Esterification of alkyl polyalkoxylates, such as $C_6H_{13}(EO)_6OH$, with (meth)acrylic acid and subsequent polymerization is claimed to give superior demulsifiers that give rapid water drop, zero residual emulsion, a sharp oil-water interface, clear water phase for disposal or reinjection, and low salt in crude content.[104]

Copolymers containing four different vinyl monomers with aromatic, oleophilic, ionizable, and hydrophilic groups have been claimed as demulsifiers.[105] A typical copolymer is made from styrene, an alkyl methacrylate, (meth)acrylic acid, and 2-hydroxyethyl(meth)acrylate. Various surfactants complete the formulation. The polymers can also be used as latex, that is, a stable dispersion of polymer microparticles in an aqueous medium. A specific class of water-soluble tetrapolymers

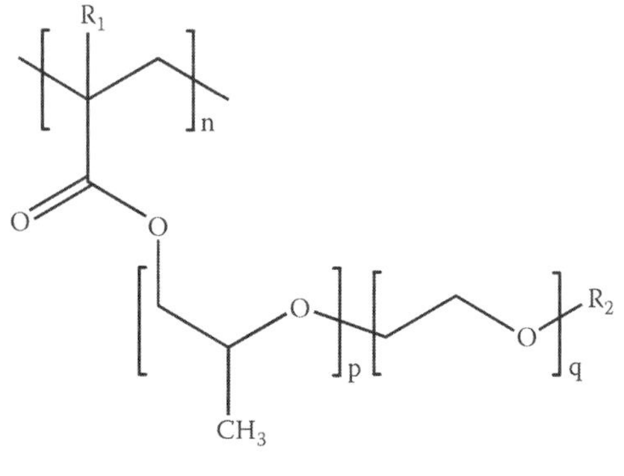

FIGURE 11.7 Alkyl polyalkoxylate (meth)acrylate ester polymers. R_1 = H or CH_3, R_2 = H or alkyl.

(M_w = 2,000–50,000) made from methyl methacrylate, butyl acrylate, acrylic acid, and methacrylic acid has been claimed as water-in-oil demulsifiers. Styrene may be tacked on to the methyl methacrylate to result in a pentapolymer.[106]

11.3.3.8 Polysilicones

Polyoxyalkylene-polysiloxane block copolymers have been claimed as primary demulsifiers.[107] Polysilicones such as dimethyl methyl(polyethylene oxide) siloxane and dimethyl siloxane, ethoxylated 3-hydroxypropyl-terminated are useful as demulsifier boosters.[108] A range of organo-modified silicone demulsifiers has been reported.[109] Polysilicones can also function as defoamers (see Chapter 12 on defoamers).

11.3.3.9 Demulsifiers with Improved Biodegradability

Since the late nineties, there has been an increasing drive towards developing more environment-friendly demulsifiers, that is, low in toxicity and high in biodegradability.[108] This has been necessary to meet the demands of the environmental authorities in environmentally strict regions such as the North Sea basin. Generally, demulsifiers are not very toxic due their high molecular weights. However, as was pointed out earlier for the alkylphenol-aldehyde resin alkoxylate demulsifiers, they may contain traces of toxic monomers or they may degrade to these or other lower molecular weight, more toxic chemicals.

Many of the traditional demulsifiers contain polyalkoxylate chains made primarily from EO and PO. Polypropoxylate chains degrade slower than polyethoxylates due to the presence of a methyl side group. Polybutoxylate chains made from ring opening of BO are even less degradable. However, straight polybutoxylate chains (polytetramethyleneglycol) made from ring opening of THF are more biodegradable. Thus, demulsifiers containing mainly polytetrahydrofurans and polyethoxylates have been claimed as biodegradable demulsifiers. Using amide or ester linkages in these polymers is a good way to introduce biodegradability. Examples are the esterified reaction products of polytetrahydrofuran, an EO/PO/EO block copolymer or a fatty amine alkoxylate, and adipic acid.[110] Twenty-eight-day BOD values of 23–52% were obtained for this class in the OECD 306 biodegradation test. Several other patents claim demulsifiers with biodegradable ester linkages.

Epoxidized fatty esters such as soya oil epoxide can also be used to synthesize potentially biodegradable demulsifiers. For example, the epoxide and a polyol containing two to six OH groups is reacted with an amine such as cocoamine or TETA and the resulting polyamine is alkoxylated with various amounts of EO, PO, or BO.[111]

Hyperbranched polymers and dendrimers discussed earlier can be significantly biodegradable if they contain ester or amide linkages. For example, polymerization of 2,2-dimethylolpropionic acid on a polyol core gives hyperbranched polyesters with many OH groups. Alkoxylation of these polymers with EO, PO, and/or BO can give a potentially biodegradable class of demulsifier.[112] Other complex ester-based demulsifiers are formed by reacting polycarboxylic acids and polyalcohols, such as the reaction of a 1:3.1 blend of esterified citric acid, triethyleneglycol, and a C12–14 alcohol.[113]

Another way of introducing more biodegradable chains into a demulsifier polymer is to use polyglycerol. Polyglycerol is available commercially as both linear polymers and hyperbranched polymers (Figure 11.8). Polyglycerols, which are subsequently

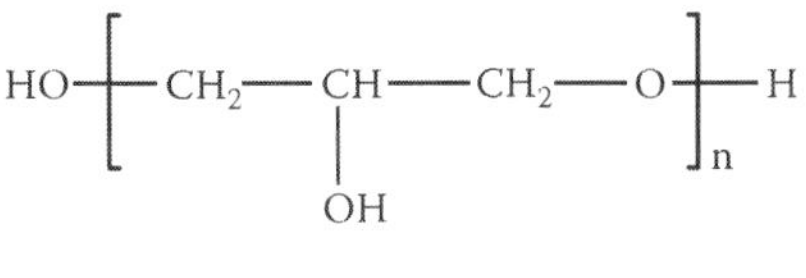

HO—[—CH$_2$—CH(OH)—CH$_2$—O—]$_n$—H

FIGURE 11.8 Linear polyglycerol.

reacted with EO and PO, have been claimed as biodegradable demulsifiers.[114] Cross-linked polyglycerols are claimed to perform even better.[115] Cross-linking can be carried out with diglycidyl ethers, dicarboxylic and polycarboxylic acids, alkylsuccinic anhydrides, alkoxyalkylsilanes, or diisocyanates.

Another class of potentially biodegradable demulsifiers is the alkoxylated alkyl polyglycosides (Figure 11.9).[116] The alkyl polyglycosides may also contain, owing to the synthesis, additional substances such as residual alcohols, monosaccharides, oligosaccharides, and oligoalkyl polyglycosides. These polymers can be optionally reacted with difunctional cross-linkers, such as diisocyanates and/or dicarboxylic acids to increase the branching and molecular weight.

Another class of demulsifier with improved biodegradability that has been claimed is that of orthoester-based polymers.[117–118] Orthoesters have the structure shown in Figure 11.10. A typical orthoester is triethyl orthoformate. These

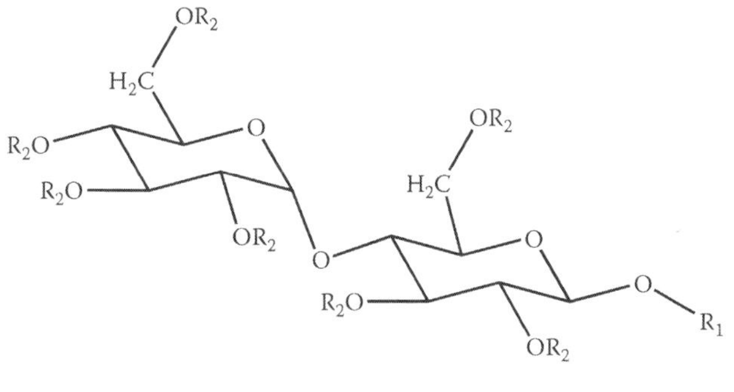

FIGURE 11.9 Example of an alkyl polyglycoside based on a β-maltose backbone. R_1 = alkyl, R_2 = polyalkylene glycol. Single six-ring molecules make up a significant percentage of the demulsifier.

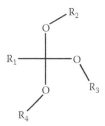

FIGURE 11.10 The structure of orthoesters.

molecule reacts with hydroxy or amino groups found in poly(ethylene glycol)s, poly(propylene glycol)s, or with some aminoalcohols to form a high-molecular-weight, potentially cross-linked polymer containing hydrophilic and hydrophobic parts.

Block, star, or branched-type amphiphilic hydroxyl polyesters have been claimed as demulsifiers with improved environmental properties.[119] The hydroxyl ester bond is made by the reaction of compounds with at least two carboxylic acid groups with a monoepoxide or the reaction of an epoxide compound, with multiepoxide groups on molecules with at least one carboxylic compound, with at least one carboxylic group. Examples of hydroxyl polyesters are the reaction products of citric acid, polyethylene glycol diacid, or carboxymethylcellulose with 2-epoxydodecane or glycidyl hexadecyl ether. Terpene alkoxylates have been claimed as demulsifiers, particularly as more environmentally friendly alternatives to nonylphenyl alkoxylates.[127]

11.3.3.10 Dual-Purpose Demulsifiers

Some demulsifiers can also have a secondary function. For example, the salt of an alkyl amine and an alkyl aryl sulfonic acid is a useful demulsifier but also has corrosion-inhibiting properties.[120] A specific example is the salt of a methyl, dicocoyl amine, and an alkyl aryl sulfonic acid. Related to these demulsifiers is the three-tailed reaction product of 2 mol of ethoxylated fatty amine and 1 mol of a sulfonated oleic acid (Figure 11.11). The product probably exists as ion pairs in nonpolar solvents.[121]

Cationic molecules and polymers designed as demulsifiers may also function as corrosion inhibitors. For example, the quaternized condensation product of 4,4′-bis(chloromethyl)diphenyl ether and tertiary dodecylamine polyglycol ether and related variations has been claimed as such.[122] Fatty amine alkoxylates have been claimed as demulsifiers, corrosion inhibitors, and/or pour-point depressants in crude oils.[123]

Sulfonic acid derivatives of alkyl aromatic compounds containing at least 2 six-rings and an alkyl chain of at least 16 carbons were first designed as asphaltene inhibitors (see Chapter 4 on asphaltene control). These molecules together with a coadditive solvent also behave as demulsifiers, possibly by stabilizing the asphaltenes, which are known to cause emulsion problems.[124] In fact, stabilizing asphaltenes can be a key to good demulsification.[125]

Compounds of the formula $H(CH_2)_z COO[C_2H_4O]_x C_y H_{2y+1}$ where z is 0–2, x is 1–5, and y is 4–9 have been claimed as demulsifiers and defoamers.[126] They may be

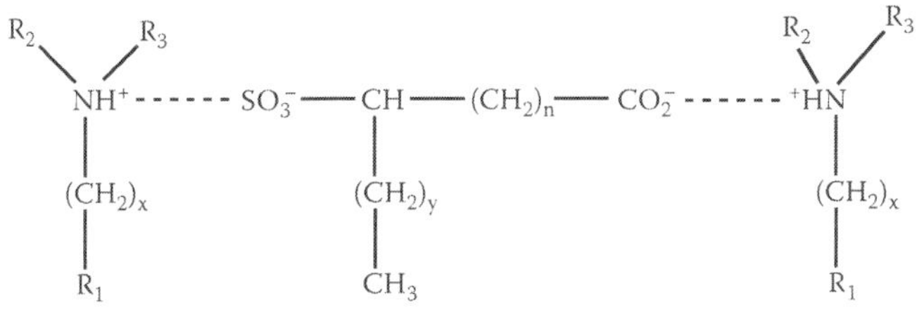

FIGURE 11.11 Three-tailed ion pair demulsifiers.

made by the simple esterification reaction of the corresponding hydroxy ethoxylated alkyl and organic acid. These compounds are also claimed to be more environmentally acceptable.

REFERENCES

1. F. Leal-Calderon, V. Schmitt, and J. Bibette, *Emulsion Science: Basic Principles*, New York: Springer, 2007.
2. L. L. Schramm, *Emulsions, Foams and Suspensions: Fundamentals and Applications*, Weinheim: Wiley-VCH, 2005.
3. P. J. Breen, D. T. Wasan, Y.-H. Kim, A. D. Nikolov, and C. S. Shetty, *Emulsions and Emulsion Stability, 2nd ed., Surfactant Science Series*, ed. J. Sjöblom, New York: Marcel Dekker, 2005, 235.
4. S. Mukherjee and A. P. Kushnick, "Effect of Demulsifiers on Interfacial Properties Governing Crude Oil Demulsification," *Oilfield Chemistry — Enhanced Recovery and Production Stimulation, ACS Symposium Series*, eds. J. K. Borchardt and T. F. Yen, Washington, DC: American Chemical Society, 1989, 364.
5. D. Arla, A. Sinquin, C. Hurtevent, and C. Dicharry, "Acidic Crude Oil Emulsions: Influence of pH and Water Cut on the Type and Stability of Emulsions" (paper presented at the 7th International Conference on Petroleum Phase Behavior and Fouling, 25–29 June 2006).
6. M. Grutters, M. van Dijk, S. Dubey, R. Adamski, F. Gelin, and P. Cornelisse, "Asphaltene Induced W/O Emulsion: False or True?," *Journal of Dispersion Science and Technology* 28 (2007): 357.
7. (*a*) J. Czarnecki and K. Moran, *Energy Fuels*, 2005, 19, 2074; (*b*) J. Czarnecki, *Energy Fuels*, 2009, 23, 1253.
8. (*a*) R. A. Rodriguez and S. J. Ubbels, "Understanding Naphthenate Salt Issues in Oil Production," *World Oil* 228(8) (2007): 143. (*b*) D. Arla, A. Sinquin, T. Palermo, C. Hurtevent, A. Graciaa, and C. Dicharry, "Influence of pH and Water Content on the Type and Stability of Acidic Crude Oil Emulsions," *Energy & Fuels* 21(3) (2007): 1337.
9. G. A. Davies, M. Yang, and A. C. Stewart, "Interactions Between Chemical Additives and Their Effects on Emulsion Separation," SPE 36617 (paper presented at the SPE Annual Technical Conference and Exhibition, Denver, CO, 6–9 October 1996).
10. D. K. Durham, S. A. Ali, and P. J. Stone, "Causes and Solutions to Surface Facilities Upsets Following Acid Stimulation in the Gulf of Mexico," SPE 29528, *SPE Production & Facilities* 12(1) (1997): 16.
11. D. Tambe, J. Paulis, and M. M. Sharma, "Factors Controlling the Stability of Colloid-Stabilized Emulsions: III. Measurements of the Rheological Properties of Colloid-Laden Interfaces," *Journal of Colloid and Interface Science* 171 (1993): 456.
12. J. J. Oren and D. M. Mackay, "Electrolyte and pH Effect on Emulsion Stability of Water-in-Petroleum Oils," *Fuel* 56 (1977): 382.
13. J. Sjöblom, P. V. Hemmingsen, A. Hannisdal, and A. Silset, "Stability Mechanisms of Crude Oil Emulsions — A Review" (paper presented at the 7th International Conference on Petroleum Phase Behavior and Fouling, 25–29 June 2006).
14. M. K. Poindexter, S. Chuai, R. A. Marble, and S. C. Marsh, "The Key to Predicting Emulsion Stability: Solid Content," SPE 93008 (paper presented at the SPE International Symposium on Oilfield Chemistry, The Woodlands, TX, 2–4 February 2005).
15. D. Graham, "Crude Oil Emulsions — Their Stability and Resolution," *Chemicals in the Oil Industry*, Royal Society of Chemistry, 1988, 155.
16. F. S. Manning and R. E. Thompson, *Oilfield Processing, Volume Two: Crude Oil*, Oklahoma City, OK: PennWell Publishing, 1995.

17. *Petroleum Extension Service, Treating Oilfield Emulsions*, 4th ed., Austin, TX: University of Texas at Austin, 1990.
18. R. Varadaraj, U.S. Patent Application 20030155307.
19. D. G. Nahmad, I. Kmiec, A. Nasir, and I. Udau, "X-O-T Technology for the Treatment of Crude Oil Emulsions SPE," SPE 115222 (paper presented at the SPE Asia Pacific Oil and Gas Conference and Exhibition, Perth, Australia, 20–22 October 2008).
20. H. H. Kartchner, U.S. Patent 6086830, 2000.
21. H. JianZhong, "Reducing the Drag Force of the Multiple Phases Flow in Gathering Lines by Injecting Demulsifiers at Wellhead," SPE 19576 (paper presented at the International Petroleum Exhibition and Technical Symposium, Beijing, China, 17–24 March 1982).
22. T. J. Jones, E. L. Neustadter, and K. P. Whittingham, "Water-in-Crude Oil Emulsion Stability and Emulsion Destabilization by Chemical Demulsifiers," *Journal of Canadian Petroleum Technology* 1978, 107.
23. S. Kokal, "Crude Oil Emulsions: A State-of-the-Art Review," SPE 77497 (paper presented at the SPE Annual Technical Conference and Exhibition, San Antonio, TX, 29 September–2 October 2002).
24. R. J. Mikula and V. A. Munoz, "Characterization of Demulsifiers," in *Surfactants: Fundamentals and Applications in the Petroleum Industry*, ed. L. L. Schramm, Cambridge, UK: Cambridge University Press, 2000, 51.
25. R. Grace, "Commercial Emulsion Breaking, in Emulsions," in *Fundamentals and Applications in the Oil Industry*, ed. L. L. Schramm, Washington, DC: ACS, 1992, 313.
26. D. Tambe, J. Paulis, and M. M. Sharma, "Factors Controlling the Stability of Colloid-Stabilized Emulsions: IV. Evaluating the Effectiveness of Demulsifiers," *Journal of Colloid and Interface Science* 171 (1995): 463.
27. Y. Wang, L. Zhang, T. Sun, S. Zhao, and J. Yu, "A Study of Interfacial Dilational Properties of Two Different Structure Demulsifiers at Oil–Water Interfaces," *Journal of Colloid and Interface Science* 270 (2004): 163.
28. G. Leopold, "Breaking Produced-Fluid and Process Stream Emulsions," in *Emulsions, Fundamentals and Applications in the Oil Industry*, ed. L. L. Schramm, Washington, DC: American Chemical Society, 1992, 341.
29. J. L. Stark and S. Asomaning, "Synergies Between Asphaltene Stabilizers and Demulsifying Agents Giving Improved Demulsification of Asphaltene-Stabilized Emulsions," *Energy & Fuels* 19 (2005): 1342.
30. H. L. Greenwald, G. L. Brown, and M. N. Fineman, "Determination of Hydrophile-Lipophile Character of Surface Active Agents and Oils by Water Titration," *Analytical Chemistry* 28(11) (1956): 1693.
31. J. Wu, Y. Xu, T. Dabros, and H. Hamza, "Development of a Method for Measurement of Relative Solubility of Nonionic Surfactants," *Colloids and Surfaces A: Physicochemical and Engineering Aspects* 232 (2004): 229.
32. M. K. Poindexter, S. Chuai, R. A. Marble, and S. C. Marsh, "Classifying Crude Oil Emulsions Using Chemical Demulsifiers and Statistical Analyses," SPE 84610 (paper presented at the SPE Annual Technical Conference and Exhibition, Denver, CO, 5–8 October 2003).
33. A. Goldszal and M. Bourrel, "Demulsification of Crude Oil Emulsions: Correlation to Microemulsion Phase Behavior," *Industrial & Engineering Chemistry Research* 39(8) (2000): 2746.
34. W. B. Allen, J. W. Harrell, and W. W. Webster, U.S. Patent 4134799, 1979.
35. J. A. Ajienka, N. O. Ogbe, and B. C. Ezeaniekwe, *Journal of Petroleum Science and Engineering* 9 (1993): 331.

36. J. H. Beetge and B. O. Horne, "Chemical Demulsifier Development Based on Critical Electric Field Measurements," SPE 93325 (paper presented at the SPE International Symposium on Oilfield Chemistry, The Woodlands, TX, 2–4 February 2005). See also SPE Journal, September 2008, 346 (same authors).
37. J. H. Beetge, "Emulsion Stability Evaluation of SAGD Product with the IPR-CEF Technique," SPE 97785 (paper presented at the SPE/PS-CIM/CHOA International Thermal Operations and Heavy Oil Symposium, Calgary, Alberta, Canada, 1–3 November 2005).
38. H.-J. Oschmann, "The DEMCON method, A New Way Forward for Evaluating Demulsifier Performance" (paper presented at the 4th International Conference on Petroleum Phase Behavior and Fouling, Trondheim, Norway, 23–26 June 2003).
39. W. Knauf, K. Oppenlander, and W. Slotman, U.S. Patent 5759409, 1998.
40. S. Radhakrishnan, S. Ananthasubramanian, and R. A. Marble, International Patent Application WO/2000/013762.
41. D. L. Gallup, P. C. Smith, J. F. Star, and S. Hamilton, "West Seno Deepwater Development Case History—Production Chemistry," SPE 92969 (paper presented at the SPE International Symposium on Oilfield Chemistry, The Woodlands, TX, 2–4 February 2005).
42. C. W. Angle, T. Dabros, and H. A. Hamza, "Demulsifier Effectiveness in Treating Heavy Oil Emulsion in the Presence of Fine Sands in the Production Fluids," *Energy & Fuels* 21(2) (2007): 912.
43. T. Balson, "The Unique Chemistry of Polyglycols," in *Chemistry in the Oil Industry VI*, Manchester, UK: Royal Society of Chemistry, 1998, 71–79.
44. P. D. Berger, C. Hsu, and J. P. Arendell, "Designing and Selecting Demulsifiers for Optimum Field Performance on the Basis of Production Fluid Characteristics," SPE 16285, *SPE Production Engineering* 3(6) (1988): 522.
45. M. A. Krawczyk, D. T. Wasan, and C. S. Shetty, "Chemical Demulsification of Petroleum Emulsions Using Oil-Soluble Demulsifiers," *Industrial & Engineering Chemistry Research* 30 (1991): 367.
46. C. R. E. Ransur, S. P. Barboza, G. Gonzalez, and E. F. Lucas, *Journal of Colloid and Interface Science* 271 (2004): 232.
47. J. Wu, Y. Xu, T. Dabros, and H. Hamza, *Colloids and Surfaces A: Physicochemical and Engineering Aspects* 252 (2005): 79.
48. G. N. Taylor and R. Mgla, U.S. Patent 5407585, 1995.
49. W. K. Langdon and R. L. Camp, U.S. Patent 4183821, 1980.
50. M. Guzmann, P. Neumann, K.-H. Büchner, and A. Oftring, International Patent Application WO/2006/134145.
51. A. M. Al-Sabagh, A. M. Badawi, and M. R. Noor El-Den, *Petroleum Science and Technology* 20 (2002): 887.
52. A. M. Al-Sabagh, N. E. Maysour, N. M. Naser, and M. R. Noor El-Din, "Synthesis and Evaluation of Some Modified Polyoxyethylene-Polyoxypropylene Block Polymer as Water-in-Oil Emulsion Breakers," *Journal of Dispersion Science and Technology* 28 (2007): 537.
53. C. W. Hahn, U.S. Patent Application 20060030491.
54. F. Staiss, R. Bohm, and R. Kupfer, "Improved Demulsifier Chemistry: A Novel Approach in the Dehydration of Crude Oil," SPE 14841, *SPE Production Engineering* 6(3) (1991): 334.
55. J. Wu, Y. Xu, T. Dabros, and H. Hamza, "Effect of Demulsifier Properties on Destabilization of Water-in-Oil Emulsion," *Energy & Fuels* 17 (2003): 1554.
56. G. Elfers, W. Sager, H.-H. Vogel, and K. Oppenlaender, U.S. Patent 5401439, 1997.
57. M. Lancaster, D. J. Moreton, and A. F. Psaila, U.S. Patent 5272226, 1993.
58. R. S. Buriks, A. R. Fauke, F. E. Mange, U.S. Patent 4032514, 1977.

59. J. Beyer, A. Skadsheim, M. A. Kelland, K. Alfsnes, and S. Sanni, "Ecotoxicology of Oilfield Chemicals: The Relevance of Evaluating Low-Dose and Long-Term Impact on Fish and Invertebrates in Marine Recipients," SPE 65039 (paper presented at the SPE International Symposium on Oilfield Chemistry, Houston, TX, 13–16 February 2001).
60. P. Jacques, I. Martin, C. Newbigging, and T. Wardell, "Alkylphenol Based Demulsifier Resins and their Continued Use in the Offshore Oil and Gas Industry," *Chemistry in the Oil Industry VII*, Royal Society of Chemistry, 2002, 30.
61. F. Holtrup, E. Wasmund, W. Baumgartner, and M. Feustel, U.S. Patent 6646016, 2003.
62. F. Holtrup, E. Wasmund, W. Baumgartner, and M. Feustel, U.S. Patent 6465528, 2002.
63. D. Leinweber and E. Wasmund, U.S. Patent Application 20040102586, 2004.
64. J. B. Byron, P. M. Lindemuth, and G. N. Taylor, U.S. Patent Application 20040266973, 2003.
65. D. Leinweber and E. Wasmund, U.S. Patent Application 20040014824, 2004.
66. K. Barthold, R. Baur, S. Crema, K. Oppenlaender, and J. Lasowski, U.S. Patent 5472617, 1995.
67. H. Diaz-Arauzo, U.S. Patent 5460750, 1995.
68. R. G. Sampson, U.S. Patent 3640894, 1972.
69. M. B. Martin, International Patent Application WO/2008/036910.
70. F. T. Lang, International Patent Application WO/2004/082604.
71. C. Myers, S. R. Hatch, and D. A. Johnson, International Patent Application WO/2006/116175.
72. K. Barthold, K. Oppenlaender, J. Lasowski, and R. Baur, U.S. Patent 4814394, 1989.
73. G. R. Meyer, U.S. Patent 20050080221.
74. W. K. Stephenson and J. D. DeShazo, U.S. Patent 5205964, 1993.
75. M. Groote and S. Kwan-Ting, U.S. Patents 2792352-6, 1957.
76. V. L. Seale, B. R. Moreland, and J. D. Shazo, U.S. Patent 3383326, 1968.
77. P. J. Breen and J. Towner, U.S. Patent 6225357, 2001.
78. A. A. Toenjes, M. R. Williams, and E. A. Goad, U.S. Patent 5102580, 1992.
79. S. Podubrin, W. Breuer, C.-P. Herold, A. Heidbreder, T. Foerster, and M. Hollenbrock, International Patent Application WO/1999/007808.
80. D. S. Treybig, D. A. Williams, and K. T. Chang, International Patent Application WO/2003/053536.
81. Y. Xu, J. Wu, T. Dabros, H. Hamza, and J. Venter, "Optimizing the Polyethylene Oxide and Polypropylene Oxide Contents in Diethylenetriamine-Based Surfactants for Destabilization of a Water-in-Oil Emulsion," *Energy & Fuels* 19 (2005): 916.
82. G. Liebold, K. Oppenlaender, E. Buettner, R. Fikentscher, and R. Mohr, U.S. Patent 3907701, 1975.
83. K. Oppenlaender, R. Fikentscher, E. Buettner, W. Slotman, E. Schwartz, and R. Mohr, U.S. Patent 4537701, 1985.
84. D. Leinweber, M. Feustel, H. Grundner, and H. Freundl, International Patent Application WO/2003/102047.
85. K. Barthold, R. Baur, R. Fikentscher, J. Lasowski, and K. Oppenlaender, U.S. Patent 4935162, 1990.
86. A. Lindert and M. S. Wiggins, U.S. Patent 6172123, 2001.
87. P. R. Hart, U.S. Patent 5250174, 1993.
88. Hoechst, British Patent 1213392, 1967.
89. R. S. Buriks, F. E. Mange, and P. M. Quinlan, U.S. Patent 3594393.
90. P. F. D. Reeve, U.S. Patent 6348509, 2002.
91. A. A. Peña, G. J. Hirasaki, and C. A. Miller, "Chemically Induced Destabilization of Water-in-Crude Oil Emulsions," *Industrial and Engineering Chemistry Research* 44 (2005): 1139.
92. B. Bruchmann, K.-H. Büchner, M. Guzmann, G. Brodt, and S. Frenzel, International Patent Application WO2006084816.
93. D. A. Tomalia and J. R. Dewald, U.S. Patent 4507466, 1985.

94. G. R. Killat, J. R. Conklin, U.S. Patent 4448708, 1984.
95. L. R. Wilson and J. R. Conklin, U.S. Patent 4457860, 1984.
96. R. S. Buriks and J. G. Dolan, U.S. Patent 4626379, 1986.
97. R. S. Buriks and J. G. Dolan, U.S. Patent 4877842, 1989.
98. K. Barthold, R. Baur, S. Crema, K. Oppenlaender, and J. Lasowski, U.S. Patent 5472617, 1995.
99. J. Fock and H. Rott, U.S. Patent 4678599, 1987.
100. D. Faul, J. Roser, H. Hartmann, H.-H. Vogel, W. Slotman, and G. Konrad, U.S. Patent 5661220, 1997.
101. C. Auschra, H. Pennewiss, U. Boehmke, and M. Neusius, U.S. Patent 6080794, 2000.
102. G. N Taylor, U.S. Patent 5609794, 1997.
103. W. K. Stephenson, U.S. Patent 4968449, 1990.
104. H. Becker, U.S. Patent 7018957, 2006.
105. J. Behles, International Patent Application WO/2007/121165.
106. B. Bhattacharyya, U.S. Patent 5100582, 1992.
107. G. Koerner and D. Schaefer, U.S. Patent 5004559, 1991.
108. C. Dalmazzone and C. Noïk, "Development of New 'Green' Demulsifiers for Oil Production," SPE 65041 (paper presented at the SPE International Symposium on Oilfield Chemistry, Houston, TX, 13–16 February 2001).
109. K. Koczo and S. Azouani, "Organomodified Silicones as Crude Oil Demulsifiers" (paper presented at the Chemistry in the Oil Industry: Oilfield Chemistry, Royal Society of Chemistry, Manchester, UK, 5–7 November 2007).
110. P. S. Newman, C. Hahn, and R. D. McClain, International Patent Application WO/2006/068702.
111. D. Leinweber, M. Feustel, H. Grundner, and H. Freundl, International Patent Application WO/2003/102047.
112. D. Leinweber, M. Feustel, E. Wasmund, and H. Rausch, International Patent Application WO/2005/003260.
113. J. Senior, L. H. Smith Sr., T. Algroy, and L. H. Smith Jr., International Patent Application WO/2004/050801.
114. D. Leinweber, F.-X. Scherl, E. Wasmund, and H. Rausch, International Patent Application WO/2002/066136.
115. D. Leinweber, F.-X. Scherl, E. Wasmund, and H. Rausch, International Patent Application WO/2004/108863.
116. R. Berkhof, H. Kwekkeboom, D. Balzer, and N. Ripke, U.S. Patent 5164116, 1992.
117. P.-E. Hellberg, "Environmentally Adapted Demulsifiers Containing Weak Links" (paper presented at the Chemistry in the Oil Industry X: Oilfield Chemistry, Royal Society of Chemistry, Manchester, UK, 5–7 November 2007).
118. P.-E. Hellberg and I. Uneback, International Patent Application WO/2007/115980.
119. W. Wang, International Patent Application WO/2008/103564.
120. R. Golden, U.S. Patent 6727388, 2004.
121. R. Varadaraj, D. W. Savage, and C. H. Brons, International Patent Application WO/2000/050541.
122. L. Heiss and M. Hille, U.S. Patent 3974220, 1976.
123. M. Hille, R. Kupfer, and R. Bohm, U.S. Patent 5421993, 1995.
124. R. Varadaraj and C. H. Brons, U.S. Patent Application 20030092779, 2003.
125. S. Kokal and J. Al-Juraid, "Reducing Emulsion Problems by Controlling Asphaltene Solubility and Precipitation," SPE 48995 (paper presented at the SPE Annual Technical Conference and Exhibition, New Orleans, LA, 27–30 September 1998).
126. K. W. Smith, J. Miller, and L. W. Gatlin, U.S. Patent Application 20050049148, 2005.
127. R. Talingting-Pabalan, G. Woodward, M. Dahanayake, and H. Adam, International Patent Application WO/2009/023724.

12 Foam Control

12.1 INTRODUCTION

Foam is a gas dispersion in a liquid or solid continuous phase.[1] Liquid foaming problems occur in many oilfield processes.[2] For example, this occurs when gas breaks out from crude oil in separators or in gas-processing plants such as amine and glycol contactors. Water systems can also foam due to chemicals or deaeration by gas or vacuum stripping. Foam in two- and three-phase separators can create several operational problems:

- poor level control that can lead to platform shutdowns
- liquid carryover in the gas outlet that can lead to flooding of downstream scrubbers and compressors
- gas carry-under in the liquid outlet that can lead to increased compression requirements

High throughput and high GOR (gas-to-oil ratio) favor foam formation.

Foams are made up of bubbles (lamellae) stabilized by surfactants, either naturally occurring surfactants (resins, asphaltenes, naphthenic acids, etc.) or added production chemicals such as some film-forming corrosion inhibitors. It is the Marangoni effect, mass transfer on or in a liquid layer due to surface tension differences, that stabilizes foams.[3] Viscosity has also been found to play a major role in determining whether a crude oil will foam under experimental conditions.[4]

12.2 DEFOAMERS AND ANTIFOAMS

Foams can be controlled by adding a defoamer or antifoam.[5] The term "antifoam" refers to a chemical that prevents or delays foam formation and "defoamer" refers to a chemical that destroys foam already formed. However, the terms often appear to be used interchangeably in the oil industry, maybe because many common defoamers are also antifoams. Thus, many antifoams not only prevent or delay foam formation but they can break already-formed foams. Oilfield defoamers and antifoams are usually used at a dosage of 1–10 ppm in crude oil separators, but there have been successful reports of their use at dosages as low as 0.1 ppm and a need for dosages as high as 100 ppm.[6-7] It is possible to overtreat with these chemicals, stabilizing the foam, so an optimum treatment level must be determined. Defoamers and antifoams displace surfactants from the gas-liquid interface of bubbles, allowing the liquid in the bubbles to coalesce and the gas to escape.[8]

The most universal characteristic of any defoamer or antifoam is the fact that it is surface active but highly insoluble in water. It is often formulated so that it will be dispersed as tiny droplets, that is, as an emulsion. There are two main classes of antifoam/defoamer used in the oil and gas industry, the first category being far the most prevalent:

- silicones and fluorosilicones
- polyglycols

Blends of the above two classes are also commercially available. They are sometimes used with hydrophobized silica particles, which have an additional destabilizing effect on foams. Fumed metal oxide particles, such as fumed silica, can also be coated with a defoamer chemical.[9] The function of such particles is to pierce the surfaces of foam bubbles, causing them to coalesce when the defoamer spreads at the interface. Some demulsifiers, such as ethoxylated fatty esters have been claimed as defoamers.[10]

As borne-out by the patent references in this chapter, defoamers and antifoams are most often laboratory tested using graduated measuring cylinders. A defoamer is tested by first making foam in the cylinder (e.g., by bubbling a gas through a liquid), adding the defoamer and then measuring the time it takes to fully break the foam. An antifoam is tested by first adding the chemical to a liquid and then bubbling gas through. The delay time in forming a certain amount of foam is recorded.

12.2.1 SILICONES AND FLUOROSILICONES

Silicones are the most commonly used antifoams for oilfield applications. Silicones are nontoxic but poorly biodegradable siloxane polymers.[11] They can be supplied as the pure product or in solvents or emulsions.[12] Mixtures with silica particles can be used for reasons discussed earlier.[13–14] The simplest silicone is poly(dimethylsiloxane) (PDMS) with the basic structure shown in Figure 12.1, although other poly(diorganosiloxane)s can be used. A polymer with peak molecular weight distribution at 15,000–130,000 Da is proposed to work best as antifoam for crude oil systems.[15] One technical disadvantage of silicone products is that there is the potential to cause damage to the surface of catalysts at the refinery, although this effect is likely to be associated with overdosing.

Derivatives of PDMS with polyoxyalkylene groups have also been claimed as antifoams.[16–18] The hydrophobicity can be varied by introducing different and varying percentages of oxyalkylene groups. Cross-linked organopolysiloxane-polyoxyalkylenes can also be used.[19] Blends of polysiloxane homopolymers and specific polysiloxane copolymers have been claimed as improved antifoams.[20] The polysiloxane copolymers are obtainable by reacting, in a first step, organopolysiloxanes, which comprise

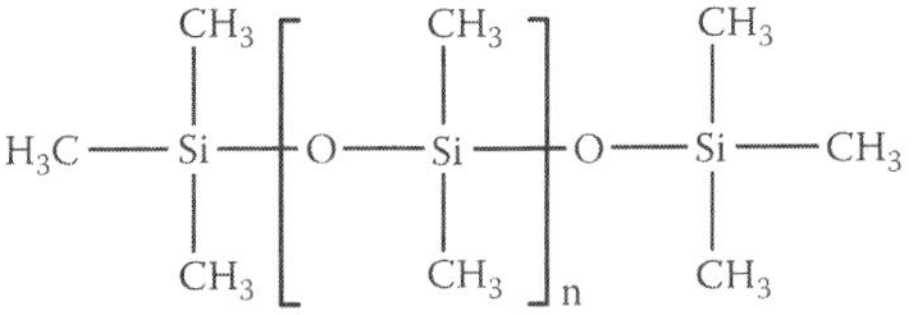

FIGURE 12.1 Structure of poly(dimethylsiloxane) (PDMS) ($n = 200$–$1,500$).

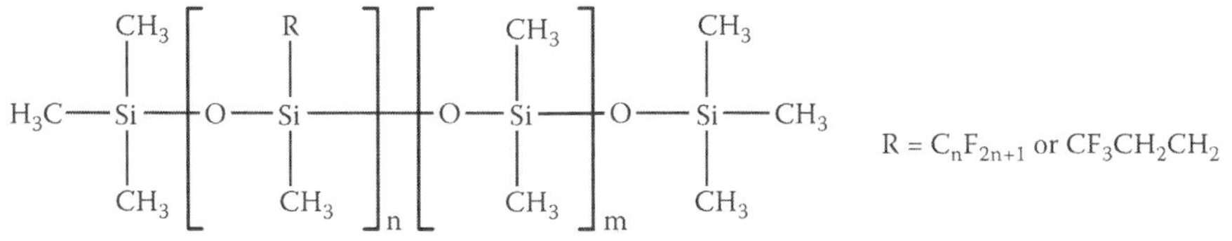

Foam Control

$$H_3C-\underset{\underset{CH_3}{|}}{\overset{\overset{CH_3}{|}}{Si}}-\left[O-\underset{\underset{CH_3}{|}}{\overset{\overset{R}{|}}{Si}}\right]_n-\left[O-\underset{\underset{CH_3}{|}}{\overset{\overset{CH_3}{|}}{Si}}\right]_m-O-\underset{\underset{CH_3}{|}}{\overset{\overset{CH_3}{|}}{Si}}-CH_3 \qquad R = C_nF_{2n+1} \text{ or } CF_3CH_2CH_2$$

FIGURE 12.2 The structure of some fluorosilicone antifoams, often containing some dimethylsiloxy groups.

at least one Si-bound hydrogen atom, with substantially linear oligomer or polymer compounds of the general formula R-(A-C_nH_{2n})$_m$-A^1-H, in which A and A^1 are groups containing oxygen and/or nitrogen atoms. Mixtures of poly(diorganosiloxane)s with polyoxyalkylene-modified silicone oils are claimed to have improved antifoaming properties, water dispersibility, and stability.[21] Other organically modified silicone oils can also be used.[14]

PDMS has some solubility in organic solvents including hydrocarbons. Its effectiveness as an antifoam in crude oil separators can often be improved by substituting some of the methyl groups for fluorinated alkyl groups.[22-25] Perfluoroalkyl groups make the fluorosilicone molecules even less soluble in hydrocarbons, thereby, improving its performance. Typical fluorosilicones are shown in Figure 12.2. 3,3,3-Trifluoropropyl siloxy groups have also been used in fluorosilicone antifoams.[26] Fluorosilicones are more expensive than simpler silicones such as PDMS, but they can usually be dosed at lower application levels.

The use of a fluorosilicone together with a nonfluorinated silicone such as PDMS has been claimed to give a synergistic blend, which performs better and at a lower dose than using either silicone alone. The blend gives both reduced liquid carry over into a gas stream and reduced gas carry under into a liquid stream in a separation process.[27]

12.2.2 Polyglycols

Polyglycols are good all-round defoamers. They are particularly useful for removing foams formed from aqueous solutions but can be applied to the crude oil separators or alternatively to glycol systems. The dose requirement is usually relatively high, for example between 10 and 200 ppm. As with all defoamers and antifoams, overdosing effects may be observed in an extreme case, such as causing foam stability and loss of control in the separator/glycol system. Nonsilicone antifoams such as the polyglycols are commonly biodegradable and low in toxicity and will not cause any damage to refinery catalyst. Polyglycols are made by adding the monomers ethylene oxide (EO), propylene oxide (PO), and/or butylene oxide (BO) to amines or alcohols. Polyglycol defoamers need to contain a high proportion of hydrophobic monomers (PO or BO) for them to have good performance. Most defoamers in this class are based on polypropoxylates, as PO is the cheaper of the two hydrophobic monomers. The polymers can be linear or branched (Figure 12.3). Some classes of polyglycols used as demulsifiers can be tailored to perform as defoamers, including alkyl phenol resin alkoxylates and diepoxides.[28]

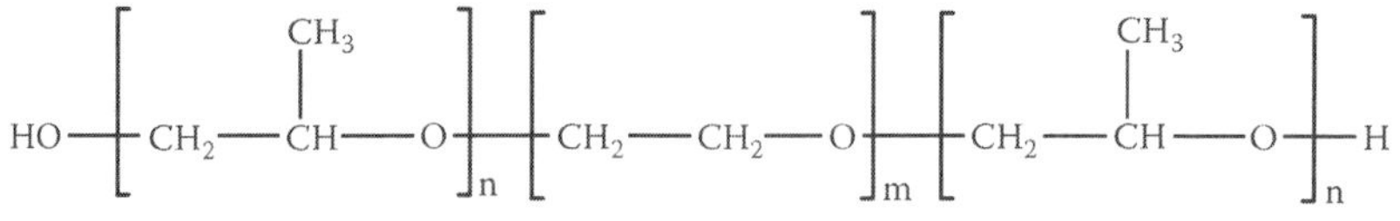

FIGURE 12.3 A linear diblock EO/PO copolymer; $n > m$ for defoamers.

Polyols can also be alkoxylated to make defoamers.[29] An example is glycerol condensed with various amounts of PO and EO, blended with aqueous 25% sodium xylene sulfonate. Other defoamer compositions are polyalkoxylated surfactants blended with a polyhydric alcohol fatty acid ester.[30]

REFERENCES

1. R. K. Prud'homme and S. A. Khan, eds., *Foams: Theory, Measurements and Applications*, Surfactant Science Series, Vol. 57, New York: Marcel Dekker, 1996.
2. L. L. Schramm, ed., *Foams: Fundamentals and Applications in the Petroleum Industry*, Advances in Chemistry Series 242, Washington, DC: ACS, 1994.
3. S. Ross, "Profoams and Antifoams," *Colloids and Surfaces A: Physicochemical and Engineering Aspects* 118 (1996): 187–192.
4. M. K. Poindexter, N. N. Zaki, P. K. Kilpatrick, S. C. Marsh, and D. H. Emmons, "Factors Contributing to Petroleum Foaming. 1. Crude Oil Systems," *Energy & Fuels* 16(3) (2002): 700–710.
5. I. C. Callaghan, "Anti-Foams for Nonaqueous Systems in the Oil Industry," in *Defoaming: Theory and Industrial Applications*, ed. P. R. Garrett, New York: Marcel Dekker, 1993, 119–150.
6. F. Cassani, P. Ortega, A. Davila, W. Rodriguez, S. A. Lagoven, and J. Seranno, "Evaluation of Foam Inhibitors at the Jusepin Oil/Gas Separation Plant, El Furrial Field, Eastern Venezuela," SPE 23681 (paper presented at the SPE Latin America Petroleum Engineering Conference, Caracas, Venezuela, 8–11 March 1992).
7. R. W. Chin, H. L. Inlow, T. Keja, P. B. Hebert, J. R. Bennett, and T. C. Yin, "Chemical Defoamer Reduction with New Internals in the Mars TLP Separators," SPE 56705 (paper presented at the SPE Annual Technical Conference and Exhibition, Houston, TX, 3–6 October 1999).
8. B. K. Jha, S. P. Christiano and D. O. Shah, "Silicone Antifoam Performance: Correlation with Spreading and Surfactant Monolayer Packing," *Langmuir* 16(26) (2000): 9947–9954.
9. D. K. Durham, J. Archer, and G. Thornton, International Patent Application WO/2007/14296.
10. K. W. Smith, J. Miller, and L. W. Gatlin, U.S. Patent Application 20050049148, 2005.
11. P. G. Pape, "Silicones: Unique Chemicals for Petroleum Processing," *Journal of Petroleum Technology* 35 (1983): 1197–1204.
12. M. J. Owen, "The Surface Activity of Silicones: A Short Review," *Industrial and Engineering Chemical Production Research and Development* 19 (1980): 97–103.
13. H. Nakahara and K. Aizawa, U.S. Patent 5153258, 1992.
14. H. Shouji and K. Aizawa, U.S. Patent 5556902, 1996.
15. I. C. Callaghan, H.-F. Fink, C. M. Gould, G. Koerner, H.-J. Patzke, and C. Weitemeyer, U.S. Patent 4557737, 1985.
16. I. C. Callaghan, C. M. Gould, and W. Grabowski, U.S. Patent 4711714, 1987.

17. I. C. Callaghan, C. M. Gould, A. J. Reid, and D. H. Seaton, "Crude-Oil Foaming Problems at the Sullom Voe Terminal," SPE 12809, *Journal of Petroleum Technology* 37(12) (1985): 2211.
18. R. A. Elms, M. A. Servinski, U.S. Patent 6512015, 2003.
19. K. C. Fey, C. S. Combs, U.S. Patent 5397367, 1995.
20. W. Burger, C. Herzig, and J. Wimmer, International Patent Application WO/2006/128624.
21. Y. Aoki and A. Itagaki, U.S. Patent 6417258, 2002.
22. H. Kobayashi and T. Masatomi, U.S. Patent 5454979, 1995.
23. R. Berger, H.-F. Fink, G. Koerner, J. Langner, and C. Weitemeyer, U.S. Patent 4626378, 1986.
24. A. S. Taylor, G.B. Patent 2244279, 1991.
25. A. S. Taylor, G.B. Patent 2244279, 1991.
26. E. R. Evans, U.S. Patent 4329528, 1982.
27. C. T. Gallagher, P. J. Breen, B. Price, and A. F. Clemmit, U.S. Patent 5853617, 1998.
28. J. H. Beetge, P. J. Venter, R. Cleary, J. Kuzyk, B. Shand, D. A. Davis, and D. W. Matalamaki, U.S. Patent Application 20060025324, 2006.
29. G. P. Sheridan, U.S. Patent 5071591, 1991.
30. J. J. Svarz, U.S. Patent 4968448, 1990.

13 Flocculants

13.1 INTRODUCTION

The water separated out at the demulsification stage at the oilfield-processing facilities contains residual oil and finely dispersed solids. The oil is present as a dispersion in water or oil-in-water emulsion (inverse emulsion). The concentration of residual oil is usually too high for discharge of the water to be allowed into the environment, plus the residual oil also has economic value.[1] For example, a new discharge standard for oil and grease in produced water in the Northeast Atlantic and North Sea area of 30 ppm (30 mg/L, previously 40 mg/L) became effective on January 1, 2007. Elsewhere, this level may be set as low as 5–10 mg/l, which can be difficult to achieve. In addition, if the water is to be reinjected, the solids may plug the pore throats in the near-well area of the injection wells, or plug filters, raising backpressures, which wastes energy, damages equipment, or can even lead to shutdown. Therefore, the water needs to be treated to remove the oil and dispersed solids. The chemical method of doing this is to add a flocculant, also called a "water clarifier," "deoiler," "oil-in-water demulsifier," "reverse emulsion breaker," or "polyelectrolyte." Flocculants can also be useful in wellbore cleanup operations.[2]

At an oil and gas production site, flocculants work in conjunction with gravity settling equipment, hydrocyclones, centrifuges, flotation devices, filtration equipment, and the like by creating a "floc" onto which oil droplets and particles are absorbed. The floc is then separated and returned to crude production. Most flocs are "sticky" and adhere to surfaces inside the equipment. After a relatively short period of time, the buildup of floc within the water-treatment system cause damage and may need to be shut down and cleaned. Preferably, the flocculant provides an "acceptable" floc, which does not cause operational problems in the system via adherence, plugging, and interface buildup. Flocculants are sometimes used upstream of hydrocyclones. In this case, they build the floc size. Hydrocyclones do not work well if the dispersed oil/solids are below about 10 μm.

Other technologies have been developed to remove oil and other contaminants from produced water.[3] One process uses a combined degassing and flotation tank for separation of a water influent containing considerable amounts of oil and gas. A rotational flow is created in the tank, which forces the lighter components such as oil and gas droplets toward an inner concentric cylindrical wall where they coalesce and rise to the surface of the liquid and are removed via the outlet whereas the heavier parts are forced down where the heavy particles sink to the lower part where they may be removed as a sludge.[4] A macroporous polymer extraction technology is now in offshore use.[5] Hydrocarbon-contaminated water is passed through a column packed with porous polymer beads that contain a specific extraction liquid. The immobilized extraction liquid removes the hydrocarbon components from the process water. Other processes use specially designed centrifuges or hydrocyclones.[6-7] Yet another process (CTour) also being used in the North Sea uses a hydrocyclone and additional liquid hydrocarbon gases (NGL), which must be available on the platform, to simultaneously extract oil including dispersed and dissolved polycyclic aromatic hydrocarbon and benzene, toluene and xylene from produced water.[8] Other nonchemical deoiling processes have been invented including an oil droplet coalescer, a cyclonic valve, and adsorbents for organic materials.[9-10] Tests on pilot-scale facility to remove polar and nonpolar organics from produced water using surfactant-modified zeolite adsorbent beds and a membrane bioreactor have been reported.[11] Organically modified clays (organoclays) can also be used for water clarification.[12]

Traditional techniques to separate oil from the water are all physical methods based on Stokes Law, prediction of the settling velocity of a sphere in a fluid, where the four main parameters are the difference of density between oil and water, the gravity force, the viscosity, and the size of the dispersed hydrocarbons. The normal equipment used with flocculants is flotation cells. Sometimes, a primary flocculant is used, the oily floc is removed, and a second treatment is carried out with a secondary flocculant. These techniques have no action on dissolved hydrocarbons. In one large Asian field, clarified water is subjected to biological oxidation for the removal of ammonia and organic carbon (BOD) in the second stage.[13]

13.2 THEORY OF FLOCCULATION

Dispersions of oil and particles in water are stabilized by a number of factors, including charge repulsion.[14] The orientation of water molecules at a hydrophobic particle surface imparts an anionic surface potential to the particle that repels similar anionic surfaces on other particles. Flocculation of these particles must overcome the charge repulsion. As the salinity of the produced water increases, the charge repulsion becomes weaker and more short-ranging. Polar molecules in crude oil, including resins and organic acids, as well as clays, scale, rust, and other polar products will be present at the oil-water interface in oil-in-water emulsions. These polar sites will be surrounded by a hydration layer that prevents oil droplets from aggregating. This is a short-range effect.

The way to get two oil droplets or two solid particles to come together is to add a flocculant and to have good mixing. The effect of the flocculant is to counterbalance the charges on the dispersed oil and particles so that they will come together (flocculate). The flocculant may also bridge particles, causing them to come together. This is usually done with a large, charged molecule, such as a high molecular weight polymer or in situ–generated polymer. Once the polymer has bridged two or more particles, the charge on the polymer is more neutralized. This causes a shift in the conformation of the polymer from being open and linear to a more coiled or globular structure. Thus, the polymer has collapsed, enveloping flocculated particles as a "floc."

13.3 FLOCCULANTS

Flocculants used in the oil industry can be summarized as follows:

- highly valent metal salts
- cationic polymers
- dithiocarbamates (DTCs; in situ–forming cationic "polymers")
- anionic polymers
- nonionic polymers
- amphoteric polymers

Surfactants or other small molecules may also be present in some formulations as enhancers. However, caution needs to be used in adding surfactants as they can just as easily stabilize as destabilize emulsions.

Highly valent metal salts have flocculation or coagulant properties due to a high positive charge density. They are not usually used alone in oilfield applications but can be added to polymer flocculants to enhance the performance. Iron(III), zinc(II), or aluminum(III) salts are the most common. They are acidic in solution due to hydrolysis and can cause corrosion problems. The hydrolyzed products, metal hydroxides, also form an unnecessary sludge. There are also restrictions on the discharge levels of some metal ions in some areas. Acids, which generate H_3O^+ ions in water, generally break oil-in-water emulsions more effectively than coagulant salts, but the resultant corrosive acidic wastewater must be neutralized after oil/water separation.

Ionic polymers (polyelectrolytes) make up the bulk of primary flocculants used in the oil industry. One service company markets DTC chemistry, which forms cationic "polymers" in situ (discussed in Section 12.3.1.5). Cationic polymers are used more than anionic polymers reflecting the charge neutralization required on usually negatively charged oily particles dispersed in water. However, anionic flocculants are known to agglomerate clays, which have a negative surface charge. In the primary oil-in-water demulsification process, a cationic polymer is usually first used. However, the charge on the dispersed oil and particles may be positive, requiring an anionic polymer, if the pH of the aqueous phase is very low, for example, after an acid stimulation treatment or if the water-in-oil demulsifier remaining in the oil-in-water emulsion is cationic. In addition, a second or even third flocculant may be needed for sufficient water clarification to meet the environmental requirements of discharged water. Nonionic polymers such as poly(ethylene oxide-b-propylene oxide), polyvinyl alcohol (PVA), and hydrophobically modified PVA have been investigated but are generally poorer flocculants than polyelectrolytes.[15–16]

The molecular weight of the ionic polymers should be fairly high (> 1,000,000 Da) for the bridging flocculation mechanism to take place effectively. However, if the molecular weight is too high, the polymer may have little mobility and impede coalescence. In addition, a high molecular weight may mean the viscosity of the aqueous polymer solution may become unmanageable. This usually limits the concentration of aqueous solutions of high molecular weight polymeric flocculants to around 5–10%.

To reduce the viscosity for injection purposes, an aqueous solution of the polymer in an invert emulsion or a latex dispersion can be used. However, invert emulsion or latex polymers add even more oil to the stream to be treated because these polymers typically include 20–30% by weight of a hydrocarbon continuous phase. Another disadvantage is that these polymer products must be inverted before use, which complicates the process of feeding the polymer into the system. Like using dry polymer, the equipment in the oil field for previous inversion is not normally available, thus forcing the direct feeding of the polymer into the system. Numerous problems associated with this feeding method have caused many customers to avoid latex polymers. In addition, the latexes generally have a very narrow treating range, often resulting in overtreatment at higher dosages. There have been many attempts to provide water-soluble, relatively high molecular weight, polymer flocculant in an aqueous composition (thereby avoiding the disadvantages of dissolving powder or dealing with the oil continuous phase) wherein the resultant composition has acceptable viscosity but much higher concentration than would be associated with high molecular weight polymer dissolved in water. These attempts involve suppressing swelling and/or dissolution of the higher molecular weight polymer by modification of the aqueous continuous phase in which it is dispersed and/or by modification of the polymer. Such products are generally referred to as "water-in-water emulsions," even though the physical state of the higher molecular weight material may not necessarily be a true emulsion. One patent claims that fairly high concentrations (at least 15%) of a high molecular weight polymer can be made by dispersing it in a solution of a low molecular weight polymer with dissolved inorganic salt.[17] A related patent uses a second polymer to disperse the polymeric flocculant in a high-concentration solution.[18] On dilution, the flocculant is made totally water-soluble. Carrying out the polymerization of the monomers in a solution of a polyvalent anionic salt, such as a phosphate or sulfate salt, is claimed to give improved handling of cationic flocculants.[19–20] A seed polymer can be added before the beginning of the polymerization for the purpose of obtaining a fine dispersion. The seed polymer is a water-soluble cationic polymer insoluble in the aqueous solution of the polyvalent anionic salt.

The charge density of the ionic polymers is also an important factor. Too much charge density can reverse the charges on the particles and restabilize them. Therefore, most ionic polymeric flocculants are copolymers, containing an ionic monomer and a neutral monomer. The ratio of the two monomers can be varied giving several potential flocculants with different charge densities in one structural class. A small amount of branching or cross-linking has been reported to improve the performance of some polymeric flocculants.[21]

13.3.1 Performance Testing of Flocculants

The bottle test (similar to the standard bottle test for crude water-in-oil emulsions) is commonly used to test flocculants. The flocculant is added to a sample of "oily" water and the capped bottle shaken. After standing for a given time, the clarity of the water is examined. Flocculant concentrations determined by this test usually exceed that required in practice in the plant. It is normal practice to start testing at a dose of 10 ppm and then increase or decrease in the range

Flocculants 323

1–50 ppm. The dose requirement is usually low compared with demulsifiers for resolving water-in-oil emulsions.

A jar test with low shear is useful for testing flocculants added to a large tank or pit. The specialized "jar test" apparatus or miniature cell attempts to duplicate the actual process when chemicals are added into a large pit, which experiences only a mild level of agitation. Stirring time and speed can varied to suit the particular application. The initial high stirring simulates the conditions close to chemical dosing but the lower stirring is representative of general plant conditions.

Many systems employ the action of gas bubbles to float both oil and solid contaminants out of the water phase. These systems usually have high amounts of agitation and create foam or froth above the water phase. Gas flotation testing is performed with a flotation test cell sometimes called a "bench Wemco" to simulate plant practice.

13.3.2 Cationic Polymers

Many types of cationic polymer (polyquaternaries) have been reported to be effective as flocculants. What the polymers have in common is that they are all made cationic by the incorporation of quaternary nitrogen atoms. They include:

- diallyldimethylammonium chloride (DADMAC) polymers
- acrylamide or acrylate-based cationic polymers
- polyalkyleneimines
- polyalkanolamines
- polyvinylammmonium chloride
- polyallylammonium chloride
- branched polyvinyl imidazoline acid salts
- cationic polysaccharides and chitosan
- condensed tannins

The first two categories are most commonly used.

13.3.2.1 Diallyldimethylammonium Chloride Polymers

The basic DADMAC polymers, such as the homopolymer polydiallyldimethylammonium chloride (poly-DADMAC), have been known, for a long time, to perform well as flocculants (Figure 13.1).[22] High molecular weight polymers (> 1,000,000 Da) are usually used. The polymers contain mainly five-ring but also some six-ring cationic groups. Various methods of making the polymers, as aqueous solutions, dispersions, or emulsions, have been claimed. An improvement on the use of the homopolymer, poly-DADMAC, is to add 5–10% by weight of residual DADMAC monomer as enhancer to the formulation.[23] DADMAC monomer can be copolymerized with acrylamide or other cheap, nonionic monomeric hydrophilic monomers to vary the charge density in the polymer.[24] A small amount of branching in some of the polymers is claimed to give improved flocculant performance.[25] Typical cross-linking agents are vinyltrimethoxysilane and methylenebisacrylamide. Hydrophobically modified DADMAC polymers are claimed to give improved

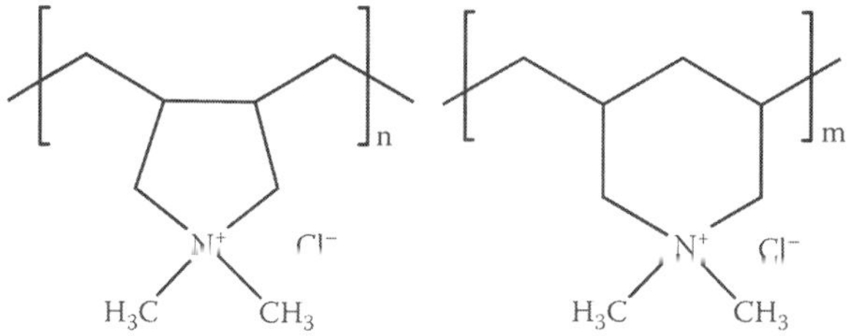

FIGURE 13.1 DADMAC polymers. The five-ring pyrolidinium monomer is the major component and the six-ring piperidinium monomer the minor component.

flocculant performance.[26] A preferred hydrophobic monomer is ethylhexylacrylate ester. The hydrophobic groups will be attracted to hydrophobic oil droplets, enhancing their flocculation. A combination of aluminum chlorohydrate and a polyamine, such as polydiallyldimethyl ammonium chloride, is claimed to give superior flocculant performance.[27]

13.3.2.2 Acrylamide or Acrylate-Based Cationic Polymers

Mannich acrylamide polymers are well-known cationic flocculants (Figure 13.2).[28–29] Generally, these polymers are homopolymers of acrylamide or copolymers thereof with comonomers such as acrylonitrile, methacrylamide, or acrylic acid in amounts up to about 50% of the resultant copolymer. The polymers have molecular weights ranging from about 10,000 to about 3,000,000 and are chemically modified by a Mannich reaction with formaldehyde and dimethylamine (or other secondary C1–8 alkylamines) to provide dimethylaminomethyl groups. These groups are modified by quaternization, for example, with dimethyl sulfate, to provide aqueous cationic polymers. Mannich acrylamide polymers, in the form of inverse microemulsions, can give superior performance.[30] It also allows the polymers to be prepared at high solids content while maintaining a very low bulk viscosity.

A variety of acrylamide or acrylate-based cationic monomers are used to make cationic polymeric flocculants (Figure 13.3). Probably the most common cationic

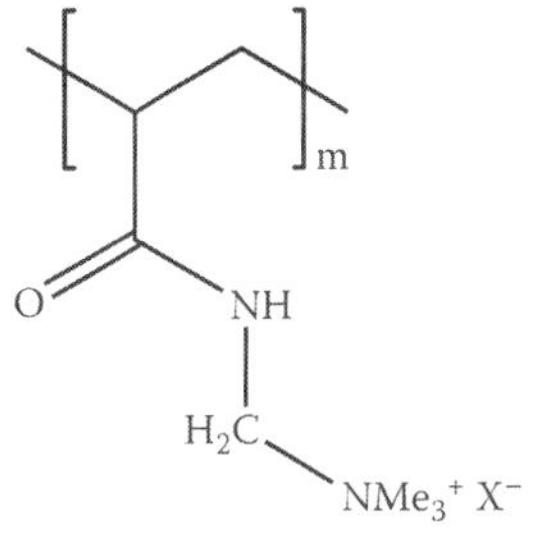

FIGURE 13.2 The active monomer structure in acrylamide Mannich copolymers.

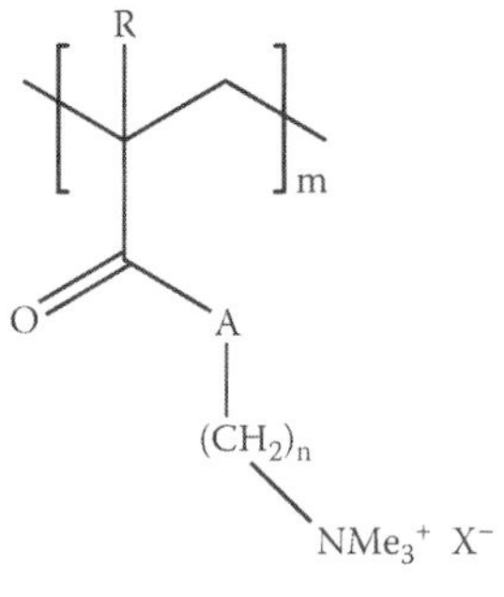

FIGURE 13.3 Structure of acrylamide or acrylate-based monomeric units found in some cationic flocculants. $A = NH$ or O, $R = H$, CH_3, $n = 2$–3.

monomers in this class of polymer are quaternary salts of dimethylaminoethyl(meth)acrylate and methylacrylamidopropyltrimethylammmonium chloride.[31-32] Another cationic polymer is based on quaternary ammonium salts of 1-acryloyl-4-methyl piperazine.[33] As with DADMAC polymers, a small amount of branching obtained by using a cross-linker can improve the performance.[25]

The cation density and effectiveness as a flocculant can be varied by copolymerization with neutral hydrophilic monomers such as acrylamide.[34] For example, a 20:80 mol% copolymer of acryloxyethyltrimethyl ammonium chloride and acrylamide with molecular weight above about 2,000,000 Da has been claimed.[35] At these high molecular weights, the viscosity of concentrated polymer solutions is very high. As explained earlier, invert emulsions or latex dispersions can be used to overcome the high viscosity. A novel method to reduce viscosity is to polymerize neutral, hydrophobic monomers such as dimethylaminoethylmethacrylate and ethyl acrylate in a water-external latex emulsion. The polymer becomes cationic and water-soluble on addition to saline solutions.[36] Dispersions of hydrophilic cationic copolymers of acrylamide in a salt media have also been claimed as easier ways of handling these otherwise high viscosity polymer solutions.[37] Di-quaternary acrylic monomers, useful to make cationic polymers, can be prepared from a vinylic tertiary amine, such as dimethylaminopropylmethacryamide, by reacting it with (3-chloro-2-hydroxypropyl)trialkylammonium chloride.[38]

Cationic polymers with a percentage of hydrophobic monomers have been claimed to give improved flocculation compared with the homopolymer. Examples are DADMAC or acrylate cationic polymers with vinyl trimethoxysilane or dimethylaminoethylacrylate benzyl chloride quaternary salt comonomers.[39-40] Other patents claim the use of lipophilic alkylacrylate comonomers with cationic acrylate or acrylamide monomers.[41] A water-dispersible terpolymer formed by polymerization of an acrylamide monomer, a water-soluble cationic monomer, and a water insoluble, hydrophobic monomer such as an alkyl(meth)acrylamide or alkyl(meth)acrylate is claimed as a superior flocculant.[42] The authors describe a possible mechanism by which the introduction of a hydrophobic monomer improves the performance of the cationic polymer. Thus, while conventional polymers can attach themselves to oil droplets by coulombic attraction, hydrogen bonding, and other undefined or not clearly understood mechanisms, the hydrophobic groups of these novel terpolymers can also be attached by a hydrophobic group–hydrophobic oil droplet association. Additionally, it may be possible that hydrophobic groups on different polymer molecules interact to form a bridge or network, which may aid in floc formation and oil flotation. While coulombic attraction still appears to be the strongest type of attraction, the hydrophobic association, or hydrophobic effect, appears to add significant strengthening to this attraction, as evidenced by improved emulsion breaking and wastewater cleanup.

13.3.2.3 Other Cationic Polymers

Diethanolamine, triethanolamines, and isopropanolamines are cheap starting materials, which can be condensed to form branched polyalkanolamines, which are useful as oil-in-water demulsifiers (Figure 13.4).[43] Triethanolamine polymers contain peripheral

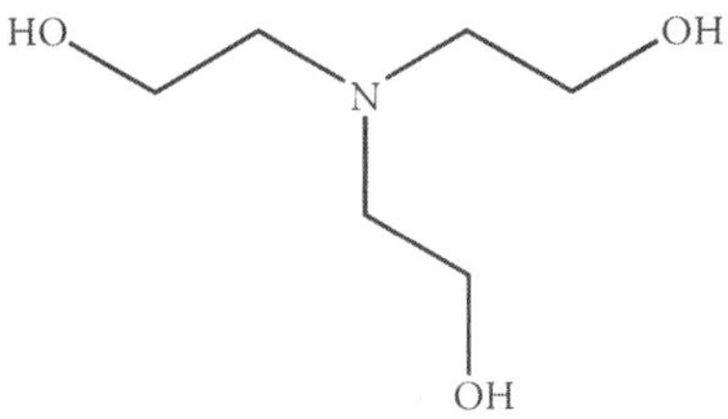

FIGURE 13.4 The structure of triethanolamine. Polymers of this molecule contain peripheral OH groups. Substituted dioxane groups are also found in the polymer.

OH groups. Substituted dioxane groups are also found in the polymer. The polymers can also be quaternized. Urea can be added to confer branching or cross-linking.[44]

Another cationic polymer class claimed as flocculants is based on the reaction of 1,2-dichloroethane with small polyamines such as 1,2-diaminoethane or diethylenetriamine (Figure 13.5).[45] These polymers used alone only work at fairly high concentrations. These polymers, and the ethanolamine polymers described above, are unusual in that the cationic charge is on the backbone and not pendant side chains. They perform well when blended with poly-DADMAC.[46]

Similar structures with the quaternary nitrogen in the backbone can be obtained by reacting epichlorohydrin and dimethylamine or larger polyamines (Figure 13.6).

Quaternized polyalkylene polyamines, such as the adduct of 2-hydroxy-3-chloropropyl trimethylammonium chloride and a polyethylene polyamine, have been claimed as flocculants.[47] Of these polyalkylene polyamines, the higher molecular weight polyethylene polyamines and polypropylene amines such as those having a number average of 100 to 15,000 Da are preferred. Of particular interest are the polyalkylene polyamines that are cross-linked with 1,2-dichloroethane or the like, as well as mixtures of such cross-linked polyamines with other polyalkylene polyamines.

Polymerized or condensed tannins have been claimed as flocculants. One such class is made from tannin, an amino compound, and an aldehyde wherein the amino compound is monoethanolamine and the aldehyde is formaldehyde.[48] Tannins consist mainly of gallic acid residues that are linked to glucose via glycosidic bonds. Condensed tannins reacted with cationic monomers such as methyl chloride quaternary salt of dimethylaminoethyl acrylate have also been claimed.[49]

Polyvinylammonium salts are cationic polymers and have been claimed as flocculants.[50–51] Polyvinylamine is commercially available and made via hydrolysis

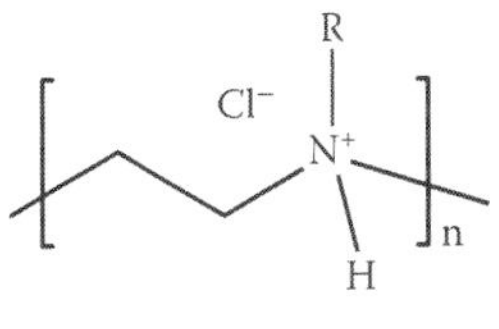

FIGURE 13.5 The polymeric reaction product of 1,2-dichloroethane with 1,2-diaminoethane.

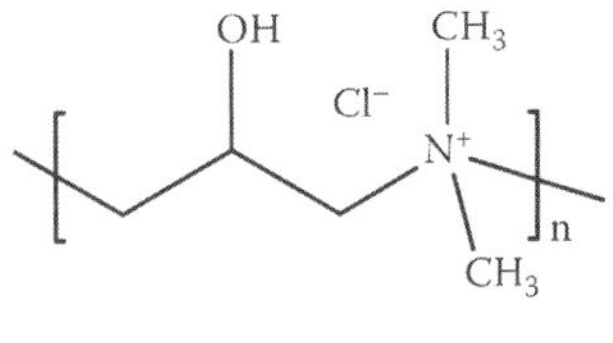

FIGURE 13.6 The polymeric reaction product of epichlorohydrin and dimethylamine.

of polyvinylformamide. Polyallylammonium chloride has also been claimed as a cationic flocculant.[52]

Various other cationic polymers have been investigated as flocculants. They include quaternized polymerized pyridines and quinolines, which can also be used as corrosion inhibitors and biocides,[53] N-diallyl-3-hydroxy azetidinium salt polymers,[54] thiazine quaternary ammonium salts of polyepihalohydrin, which can also function as biocides or water-in-oil demulsifiers,[55] branched polyvinyl imidazoline acid salts have been shown to be better flocculants than linear polymers.[21] Cationic dendritic polymers such as dendrimers and hyperbranched polymers can also be used as flocculants. These include dendritic polyamines, dendritic polyamidoamines, and hyperbranched polyethyleneimines and the reaction products thereof with gluconolactone, alkylene oxides, salts of 3-chloro-2-hydroxypropanesulfonic acid, alkyl halides, benzyl halides, and dialkyl sulfates.[56] The flocculation behaviors of three series of polycations with narrow-molecular-weight distributions carrying hydrophobic substituents on their backbones, [poly(N-vinylbenzyl-N,N,N-trimethylammonium chloride), poly(N-vinylbenzyl-N,N-dimethyl-N-butylammonium chloride), and poly(N-vinylbenzylpyridinium chloride)] has been reported. When the substrate has a low charge density, the hydrophobic interactions play a much more significant role in the flocculation process.[57]

13.3.2.4 Environment-Friendly Cationic Polymeric Flocculants

Very little seawater environmental data are available on cationic polymeric flocculants (DTCs are discussed later). Due to their large size and water solubility, they would be expected to be low in toxicity and bioaccumulation, respectively. However, any residual monomers and oligomers in the formulations will increase the toxicity. For example, residual cationic monomers will exhibit acute toxicity and residual formaldehyde and acrylamide monomer (which could occur in Mannich polymer products) are carcinogenic.[58] The DADMAC and vinyl polymers are poorly biodegradable, which is typical of polymers containing an all-carbon backbone. Polyamides made from piperazine derivatives and amines have been claimed as flocculants and may be more biodegradable due to the amide linkages although this is not claimed.[59] An example is the reaction product of N,N'-bis-(methoxycarbonylethyl)-piperazine and N,N-bis(3-aminopropyl)-methylamine or diethylenetriamine or higher polyalkyleneamine. The higher cost of the piperazine derivatives, compared with DADMAC and certain cationic vinyl monomers, may prohibit their application.

Graft polymers based on polyalkylene glycols grafted with a water-soluble ethylenically unsaturated monomer have been claimed as flocculants.[60] With a polyethyleneglycol backbone, these polymers could exhibit substantial biodegradability depending on the degree of grafting. An example is polyethyleneglycol grafted with the cationic monomer 2-(acryloyloxy)ethyltrimethylammonium chloride graft polymer. The graft polymers should have an average molecular weight of above 100,000 Da.

A class of cationic polymers that should exhibit substantial biodegradability is the cationic polysaccharides. No applications of these polymers as flocculants appear to have been reported in the oil industry but they are used in onshore wastewater treatment and are widely use in the pulp and paper industry. Polysaccharides such as starch, glycogen, glucomannan, and xanthan can be derivatized at the OH groups with

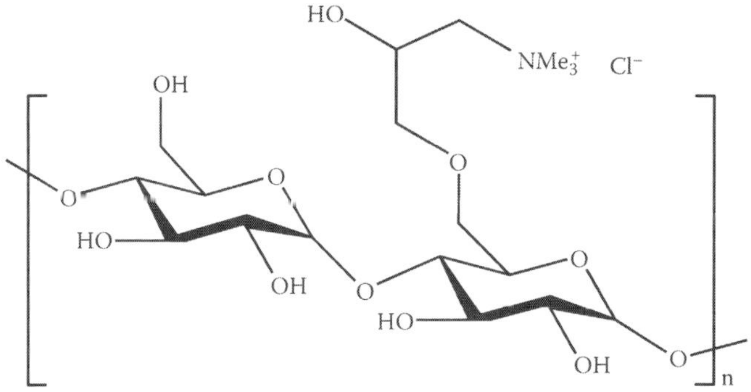

FIGURE 13.7 The structure of a cationic starch.

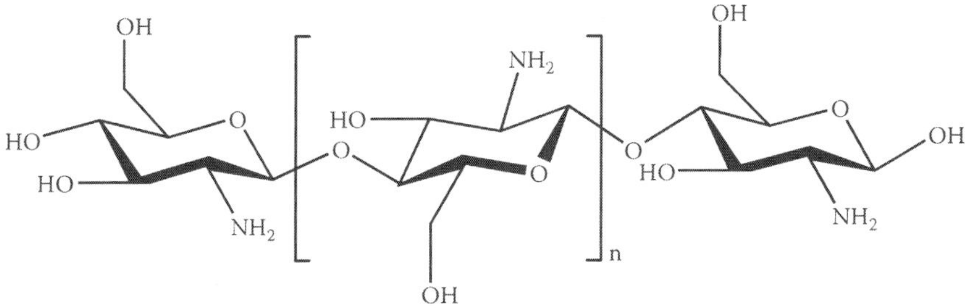

FIGURE 13.8 The structure of chitosan, fully deacetylated.

quaternary molecules such as *N*-(3-chloro-2-hydroxypropyl) trimethyl ammonium chloride or the epoxy equivalent to form cationic polysaccharides (Figure 13.7).[61–65] Alternatively, polysaccharides can be grafted with vinyl cationic monomers.[66]

Chitosan contains pendant, primary amine groups and behaves as a cationic polymer when acidified due to nitrogen quaternization (Figure 13.8).[67] Hydrophobically modified chitosan derivatives have been reported to work as improved flocculants of oil-in-water emulsions, although their cost may prohibit widespread application.[68] Chitosan is produced commercially by deacetylation of chitin, obtained from the exoskeleton of crustaceans (crabs, shrimp, etc.).

13.3.2.5 Dithiocarbamates: Pseudocationic Polymeric Flocculants with Good Environmental Properties

A class of flocculants that has advantages over the high molecular weight cationic polymers is the DTCs.[14,69] These chemicals are made by the reaction between polymeric/oligomeric primary or secondary amines with CS_2 and base in aqueous or alcoholic solution. They are fairly biodegradable and claimed to have lower acute toxicity than the cationic polymers. Being lipophilic, they will not end up in overboard water. The DTCs are low molecular weight water-soluble anionic polymers that form a high molecular weight cationic and lipophilic polymer in situ by complexing with iron(II) (ferrous) ions. The ferrous ions are usually present in sufficient

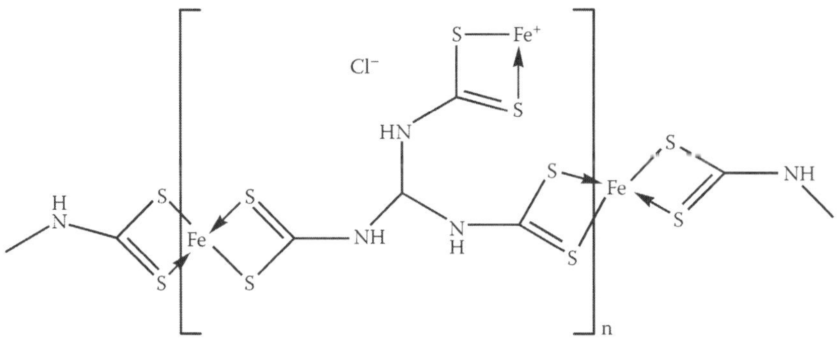

Flocculants

FIGURE 13.9 A simple DTC, cross-linked and cationized with ferrous ion.

concentration in the produced water, or they can be added.[70] A terminating agent for increasing water solubility of the polymeric matrix so-formed can be added. The terminating agent is selected from the group consisting of nonemulsifying hydrotropes containing a sulfonate or sulfate group.[71] A simple DTC structure illustrating how the polymer becomes cationic by complexing with ferrous ions is given in Figure 13.9. Since the injected polymers have low molecular weight, they can be applied at fairly high concentrations without being too viscous.[72]

The DTC flocculant technology has been improved over the years. The DTCs can be made from any suitable amine, including, but not limited to, bis(hexamethylene) triamine, hexamethylenediamine pentaethylenehexamine, polyoxyalkylenetriamines, aminoethylethanolamine, and blends of primarily triethylenetetraamine and aminoethylpiperazine.[73] Epoxy-modified DTC flocculants have also been claimed as an improvement on earlier DTC flocculants.[74] In this patent, the polyamines required to react with CS_2 are made from bisphenol A epichlorohydrin–based epoxy resins reacted with ethylenediamine. The polyamine reaction products have the structure shown in Figure 13.10. The dithiocarbamic salts made from these or other polyamines also find use as scale or corrosion inhibitors or biocides.[75]

Materials such as amines, alcohols, aminoalcohols, ethers, and mixtures thereof, including halogenated adducts thereof, have been found to be useful floc modifiers for water clarifiers to form an effective overall water clarifier composition. An improved floc is one that is easily skimmed and does not build up in the system—essentially a floc, which is easier to handle. An example of a floc modifier is ethanolamine.[76]

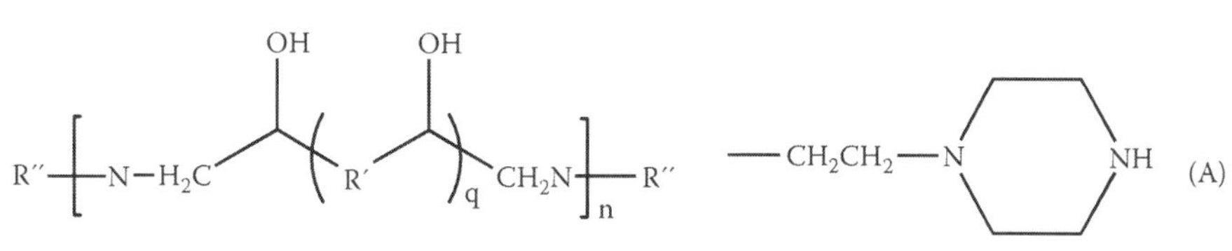

FIGURE 13.10 Epoxy-modified polyamine precursors for making DTC flocculants. R″ is selected from the group consisting of the structure: –R-NH_2 and (A), where R is a hydrocarbyl group and where R′ is –$(CH_2)_m$-O-R-O-$(CH_2)_m$–, where n and m independently range from 1 to 5 and q is 0 or 1.

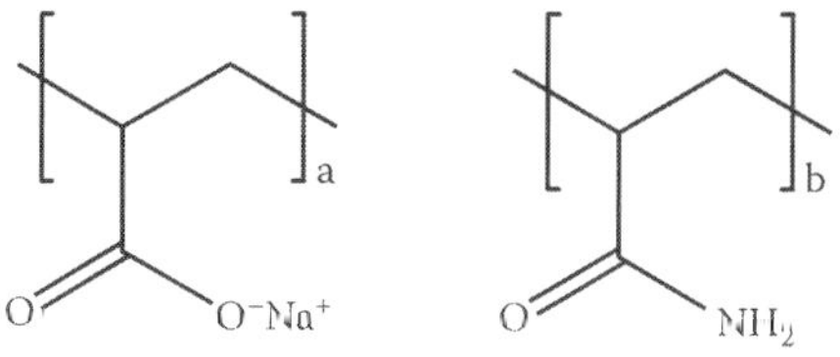

FIGURE 13.11 The structure of sodium acrylate/acrylamide copolymer, two monomer groups often found in commercial anionic flocculants.

13.3.3 ANIONIC POLYMERS

Anionic polymeric flocculants are needed if particles and oily droplets have a positive charge. This can be after an acidizing operation, or after treatment with a primary cationic flocculant to promote polymer bridging between droplets and accelerate floc formation. The market for anionic polymers is dominated by salts of high molecular weight acrylic polymers (Figure 13.11). They are most often acrylic acid copolymers with acrylamide that have been neutralized with base. Alternatively, partially hydrolyzed polyacrylamide can be used. Unhydrolyzed polyacrylamide can be used if a pure, nonionic polymer is required. An acrylamide copolymer in combination with an ionic, hydrophilic surfactant is claimed to have superior flocculant properties for oil-in-water emulsions.[77] Anionic and neutral polyacrylamides with good environmental properties have been used as flocculants in wellbore cleanup operations.[78] Polyacrylamide can be derivatized with phosphorous acid to give a phosphono-containing polymeric flocculant.[79] When fully deprotonated, the phosphono pendants are transformed into phosphonate pendants, each having two negative sites, that is, twice the anionic charge per pendant as carboxylated flocculants. These polymers ought to function also as scale inhibitors.

Poly-γ-glutamic acid has been investigated as a biodegradable anionic flocculant but no reports of its use in the oil industry exist.[80] Unlike polyaspartic acid, poly-γ-glutamic acid can be made biosynthetically in high molecular weights of up to 2,000,000 Da.[81] Phosphate modification of konjac (glucomannan), a polysaccharide, gives a biodegradable anionic flocculant.[82] Other potentially biodegradable anionic flocculants are partially hydrolyzed sodium alginate–grafted polyacrylamide and hydrolyzed polyacrylamide–grafted xanthan gum.[83–84] Other grafted polysaccharides can be used.

REFERENCES

1. E. Garland, "Discharge of Produced Water in the North Sea: Where We Are, Where We Go," SPE 97048 (paper presented at the Offshore Europe, Aberdeen, UK, 6–9 September 2005).
2. B. Holland, T. Cooksley, and C. A. Malbrel, "Polymeric Flocculants Prove Most Effective Chemicals for Wellbore Cleanup Operations," SPE 27408 (paper presented at the SPE Formation Damage Control Symposium, 7–10 February 1994, Lafayette, LA).

3. (a) J. Robinson and J. Veil, "An Overview of Offshore and Other Onshore Produced Water Treatment Technologies" (paper presented at the 14th Annual International Petroleum Environmental Conference, Houston, TX, 8 November 2007). (b) M. J. Plebon, Xuejun Chen, and Marc A. Saad, "Adapting a De-Oiling Solution to the Unique Characteristics of Produced Water" (paper presented at the 14th Annual International Petroleum Environmental Conference, Houston, TX, 8 November 2007), (c) M. Davies and P. J. B. Scott, *Oilfield Water Technology*, Houston, TX: National Association of Corrosion Engineers (NACE), 2006.
4. S. E. Oseroed, International Patent Application WO/2002/041965.
5. (a) D. Th. Meijer and C. A. T. Kuijvenhoven, "Field-Proven Removal of Dissolved Hydrocarbons from Offshore Produced Water by the Macro Porous Polymer-Extraction Technology," OTC 13217 (paper presented at the 2001 Offshore Technology Conference, Houston, TX, 30 April–3 May 2001). (b) D. Meijer, "The Removal of Toxic Dissolved and Dispersed Hydrocarbons from Oil and Gas Produced Water with the Macro Porous Polymer Extraction Technology" (paper presented at the Offshore Mediterranean Conference and Exhibition, Ravenna, Italy, 28–30 March 2007).
6. (a) Z. I. Khatib, M. S. Faucher, and E. L. Sellman, "Field Evaluation of Disc-Stack Centrifuges for Separating Oil/Water Emulsions on Offshore Platforms," SPE 30674 (paper presented at the SPE Annual Technical Conference and Exhibition, Dallas, TX, 22–25 October 1995). (b) S. E. Rye, "A New Method for Removal of Oil in Produced Water," SPE 26775 (paper presented at the Offshore Europe, 7–10 September 1993, Aberdeen, UK).
7. (a) J. C. Ditria and M. E. Hoyack, "The Separation of Solids and Liquids with Hydrocyclone-Based Technology for Water Treatment and Crude Processing," SPE 28815 (paper presented at the SPE Asia Pacific Oil and Gas Conference, 7–10 November 1994, Melbourne, Australia). (b) L. Nnabuihe, "Novel Compact Oil/Water Separator Tested in Nimr," SPE 68150 (paper presented at the SPE Middle East Oil Show, 17–20 March 2001, Bahrain). (c) A. Sinker, "Produced Water Treatment Using Hydrocyclones: Theory and Practical Application" (paper presented at the 14th Annual International Petroleum Environmental Conference, Houston, TX, 8 November 2007).
8. I. B. Henriksen, International Patent Application WO/2005/123213.
9. H. Goksoeyr, N. Henriksen, International Patent Application WO/2004/069753.
10. (a) T. Husveg, International Patent Application WO/2007/024138. (b) B. L. Knudsen, M. Hjelsvold, T. K. Frost, M. B. E. Svarstad, P. G. Grini, C. F. Willumsen, and H. Torvik, "Meeting the Zero Discharge Challenge for Produced Water," SPE 86671 (paper presented at the SPE International Conference on Health, Safety, and Environment in Oil and Gas Exploration and Production, Calgary, Alberta, Canada, 29–31 March 2004). (c) T. Arato, H. Iizuka, A. Mochizuki, T. Suzuki, H. Honji, S. Komatsu, H. Isogami, and Sasaki, U.S. Patent Application 20080023401.
11. S. Kwon, E. J. Sullivan, L. Katz, K. Kinney, and R. Bowman, "Pilot Scale Test of a Produced Water-Treatment System for Initial Removal of Organic Compounds," SPE 116209 (paper presented at the SPE Annual Technical Conference and Exhibition, Denver, CO, 21–24 September 2008).
12. G. Alther and T. Wilkinson, "Organoclays Can Cut the Cost of Cleanup of Produced Water, Wastewater and Groundwater by 50%" (paper presented at the 13th Annual International Petroleum Environmental Conference, San Antonio, TX, October 17–20, 2006).
13. E. S. Madian and R. J. Jan, "Treating of Produced Water at the Giant Arun Field," SPE 27130 (paper presented at the SPE Health, Safety and Environment in Oil and Gas Exploration and Production Conference, Jakarta, Indonesia, 25–27 January 1994).
14. P. R. Hart, "The Development and Application of Dithiocarbamate (DTC) Chemistries for Use as Flocculants by North Sea Operators," Proceedings of the Chemistry in the Oil Industry VII Symposium, Royal Society of Chemistry, Manchester, UK, 2002, p. 149.

15. R. S. Fernandes, G. Gonzalez, and E. F. Lucas, *Colloid and Polymer Science* 283 (2005): 375.
16. A. L. Feder, G. Gonzalez, C. L. S. Teixeira, and E. F. Lucas, *Journal of Applied Polymer Science* 94(4) (2004): 1473.
17. M. S. Ghafoor, M. Skinner, and I. M. Johnson, U.S. Patent 6001920, 1999.
18. J. W. Sparapany and J. R. Hurlock, U.S. Patent 5938937, 1999.
19. J. R. Hurlock, U.S. Patent 6025426, 2000.
20. N. E. Byrne, R. A. Marble, and M. Ramesh, U.S. Patent 5330650, 1994.
21. T. Higashino and S. Shimosato, U.S. Patent 6890996, 2005.
22. W. E. Hunter and T. P. Sieder, U.S. Patent 4151202, 1979.
23. M. Hofinger, M. Hille, and R. Bohm, U.S. Patent 4686066, 1987.
24. A. Sivakumar and P. G. Murray, U.S. Patent 6036868, 2000.
25. W. L. Whipple, C. Maltesh, C. C. Johnson, A. Sivakumar, T. M. Guddendorf, and A. P. Zagala, U.S. Patent 6753388, 2004.
26. M. Ramesh and A. Sivakumar, U.S. Patent 5635112, 1997.
27. P. R. Hart, U.S. Patent 5607574, 1997.
28. A. T. Coscia and M. N. D. O'Connor, U.S. Patent 4137164, 1979.
29. T. V. Vyshkina, U.S. Patent, 5744563, 1998.
30. J. J. Kozakiewicz and S.-Y. Huang, U.S. Patent 5723548, 1998.
31. M. S. Raman, U.S. Patent 4160742, 1979.
32. P. Flesher, D. Farrar, M. Hawe, and J. Langley, U.S. Patent 4702844, 1987.
33. D. W. Fong and A. M. Halverson, U.S. Patent 4802992, 1989.
34. R. S. Buriks, A. R. Fauke, and D. W. Griffiths, U.S. Patent 4224150, 1980.
35. P. R. Hart, J. M. Brown, and E. J. Connors, U.S. Patent 5730905, 1995.
36. M. L. Braden and S. J. Allenson, U.S. Patent 4931191, 1990.
37. (*a*) J. R. Hurlock U.S. Patent 6025426, 2000. (*b*) S.-Y. Huang, L. Rosati, and J. J. Kozakiewicz, U.S. Patent 6702946, 2004.
38. L. Z. Liu, J. D. Kiplinger, and D. Radtke, International Patent Application WO/2008/118315.
39. M. Ramesh and A. Sivakumar, U.S. Patent 5635112, 1997.
40. A. Sivakumar and M. Ramesh, U.S. Patent 5560832, 1996.
41. P. R. Hart, F. Chen, W. P. Liao, and W. J. Burgess, U.S. Patent 5921912, 1999.
42. J. Bock, P. L. Valint, T. J. Pacansky, and H. W. H. Yang, U.S. Patent 5362827, 1994.
43. T. J. Bellos, U.S. Patent 4459220, 1984.
44. R. Fikentscher, K. Oppenlaender, J. P. Dix, W. Sager, H.-H. Vogel and G. Elfers, U.S. Patent 5234626, 1993.
45. C. W. Burkhardt, U.S. Patent 4411814, 1983.
46. B. Lehmann and U. Litzinger, U.S. Patent 5707531, 1998.
47. B. S. Fee, U.S. Patent 4387028, 1983.
48. J. E. Quamme and A. H. Kemp, U.S. Patent 4558080, 1985.
49. P. R. Hart, J.-C. Chen, F. Chen, and T. H. Duong, U.S. Patent 5851433, 1998.
50. P. L. Dubin, U.S. Patent 4217214, 1980.
51. A. G. Sommese and A. Sivakumar, U.S. Patent 5702613, 1997.
52. D. N. Roark, U.S. Patent 4614593, 1986.
53. P. M. Quinlan, U.S. Patent 4339347, 1982.
54. R. S. Buriks and E. G. Lovett, U.S. Patent 4383926, 1983.
55. P. M. Quinlan, U.S. Patent 4331554, 1982.
56. M. B. Manek, M. J. Howdeshell, K. E. Wells, H. A. Clever, and W. K. Stephenson, U.S. Patent Application 20060289359, 2006.
57. S. Schwarz, W. Jaeger, B.-R. Paulke, S. Bratskaya, N. Smolka, and J. Bohrisch, *J. Phys. Chem. B*, 2007, 111(29), 8649.
58. J. M. Rice, "The Carcinogenicity of Acrylamide," *Mutation Research/Genetic Toxicology and Environmental Mutagenesis* 580 (2005): 3–20.

59. U.-W. Hendricks, B. Lehmann, and U. Litzinger, European Patent Application EP691150, 1996.
60. M. Singh, B. Dymond, A. Hooley, and K. Symes, International Patent Application WO/2006/050811.
61. S. Pal, D. Mal, and R. P. Singh, *Colloids and Surfaces A: Physicochemical and Engineering Aspects* 289 (2006): 193,
62. S. Bratskaya, S. Schwarz, T. Liebert, and T. Heinze, "Starch Derivatives of High Degree of Functionalization: 10. Flocculation of Kaolin Dispersions,"*Colloids and Surfaces A: Physicochemical and Engineering Aspects* 254 (2005): 75–80.
63. S. Pal, D. Mal, and R. P. Singh, "Cationic Starch: An Effective Flocculating Agent,"*Carbohydrate Polymers* 59 (2005): 417.
64. L. Järnström, L. Lason, and M. Rigdahl, "Flocculation in Kaolin Suspensions Induced by Modified Starches 1. Cationically Modified Starch—Effects of Temperature and Ionic Strength," *Colloids and Surfaces A: Physicochemical and Engineering Aspects* 104 (1995): 191.
65. E. Gunn, A. Gabbianelli, R. Crooks, and K. Shanmuganandamurthy, International Patent Application WO/2006/055877.
66. D. W. Fong and A. M. Halverson, U.S. Patent 4568721, 1986.
67. A. Pinotti, A. Bevilacqua, and N. Zaritzky, *Journal of Surfactants and Detergents* 4 (2001): 57.
68. S. Bratskaya, V. Avramenko, S. Schwarz, and I. Philippova, *Colloids and Surfaces A: Physicochemical and Engineering Aspects* 275 (2006): 168.
69. N. E. S. Thompson and R. G. Asperger, U.S. Patent 4689177, 1987.
70. T. J. Bellos, U.S. Patent 6019912, 2000.
71. T. J. Bellos, U.S. Patent 6130258, 2000.
72. D. K. Durham, "Advances in Water Clarifier Chemistry for Treatment of Produced Water on Gulf of Mexico and North Sea Offshore Production Facilities," SPE 26008 (paper presented at the SPE/EPA Exploration and Production Environmental Conference, San Antonio, TX, 7–10 March 1993).
73. D. K. Durham, U. C. Conkle, and H. H. Downs, U.S. Patent 5006274, 1991.
74. G. T. Rivers, U.S. Patent 5247087, 1993.
75. N. E. S. Thompson and R. G. Asperger, U.S. Patent 5089619, 1992.
76. E. J. Evain, H. H. Downs, and D. K. Durham, U.S. Patent 5302296, 1994.
77. P. R. Hart, J. M. Brown, and E. J. Connors, Canadian Patent CA2156444, 1996.
78. M. N. M. Yunus, A. D. Procyk, C. A. Malbrel, and K. L. C. Ling, "Environmental Impact of a Flocculant Used to Enhance Solids Transport during Well Bore Clean-Up Operations," SPR 29736 (paper presented at the SPE/EPA Exploration and Production Environmental Conference, Houston, TX, 27–29 March 1995).
79. L. E. Nagan, U.S. Patent 5393436, 1995.
80. H. Yokoi, T. Arima, J. Hirose, S. Hayashi, and Y. Takasaki, *Journal of Fermentation and Bioengineering* 82 (1996): 84.
81. I.-L. Shih and Y.-T. Van, "The Production of Poly-(Gamma-Glutamic Acid) from Microorganisms and Its Various Applications," *Bioresource Technology* 79 (2001): 207–225.
82. C. Xie, Y. Feng, W. Cao, Y. Xia, and Z. Lu, "Novel Biodegradable Flocculating Agents Prepared by Phosphate Modification of Konjac," *Carbohydrate Polymers* 67 (2007): 566.
83. T. Tripathy and R. P. Singh, *European Polymer Journal* 36 (2000): 1471.
84. P. Adhikary and R. P. Singh, *Journal of Applied Polymer Science* 94 (2004): 1411.

14 Biocides

14.1 INTRODUCTION

Biocides, also called "bactericides" or "antimicrobials," are used in oil and gas production. Their aim is to kill microorganisms, especially bacteria, or interfere with their activity. Microorganisms in oilfields or in injection water are generally classified by their effect. Sulfate-reducing bacteria (SRBs), denitrifying bacteria (hNRB), slime-forming bacteria, iron-oxidizing bacteria, and miscellaneous organisms such as algae, sulfide-oxidizing bacteria (NR-SOB), yeast and molds, and protozoa can be encountered in bodies of water of oilfields to be treated.[1] Even carbonate-scale–forming bacteria have been observed in a Middle East field.[2] Bacteria can be found in solution (planktonic), as dispersed colonies or immobile deposits (sessile bacteria plus their waste products). Bacteria can utilize a wide variety of nitrogen, phosphorus, and carbon compounds (such as organic acids) to sustain growth. Nitrogen and phosphorus are usually sufficiently present in the formation water to sustain bacterial growth but injection of organic nitrogen- and phosphorus-containing chemicals can increase the growth potential.

Bacteria can be aerobic or anaerobic bacteria such as the SRB (desulfovibrio), which are present in all waters handled in oilfield operations. SRB convert sulfate ions to hydrogen sulfide leading to reservoir souring. Hydrogen sulfide is acidic and can in turn cause sulfide scales, most importantly, iron sulfides.[3-4] Solid deposits of bacterial colonies are called "biofilms" or "biofouling." The presence of iron sulfide or an increase in the water-soluble sulfide concentrations in a flowline is a strong indicator for microbially induced corrosion (MIC).[5-6] Therefore, it is very important to prevent the formation of biofilms on the surfaces of pipelines and vessels and to have viable treatment strategies for both planktonic and sessile bacteria numbers. The potential for SRB activity is greater in the case of produced water reinjection (PWRI). Water that is reinjected can be a mixture of produced water and seawater. In such cases, one has a mixture of SRB nutrients including sulfate ions, organic carbon, ammoniacal nitrogen, and low temperatures. There are SRB that can survive extremes of temperature, pressure, salinity, and pH but their growth is particularly favored in the temperature range of 5–80°C (41–176°F).

Bacteria are very small (approximately 1.5 µm^3), but have the largest surface area to volume ratio of any life form. As a result, by providing interfaces for sorption of metal cations, bacteria are efficient scavengers of dilute metals and can concentrate them from the surrounding aqueous environment.[7] This is mainly due to the overall anionic charge of bacterial surfaces imparted by the macromolecules, which make up their fabric. Once metal ions have interacted with the electronegative sites on these molecules, they nucleate the formation of fine-grained minerals using anions from the external milieu as counterions for additional metal complexation. Diverse mineral types, including carbonates and sulfates/sulfides are commonly found in oilfield biofilms. Even radioactive uranium salts have been found in biofilms in water injectors in the Middle East together with iron sulfide.[8]

Biocides are used both upstream and downstream of the separator during oil and gas production.[9] Other upstream uses include drilling and fracturing operations as well as well treatments to reduce hydrogen sulfide production and sulfide scales, reduce biofouling, and corrosion, and thus improve well productivity.

The greatest oilfield use of biocides is in raw seawater injection projects used for pressure support and enhanced oil recovery.[9] SRBs present in the seawater, and hydrogen sulfide formed by them, can cause significant damage to water injection systems by corroding deep pits and holes that can completely penetrate the pipe walls. Oxygen must also be removed from the seawater to prevent oxygen corrosion. This is done by first passing the seawater through a deaerator tower and then adding oxygen scavengers to remove the last traces of oxygen. Upstream of the deaerator, the seawater is treated with an oxidizing biocide. This is usually chlorine, generated electrochemically from the seawater, and present mainly as the hypochlorite ion at pH 8–9. The electrochlorinator controls bacterial growth upstream of the deaerator, but surviving bacteria will pass through to contaminate the system downstream. In addition, the oxygen scavenger, which is usually a bisulfite salt, will react with residual chlorine leaving no biocide in the water. Therefore, a nonoxidizing organic biocide is injected, usually in batch doses.[10] If the organic biocide is known to cause foam problems, it will be injected downstream of the deaerator tower. However, this will leave the deaerator uninhibited.

SRBs were once thought to only be able to metabolize organic acids and alcohols besides sulfate ions. However, more recent studies have shown that saturated hydrocarbons and even toluene can be metabolized by some strains of SRB. Hence, the degree of SRB production and reservoir souring may be greater than first anticipated based on available organic acids/alcohol SRB nutrients. The occurrence of different strains of SRB may also mean that not all biocide classes may work against all strains of SRB.[11–12]

There are five basic methods to minimize reservoir souring:

1. Add a biocide to kill SRB.
2. Treat the SRB with a biostat (control biocide or metabolic inhibitor) that controls SRB growth.
3. Stimulate the formation of nitrate-reducing sulfide-oxidizing bacteria (NR-SOB) by adding nutrients such as nitrate ions. This uses up the carbon-based nutrients, forms nitrite control biocide, and thereby inhibits SRB growth.
4. Use unsulfated aquifer or desulfated seawater in water injector wells.
5. Use a H_2S scavenger (see chapter on H_2S scavengers).

Biocides

The concentration of sulfate ions in seawater can be greatly reduced using membrane technology. This will also reduce potential sulfate scaling in the production wells. Phosphorus compounds, such as phosphates, are also needed as nutrients for SRB growth but are not specifically used by the SRB to generate hydrogen sulfide. It has been claimed to remove water-soluble phosphorus compounds from injected seawater using membrane technology to minimize SRB growth.[13] Each of the one or more separation membranes is preferably either a reverse osmosis membrane or a nanofiltration membrane.

Ultraviolet radiation is another bactericidal method that has been investigated. However, it is not proven that it kills all SRBs, so a secondary chemical bactericidal treatment is needed. Methods 1–4 above are usually carried out at the injector wells while H_2S scavengers are injected into the production stream either downhole or topside. Methods 2 and 3 can be carried out simultaneously using nitrate treatment and are, therefore, discussed together under the section on control biocides.

14.2 CHEMICALS FOR CONTROL OF BACTERIA

Chemicals for control of bacteria in oilfield applications can be divided into two main classes:

- Biocides (oxidizing and nonoxidizing/organic)
- Biostats (control "biocides" or metabolic inhibitors)

Biocides kill bacteria at normal use concentrations. Biostats do not kill bacteria but interfere with their activity (metabolism) so that the formation of sulfide species is minimized. As will become apparent in this chapter, some combinations of two organic biocides can work synergistically, being better than either biocide alone. Also, combinations of biocides and biostats can also have advantages over single-product treatments. In addition, many biocides appear to work best on planktonic organisms while fewer biocides are able to reduce sessile populations (biofouling) at the same dosage.[14]

The evaluation of biocides is a detailed process as there are a number of factors that can affect the performance besides the concentration of biocide. A rough screening of biocides can be carried out on planktonic bacteria but it is generally accepted that the final selection of biocide must be made from tests on sessile bacteria. Test procedures for determining the efficacy of biocides can be done in the laboratory for R&D purposes, or in field trials in water systems with monitoring of sessile SRB growth (biofouling), sulfide production, microbial corrosion, and iron sulfide production. Physical appearance, microorganism count, microscopic analysis, pressure fluctuations, and heat transfer are all methods of monitoring SRB growth. Epifluorescent microscopy has also been used effectively on field samples of bacteria.[15] Laboratory tests on biofilm (sessile bacteria) reduction by biocides are most often done in loops or rotating biofilm cylinder autoclaves.[16–18] Laboratory tests with

SRBs are less easy, so other bacteria such as *Pseudomonas fluorescens* have been used instead, often as mixed cultures with SRBs. However, SRBs can grow beneath a biofilm in an aerobic environment. It is important to measure reduction in sessile populations (biofilms), rather than planktonic organisms.[19–20] One laboratory study using a biofouling loop showed that once the slug dose of biocide was complete, the biofilm population recovered rapidly.[21]

Oxidizing biocides, such as chlorine/hypochlorite generally require a longer residence time (up to 30 min) than organic biocides in order to get a complete kill of bacteria. Organic biocides are characterized by high "speed-of-kill" or "knockdown" properties, usually requiring relatively high-dosage concentrations, often in the range 400–1000 ppm.[22] Organic biocides are usually added to the water-injection system as slug doses, that is, addition in one portion or shot over a relatively short period of time by measured delivery. Typically, the organic biocide is added once every few days for a few hours at a time. High population levels of biofilm bacteria are quickly reduced by application of the biocide. As soon as the biocide addition is completed, however, the bacteria in the biofilm commence a period of regrowth. The biofilm bacteria can quickly repopulate to levels equal to those present before the biocide slug dose was added. It is during the growth period that the bacteria are most active and can cause the most damage. Oxidizing biocides such as chlorine/hypochlorite are not added by the method described.

In many injector wells, MIC has occurred despite the application of biocides into the affected system. Besides a possible ineffectual biocide, another reason for increased MIC is that the application of the chemical is such that it is not contacting the target bacteria with sufficient concentration, retention time, and/or frequency to achieve a kill.[23] Another study has shown that some water-treatment additives are actually nutrients promoting bacterial growth and should therefore be investigated prior to chemical selection.[24] Another and more common reason for MIC is due to underdosing of the biocide.

Biostats inhibit further growth of a microorganism without killing it. As long as the microorganism is exposed to the biostatic agent, the microorganism will not be able to proliferate. However, once the microorganism is not exposed to the biostatic material, it will be able to proliferate. Biostats for the oil and gas industry are very effective in preventing the formation of hydrogen sulfide by maintaining a low level of SRB and inhibiting their metabolic activity. They are added at much lower concentrations (2–10 ppm) than the organic biocide. Some organic biocides also function as biostats but not all biostats are biocides. Typically an organic biocide may be dosed in periodic slugs with a low, continuous, or slug dose of a biostat. Test methods for biocides and biostats have been compared.[25]

14.3 BIOCIDES

There are two classes of biocides:

- oxidizing biocides
- nonoxidizing organic biocides

Biocides

The primary means of treating water-injection systems is usually with an oxidizing biocide and the secondary means with a nonoxidizing organic biocide. Chlorine/hypochlorite is generally used as the oxidizing biocide for seawater lift systems. This and other oxidizing biocides are discussed in the next section.

14.3.1 Oxidizing Biocides

Oxidizing biocides cause irreversible oxidation/hydrolysis of protein groups in the microorganism and in the polysaccharides that bind the microorganisms to the surfaces of the equipment. The result of this process is a loss of normal enzyme activity and cell death. Thus, oxidizing biocides will work against all strains of SRB, whereas some SRB may be resistant to some nonoxidizing organic biocides.

A summary of oxidizing biocides is as follows:

1. Electrochemically generated chlorine/hypochlorite (and bromine/hypobromite)
2. Hypochlorite and hypobromite salts
3. Stabilized bromine chloride
4. Hydroxyl radicals
5. Chloramines
6. Chlorine dioxide
7. Chloroisocyanurates
8. Halogen-containing hydantoins
9. Hydrogen peroxide and peracetic acid

A review of the factors to consider when applying oxidizing biocides in the field has been published.[26] Only chlorine/hypochlorite is commonly deployed in the oil-producing industry, primarily for water-injection systems.

Chlorine gas and bromine liquid are toxic, corrosive, and difficult to handle. However, chlorine is easily generated on site by electrolytic oxidation of the chloride ions in seawater using a DC current:[27]

$$2Cl^- \rightarrow Cl_2 + 2e^- \text{ (anode reaction)}$$

At the same time, a reduction of H^+ ions to hydrogen (H_2) gas takes place at the cathode, resulting in a more alkaline (basic) solution being produced. The chlorine reacts with water to produce hypochlorous acid and hydrochloric acids:

$$Cl_2 + H_2O \rightarrow HOCl + HCl$$

However, since the seawater is now alkaline, the acids react with hydroxide ions. Thus, hypochlorite ions (OCl^-) are produced, which are also biocidal:

$$OH^- + HOCl \rightarrow OCl^- + H_2O$$

Hypochlorous acid is a more powerful biocide than hypochlorite ions, and their mode of action in causing oxidative protein unfolding in vitro has been reported.[28]

Therefore, the efficiency of the treatment increases as the pH decreases, particularly below pH 6. In practice, at typical pH values of 8–9, there is little free hypochlorous acid in solution so the performance is not optimal. However, seawater also contains bromide ions, which will be oxidized by the chlorine to bromine, which reacts further to form hypobromite ions. The hypobromite ion is a much more powerful oxidizing biocide than hypochlorite. Thus, the hypobromite ions may also be making a significant contribution to preventing bacterial growth. A problem with the use of electrochlorination is that the generated chlorine will react with any organic material in the water, reducing the concentration available for bacterial control. Chlorination downstream of the deaerator has occasionally been used as the secondary means of bacterial control, rather than using a nonoxidizing biocide. However, chlorine is corrosive to steels and will remove bisulfite oxygen scavengers from the water, which is needed to prevent oxygen corrosion.

Hydroxyl radicals (OH×) are powerful biocides. On-site electrocatalytic hydroxyl radical generation could also be used for bacterial control but is currently not used in the oil industry.

Commercial alkaline solutions of sodium hypochlorite (NaOCl) or solid calcium hypochlorite (bleaching powder) could also be used as oxidizing biocides. However, the use of these salts in seawater injection systems has logistic and storage-tank limitations as well as high associated transportation costs. They are mostly used in closed or semiclosed water systems. Stable hypobromite or bromine compositions superior to hypochlorite salts have been claimed.[29]

Bromine chloride (BrCl) has the same disadvantages as the hypochlorite salts but has been made easier to handle by stabilizing it with sulfamate salts. It effectively gives a dose of the powerful biocide hypobromous acid, HOBr, at near-neutral pH.[30–32] Addition of C8–14 alkylamines as synergists to these blends is claimed to improve the biocidal performance further.[33] Alkylamines by themselves are fairly poor nonoxidizing biocides.[34]

A source of high-valence chlorine can be found in the chloramines (N–Cl bonds). Chloramines are used in the papermaking industry as superior biocides to the hypochlorites, but they are difficult to handle. However, methods whereby chloramines are generated in situ, for example, from ammonium bromide activated with sodium hypochlorite, have been claimed.[35–37] This method does not appear to have been tried in the oil industry yet.

Chlorine dioxide (ClO_2) as chlorite ions (ClO_2^-, ClO_2 is an explosive gas in air but only at high concentrations) is usually classified as an oxidizing biocide, although its kill mechanism is not oxidation. It is more effective at a higher pH, in nitrogen, or in organic-contaminated systems than chlorine. Chlorine dioxide can be generated on site using chlorite and an acid.[38] The problem in stabilizing chlorite solutions to give a high active dose of chlorine dioxide has been overcome by using an aqueous solution of a chlorite (e.g., sodium chlorite), a chlorine-generating agent (e.g., sodium hypochlorite), and a base (e.g., sodium hydroxide).[39] Chlorine dioxide solutions are corrosive. Injection of sodium dichromate can provide some corrosion inhibition. Less toxic corrosion inhibitors, such as a blend of glycol, acetic acid, a fatty imidazoline, and an ethoxylated fatty diamine, have been investigated but have very limited performance.[40]

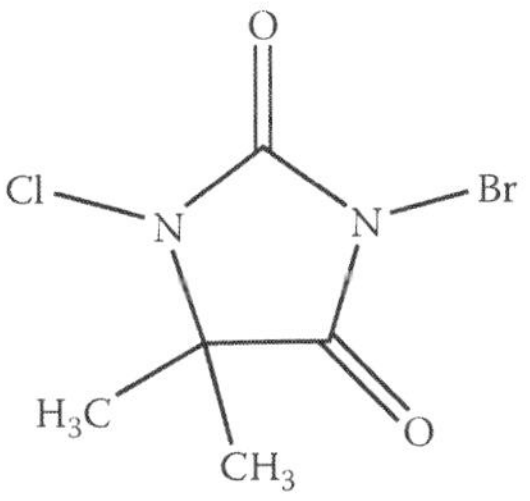

Biocides

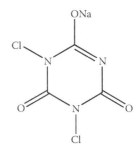

FIGURE 14.1 1-Bromo-3-chloro-5,5-dimethylhydantoin.

FIGURE 14.2 Sodium dichloroisocyanuric acid.

1-Bromo-3-chloro-5,5-dimethylhydantoin (BCDMH, Figure 14.1) is a solid and an excellent source of both chlorine and bromine, as it reacts slowly with water-releasing hypochlorous acid and hypobromous acid. It is more active than sodium hypochlorite, at the same dosage, on biofilms and is less corrosive.[17] Dichloro- and dibromo-hydantoins have also been used as oxidizing biocides. The hydantoins are widely used as biocides in a variety of industries. Pumpable, liquid BCDMH products have been developed.[20,41–42]

Chloroisocyanurates are easily handled powdered compounds, which hydrolyze in water to slowly release chlorine and cyanuric acid. Sodium dichloroisocyanuric acid is an example in this class (Figure 14.2). However, this class suffers all the drawbacks of the other chlorine-containing products in pH effectiveness ranges and potential corrosion problems.

14.3.2 NONOXIDIZING ORGANIC BIOCIDES

In general, the nonoxidizing organic biocides function primarily by altering the permeability of the cell walls of the microorganisms and interfering with their biological processes. Nonoxidizing organic biocides are less prone to cause corrosion than oxidizing biocides: in fact, some can inhibit corrosion. They include the following:

1. aldehydes
2. quaternary phosphonium compounds
3. quaternary ammonium surfactants
4. cationic polymers
5. organic bromides
6. metronidazole
7. isothiazolones (isothiazolinones) and thiones
8. organic thiocyanates
9. phenolics
10. alkylamines, diamines, and triamines
11. dithiocarbamates
12. 2-(decylthio)ethanamine (DTEA) and its hydrochloride
13. triazine derivatives

14. oxazolidines
15. other specific surfactant classes

Only some appear to have been used in oilfield applications mainly due their low cost or environmental acceptability, others have, for example, been used in closed-loop water systems or cooling towers. The most common nonoxidizing organic biocides in the oil industry are glutaraldehyde (glut) and tetrakis-hydroxymethyl-phosphonium (THPS), with smaller amounts of formaldehyde and acrolein being used (the latter two aldehydes are suspected carcinogens). These biocides are sometimes used in combination with quaternary ammonium surfactants and other synergists. These and several other classes of organic biocides will be discussed in the next sections.

14.3.2.1 Aldehydes

Aldehyde biocides include (Figure 14.3):

- C3–C7 alkanedials, especially glutaraldehyde (1,5-pentanedial)
- formaldehyde
- acrolein
- *ortho*-Phthalaldehyde

A potential advantage with using aldehydes to kill SRB populations is that they are also H_2S scavengers (see Chapter 15). By tonnage, glutaraldehyde (1,5-pentanedial) is, by far, the most commonly used oilfield nonoxidizing biocide.[43] The kill mechanism is by cross-linking outer proteins of cells and preventing cell permeability. If not stored properly, glutaraldehyde is unstable on storage. In one laboratory study, the use of glutaraldehyde alone led to decreased bacterial populations, decreased metabolic activity, and some decrease in biofilm accumulation, although this was dosage dependent.[19,44] Glutaraldehyde is often combined with other surface-active agents

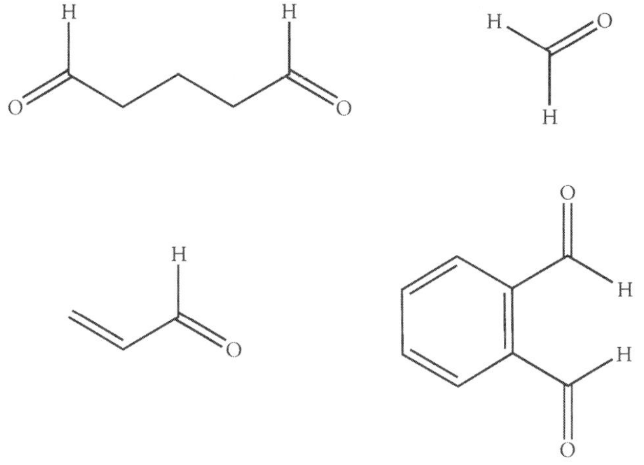

FIGURE 14.3 Glutaraldehyde (top left), formaldehyde (top right), acrolein (bottom left), and *ortho*-phthalaldehyde (bottom right).

Biocides

and biocides such as quaternary ammonium or phosphonium compounds so as to increase the speed with which it kills bacteria and to reduce the dosage of glutaraldehyde, which by itself would be quite high. Blending also reduces the necessary contact time. Glutaraldehyde has also been shown to work synergistically with methylene bis(thiocyanate).[45] Glutaraldehyde is pH-sensitive: it works well in neutral to alkaline water, that is, it may be effective for injection waters but less effective for produced water treatments.

Formaldehyde is another low-cost aldehyde that has been documented as being effective in killing sessile microorganisms in established biofilms.[46] One study showed that more biocide was needed in the field than in the laboratory study.[47] However, the European Union decided, in 2007, to ban formaldehyde use throughout Europe due to its carcinogenic properties. Formaldehyde is classified as a probable human carcinogen by the U.S. Environmental Protection Agency (EPA). Polyoxymethylene polymers have been shown to offer significant advantages over aqueous formaldehyde solutions and solid paraformaldehyde as sources of biocidal formaldehyde. Polyoxymethylenes are water-insoluble materials, wherein the decomposition rate to formaldehyde can be controlled through the use of pH, temperature, and certain decomposition catalysts.[48] Ethyleneglycol hemiformals are also formaldehyde-releasing compounds.

Acrolein (2-propenal) enables a threefold approach to oilfield problems stemming from SRB activity. Firstly, it is an effective biocide, secondly, like the other small aldehydes, it scavenges H_2S, and, thirdly, it dissolves iron sulfide.[49-51] However, it should be noted that acrolein has high acute toxicity and is a suspected carcinogen, so it needs to be handled carefully.

ortho-Phthalaldehyde has been proposed as an improvement on glutaraldehyde as it can remove sessile populations at lower doses with less contact time.[52]

14.3.2.2 Quaternary Phosphonium Compounds

Salts of the THPS ion have been shown to be excellent nonoxidizing, nonfoaming biocides, and are now widely used in the oil and gas industry.[53-55] These phosphonium salts kill bacteria by a number of mechanisms, but mainly by cross-linking of proteins, which leads to collapse of cell membranes (cell lysis). It is usually sold as the sulfate salt THPS, which has a good environmental profile and does not adsorb significantly on reservoir rock (Figure 14.4). THPS has a wide application as it appears to kill all types of anaerobic strains of SRB at pH 3–10. According to the manufacturers, once discharged from a treated system, THPS loses its antimicrobial properties almost immediately and degrades to a nontoxic substance, *tris*-hydroxymethyl phosphine oxide. Under certain conditions, THPS is a corrosion aggravator,

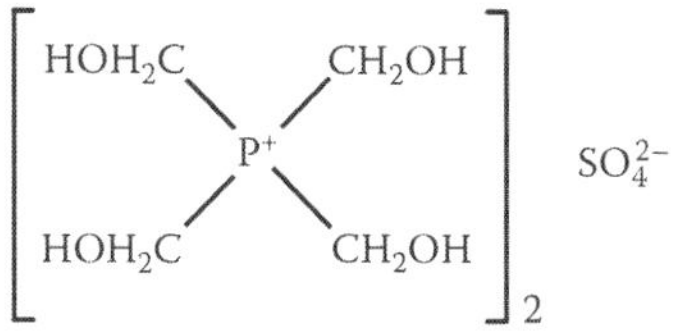

$$\left[\begin{array}{c} HOH_2C \diagdown \diagup CH_2OH \\ P^+ \\ HOH_2C \diagup \diagdown CH_2OH \end{array} \right]_2 SO_4^{2-}$$

FIGURE 14.4 Tetrakis-hydroxymethylphosphonium sulfate.

under others, it is an effective inhibitor. Thus, sometimes, a film-forming corrosion inhibitor needs to be injected also.[56] THPS also works synergistically with a range of surfactants to kill biofilms.[57] New-generation higher performing THPS formulations that do not contain surfactants have been reported.[58] These new formulations also minimize the risk of calcium or barium/strontium sulfate scaling, which might be induced by standard THPS formulations. THPS also works synergistically as a biocide with certain aldehydes such as formaldehyde.[59]

THPS is conventionally supplied as a liquid-based product, but solid forms adsorbed on adipic acid can be made, but are not currently commercially available. THPS adsorbed onto silica is used in fracturing fluids. Like many biocides, liquid-based phosphonium compounds such as THPS react or interfere with the performance of commonly used oxygen scavengers, for example, sulfite-based scavengers and erythorbic acid, with the result being that complete deaeration of systems is difficult to achieve. A way around this has been patented, whereby a phosphonium compound is embedded in a matrix substrate wherein the phosphonium compound is selected from the group consisting of *tris*-hydroxyorganophosphine (THP), a THP+ salt (tetrakis-hydroxyorganophosphonium salt) or a condensate of THP and a nitrogen-containing compound, most preferably urea.[60]

THPS has also been shown to remove iron sulfide scale both downhole and topside when coinjected with an ammonium salt.[61-62] In many fields, sufficient ammonium ions exist in the produced waters to give the desired effect. The iron ends up being chelated to a nitrogen-phosphorus ligand, giving the water a red color.[63] In one field, use of THPS with a surfactant killed the SRB, removed iron sulfide scale, and, surprisingly, increased well production in several wells by as much as 300%.[64] Removing particulate iron sulfide from produced fluids also facilitates demulsification processes.

The biocidal activity of THPS can be synergistically enhanced by using it in conjunction with wetting agents (various classes of surfactants), hydrotropes (mutual solvents), or biopenetrants such as poly[oxyethylene(dimethyliminio)ethylene (dimethyliminio)ethylene dichloride].[65-66]

Another synergist blend comprises an anionic scale inhibitor such as a mixture of 1-hydroxyethane-1,1-diphosphonic acid and diethylenetriaminepenta(methyle nephosphonic acid) blended with a cationically charged biocide such as THPS or poly(oxyethylene(dimethyliminio)ethylene(dimethyliminio)ethylene dichloride).[67]

Reaction products of acrylic acid and THPS salts have been shown to perform well as biocides and also dissolve iron sulfide but they are not currently commercially available (Figure 14.5).[68]

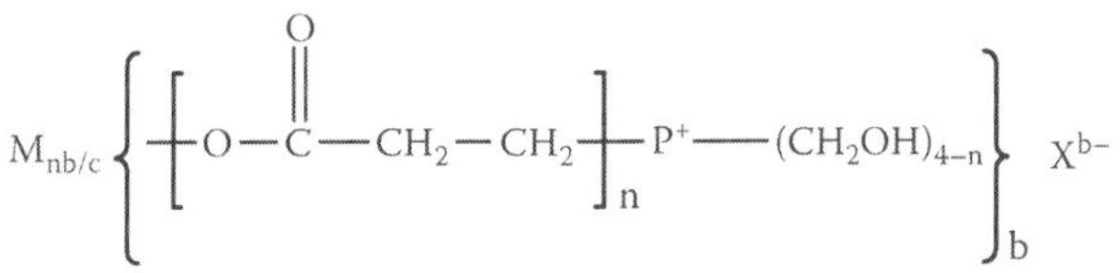

$$M_{nb/c} \left\{ \left[O - \overset{O}{\underset{\|}{C}} - CH_2 - CH_2 \right]_n P^+ - (CH_2OH)_{4-n} \right\}_b X^{b-}$$

FIGURE 14.5 Reaction products of acrylic acid and tetrakis-hydroxymethylphosphonium salts.

A new quaternary phosphonium biocide surfactant based on tributyl tetradecyl phosphonium chloride (TTPC) has been reported.[69] Laboratory and field data have shown that TTPC is effective at low concentrations, is fast-acting, and is effective against both acid-producing and sulfate-reducing bacteria. It has outperformed both glutaraldehyde and THPS in comparative biocidal tests. TTPC is compatible with oxidizing biocides, hydrogen sulfide, and oxygen scavengers and has excellent thermal stability. However, unlike THPS, TTPC can cause some low foaming, adsorbs onto surfaces, cannot be easily deactivated, and will not dissolve iron sulfide scale.

14.3.2.3 Quaternary Ammonium Compounds

Quaternary ammonium surfactants (quats) biocides are very surface-active and because of this property, they are sometimes used in blends with other biocides to improve the performance. The most common "quats" in this class are long chain n-alkyldimethylbenzylammonium chlorides where the alkyl group is 12 carbons or more (Figure 14.6). The quaternary salts are generally most effective against algae and bacteria in alkaline pH ranges but lose their activity in systems fouled with dirt, oil, and other debris. Low concentrations (< 250 ppm) of the alkyldimethylbenzylammonium surfactants have been shown to reduce the growth of numerous bacterial strains, including the SRBs. The kill mechanism is due to the cationic nature, whereby an electrostatic bond is formed with the cell wall, which affects permeability and protein denaturing. Used alone, quat biocides can take up to 10 min to kill bacteria. The use of a cellular membrane disruptor, such as ascorbic acid or glycolic acid, to improve the kill rate has been proposed for bactericide applications outside the oil industry.[70] As the quats are surfactants, they have film-forming corrosion-inhibiting properties. The quats are often formulated with amines such as cocodiamine. Quaternary ammonium surfactants are often foamers, but nonfoaming quaternary biocidal surfactants have been discovered.[71]

Nonpolymeric quaternary ammonium polyhalides such as N,N-dimethyl-N-ethyl-N-propylammonium tribromide have also been shown to behave as biocides. The tribromide ion is a source of bromine, so they may really be oxidizing biocides.[72] Other cationic biocides include long chain alkylguanidinium salts and 1-(3-chloroallyl)-3,5,7-triaza-1-azoniaadamantane (Figure 14.7).

14.3.2.4 Cationic Polymers

A composition useful as a biodegradable corrosion inhibitor and a biocide has been reported comprising a polymeric quaternary ammonium salt prepared by a reaction of a polyepihalohydrin with a tertiary amine.[73]

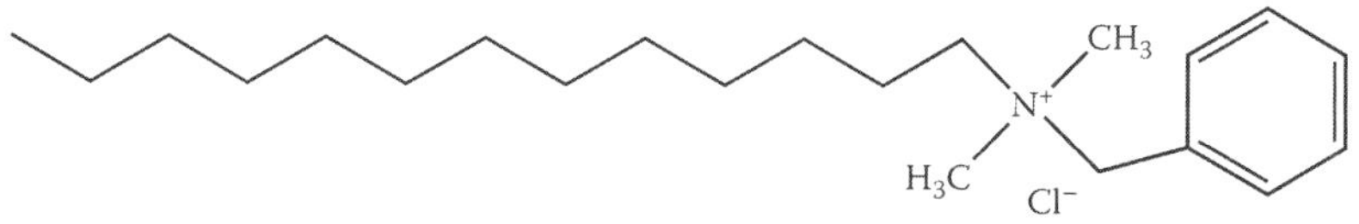

FIGURE 14.6 A typical n-alkyldimethylbenzylammonium chloride surfactant biocide.

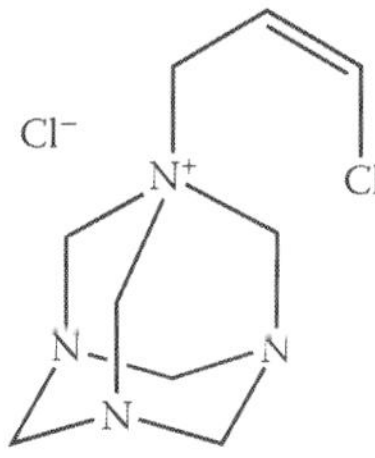

FIGURE 14.7 1-(3-Chloroallyl)-3,5,7-triaza-1-azoniaadamantane.

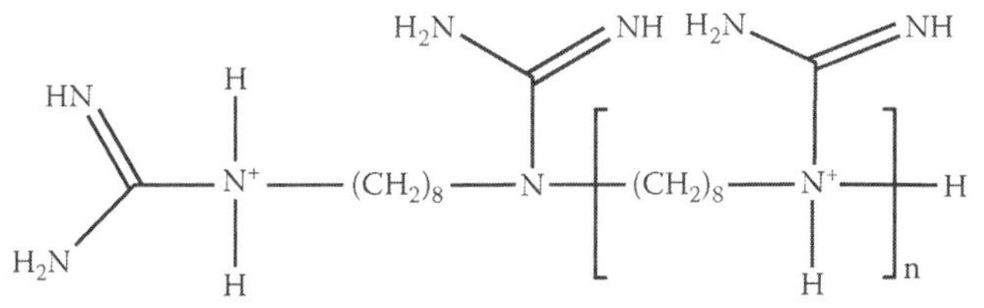

FIGURE 14.8 Poly(hexamethylene)biguanide hydrochloride ($n = 1–3$).

Another class of cationic polymeric biocides are the biguanides, for example, poly(hexamethylene)biguanide hydrochloride (Figure 14.8). They are very short polymers or oligomers and have been mainly used as disinfectants.[74–75]

Quaternary ammonium polymers made from epichlorohydrin, diamines, and tertiary have been proposed not only as nonsurfactant biopenetrant biocides but also as corrosion inhibitors.[76] Related quaternary polymers such as poly[oxyethylene-(dimethyliminio)ethylene (dimethyliminio)ethylene dichloride] are also used as biocides. This copolymer is made from *NNN'N'*-tetramethyl-1,2-diamino ethane with bis(2-chloroethyl) ether. Poly[oxyethylene(dimethyliminio)ethylene(dimethyliminio) ethylene dichloride] blended with an anionic scale inhibitor such as a mixture of 1-hydroxyethane-1,1-diphosphonic acid and diethylenetriaminepenta(methylenephosphonic acid) has been claimed as a superior biocide.[67]

14.3.2.5 Organic Bromides

The most common organic bromide biocides are 2-bromo-2-nitropropanediol (BNPD), 2,2-dibromo-3-nitrilopropionamide (DBNPA), and 1-bromo-1-(bromomethyl)-1,3-

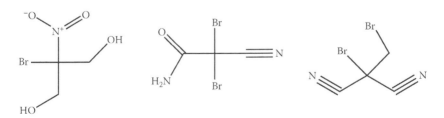

FIGURE 14.9 2-Bromo-2-nitropropanediol (left), 2,2-dibromo-3-nitrilopropionamide (middle), and 1-bromo-1-(bromomethyl)-1,3-propanedicarbonitrile (right).

Biocides

propanedicarbonitrile (Figure 14.9). 2,2-Dibromo-2-nitroethanol is also a useful biocide. BNPD was invented in the sixties and has been used as a biocide in oilfield operations.[77–78] DBNPA is pH-sensitive, quickly hydrolyzing under both acid and alkaline conditions. It is preferred for its instability in water as it quickly kills and then quickly degrades to ammonia and bromide ions. Methods to get around its low water solubility have been proposed.[79] DBNPA is proposed to act synergistically with a number of oxidizing (e.g., hypochlorite) and nonoxidizing biocides.[80]

14.3.2.6 Metronidazole

Metronidazole is an imidazole derivative with additional weak corrosion-inhibiting properties (Figure 14.10). Metronidazole is selectively taken up by anaerobic bacteria. The nitro group of metronidazole is chemically reduced by ferredoxin (or ferredoxin-linked metabolic process) and the products are responsible for disrupting the DNA helical structure, thus inhibiting nucleic acid synthesis. It has been shown to be effective in controlling biogenic sulfide production in oilfield water injectors.[81]

A more surface-active biocide was made by derivatizing metronidazole to make a quaternary surfactant. This molecule also had good corrosion inhibition properties (see Chapter 8 on corrosion control for the structure).[82]

14.3.2.7 Isothiazolones (or Isothiazolinones) and Thiones

Products in this class are chloro-2-methyl-4-isothiazolin-3-one, 5- and 2-methyl-4-isothiazolin-3-one, and 4,5-dichloro-2-(n-octyl)-4-isothiazolin-3-one (Figure 14.11).[83]

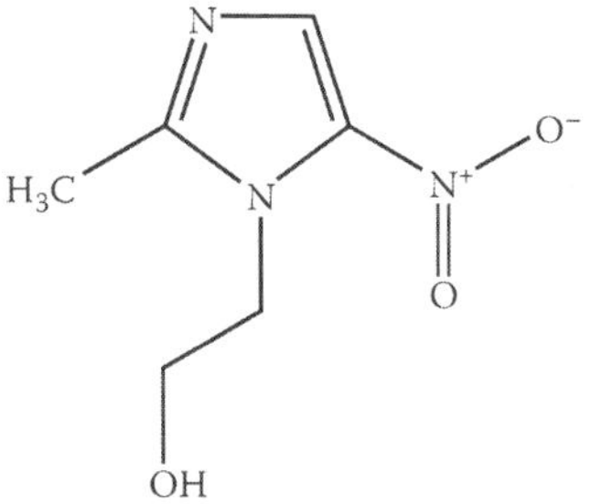

FIGURE 14.10 Metronidazole.

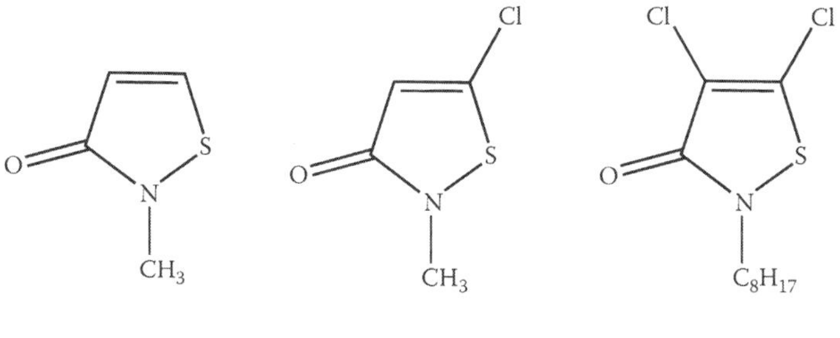

FIGURE 14.11 Common isothiazolone biocides.

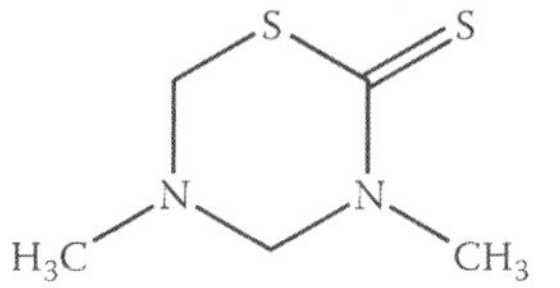

FIGURE 14.12 3,5-Dimethyl-1,3,5-thiadiazinane-2-thione.

Less water-soluble benzisothiazolones are also commercially available. The isothiazolones are broad-spectrum antimicrobials with good biodegradability, effective against sessile organisms (biofilms).[14,84] Isothiazolones kill by inhibiting microbial respiration and food transport through the cell wall.

Although isothiazolones may be used as biocides, their effectiveness is reduced by hydrogen sulfide formed by the existing SRB. Isothiazolones are however very effective biostats in preventing the formation of hydrogen sulfide by maintaining a low level of SRB and inhibiting their metabolic activity. Hence, they are best used in combination with other organic biocides.[85] For example, one method of reducing sessile SRB comprises adding periodic slug doses of an alkanedial such as glutaraldehyde as biocide, together with a continuous dose of an isothiazolone as biostat.[86–87] An improvement on this method is to use a slug dose of a biologically effective amount of a biocide simultaneously with or followed by intermittent addition of a biologically effective amount of a biostat such as an isothiazolone.[88] An improved isothiazolone-based composition contains 3-iodo-2-propynyl-*N*-butylcarbamate as a second biocidal agent.[89] Another patent claims the use of zinc salts to enhance the performance of isothiazolinone biocides.[90] The antimicrobial composition can also contain cobiocides, such as pyrithiones, including zinc pyrithione or copper pyrithione. Six-ring thiones are also useful oilfield biocides. A preferred thione is 3,5-dimethyl-1,3,5-thiadiazinane-2-thione (Figure 14.12).[91] Isothiazolones are not in current use in oilfield production applications due to their limited performance as biocides and the availability of other more cost-effective treatments.

14.3.2.8 Organic Thiocyanates

The only product in this class that has been widely used is methylene bis(thiocyanate) (Figure 14.13). Methylene bis(thiocyanate) suffers from rapid decomposition above approximately pH 8.0, eventually releasing toxic hydrogen cyanide. It is relatively expensive and generally requires a dispersant to obtain effective penetration of the biocide into the biofouling. It needs to be added at a fairly high concentration if used by itself. At low concentrations, methylene-bis(thiocyanate) tends to have a narrow antimicrobial spectrum and fails to completely prevent the growth of

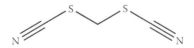

FIGURE 14.13 Methylene bis(thiocyanate).

Biocides 349

microorganisms. It has been shown to work synergistically with several other biocides such as glutaraldehyde, DTEA, and organic bromides.[92–93] The kill mechanism is to block the transfer of electrons in the microorganism, preventing oxidation/reduction mechanisms.

14.3.2.9 Phenolics

An example of a phenolic biocide is sodium pentachlorophenate. Although they are low-cost and powerful biocides, this class has been withdrawn from most markets because of environmental concerns. Parabens such as 4-hydroxybenzoic acid are also phenolic-type biocides but do not appear to be have been used widely in the oil and gas industry. Sodium phenyl phenate, *ortho*-phenyl phenol, and dichloro-*m*-xylenol are also sold as industrial biocides.

14.3.2.10 Alkylamines, Diamines, and Tramines

Cocodiamine surfactant has long been used as a biocide in the oil industry (Figure 14.14).[49,94] Triamines such as *N,N*-bis(3-aminopropyl)dodecylamine or bis(3-aminopropyl)octylamine have also been shown to act as biocides.[95] These amine-based surfactant molecules probably also exhibit some corrosion inhibition properties.

Nonsurfactant tertiary amine acrylamide monomers such as dimethylaminopropylmethacrylamide or dimethylaminopropylacrylamide have been shown to inhibit microbiological growth (Figure 14.15). In the presence of corrosion inhibitors, it is found that these biocides will permit attainment of 100% biocidal kill when present in lesser amounts than are required when no corrosion inhibitor is present. They also improve the corrosion inhibition.[96]

14.3.2.11 Dithiocarbamates

An example of a dithiocarbamate biocide is disodium ethylene-1,2-bisdithiocarbamate (Figure 14.16). They are used more in the paper industry, sometimes in blends with aldehydes.[97]

FIGURE 14.14 Cocodiamine.

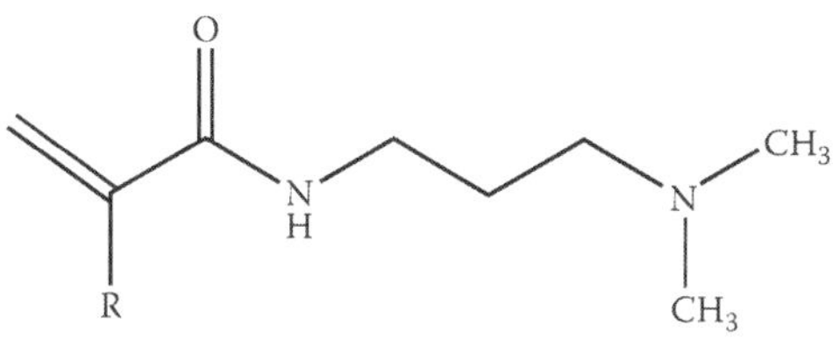

FIGURE 14.15 Dimethylaminopropylmethacrylamide ($R = CH_3$) or dimethylaminopropylacrylamide ($R = H$).

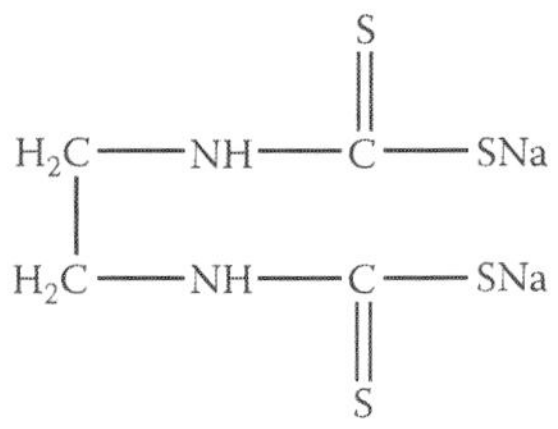

FIGURE 14.16 Disodium ethylene-1,2-bisdithiocarbamate.

FIGURE 14.17 2-(Decylthio)ethanamine hydrochloride.

14.3.2.12 2-(Decylthio)ethanamine and Its Hydrochloride

At low pH, DTEA exists as the hydrochloride and as the free amine at high pH (Figure 14.17). Both the amine and the hydrochloride have been shown to have biocidal properties. The hydrochloride is a quaternary surfactant and has been shown to have additional corrosion inhibition properties. The use of glutaraldehyde in combination with DTEA is recommended and is in use today in some oilfields in the United States.[98–99] 2-(Decylthio)ethanamine hydrochloride has been shown to work synergistically with methylene bis(thiocyanate).[100]

14.3.2.13 Triazine Derivatives

Certain triazine derivatives have been shown to have antimicrobial properties. Examples are sulfur-containing triazines and 2-(*tert*-butylamino)-4-chloro-6-(ethylamino)-*S*-triazine (Figure 14.18). Some triazine chemicals that are used as H_2S scavengers are also commercially available as biocides[101–102] (see Chapter 15 for details).

14.3.2.14 Oxazolidines

Oxazolidines, such as 7-ethyl bicyclooxazolidine, 4,4-dimethyloxazolidine, and methylene bis-oxazolidine, as well as some halooxazolidinones, are broad-spectrum commercial biocides (Figure 14.19).[103]

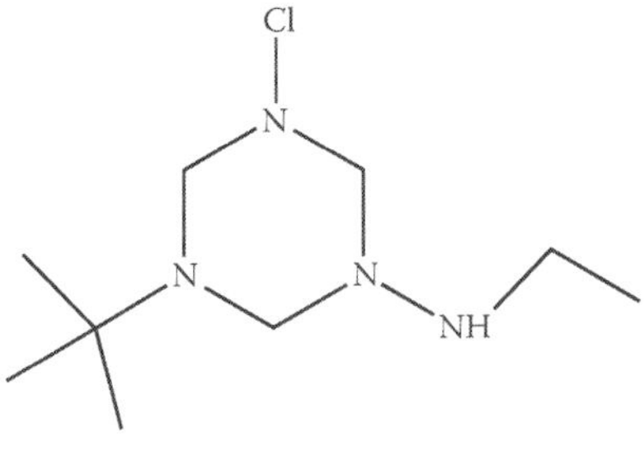

FIGURE 14.18 2-(*tert*-Butylamino)-4-chloro-6-(ethylamino)-*S*-triazine.

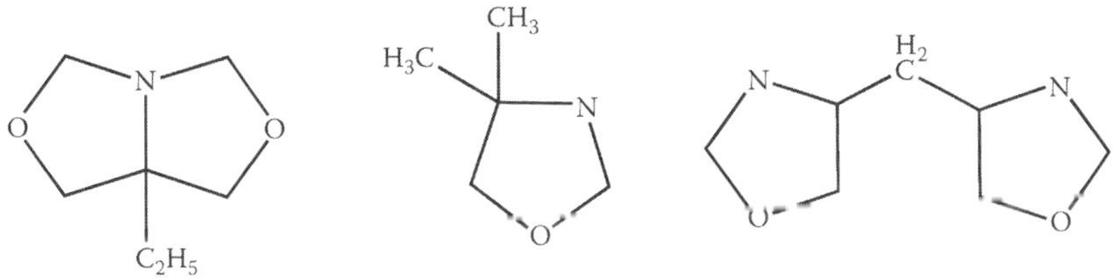

FIGURE 14.19 7-Ethyl bicyclooxazolidine (left), 4,4-dimethyloxazolidine (middle), and methylene bis-oxazolidine (right).

14.3.2.15 Specific Surfactant Classes

Alkylaminomethylenephosphonic acid amphiphilic compounds having the general formula $R'R_2NCH_2P(O)(OH)_2$, such as octylaminomethylenephosphonic acid, have been claimed as biocides. It is possible that hydrophobically modified phosphonate scale inhibitors with several phosphonate groups might also show biocidal properties but this was not claimed.[104] Sulfamic acid surfactants such as dodecyl sulfamic acid have also been claimed by the same group as biocides.[105]

14.4 BIOSTATS (CONTROL "BIOCIDES" OR METABOLIC INHIBITORS)

Biostats do not necessarily kill bacteria but interfere with their metabolic processes, controlling their growth. As discussed in the section on organic biocides, isothiazolones are one class of biostats, preventing the formation of iron sulfide scale (via hydrogen sulfide) by maintaining a low level of SRB and inhibiting their metabolic activity. Metabolic inhibitors deprive SRB of the ability to produce ATP, and as a result, cells are unable to grow and/or divide. This inability to grow or divide may eventually cause the death of some of the SRB; however, the cell death is not a direct result of exposure to the metabolic inhibitors as it would be for biocides. Alkylbenzyldimethylammonium salts are also a commercially significant class of biostats. A combination of biocide and biostat has been shown in laboratory studies to inhibit biogenic sulfide production at significantly lower concentrations than would be required if the biocide or biostat was used alone.[106]

Examples of biostats that are not biocides are discussed in this section. They include:

- anthraquinone
- azide ions
- nitrite and nitrate ions
- molybdate or tungstate ions
- selenate ions

14.4.1 ANTHRAQUINONE AS CONTROL BIOCIDE

Anthraquinone has been used as a biostat in a number of projects since the late nineties (Figure 14.20). Anthraquinone is not water-soluble, but 9,10-anthracenediol

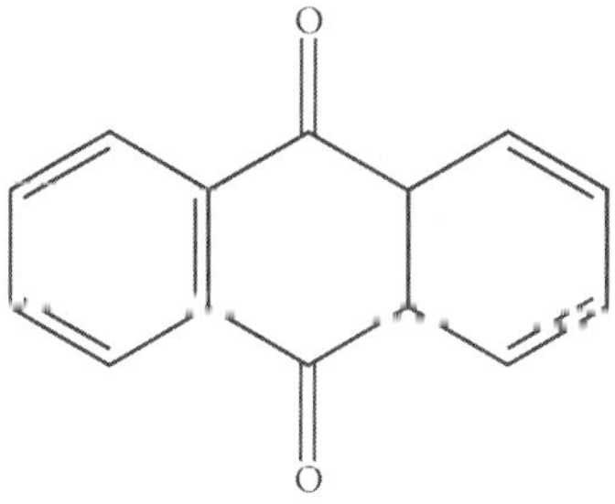

FIGURE 14.20 Anthraquinone.

disodium salt is water-soluble and works as if it were anthraquinone.[107] Anthraquinones have been shown to inhibit sulfate respiration in SRBs effectively shutting down the sulfide-producing mechanism but having little effect on other classes of bacteria.[108–109] One water injection program managed to eliminate continuous injection of a quaternary biocide with slug doses of the easier-to-handle anthraquinone product. Slug doses of synergistic acrolein biocide were used in addition.[110–111] Batch treatment of anthraquinone together with THPS biocide has also been shown to be effective in stopping biogenic sulfide production in produced water tanks in slop-handling systems.[112–113]

14.4.2 Nitrate and Nitrite Treatment

Nitrite (NO_3^-) and nitrate ions (NO_2^-) as found in calcium or sodium nitrate/nitrite salts are cheap, easy to handle, environment-friendly inorganic chemicals, and becoming more and more used in the oil industry to inhibit SRB. The corrosion consequences of using nitrate or nitrite in oilfield brines have been reviewed.[114]

Nitrite directly inhibits sulfate reduction by SRB because it is reduced more slowly than sulfite by the final enzyme in the sulfate reduction pathway, dissimilatory sulfite reductase. Thus, addition of nitrite ions to produced water or seawater injection systems can control biogenic sulfide formation if present in high-enough concentration (depending on pH, nitrite, but not nitrate, ions also react directly with any H_2S already present to form sulfur and reduced nitrogen compounds). In one field study, pulses of nitrite were more effective than the same amount of nitrite added continuously. Nitrite was more effective at inhibiting souring than was glutaraldehyde, and SRB recovery was delayed longer with nitrite than with glutaraldehyde.[115] Nitrite injection has also been used successfully in producer wells to scavenge H_2S and prevent SRB activity.[116] In one case, oil production increased immediately following the treatment, probably due to the dissolution of precipitated iron sulfides in the zone surrounding the wellbore.[117]

There exist nitrate-reducing and sulfide-oxidizing bacteria (hNRB and NR-SOB) in most oilfields that can reduce nitrate ions to nitrite ions.[118–119] Thus, the upstream petroleum industry has introduced a nitrate-based microbial treatment technology, useful for both the prevention and removal of biogenic sulfide from reservoirs, produced water, surface facilities, pipelines, and gas-storage reservoirs, as well as

Biocides

increasing oil recovery. This reservoir treatment technology works by replacing SRB with a naturally occurring suite of beneficial microorganisms enhanced by the introduction of an inorganic nitrate-based formulation.[120] If the reservoir is a poor source of carbon for SRB (low in small organic acids), injection of nitrate ions will stimulate growth of "hardier" denitrifying bacteria (nutrient augmentation), thereby, dominating the system and inhibiting the growth of SRB. Production of nitrite and sometimes nitrous oxide (NO) from nitrate ions by the hNRB and NR-SOB will also directly inhibit SRB growth by acting as toxins. In addition, hNRB and NR-SOB may produce compounds that raise the oxidation-reduction potential of the environment to a level that is inhibitory to the growth of SRB.[121-129]

Calcium or sodium nitrate are environment-friendly and complement the naturally occurring organic acids in the reservoir, selectively stimulating and increasing the targeted nitrate-reducing bacteria. Many North American gas fields have been treated successfully this way and several fields in the North Sea have already successfully used this treatment strategy for reducing, but not totally eliminating, reservoir souring, both by reactive and proactive strategies. It is a simple method for preventing biogenic sulfide formation, which can mostly eliminate the use of organic biocides.[130-134]

For the reasons discussed above, the use of a mixture of nitrate and nitrite ions may perform superior to a simple nitrate treatment.[135] One study showed that nitrite treatment alone may be preferable in reservoirs with only SRB present.[136] In addition, the use of a molybdenum compound (molybdate) may also enhance the treatment effect. Molybdenum is a known enzymatic inhibitor of the hydrogenase enzyme found in SRB.[137] It should be noted that nitrate/nitrite treatment may not totally eliminate the use of biocides if they are injected very late in the system. For example, a deepwater field offshore Nigeria that has been treated with calcium nitrate still uses a biocide (THPS) to prevent biofilms in the injection facilities.[138-139] A potentially large advantageous side effect of using nitrate-based water injection treatments is microbial-enhanced oil recovery (MEOR). Formation of NR-SOB biofilms in the reservoir may help to release more oil from the rock surface, which can then migrate to the producer wells. The use of nitrate, together with other nutrients such as vitamins and phosphate, has been suggested as an MEOR improvement.[140]

Instead of relying on indigenous bacteria, injection of non-SRB bacteria with nutrients such as nitrate ions has also been carried out, eliminating, in one case, the use of hazardous acrolein biocide.[141] A novel class of bacteria that oxidize sulfide as well as oil organics with nitrate has been reported.[142] There is still a lot to learn about nitrite/nitrate treatments, such as the relative effectiveness of nitrate versus nitrite ions and determining which of the several mechanisms of reducing biogenic sulfide production is dominant.

14.4.3 OTHER BIOSTATS

Azide salts such as sodium azide (NaN_3) have long been known as biostats and have been suggested for use in preventing biofouling of wells.[143] Sodium azide acts as a biostat by inhibiting cytochrome oxidase in gram-negative bacteria. Its effect on SRB has not been reported.

Other more expensive biostats that inhibit SRB growth include selenate and some transition metal oxyanions such as vanadate, molybdate, permanganate, and tungstate ions.[144–146] Molybdate ions are also useful in some corrosion inhibitor blends. The metal oxyanion SRB inhibitors are used to deplete ATP pools in the SRB, thereby, resulting in the death of the SRB.

14.5 SUMMARY

In summary, the use of biocides and biostats are the two most common methods of preventing biogenic sulfide formation (reservoir souring) and MIC. If produced, toxic hydrogen sulfide (H_2S) can be converted to benign chemicals using hydrogen sulfide scavengers, which are discussed in Chapter 15.

REFERENCES

1. M. Magot, "Indigenous Microbial Communities in Oil Fields," in *Petroleum Microbiology*, eds. B. Ollivier and M. Magot, Washington, DC: ASM Press, 2005, 21.
2. P. A. Lapointe, M. A. Muhsin, and A. F. Maurin, "Microbial Corrosion and Biologically Induced New Products Example from a Seawater Injection System, Umm Shaif Field, U.A.E." SPE 21367 (paper presented at the SPE Middle East Oil Show, Bahrain, 16–19 November 1991).
3. J. L. Lynch and R. G. J. Edyvean, "Biofouling in Oilfield Water Systems—A Review," *Biofouling* 1 (1988): 147–162.
4. R. CordRuwisch, W. Kleintz, and F. Widdel, "Sulfate-Reducing Bacteria and Their Activities in Oil Production," SPE 13554, *Journal of Petroleum Technology* 39 (1987): 97.
5. P. F. Sanders and P. J. Sturman, "Biofouling in the Oil Industry," in *Petroleum Microbiology*, eds. B. Ollivier and M. Magot, Washington DC: ASM Press, 2005, 171.
6. J.-L. Crolet, "Microbial Corrosion in the Oil Industry: A Corrosionist's View," in *Petroleum Microbiology*, eds. B. Ollivier and M. Magot Washington, DC: ASM Press, 2005, 143.
7. S. Schultze-Lam, D. Fortin, B. S. Davis, and T. J. Beveridge, "Mineralization of Bacterial Surfaces," *Chemical Geology* 132 (1996): 171.
8. A. F. Bird, H. R. Rosser, M. E. Worrall, K. A. Mously, and O. I. Fageeha, "Sulfate Reducing Bacteria Biofilms in a Large Seawater Injection System," SPE 73959, SPE (paper presented at the International Conference on Health, Safety and Environment in Oil and Gas Exploration and Production, Kuala Lumpur, Malaysia, 20–22 March 2002).
9. (*a*) M. Davies and P. J. B. Scott, "Oilfield Water Technology," Houston, TX: National Association of Corrosion Engineers (NACE), 2006. (*b*) E. A. Morris, R. Gomez, and R. Peterson, "Application of Chemical and Microbiological Data for Sulfide Control," SPE 52705 (paper presented at the SPE/EPA Exploration and Production Environmental Conference, Austin, TX, February–3 March 1999).
10. P. A. Lapointe, M. A. Muhsin, and A. F. Maurin, "Microbial Corrosion and Biologically Induced New Products: Example from a Seawater Injection System," Umm Shaif Field, UAE, SPE 21367 (paper presented at the SPE Middle East Oil Show, Bahrain, 16–19 November 1991).

11. F. Akersburg, F. Bak, and F. Widdel, "Anaerobic Oxidation of Saturated Hydrocarbons to CO_2 by a New Type of Sulfate-Reducing Bacterium," *Archives of Microbiology*, Feb 1991.
12. H. Beller, P. Spormann, and J. Cole, "Isolation and Characterization of a Novel Toluene-Degrading Sulfate-Reducing Bacterium," *Applied and Environmental Microbiology* 62 (1996); 1188.
13. J. E. McElhiney, International Patent Application WO/2007/106691.
14. (*a*) L. Watkins and J. W. Costerton, Paper 246 (paper presented at the NACE CORROSION Conference, Anaheim, CA, 1983). (*b*) I. Ruseska, J. Robbins, and J. W. Costerton, "Biocide Testing against Corrosion-Causing Oil-Field Bacteria Helps Control Plugging," *Oil & Gas Journal* 80 (1982): 253. (*c*) V. Keasler, B. Bennett, R. Diaz, P. Lindmuth, D. Kasowski, C. Adelizzi, Nalco, L. Santiago-Vazquez, Identification and Analysis of Biocides Effective Against Sessile Organisms SPE 121082, SPE International Symposium on Oilfield Chemistry, The Woodlands, TX, April 20–22, 2009.
15. T. Thorstenson, G. Boedtker, B.-L. P. Lilleboe, T. Torsvik, E. Sunde, and J. Beeder, "Biocide Replacement by Nitrate in Sea Water Injection," Paper 02033 (paper presented at the NACE CORROSION Conference, 2002).
16. D. B. McIlwaine, J. Diemer, and L. Grab, "Determining the Biofilm Penetrating Ability of Various Biocides Utilizing an Artificial Biofilm Matrix," Paper 97400 (paper presented at the NACE CORROSION Conference, 1997).
17. C. J. Nalepa, H. Ceri, and C. A. Stremick, "A Novel Technique for Evaluating the Activity of Biocides against Biofilm Bacteria," Paper 00347 (paper presented at the NACE CORROSION Conference, 2000).
18. B. Yin, J. Yang, U. Bertheas, and J. Adams, "A High Throughput Evaluation of Biocides for Biofouling Control in Oilfields," *Chemistry in the Oil Industry X: Oilfield Chemistry*, Royal Society of Chemistry, Manchester, UK, 5–7 November, 2007.
19. R. G. Eager, J. Leder, J. P Stanley, and A. B. Theis, "The Use of Glutaraldehyde for Microbiological Control in Waterflood Systems," *Materials Performance* 27 (1988): 40.
20. N. T. Macchiarolo, B. McGuire, and J. M. Scalise, U.S. Patent 4297224, 1981.
21. T. K. Haack, E. S. Lashen, and D. E. Greenly, *Developments in Industrial Microbiology* (*Journal of Industrial Microbiololgy* Suppl. 3) (1988): 247.
22. T. K. Haack, D. E. Greenley, U.S. Patent 5026491, 1991.
23. S. Maxwell, "Controlling Corrosive Biofilms by the Application of Biocides," SPE 93172 (paper presented at the SPE International Symposium on Oilfield Corrosion, Aberdeen, UK, 13 May 2005).
24. E. Sunde, T. Thorstenson, and T. Torsvik, "Growth of Bacteria on Water Injection Additives," SPE 20690 (paper presented at the SPE Annual Technical Conference and Exhibition, New Orleans, LA, 23–26 September 1990).
25. B. Yin, J. Yang, U. Bertheas, and J. Adams, "High Throughput Evaluation of Biocides for Biofouling Control in Oilfields" (paper presented at teh Chemistry in the Oil Industry: Oilfield Chemistry, Royal Society of Chemistry, Manchester, UK, 5–7 November 2007).
26. C. J. Nalepa, J. N. Howarth, and F. D. Azarnia, "Factors to Consider when Applying Oxidizing Biocides in the Field," Paper 02223 (paper presented at the NACE CORROSION Conference, 2002).
27. K. B. Flatval, S. Sathyamorrthy, C. Kuijvenhoven, and D. Ligthelm, "Building the Case for Raw Seawater Injection Scheme in Barton," SPE 88568 (paper presented at the SPE Asia Pacific Oil and Gas Conference and Exhibition, Perth, Australia, 18–20 October 2004).

28. (a) D. M. Clementz, D. E. Patterson, R. J. Aseltine, and R. E. Young, "Stimulation of Water Injection Wells in the Los Angeles Basin by Using Sodium Hypochlorite and Mineral Acids," SPE 10624, *Journal of Petroleum Technology* 34 (1982): 2087. (b) J. Winter, M. Ilbert, P. C. F. Graf, D. Özcelik, and U. Jakob, *Cell*, 2008, 135, 691.
29. S. Yang, World Patent Application WO/2004/026770.
30. R. M. Moore Jr. and C. J. Nalepa, U.S. Patent 6322822, 2001.
31. C. J. Nalepa, U.S. Patent 7087251 2006.
32. J. F. Carpenter and C. J. Nalepa, "Bromine-Based Biocides for Effective Microbiological Control in the Oil Field," SPE 92702 (paper presented at the SPE International Symposium on Oilfield Chemistry, Houston, TX, 2–4 February 2005).
33. C. J. Nalepa, U.S. Patent 6419838, 2002.
34. J. A. Findlay, "The Potential of Alkyl Amines as Antifouling Biocides I: Toxicity and Structure Activity Relationships," *Biofouling* 9(4) (1996): 257.
35. M. J. Mayer and F. L. Singleton, U.S. Patent Application 20070045199.
36. H. N. Cheng, D. Sharoyan, M. J. Mayer, and F. L. Singleton, International Patent Application, WO/2008/091678.
37. A. Gupta, M. Ramesh, and R. Elliott, International Patent Application WO/2008/083182.
38. G. H. Zaid and D. W. Sanders, U.S. Patent 6431279, 2002.
39. T. J. Parkinson and A. T. Harris, U.S. Patent 6325970, 2001.
40. J. R. Ohlsen, J. M. Brown, G. F. Brock, and V. K. Mandlay, U.S. Patent 5459125, 1995.
41. M. L. Ludyanakiy and F. J. Himpler, "The Effect of Halogenated Hydantoins on Biofilms," Paper 97405 (paper presented at the NACE CORROSION Conference, 1997).
42. B. R. Sook, T. F. Ling, and A. D. Harrison, "A New Thixotropic Form of Bromo-chlorodimethylhydantoin: A Case Study," Paper 03715 (paper presented at the NACE CORROSION Conference, 2003).
43. (a) D. M. Brandon, J. P. Fillo, A. E. Morris, and J. M. Evans, "Biocide and Corrosion Inhibition Use in the Oil and Gas Industry: Effectiveness and Potential Environmental Impacts," SPE 29735 (paper presented at the SPE/EPA Exploration and Production Environmental Conference, Houston, TX, 27–29 March 1995). (b) H. R. Mcginley, M. Enzien, G. Hancock, S. Gonisor, Maureen Mikoztal, Glutaraldehyde: An Understanding of its Ecotoxicity Profile and Environmental Chemistry, paper 09405, NACE, Corrision 2009 Conference and Exposition, Atlanta, GA, March 22–26, 2009.
44. R. G. Eager et al., "Glutaraldehyde: Impact of Corrosion Causing Biofilms," Paper 86-125 (paper presented at the NACE CORROSION Conference, 1986).
45. T. M. LaMarre and C. H. Martin, U.S. Patent 4616037, 1986.
46. W. G. McLelland, "Results of Using Formaldehyde in a Large North Slope Water Treatment System," SPE 35675 (paper presented at the SPE Computer Applications, April 1997, 55).
47. B. G. Kriel, A. B. Crews, E. D. Burger, E. Vanderwende, and D. O. Hitzman, "The Efficacy of Formaldehyde for the Control of Biogenic Sulfide Production in Porous Media," SPE 25196 (paper presented at the SPE International Symposium on Oilfield Chemistry, New Orleans, LA, 2–5 March 1993).
48. J. T. Fenton and J. F. Miller, U.S. Patent 4911923, 1990.
49. R. M. Jorda, "Aqualin Biocide in Injection Waters," SPE 280 (paper presented at the SPE Production Research Symposium, Tulsa, OK, 12–13 April 1962).
50. C. Reed, J. Foshee, J. E. Penkala, and M. Roberson, "Acrolein Application to Mitigate Biogenic Sulfides and Remediate Injection Well Damage in a Gas Plant Water Disposal System," SPE 93602 (paper presented at the SPE International Symposium on Oilfield Chemistry, The Woodlands, TX, 2–4 February 2005).

51. J. Penkala, M. D. Law, D. Horaska, and A. L. Dickinson, "Baker Petrolite, Acrolein 2-Propenal: A Versatile Microbiocide for Control of Bacteria in Oilfield Systems," Paper 04749 (paper presented at the NACE CORROSION Conference, 2004).
52. A. B. Theis and J. Leder, U.S. Patent 5128051, 1992.
53. K. P. Davis and R. E. Talbot, K Patent Application GB2145798, 1984.
54. N. Macleod, T. Bryan, A. J. Buckley, R. F. Talbot, and M. A. Veale, SPE 30171, Society of Petroleum Engineers eLibrary, 1994.
55. B. L. Downward, R. E. Talbot, and T. K. Haack, "Tetrakishydroxymethylphosphonium sulfate (THPS), A New Industrial Biocide with Low Environmental Toxicity," Paper 97401 (paper presented at the NACE CORROSION Conference, 1997).
56. R. L. Martin, "Unusual Oilfield Corrosion Inhibitors," SPE 80219 (paper presented at the SPE International Symposium on Oilfield Chemistry, Houston, TX, 5–7 February 2003).
57. K. G. Cooper, R. E. Talbot, and M. J. Turvey, U.K. Patent Application GB2178960, 1986.
58. C. R. Jones, G. Collins, B. L. Downward, and K. Hernandez, "THPS: A Holistic Approach to Treating Sour Systems," Paper 08659 (paper presented at the NACE CORROSION Conference & Exposition, New Orleans, LA, 16–20 March 2008).
59. E. Bryan, M. A. Veale, R. E. Talbot, K. G. Cooper, and N. S. Matthews, European Patent EP0385801, 1990.
60. C. R. Jones and R. Diaz, International Patent Application WO/2005/079578.
61. P. D. Gilbert, J. M. Grech, R. E. Talbot, M. A. Veale, and K. A. Hernandez, "Tetrakis-hydroxymethylphosphonium Sulfate (THPS), for Dissolving Iron Sulfides Downhole and Topside—A Study of the Chemistry Influencing Dissolution," Paper 02030 (paper presented at the NACE CORROSION Conference, 2002).
62. H. A. Nasr-El-Din, A. M. Al-Mohammad, M. A. Al-Hajri, and J. B. Chesson, "A New Chemical Treatment to Remove Multiple Damages in a Water Supply Well," SPE 95001 (paper presented at the SPE European Formation Damage Conference, Sheveningen, The Netherlands, 25–27 May 2005).
63. J. C. Jeffery, B. Odell, N. Stevens, and R. E. Talbot, "Self-Assembly of a Novel Water-Soluble Iron(II) Macrocylic Phosphine Complex from Tetrakis(Hydroxymethyl) Phosphonium Sulfate and Iron(II) Ammonium Sulfate: Single Crystal X-ray Structure of the Complex $[Fe(H_2O)_2\{RP(CH_2N(CH_2PR_2)CH_2)_2PR\}]SO_4 \times 4H_2O$ ($R = CH_2OH$)," *Chemical Communications* (2000): 101.
64. P. R. Rincon, J. P. McKee, C. E. Tarazon, L. A. Guevara, B. Vinccler, "Biocide Stimulation in Oilwells for Downhole Corrosion Control and Increasing Production," SPE 87562 (paper presented at the SPE International Symposium on Oilfield Corrosion, Aberdeen, United Kingdom, 28 May 2004).
65. K. G. Cooper, R. E. Talbot, and M. J. Turvey, U.S. Patent 5741757, 1998.
66. C. R. Jones and R. E. Talbot, U.S. Patent 6784168, 2004.
67. T. F. McNeel, D. L. Comstock, M. Z. Anstead, and R. A. Clark, U.S. Patent 6180056, 2001.
68. G. P. Otter, S. G. Breen, G. Woodward, R. E. Talbot, R. S. Padda, K. P. Davis, S. D'Arbeloff-Wilson, and C. R. Jones, International Patent Application WO/2003/021031.
69. J. F. Kramer, F. O'Brien, and S. F. Strba, "A New High Performance Quaternary Phosphonium Biocide for Microbiological Control in Oilfield Water Systems," Paper 08660 (paper presented at the NACE CORROSION Conference & Exposition, New Orleans, LA, 16–20 March 2008).
70. V. Vlasaty and D. Q. Cao, International Patent Application WO/2008/019320
71. D. T. Murray, "A New Quat Demonstrates High Biocidal Efficacy with Low Foam," Paper 97406 (paper presented at the NACE CORROSION Conference, 1997).

72. J. E. Gannon and S. Thornburgh, World Patent Application WO/1988/002351.
73. A. Naraghi and N. Obeyesekere, International Patent Application WO/2006/034101.
74. V. L. Colclough, U.S. Patent 6303557, 2001.
75. R. F. Stockel, U.S. Patent 4891423, 1990.
76. J. G. Fenyes and J. D. Pera, U.S. Patent 4778813, 1988.
77. L. G.Kleina, M. H. Czechowski, J. S. Clavin, W. K. Whitekettle, and C. R. Ascolese, "Performance and Monitoring of a New Non-Oxidizing Biocide—The Study of BNPD/ISO and ATP," Paper 97403 (paper presented at the NACE CORROSION Conference, 1997).
78. D. M. Bryce, B. Croshaw, J. E. Hall, V. R. Holland, and B. Lessel, "The Activity and Safety of the Antimicrobial Agent Bronopol (2-Bromo-Nitropan-1,3-Diol)," *Journal of the Society of Cosmetic Chemists* 29 (1978): 3.
79. C. D. Gartner, U.S. Patent 5627135, 1997.
80. J. M. Cronan Jr. and M. J. Myer, U.S. Patent 7008545, 2006.
81. E. S. Littmann and T. L. McLean, "Chemical Control of Biogenic H_2S in Producing Formations," SPE 16218 (paper presented at the SPE Production Operations Symposium, Olahoma City, OK, 8–10 March 1987).
82. H. Jin-Ying, Z. Jia-Shen, F. Chao-Yang, Q. Jun-e, and L. Jian-Guo, "The Inhibition Effects of a New Heterocyclic Bisquaternary Ammonium Salt in Simulated Oilfield Water," *Anti-Corrosion Methods and Materials* 51 (2004): 272.
83. T. M. Williams, R. Levy, and B. Hegarty, "Control of SRB Biofouling and MIC by Chloromethyl-Methyl-Isothiazolone," Paper 01273 (paper presented at the NACE CORROSION Conference, 2001).
84. S. N. Lewis, G. A. Miller, and A. B. Law, U.S. Patent 4322475.
85. R. P. Clifford and G. A. Birchall, U.S. Patent 4539071, 1985.
86. T. M. Williams, B. M. Hegarty, and R. Levy, "Control of Oilfield Biofouling," Paper 01273 (paper presented at the NACE CORROSION Conference, 2001).
87. T. K. Haack, European Patent P 337624A, 1989.
88. B. M. Hegarty and R. Levy, U.S. Patent 5827433, 1998.
89. D. Antoni-Zimmermann, R. Baum, T. Wunder, and H.-J. Schmidt, U.S. Patent Application 20050124674.
90. N. E. Thompson and M. Greenhalgh, International Patent Application WO/2007/139950.
91. R. J. Starkey, G. A. Monteith, and C. W. Aften, International Patent Application WO/2008/016662.
92. J. H. Payton, U.S. Patent 3996378, 1976.
93. T. M. LaMarre and C. H. Martin, U.S. Patent 4616037, 1986.
94. A. J. Telang, S. Ebert, J. M. Foght, D. W. S. Westlake, and G. Voordouw, "Effects of Two Diamine Biocides on the Microbial Community from an Oil Field," *Canadian Journal of Microbiology/Review* 44(11) (1998): 1060.
95. M. Ludensky, C. Hill, and F. C. A. Lichtenberg, U.S. Patent Application 20030228373.
96. F. W. Valone, U.S. Patent 4647589, 1987.
97. J. G. La Zonby, U.S. Patent 5209824, 1993.
98. R. W. Walter, A. G. Relenyi, and R. L. Johnson, U.S. Patent 4816061, 1989.
99. R. W. Walter and L. M. Cooke, Paper 410 (paper presented at the NACE CORROSION Conference, 1997).
100. W. K. Whitekettle and D. K. Donofrio, U.S. Patent 4916158, 1990.
101. B. Heer, G. Tiedtke, and B. M. Hegarty, U.S. Patent Application 20040198713.
102. J. S. Gill and A. Gupta, U.S. Patent Application 20030200997.
103. H. Eggensperger and K.-H.Diehl, U.S. Patent 4148905, 1979.
104. T. E. McNeel, M. S. Whittlemore, S. D. Bryant, and G. H. Vunk, World Patent Application WO/1999/006325.

105. T. E. McNeel, M. S. Whittlemore, S. D. Bryant, and G. H. Vunk, World Patent Application WO/1999/005912.
106. G. E. Jenneman, A. Greene, and G. Voordouw, U.S. Patent Application 20050238729.
107. E. D. Burger, A. B. Crews, and H. W. Ikerd II, Paper 01274 (paper presented at the NACE CORROSION Conference, 1999).
108. E. D. Burger and J. M. Odom, "Mechanisms of Anthraquinone Inhibition of Sulfate-Reducing Bacteria," SPE 50764 (paper presented at the SPE International Symposium on Oilfield Chemistry, Richardson, TX, 1999).
109. F. B. Cooling, C. L. Maloney, E. Nagel, J. Tabinowski, and J. D. Odom, "Inhibition of Sulfaterespiration by 1,8-Dihydroxyanthraquinone and Other Anthaquinone Derivatives," *Applied and Environmental Microbiology* 62 (1996): 2999.
110. M. D. Law, M. B. Kretsinger, E. D. Burger, J. G. Schoenenberger, and M. A. Ulman, "A Field Case History: Chemical Treatment of a Produced-Water Injection System Using Anthraquinone Improves Water Quality and Reduces Costs," SPE 65023 (paper presented at the SPE International Symposium on Oilfield Chemistry, Houston, TX, 13–16 February 2001).
111. M. D. Johnson, M. L. Harless, A. L. Dickinson, and E. D. Burger, "A New Chemical Approach to Mitigate Sulfide Production in Oilfield Water Injection Systems," SPE 50741 (paper presented at the SPE International Symposium on Oilfield Chemistry, Houston, TX, 16–19 February 1999).
112. A. Dickinson, G. Peck, and B. Arnold, "Effective Chemistries to Control SRB, H_2S, and FeS Problems," SPE 93007 (paper presented at the SPE Western Regional Meeting, Irvine, CA, March 30–April 1 2005).
113. E. D. Burger, C. A. Andrade, M. Rebelo, and R. Ribeiro, "Flexible Treatment Program for Controlling H_2S in FPSO Produced-Water Tanks," SPE 106106 (paper presented at the SPE International Symposium on Oilfield Chemistry, Houston, TX, 28 February–2 March 2007).
114. R. L. Martin, "Corrosion Consequences of Nitrate/Nitrite Additions to Oilfield Brines," SPE 114923 (paper presented at the Annual Technical Conference and Exhibition, Denver, CO, 22–24 September 2008).
115. M. A. Reinsel, J. T. Sears, P. S. Stewart, and M. J. McInerney, "Isolation and Characterization of Strains CVO and FWKO B, Two Novel Nitrate-Reducing, Sufide-Oxidizing Bacteria Isolated from Oil Field Brine," *Journal of Industrial Microbiology* 17(2) (1996): 128.
116. D. O. Hitzman and D. M. Dennis, "Sulfide Removal and Prevention in Gas Wells," SPE 50980, *SPE Reservoir Evaluation & Engineering* 1(4) (1998): 367.
117. P. J. Sturman and D. M. Goeres, "Control of Hydrogen Sulfide in Oil and Gas Wells with Nitrite Injection," SPE 56772 (paper presented at the SPE Annual Technical Conference and Exhibition, Houston, TX, 3–6 October 1999).
118. C. Hubert, M. Nemati, G. Jenneman, and G. Voordouw, "Containment of Biogenic Sulfide Priduction in Continuous Up-Flow Packed-Bed Bioreactors with Nitrate or Nitrite," *Biotechnology Progress* 19(2) (2003): 338.
119. G. Voordouw, M. Nemati, and G. E. Jenneman, "Use of Nitrate-Reducing, Sulfide-Oxidizing Bacteria to Reduce Souring in Oil Fields," Paper 02034 (paper presented at the NACE CORROSION Conference, 2002).
120. E. A. Morris, R. M. Derr, T. M. Kenney, and D. H. Pope, "Field and Laboratory Tests on Nitrate Treatment for Potential Use in Natural Gas Operations—Stimulate Non-SRB Bacteria," SPE 29738 (paper presented at the SPE/EPA Exploration and Production Environmental Conference, Houston, TX, 27–29 March 1995).

121. J. F. D. Scott, "Modern Concepts of Chemical Treatment for the Control of Microbially-Induced Corrosion in Oilfield Water Systems," Proceedings of Chemistry in the Oil Industry IX Symposium, 31 October–2 November 2005, Manchester, UK: Royal Society of Chemistry Publications.
122. C. Hubert, G. Voordouw, M. Nemati, and G. E. Jenneman, "Is Souring and Corrosion by Sulfate-Reducing Bacteria in Oil Fields Reduced More Efficiently by Nitrate or by Nitrate?," Paper 04762 (paper presented at the NACE CORROSION Conference, 2004).
123. E. A. Greene, C. Hubert, M. Nemati, G. E. Jenneman, and G. Voordouw, "Nitrite Reductase Activity of Sulfate-Reducing Bacteria Prevents Their Inhibition by Nitrate-Reducing, Sulfide-Oxidizing Bacteria," *Environmental Microbiology* 5 (2003): 607.
124. D. O. Hitzman and G. T. Sperle, "A New Microbial Technology for Enhanced Oil Recovery and Sulfide Prevention and Reduction," SPE 27752 (paper presented at the SPE/DOE Enhanced Oil Recovery Symposium, Tulsa, OK, 17–20 April 1994).
125. K. A. Sandbeck and D. O. Hitzman, "Biocompetitive Exclusion Technology: A Field System to Control Reservoir Souring and Increase Production" (paper presented at the 5th Int. Conf. on Microbial Enhanced Oil Recovery and Related Biotechnology for Solving Environmental Problems, Sponsored by U.S. DOE, 1995).
126. D. O. Hitzman and D. M. Dennis, "New Technology for Prevention of Sour Oil and Gas," SPE 37908 (paper presented at the SPE/EPA Exploration and Production Environmental Conference, Dallas, TX, 3–5 March 1997).
127. D. O. Hitzman, G. T. Sperl, and K. A. Sandbeck, U.S. Patent 5750392, 1998.
128. M. A. Reinsel, J. T. Sears, P. S. Stewart, and M. J. McInerney, "Control of Microbial Souring by Nitrate, Nitrite or Glutaraldehyde Injection in a Sandstone Column," *Journal of Industrial Microbiology* 17 (1996): 128.
129. D. O. Hitzman, M. Dennis, and D. C. Hitzman, "Recent Successes: MEOR Using Synergistic H_2S Prevention and Increased Oil Recovery Systems," SPE 89453 (paper presented at the SPE/DOE Symposium on Improved Oil Recovery, Tulsa, OK, 17–21 April 2004).
130. M. Collison, "Biological H_2S Removal Is Gaining a Toehold in the Sour Gas Fields of Western Canada," *Oilweek Magazine*, April 2006.
131. A. Anchliya, "New Nitrate-Based Treatments—A Novel Approach to Control Hydrogen Sulfide in Reservoir and to Increase Oil Recovery," SPE 100337 (paper presented at the SPE Europec/EAGE Annual Conference and Exhibition, Vienna, Austria, 12–15 June 2006).
132. J. Larsen, "Downhole Nitrate Applications to Control Sulfate Reducing Bacteria Activity and Reservoir Souring," Paper 02025 (paper presented at the NACE CORROSION Conference, 2002).
133. T. Thorstenson, G. Bodtker, E. Sunde and J. Beeder, "Biocide Replacement by Nitrate in Sea Water Injection Systems," Paper 02033 (paper presented at the NACE CORROSION Conference, 2002).
134. E. Sunde and T. Torsvik, "Microbial Control of Hydrogen Sulfide Production in Oil Reservoirs," *Petroleum Microbiology*, eds. B. Ollivier and M. Magot, Washington, DC: ASM Press, 2005, 201.
135. D. O. Hitzman, G. T. Sperl, and K. A. Sandbeck, U.S. Patent 5405531, 1995.
136. G. Voordouw, B. Buziak, S. Lin, A. Grigoryan, K. M. Kaster, G. E. Jenneman, and J. J. Arensdorf, "Use of Nitrate or Nitrite for the Management of the Sulfur Cycle in Oil and Gas Fields," SPE 106288 (paper presented at the SPE International Symposium on Oilfield Chemistry, Houston, TX, 28 February–2 March 2007).
137. D. M. Dennis and D. O. Hitzman, "Advanced Nitrate-Based Technology for Sulfide Control and Improved Oil Recovery," SPE 106154 (paper presented at the SPE International Symposium on Oilfield Chemistry, Houston, TX, 28 February–2 March 2007).

138. C. Kuijvenhoven, A. Bostock, D. Chappell, J. C. Noirot, and A. Khan, "Use of Nitrate to Mitigate Reservoir Souring in Bonga Deepwater Development Offshore Nigeria," SPE 92795 (paper presented at the SPE International Symposium on Oilfield Chemistry, The Woodlands, TX, 2–4 February 2005).
139. C. Koijvenhoven, J. C. Noirot, P. Hubbard, and L. Oduola, "One Year Experience with the Injection of Nitrate to Control Souring in Bonga Deepwater Development Offshore Nigeria," SPE 105784 (paper presented at the SPE International Symposium on Oilfield Chemistry," Houston, TX, 28 February–2 March, 2007).
140. E. Sunde and T. Torsvik, U.S. Patent 6758270, 2004.
141. R. B. Cassinis, W. A. Farrone, and J. H. Portwood, "Microbial Water Treatment: An Alternative Treatment to Manage Sulfate Reducing Bacteria (SRB) Activity, Corrosion, Scale, Oxygen, and Oil Carry-Over at Wilmington Oil Field, Wilmington, CA," SPE 49152 (paper presented at the SPE Annual Technical Conference and Exhibition, New Orleans, LA, 27–30 September 1998).
142. C. Hubert, G. Voordouw, J. Arensdorf, and G. E. Jenneman, "Control of Souring Through a Novel Class of Bacteria That Oxidize Sulfide as Well as Oil Organics with Nitrate," Paper 06669 (paper presented at the NACE CORROSION Conference, 2006).
143. D. R. Grimshaw, U.S. Patent Application, 20060185851.
144. S. L. Percival, "The Effect of Molybdenum on Biofilm Development," *Journal of Industrial Microbiology and Biotechnology* 23 (1999): 112.
145. P. Angell and D. C. White, "Is Metabolic Activity by Biofilms with Sulfate-Reducing Bacterial Consortia Essential for Long-Term Propagation of Pitting Corrosion of Stainless Steel?," *Journal of Industrial Microbiology and Biotechnology* 15(4) (1995): 329.
146. D. Stepan and D. Ye, International Patent Application WO/2008/076928.

15 Hydrogen Sulfide Scavengers

15.1 INTRODUCTION

Hydrogen sulfide (H_2S) is a very toxic and pungent gas that causes problems in both the upstream and downstream oil and gas industry. Exposure to this gas even at fairly low concentrations can cause serious injury or death. Natural gas for sale often requires the concentration of H_2S to be below about 4 ppm. H_2S is often accompanied by smaller amounts of mercaptans (RSH or R_2S) such as methyl mercaptan CH_3SH, aromatic sulfide species, polysulfides, and carbonyl sulfide (COS).

H_2S is known as a sour gas, which is appreciably soluble in water. It behaves as a weak acid partially dissociating into hydrosulfide and sulfide ions:

$$H_2S + H_2O = H_3O^+ + HS^- \qquad pK_a = 6.9$$

$$HS^- + H_2O = H_3O^+ + S^{2-} \qquad pK_a = 19$$

The concentrations of the anionic species are dictated by the pH, particularly by the presence of another acid gas, CO_2. Being a weak acid, H_2S is corrosive, reacting with steels in wells and pipelines, causing pitting and stress cracking corrosion and deposition of iron sulfide scales.[2] Other sulfide scales that have been observed include zinc sulfide and lead sulfide (see Chapter 3 on scale control for further information).

There are several natural processes that can generate H_2S in reservoirs. They include bacterial sulfate reduction by indigenous sulfate-reducing bacteria (SRBs), thermal cracking, and thermochemical sulfate reduction (TSR) by hydrocarbons. It has been proposed that it is the TSR that leads to the largest amount of H_2S; however, other studies suggest that TSR by hydrocarbons only takes place significantly above 140°C (284°F) in the reservoir.[3] This phenomenon involves hydrocarbon oxidation and sulfate reduction (from anhydrite either naturally occurring or formed from injected sulfate anion in seawater) and produces as by-products, hydrogen sulfide, carbon dioxide, carbonate minerals, and heavy organosulfur compounds.

H_2S production is usually significantly higher in reservoirs that are seawater-flooded for secondary oil recovery.[4] Seawater contains about 2,800 ppm sulfate ions. These ions are reduced to H_2S by indigenous SRBs in the reservoir and by the TSR process, which eventually reaches the production wells. The growth of SRBs also requires the presence of an easily metabolized carbon source such as organic acids. These are usually found in high-enough quantities in the reservoir fluids to sustain

SRB growth.[5-6] Thus, one way to reduce reservoir souring is to treat the injection water with sufficient biocide to prevent growth of SRB. Other preventive methods include controlling SRB metabolism or promoting the growth of nonindigenous SRB. Chapter 14 on biocides should be consulted for these methods. A final method of preventing biogenic sulfide formation is to inject nonsulfated aquifer water, if it is available, or inject desulfated seawater using membrane technology. In the latter case, not all the sulfate ions can be removed from seawater, but enough are removed to reduce reservoir souring as well as sulfate scaling considerably.

H_2S must be removed from produced natural gas to meet gas sales specifications, which require a maximum H_2S level of a few ppm. Batch treatments to adsorb H_2S and other sulfur compounds can be used to remove very small amounts, that is, low gas-flow rate and/or small concentrations of hydrogen sulfide. Batch treatments can use solids or liquid slurries, nitrite, aqueous ClO_2, zinc or iron oxide slurries, formaldehyde/methanol/water, molecular sieves, iron sponge, and other proprietary metal-based processes. In large production facilities, the most economic solution to remove H_2S in the gas process stream is to install a regenerative system for treating the sour gas.[7-8] These systems typically employ a compound used in an absorption tower to contact the produced fluids and selectively absorb the H_2S and possibly other toxic materials such as mercaptans. The process is known as gas sweetening. The absorption compound and H_2S is then regenerated, usually by heating. The absorption material is reused in the system and the separated H_2S treated by a modified Claus process to form elemental sulfur.

Concentrated solutions of aqueous amines (actually alkanolamines), mixed with activators, are by far the most common chemicals for removing H_2S from produced natural gas. These solutions undergo reversible reactions with acid gases and can be regenerated (usually by applying heat) in a cyclic process to remove rather large amounts of sulfur, and CO_2, when needed. Several types of amine solutions may be used depending on the sour gas specifications. Typical amines that have been used include:

- monoethanolamine (MEA)
- diethanolamine (DEA)
- *N*-methyldiethanolamine (MDEA)
- Diglycolamine (DGA), also known as 2-(2-aminoethoxy)ethanolamine

MEA and DEA have been used in the past to sweeten gas streams. These alkanolamines absorb both H_2S and CO_2. DEA is a secondary amine that reacts rather fast with CO_2 to form a carbamate that is partially hydrolyzed into bicarbonate. MDEA is selective to H_2S. Most modern amine gas–sweetening processes are MDEA-based. MDEA is a tertiary amine that cannot react with CO_2 (as there is no free H atoms on the N atom), so that the absorption of CO_2 is only possible via bicarbonate/carbonate formation. This is a slow reaction compared with the one between the alkanolamine and H_2S. However, activators can be added to improve the CO_2 absorption if desired, such as polyalkyleneamines, alkoxypropylamines, aminopiperazines, aminopiperidines, butyldiethanolamine, and aminoethylethanolamine.[9] DGA (also known as [2-(2-aminoethoxy)]ethanol) in an aqueous solution has been used in both natural and refinery gas applications. DGA is a primary amine capable of removing not

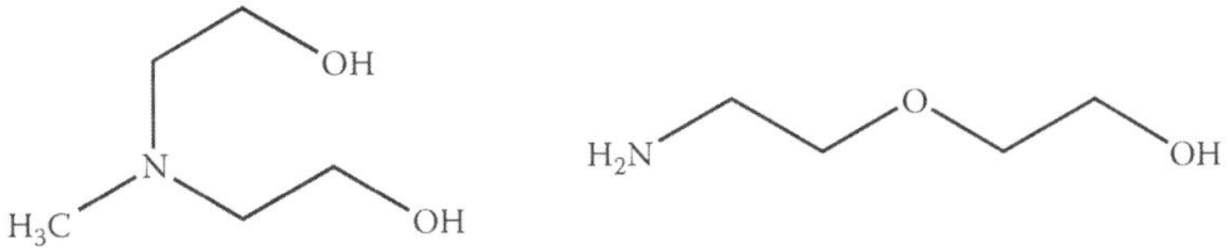

Hydrogen Sulfide Scavengers

FIGURE 15.1 *N*-methyldiethanolamine and 2-(2-aminoethoxy)ethanolamine (DGA).

only H_2S and CO_2, but also COS and mercaptans from gas and liquid streams. Triethanolamine and diisopropanolamine have also been used to a lesser extent for H_2S absorption.

Other commercial solvent processes to remove H_2S include the use of polyethyleneglycol, *N*-methyl pyrrolidone (NMP), propylene carbonate, and methanol. Potassium carbonate, alone or mixed with amine solutions with other activators, has been used for removal of H_2S and carbon dioxide (CO_2) from gas streams in petrochemical and crude oil refining industries for many years. Sulfolane, optionally formulated with amines, is also a commercial H_2S solvent. Iron (or vanadium) chelate sulfur removal processes belong to a commercial process group referred to as liquid redox sulfur recovery. The reaction is very efficient but best for low sulfur applications, that is, under roughly 10–12 tons per day.

In cases where the concentration of H_2S is only a few hundred parts per million or less, it is most economic to treat the sour hydrocarbon production stream with nonregenerative H_2S scavengers. This is most common for wet gas lines but is also carried out in oil-production lines.

15.2 NONREGENERATIVE H_2S SCAVENGERS

Nonregenerative H_2S scavengers that have been investigated can be divided into the following categories:

1. solid, basic metallic compounds
2. oxidizing chemicals
3. aldehydes and aldehyde-related products
4. reaction products of aldehydes and amines, including triazines
5. metal carboxylates and chelates (some of these are regenerative)
6. other amine-based products

Triazine scavengers are the most commonly used today. Some transition metal complexes and chelates such as ferric chelates are regenerative, selective H_2S scavengers and are discussed below.

The scavenger reacts with the H_2S to form a nontoxic compound, which can be removed from the hydrocarbons or can be discharged in the water phase. Products in categories 2–6 are mostly water-soluble formulations. Laboratory equipment for testing H_2S scavengers have been described.[11–12] Liquid scavengers can be injected anywhere along the production line, for example, at the wellhead or further downstream. A patented injection equipment that atomizes the H_2S scavenger is claimed to reduce scavenger consumption by 30–35%.[13]

15.2.1 SOLID SCAVENGERS

Solid scavengers cannot be injected and are therefore only useful in treating sour gas in process facilities. Solid scavengers are generally zinc- or iron-based materials. A bed of zinc oxide, ZnO, was used successfully to remove H_2S from the produced gas for one North Sea operator.[14] Improved solid iron oxide scavengers used in the industry today form innocuous iron pyrite (FeS_2) as the reaction product. The catalyst can be impregnated on an inert ceramic material, solving the pyrophoricity problem of the earlier iron-sponge type scavengers.

15.2.2 OXIDIZING CHEMICALS

Examples of water-soluble salts with oxidizing anions include chlorites (e.g., $NaClO_2$), bromates/iodates (e.g., $NaBrO_3$), nitrites (e.g., $NaNO_2$), and persalts.[15–16] The reaction of these oxidizing agents, except the persalts, with H_2S is complicated, but elemental sulfur is usually one of the products.

Chlorite oxidizing agents react very rapidly with H_2S, but their use is restricted, as they give handling and operational problems such as sulfur deposition and corrosion problems. They have, however, been used very effectively in scavenging H_2S from disused storage tanks where further corrosion was not a major issue.[17]

Sodium nitrite solution has been used in downhole squeeze injection into sour oil and gas wells and has effectively removed H_2S from the aqueous and gas phases.[18] This method also reduced corrosion due to H_2S and removed iron sulfide scale from the near-wellbore. H_2S removal from topsides water separations equipment was also facilitated using sodium nitrite.

The common use of triazine H_2S scavengers (discussed in Section 15.2.4) has sometimes caused environmental discharge problems in one sector of the North Sea because the overboard water contained unused triazine or toxic amines formed from reaction of triazines with H_2S. An environment-friendly H_2S scavenger that is peroxide-based has been developed to overcome this problem.[19] The simplest product is hydrogen peroxide, although one could imagine persalts such as perborate or persulfate, which also contain the O–O peroxide linkage, might be equally effective. Organic peroxides have also been field-tested but led to corrosion problems. Peroxide chemicals convert sulfide to sulfate. For example, the reaction of hydrogen peroxide with H_2S will generate sulfate ions and protons:

$$H_2S + 4H_2O_2(aq) \rightarrow SO_4^{2-}(aq) + 4H_2O(l) + 2H^+(aq)$$

Offshore experience suggests that H_2S must be present as ions (HS^- or S^{2-}) to react quickly and that the pH of the peroxide product must be high. However, a high pH usually destabilizes hydrogen peroxide especially in the presence of contaminants such as trace transition metals. In collaboration with an operator, one service company has managed to formulate a stable, high-pH peroxide product and has successfully used this in the laboratory and pilot plants (imidazoles or triazoles, which prevent the peroxide from binding to metal sites, where it is catalytically decomposed, are possible stabilizers). However, the stabilizer did not fulfill the OSPAR

Hydrogen Sulfide Scavengers

environmental requirements for use in the North Sea and new stabilizers were sought. Corrosion issues due to the strongly oxidizing peroxide were not addressed.

15.2.3 Aldehydes

Aldehydes react with H_2S to form various sulfur products. Typical aldehydes that have been used include formaldehyde, glutaraldehyde, acrolein, and glyoxal (Figure 15.2).[20–23]

Formaldehyde forms mainly the ring compound 1,2,3-trithiane with H_2S (Figure 15.3). Formaldehyde is a suspected carcinogen, banned in Europe and is, therefore, not much used today. However, the safer, water-soluble chemical hexamethylenetetramine (HMTA) can be used in acid stimulation fluids since it decomposes on reaction with strong acids to give formaldehyde in situ.[11,34] However, there is a possibility of formation damage from oily polymeric substances formed by reaction of aldehydes and acids (see below). Further, the reaction products of H_2S and formaldehyde are water-insoluble solids, and these can cause processing issues.

Glyoxal reacts with an excess of H_2S to yield a partially water-soluble *trans,trans*-4,4,5,5′-tetrahydroxy-2,2′-bi(1,3-thioxolane).[24]

The larger the aldehyde, the greater the fraction that will partition into the liquid hydrocarbon phase. Thus, formaldehyde, glutaraldehyde, and glyoxal partition about 98–99% into the brine phase, whereas acrolein partitions over 50% into the liquid hydrocarbon phase. Acrolein is a very effective scavenger of sulfides but is not widely used because of its high toxicity and handling problems.

One disadvantage with the aldehydes is that most of the products from reaction with H_2S are poorly soluble in water. However, aldehydes have the additional benefit that they are biocides (see Chapter 14). Thus, they can kill SRBs preventing them from forming H_2S.[25] This makes them good candidates for seawater injection projects, near-well, or squeeze treatments as they can control SRB formation and remove any H_2S that is still formed.[26] Acrolein can actually perform three tasks. First, it is a H_2S scavenger, a biocide, and it also dissolves iron sulfide scale.[27–28] Acrolein reacts with H_2S to form first a thiolaldehyde, which reacts further with acrolein to form a water-soluble thiopyran (Figure 15.4).[29–30]

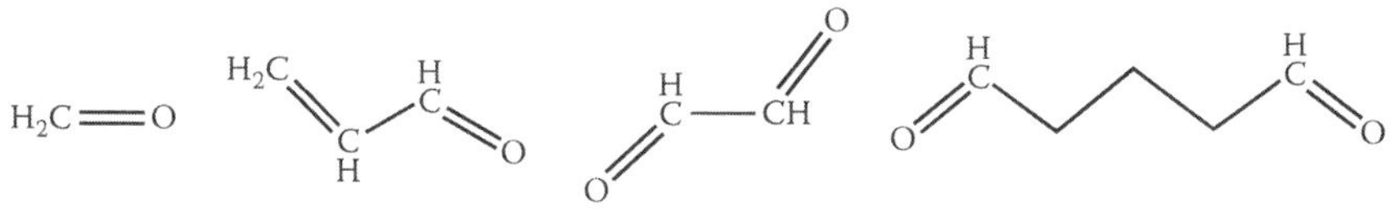

FIGURE 15.2 From left to right, formaldehyde, acrolein, glyoxal, and glutaraldehyde.

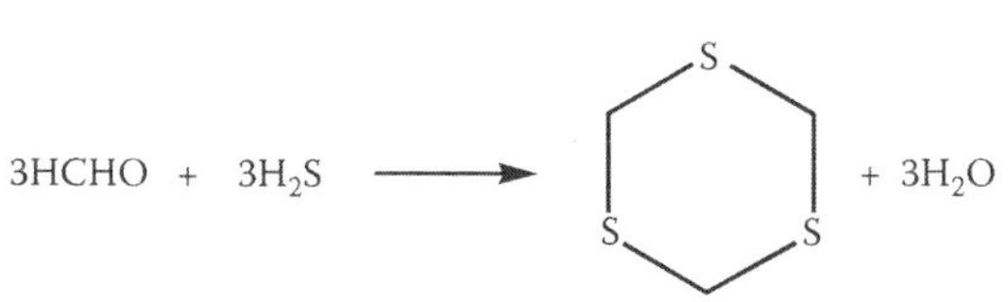

$$3HCHO + 3H_2S \longrightarrow \text{(1,2,3-trithiane ring)} + 3H_2O$$

FIGURE 15.3 Reaction of formaldehyde with H_2S.

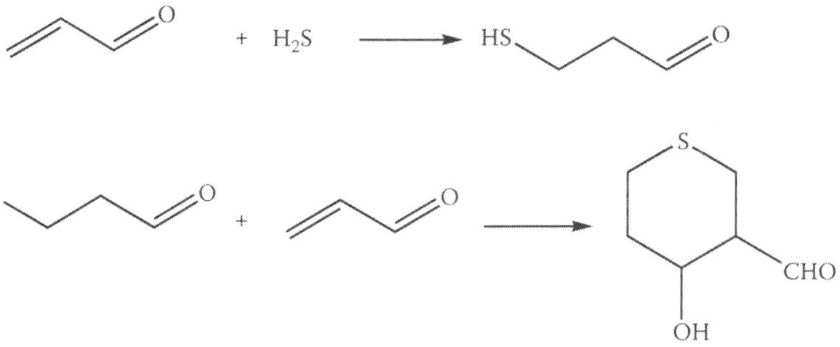

FIGURE 15.4 Reaction of acrolein with H_2S.

Acrolein has also been used as a H_2S scavenger in multiphase production.[31] Water-soluble low-molecular-weight polycondensation products produced by the condensation of acrolein and formaldehyde have been proposed for use in the elimination of hydrogen sulfide and iron sulfide present in aqueous systems.[32]

One drawback of the aldehydes is that the reaction with H_2S is somewhat slow, especially at low temperatures. In comparison, triazine-based scavengers (see below) react significantly faster with H_2S (but not at low temperature). However, aldehydes do not raise the pH of the produced water as do amine and triazine products. A raised pH can cause possible carbonate scaling and exacerbate emulsion tendencies.

It has been found that the reaction products of aldehydes and polyhydroxyl or urea-based compounds are excellent H_2S scavengers. This technology is amine-free, although the inventors state that a blend with alkanolamines and/or triazines is preferred. Use of this or any aldehyde-based scavenger avoids raising the pH of the water, which occurs with triazine H_2S scavengers, leading to carbonate scaling (see Section 15.2.4).[10] A typical product is made by reacting formaldehyde with ethylene glycol or glycerol. The product using ethylene glycol is ethylene glycol hemiformal, also known as [1,2-ethanediylbis(oxy)]-bis-methanol or 1,6-dihydroxy-2,5-dioxahexane (Figure 15.5). This product can be used in combination with dimethylolurea (also known as N,N-bis-(hydroxymethyl)urea) formed from the reaction of urea and formaldehyde. The heterocyclic acetal, 1,3-dioxolane, can also be used, made by a 1:1 reaction of ethylene glycol and formaldehyde. This will act like formaldehyde in its reaction with H_2S, producing 1,3,5-trithiane and regenerating ethylene glycol. However, dioxolanes and trioxanes do not react with the HS^- ion.[35] Butylformal (butoxymethanol) can also be used to scavenge H_2S.

Aldehyde-based H_2S scavengers can be used in acid-stimulation packages for stimulation of the near-wellbore of sour carbonate reservoirs. An early study in the Middle East used aldehydes as H_2S scavengers in acid stimulation of injector wells,

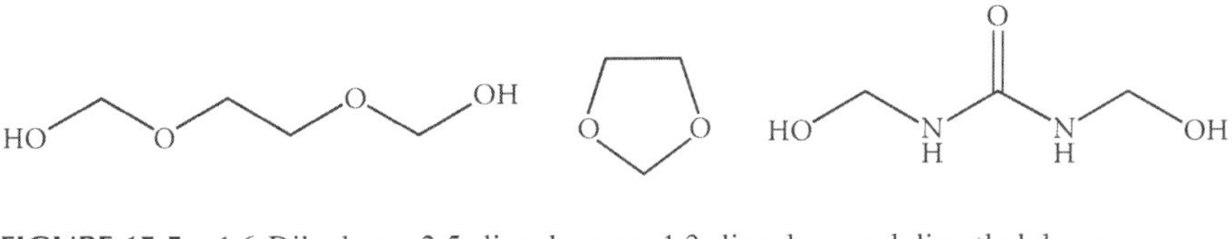

FIGURE 15.5 1,6-Dihydroxy-2,5-dioxahexane, 1,3-dioxolane and dimethylolurea.

but it was observed that at high scavenger concentration, a polymeric material was formed, which adversely affected acid reaction with iron sulfide and could cause formation damage.[33] A later study used HMTA, which generates formaldehyde in low-pH solutions such as 7.5% HCl. However, HTMA was found to be a worse scavenger than formaldehyde. The best H_2S scavenger from the laboratory test promoting good iron sulfide dissolution, and the one used in the field, was a mixture of an aliphatic and an aromatic aldehyde.[34] However, due to concerns over formation damage with the use of aldehydes, a hydroxyalkyltriazine was successfully used in later acid stimulation field applications (see Section 15.2.4 for details about triazines).[11]

15.2.4 Reaction Products of Aldehydes and Amines, Especially Triazines

Aldehydes react with amines to form imines, but with one equivalent of formaldehyde cyclic 1,3,5-hexahydrotriazine products are formed (the IUPAC correct name is triazinane). A side product is N,N'-methylenebis-oxazolidine or this can be made the main product if the ratio of amine to formaldehyde is adjusted. The bis-oxazolidines alone have been claimed as H_2S scavengers,[36] but it is the 1,3,5-hexahydrotriazine products that are most widely used in the oil and gas industry. Although their correct name is 1,3,5-hexahydrotriazines, they are usually called just triazines by production chemists. Laboratory experiments, using a wide range of CO_2 partial pressures and H_2S-to-CO_2 ratios, indicate that CO_2 has very little effect on the H_2S scavenging performance of triazine-based H_2S scavengers.[37]

The first triazines to be investigated were halogenated triazines such as trichloro-S-triazinetrione.[38] The most common triazines used today are water-soluble and made by the reaction of alkanolamines and/or methylamine with formaldehyde. For example, with ethanolamine, the reaction product is mainly 1,3,5-tri-(2-hydroxyethyl)-hexahydro-S-triazine (Figure 15.6).[39-40] N,N'-methylenebis-oxazolidine formed as a biproduct (or main product if the reaction ratio is changed) will also react with H_2S.[41] 1,3,5-Trimethyl-hexahydrotriazine is another common H_2S scavenger.

Triazines made from other amines have also been claimed. For example, the reaction product of 3-methoxypropylamine (MOPA), methylamine, and formaldehyde is claimed to be a superior H_2S scavenger to the hydroxyethyltriazines.[42-43] It is also possible to make triazines in situ from an alkanolamine and formaldehyde as long as this is far enough upstream for the triazines to form and react with H_2S.[44]

Triazines react faster with H_2S than aldehydes, and also react with the HS^- ion.[35] Triazine liquid scavengers are economical up to about 50 kg of H_2S/day and will

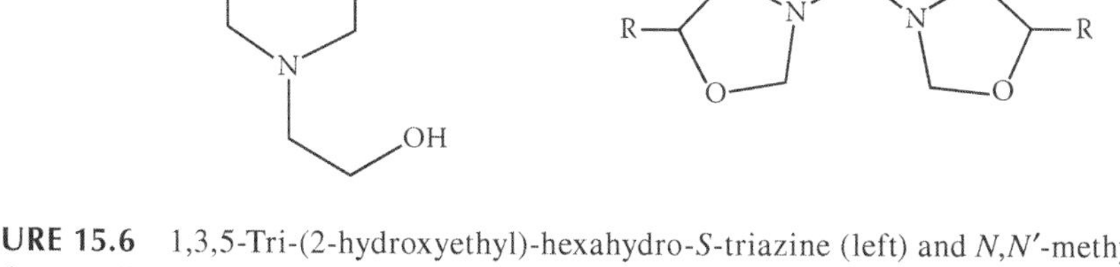

FIGURE 15.6 1,3,5-Tri-(2-hydroxyethyl)-hexahydro-S-triazine (left) and N,N'-methylenebis-oxazolidines (right). R = alkyl.

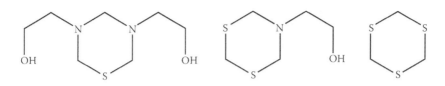

FIGURE 15.7 Sulfur reaction products of 1,3,5-tri-(2-hydroxyethyl) hexahydro-S-triazine with H_2S.

remove H_2S down to approximately 5-ppm levels in streams with relatively low concentrations of H_2S. Triazines have low toxicity characteristics and are usually biodegradable, but contain small amounts of free formaldehyde, which is a suspected carcinogen. They are usually injected as water or methanol/water solutions.[45–46] The original triazine can be regenerated by treatment with hot aqueous base, but this is not normally economic in practice.[47]

The main products from the reaction of 1,3,5-tri-(2-hydroxyethyl)-hexahydro-S-triazine with H_2S are 5-(2-hydroxyethyl)hexahydro-1,3,5-dithiazine, 3,5-bis-(2-hydroxyethyl)hexahydro-1,3,5-thiadiazine and 1,3,5-trithiazane depending on the pH (Figure 15.7).[48] The bis-hydroxyethyl compound is fairly water-soluble and does present problems but the dithiazine and other products are less water-soluble, especially at low temperatures, and this can cause a deposition problem in cold pipelines or process facilities. Dithiazine forms a separate liquid phase or layer in gas-processing equipment. At temperatures of about 20°C (68°F) or lower, solid dithiazine crystals form in this layer and precipitate out of solution. It has been claimed to use a chemical formulation consisting of triethylene glycol and the reaction product of an amine and an aldehyde (a triazine) to alleviate the problems associated with the deposition of dithiazine solids.[49] 1,3,5-Tri-(2-hydroxyethyl)-hexahydro-S-triazine has been used successfully in acid workovers of sour-water injectors to remove hydrogen sulfide and prevent further iron sulfide scale. The reaction products were water-soluble and the scavenger also gave some corrosion inhibition.[11]

There have been a number of attempts to improve the performance of triazine-based scavengers. An improvement that gives more water-soluble products after reaction with H_2S is to use a more hydrophilic triazine such as the reaction product of MEA, DGA, and formaldehyde. This gives a mixture including 1,3,5-tri-(2-hydroxyethyl)-hexahydro-S-triazine and 1,3,5-tri(2-ethoxyethanol)hexahydro-triazine.[50] Another more hydrophilic triazine is made by reacting dimethylaminopropylamine with formaldehyde. This molecule has pendant water-soluble dimethylamino groups.[51] Triazine-scavenging agents having hydroxyalkyl and alkylamine functionality are claimed to be superior scavengers than products containing just hydroxyalkyl functionality.[52] These scavenging agents are made by reacting at least one alkanolamine and at least one alkyl amine with an aldehyde, such as formaldehyde.

A scavenger composition, which comprises a triazine reaction product of formaldehyde with the very hydrophilic molecule aminoethylpiperazine (AEP) and a second or "enhancing" amine, such as n-butylamine or MEA, has also been claimed to avoid any organic sulfur solids deposition problems. N.A.[53] Several case histories have been reported.[54]

Another improvement is the use of a quaternary ammonium compound to accelerate the reaction of the triazine with H_2S.[55] The preferred hexahydro-triazine is 1,3,5-tri-methoxypropyl-hexahydro-1,3,5-triazine (MOPA hexahydro-triazine) and an example of a quaternary ammonium compound is benzyl cocoalkyl dimethyl quaternary ammonium chloride, which is also a film-forming corrosion inhibitor.

Another improvement is the use of oligomers of two or more triazine rings linked by CH_2 groups. The benefits of these compounds is that they are less liable to produce free formaldehyde and the reaction products with H_2S are less pungent.[9] Bis-triazines are formed by reacting ethylenediamine with formaldehyde.[56] Other complex polytriazines have been made by reacting an alkylenepolyamine, such as diethylenetriamine with formaldehyde.[57] Triazines, which are the product of the reaction of dimethylaminopropyl amine with a molar excess of formaldehyde, are also useful H_2S scavengers.[58] Triazines have also been used in foam-well treatments with aldehydes that have a dual function as biocides and H_2S scavengers.[59]

Triazines can also be made oil-soluble from the reaction of fairly hydrophobic alkylamines with formaldehyde. For example, the oil-soluble 1,3,5-trihexahydro-1,3,5-*tert*-butyltriazine has been claimed.[60] However, one factor demanding attention in scavenger selection for hydrocarbon liquids is the downstream processing of the condensate or crude, which often requires maintaining a low-nitrogen content to avoid poisoning of reforming catalysts. Water-soluble scavengers have been shown to work equally as well for removing H_2S from a condensate as the scavenger does not follow the condensate phase due to solubility and density differences.[46]

One problem with the triazine H_2S scavengers is that they contain tertiary amine groups and are, therefore, basic in aqueous solution raising the pH of the water. Amines formed after reaction of the triazine with H_2S are also basic. The increase in pH when using triazine H_2S scavengers has been shown to lead to new or increased calcium carbonate scaling problems:[61–62]

$$Ca^{2+}(aq) + 2HCO_3^-(aq) \leftrightarrow CaCO_3(s) + H_2O(l) + CO_2(g)$$

$$HCO_3^-(aq) + R_2R'NH(aq) \rightarrow CO_3^{2-}(aq) + R_2R'NH^+(aq)$$

$$CO_3^{2-}(aq) + Ca^{2+}(aq) \rightarrow CaCO_3(s)$$

To overcome this problem, one could use a H_2S-scavenging aldehyde, which is not basic, or treat with a carbonate-scale inhibitor. Combined scale inhibitor–H_2S scavenger blends have also been formulated that gave good field results.[63] A phosphonate-scale inhibitor was found to perform best in one combined treatment package.[64] Aromatic imine compounds, containing no free formaldehyde have been proposed as improved, less toxic alternatives to triazines.[79]

15.2.5 Metal Carboxylates and Chelates

Both water- and oil-soluble high-valence metal chelates have been used as H_2S scavengers. They have been used both for treating drilling fluids and contaminated water and oil streams.[65] The metal ion is usually zinc(II) or iron(III). The chelate contains carboxylate groups and can be based on nitrilotrisacetic acid, EDTA, polyaminodisuccinic acid, or a more biodegradable chelate such as a gluconate.[66–69] Normally, metal sulfides are formed, but if the metal is in a high oxidation state, it can be used to oxidize H_2S to sulfur.[70] An example is the ferric chelate of N-(2-hydroxyethyl) EDTA, which is reduced to the ferrous chelate after reaction with H_2S.[71] Ferric chelates can be regenerated, if necessary, by reaction of the ferrous chelate with O_2 at elevated temperature. However, an antioxidant stabilizer such as thiosulfate ion or enzyme catalase, which destroys hydroxyl radicals, is needed to prevent the chelate from degrading. This and other more common chelates such as EDTA are degraded.[72]

Fast-acting, oil-soluble H_2S scavengers based on carboxylates of zinc and iron have been developed. This technology's efficacy was demonstrated in a plant trial treating a 1,200-ppm hydrogen sulfide–contaminated oil stream.[66] A typical product mentioned is zinc carboxylate made from a long-chain fatty acid, which gives it the product its oil solubility. It was shown to react with H_2S to give fine transportable dispersions of ZnS in the oil phase:

$$Zn(OOCR)_2 + H_2S \leftrightarrow ZnS + 2HOOCR$$

15.2.6 Other Amine-Based Products

Besides the nonselective regenerative alkanolamines, several other more selective classes of amine-based products have been investigated. Small water-soluble amidines ($RC(=NH)NH_2$) or oligomeric amidines have been shown to have good H_2S scavenger properties.[73] Mixtures of amine oxides and specific enzymes have been patented, but no field data are available.[74] Piperazinone or an alkyl-substituted derivative such as 1,4-dimethylpiperazinone have also been claimed as good H_2S scavengers.[75] Maleimides, which also have biocidal properties, have also been shown to be good H_2S scavengers.[76] Another patent application claims a H_2S scavenger selected from a 1,3,5-trisalkanylamino hexahydro-1,3,5-triazine derivative, a morpholine or piperazine derivative, an amine oxide, an alkanolamine, or an aliphatic or aromatic polyamine.[77]

H_2S and small mercaptans (thiols) in hydrocarbons may be scavenged using a formulation of quaternary ammonium alkoxide or hydroxide in the presence of a high-oxidative-state metal such as cobalt, iron, chromium, and/or nickel.[78] The high-oxidative-state metal, being an oxidizer, probably acts as a catalyst for improved H_2S and mercaptan-scavenging performance. The exact mechanism by which the method operates is not known, but the product of the reaction of a mercaptan with the scavenger is believed to be a disulfide. Moreover, it has been found that introducing oxygen, such as by sparging the treated fluid with air, increases the scavenging activity dramatically. The performance of the scavenger has been found to be improved at higher temperatures such as about 50–70°C. A typical scavenger that was investigated on a sulfurous crude oil was a quaternary ammonium hydroxide prepared from dimethyl soya amine and ethylene oxide, with Co^{3+} ions.

15.3 SUMMARY

The choice of H_2S control strategy depends on many factors including H_2S concentration, temperature, produced fluid chemistry, environmental requirements, and where in the system the treatment is to be made. To summarize, there are five basic methods to treat reservoir souring and H_2S production:

1. Add a biocide to kill SRB. The biocide may also be a H_2S scavenger.
2. Treat the SRB with a metabolic inhibitor that prevents them from reducing sulfate to sulfide.
3. Stimulate the formation of non-SRB bacteria by adding nutrients such as nitrate ions for stimulating denitrifying bacteria.
4. Use unsulfated aquifer or desulfated seawater in water injector wells.
5. Use a H_2S scavenger.

Methods 1–4 are usually carried out at the injector wells, but biocides can also be injected into production streams. H_2S scavengers are injected either downhole or, most usually, topside in production. Methods 1–4 are discussed in Chapter 14.

REFERENCES

1. *Hydrogen Sulfide in Production Operations (Petroleum)*, 2nd ed., Austin, TX: University of Texas at Austin, 1996.
2. N. P. Tung, P. V. Hung, and H. D. Tien, "Study of Corrosion Control Effect of H 2 S Scavenger in Multiphase System," SPE 65399 (paper presented at the SPE International Symposium on Oilfield Chemistry, Houston, TX, 13–16 February 2001).
3. (a) P. Mougin, V. Lamoureux-Var, A. Bariteau, and A. Y. Huc, *Journal of Petroleum Science and Technology* 58 (2007): 413. (b) R. H. Worden, P. C. Smalley, and M. M. Cross, "The Influence of Rock Fabric and Mineralogy on Thermochemical Sulfate Reduction: Khuff Formation, Abu Dhabi," *Journal of Sediment Research* 70 (2000): 1210–1211.
4. I. Vance and D. R. Thrasher, "Reservoir Souring: Mechanism and Prevention," in *Petroleum Microbiology*, B. Ollivier, and M. Magot, eds., Washington, DC: ASM Press, 2005, 123.
5. (a) E. D. Berger, I. Vance, G. F. Gammack, and S. E. Duncan, The 5th International Conference on Microbial Enhanced Oil Recovery and Related Biotechnology for Solving Environmental Problems, sponsored by U.S. DOE, 1995. (b) D. O. Hitzman and D. M. Dennis, "New Technology for Prevention of Sour Oil and Gas," SPE 37908 (paper presented at the SPE/EPA Exploration and Production Environmental Conference, Dallas, TX, 3–5 March 1997).
6. E. A. Morris, R. Gomez, and R. Peterson, "Application of Chemical and Microbiological Data for Sulfide Control," SPE 52705 (paper presented at the SPE/EPA Exploration and Production Environmental Conference, Austin, TX, 1–3 March 1999).
7. A. L. Kohl and R. Nielsen, *Gas Purification*, 5th ed., Houston, TX: Gulf Professional Publishing, 1997.
8. M. Abedinzadegan Abdi, "Design and Operations of Natural Gas Sweetening Facilities Course for the National Iranian Gas Company Workshop," 2nd Iranian Gas Forum, Memorial University of Newfoundland (MUN), June 2008 (http://www.engr.mun.ca/people/mabdi.php).

9. (*a*) J.-L. Peytavy, S. Capdeville, and H. Lacamoire, U.S. Patent 6290754, 2001. (*b*) J.-L. Peytavy, P. Le Coz, and O. Oliveau, U.S. Patents 5348714 and 5209914, 1993.
10. (*a*) C. J. N. Buisman, A. J. H. Janssen, and R. J. Van Bodegraven, U.S. Patent 6656249, 2003. (*b*) C. Cline, A. Hoksberg, R. Abry, and A. Janssen, "Biological Process for H_2S Removal from Gas Streams: The Shell-Paques/THIOPAQ Gas Desulfurization Process" (paper presented at the Laurance Reid Gas Conditioning Conference, LRGCC, Norman, OK, 23–26 February 2003).
11. (*a*) H. L. Smith, A. F. Johnsen, and B. Knudsen, U.S. Patent 7078005, 2006. (*b*) J. van Dijk, and A. Bos, "An Experimental Study of the Reactivity and Selectivity of Novel Polymeric 'Triazine-Type' H_2S Scavengers," Proceedings of the Chemicals in the Oil Industry VI, 14–17 April 1997, p. 170.
12. H. A. Nasr-El-Din, M. Zabihi, S. K. Kelkar, and M. Samuel, "Development and Field Application of a New Hydrogen Sulfide Scavenger for Acidizing Sour-Water Injectors," SPE 106442 (paper presented at the SPE International Symposium on Oilfield Chemistry, Houston, TX, 28 February–2 March 2007).
13. H. Linga, F. P. Nilsen, R. Abiven, and B. H. Kalgraff, International Patent Application WO/2006/038810.
14. D. R. Wilson, "Hydrogen Sulfide Scavengers: Recent Experience in a Major North Sea Field," SPE 36943 (paper presented at the European Petroleum Conference, Milan, Italy, 22–24 October 1996).
15. E. E. Burnes and K. Bhatia, U.S. Patent 4515759, 1985.
16. D. Geverte and G. E. Jenneman, U.S. Patent 5820766, 1998.
17. A. G. Hunton, P. A. Read, and R. D. Wilson, "Evaluation and Field Application of a New Hydrogen Sulfide Scavenger" (paper presented at the 10th International Oil Field Chemical Symposium, Fagernes, Norway, 1–3 March 1999).
18. P. J. Sturman and D. M. Goeres, Center for Biofilm Engineering, Montana State University, and M. A. Winters, "Control of Hydrogen Sulfide in Oil and Gas Wells with Nitrite Injection," SPE 56772 (paper presented at the SPE Annual Technical Conference and Exhibition, Houston, TX, 3–6 October 1999).
19. B. Knudsen, S. Tjelle, and H. Linga, "A New Approach Towards Environmentally Friendly Desulfurization," SPE 73957 (paper presented at the SPE International Conference on Health, Safety and Environment in Oil and Gas Exploration and Production, Kuala Lumpur, Malaysia, 20–22 March 2002).
20. J. G. Edmondson, U.S. Patent, 4680127, 1987.
21. R. Roehm, U.S. Patent 3459852, 1969.
22. J. A. Hardy and J. W. Georgie, U.S. Patent Application 20040074813.
23. J. R. Elliott, M. B. Raymond, B. Kalpakci, and N. F. Magri, "Theory and Measurements of Fates of H_2S Scavengers," SPE 28949 (paper presented at the SPE International Symposium on Oilfield Chemistry, San Antonio, TX, 14–17 February 1995).
24. C. T. Bedford, A. Fallah, E. Mentzer, and F. A. Williamson, "The First Characterisation of a Glyoxal-Hydrogen Sulfide Adduct," *Journal of the Chemical Society, Chemical Communications* (1992): 1035.
25. B. G. Kriel, A. B. Crews, E. D. Burger, E. Vanderwende, and D. O. Hitzman, "The Efficacy of Formaldehyde for the Control of Biogenic Sulfide Production in Porous Media," SPE 25196 (paper presented at the SPE International Symposium on Oilfield Chemistry, New Orleans, LA, 2–5 March 1993).
26. C. L. Kissel, J. L. Brady, H. Gottry, H. N Clifton, M. J. Meshishnek, and M. W. Preus, "Factors Contributing to the Ability of Acrolein to Scavenge Corrosive Hydrogen Sulfide," SPE 11749, *SPE Journal* 25 (1985): 647–655.

27. J. E. Penkala, C. Reed, and J. Foshee, "Acrolein Application to Mitigate Biogenic Sulfides and Remediate Injection-Well Damage in a Gas-Plant Water-Disposal System," SPE 98067 (paper presented at the International Symposium and Exhibition on Formation Damage Control, Lafayette, LA, 15–17 February 2006).
28. T. Salma, "Cost Effective Removal of Iron Sulfide and Hydrogen Sulfide from Water Using Acrolein," SPE 59708 (paper presented at the SPE Permian Basin Oil and Gas Recovery Conference, Midland, TX, 21–23 March 2000).
29. J. Penkala, M. D. Law, D. D. Horaska, and A. L. Dickinson, "Baker Petrolite, Acrolein 2-Propenal: A Versatile Microbiocide for Control of Bacteria in Oilfield Systems," Paper 04749 (paper presented at the NACE CORROSION 2004).
30. C. Reed, J. Foshee, J. E. Penkala, and M. Roberson, "Acrolein Application to Mitigate Biogenic Sulfides and Remediate Injection Well Damage in a Gas Plant Water Disposal System," SPE 93602 (paper presented at the SPE International Symposium on Oilfield Chemistry, The Woodlands, TX, 2–4 February 2005).
31. J. J. Howell and M. B. Ward, "The Use of Acrolein as an H_2S Scavenger in Multiphase Production," SPE 21712 (paper presented at the SPE Production Operations Symposium, Oklahoma City, OK, 7–9 April 1991).
32. W. Merk and K.-H. Rink, U.S. Patent 4501668, 1985.
33. A. Y. Al-Humaidan and H. A. Nasr-El-Din, "Optimization of Hydrogen Sulfide Scavengers Used during Well Stimulation," SPE 50765 (paper presented at the SPE International Symposium on Oilfield Chemistry, Houston, TX, 15–19 February 1999).
34. H. A. Nasr-El-Din, A. Y. Al-Humaidan, B. A. Fadhel, W. W. Frenler, and D. G. Hill, "Investigation of Sulfide Scavengers in Well-Acidizing Fluids," SPE 80289, *SPE Production and Facilities* (2002): 229.
35. (*a*) J. M. Bakke and J. B. Buhaug, "Hydrogen Sulfide Scavenging by 1,3,5-Triazinanes. Comparison of the Rates of Reaction," *Industrial & Engineering Chemistry Research* 43(9) (2004): 1962–1965. (*b*) L. Zea, P. Jepson, and R. Kumar, "Role of Pressure and Reaction Time on Corrosion Control of H_2S Scavenger," SPE 114175 (paper presented at the SPE International Oilfield Corrosion Conference, Aberdeen, UK, 27 May 2008).
36. (*a*) G. T. Rivers, U.S. Patent 6117310, 2000. (*b*) G. T. Rivers and J. T. Hackerott, U.S. Patent 6339153, 2002.
37. T. Salma, "Effect of Carbon Dioxide on Hydrogen Sulfide Scavenging," SPE 59765 (paper presented at the SPE/CERI Gas Technology Symposium, Calgary, Alberta, 3–5 April 2000).
38. J. D. Allison and J. W. Wimberley, U.S. Patent 4710305, 1987.
39. E. T. Dillon, World Patent Application WO9007467, 1990.
40. E. T. Dillon, *Hydrocarbon Process, International Edition* 70(12) (1991): 65.
41. E. T. Dillon, U.S. Patent 4978512, 1990.
42. G. T. Rivers and R. L. Rybacki, U.S. Patent 5347004, 1994.
43. G. T. Rivers and R. L. Rybacki, U.S. Patent 5554349, 1996.
44. A. J. Galloway, U.S. Patent 5405591, 1995.
45. G. J. Nagl, "Removing Hydrogen Sulfide," *Hydrocarbon Engineering* 6(2) (2001): 35.
46. T. Salma, M. L. Briggs, D. T. Herrmann, and E. K. Yelverton, "Hydrogen Sulfide Removal from Sour Condensate Using Non-Regenerable Liquid Sulfide Scavengers: A Case Study," SPE 71078 (paper presented at the SPE Rocky Mountain Petroleum Technology Conference, Keystone, CO, 21–23 May 2001).
47. E. A. Trauffer and R. D. Evans, U.S. Patent 5347003, 1994.
48. J. Bakke, J. Buhaug, and J. Riha, "Hydrolysis of 1,3,5-Tris(2-Hydroxyethyl)Hexahydro-*s*-Triazine and Its Reaction with H_2S," *Industrial & Engineering Chemistry Research* 40 (2001): 6051–6054.
49. T. R. Owens, International Patent Application WO/2008/049188.

50. C. W. Titley and P. H. Wieninger, U.S. Patent 7115215, 2006.
51. L. W. Gatlin, World Patent Application WO2004043938, 2004.
52. T. Salma, A. A. Lambert III, and G. T. Rivers, International Patent Application WO/2008/027721.
53. J. C. Warrender, World Patent Application WO9819774, 1998.
54. S. R. Schieman, "Solids-Free H_2S Scavenger Improves Performance and Operational Flexibility," SPE 50788 (paper presented at the SPE International Symposium on Oilfield Chemistry, Houston, TX, 16–19 February 1999).
55. D. Sullivan III, A. R. Thomas, J. M. Garcia, and P Yon-Hin, U.S. Patent 5744024, 1998.
56. J. F. Vasil, U.S. Patent 5314672, 1994.
57. J. J. Weers and T. J. O'Brien, U.S. Patent 6024866, 2000.
58. L. W. Gatlin, U.S. Patent Application 20030089641.
59. T. Arnold, W. R. Graham, and J. D. Cranmer, U.S. Patent, 6942037, 2005.
60. D. S. Sullivan III, A. R. Thomas, M. A. Edwards, G. N. Taylor, P. Yon-Hin, and J. M. Garcia III, U.S. Patent 5674377, 1997.
61. J. C. Millan, S. Dubey, and W. Koot, "Accelerated Mechanism of Scale Deposition in UW Production Operation," SPE 87446 (paper presented at the SPE International Symposium on Oilfield Scale," Aberdeen, UK, 26–27 May 2004).
62. M. M. Jordan, K. Mackin, C. J. Johnston, and N. D. Feasey, "Control of Hydrogen Sulfide Scavenger Induced Scale (Due to Raised pH) and the Associated Challenge of Sulfide Scale Formation within a North Sea High Temperature/High Salinity Field Production Wells: Laboratory Evaluation to Field Application," SPE 87433 (paper presented at the SPE International Symposium on Oilfield Scale, Aberdeen, UK, 26–27 May 2004).
63. C. D. Sitz, D. K. Barbin, and B. J. Hampton, "Scale Control in a Hydrogen Sulfide Treatment Program," SPE 80235 (paper presented at the International Symposium on Oilfield Chemistry, Houston, TX, 5–7 February 2003).
64. J. G. R. Eylander, H. A. Holtman, T. Salma, M. Yuan, M. Callaway, and J. R. Johnstone, "The Development of Low-Sour Gas Reserves Utilizing Direct-Injection Liquid Hydrogen Sulfide Scavengers," SPE 71541 (paper presented at the SPE Annual Technical Conference and Exhibition, New Orleans, LA, 30 September–3 October 2001).
65. E. Davidson, J. Hall, and C. Temple, "A New Iron-Based, Environmentally Friendly Hydrogen Sulfide Scavenger for Drilling Fluids," SPE 84313, *SPE Drilling & Completion* 19(4) (2004): 229–234.
66. J. Buller and J. F. Carpenter, "H_2S Scavengers for Non-Aqueous Systems," SPE 93353 (paper presented at the SPE International Symposium on Oilfield Chemistry, The Woodlands, TX, 2–4 February 2005).
67. A. S. Deshpande, N. V. Sankpal, and B. D. Kulkarni, U.S. Patent Application 20040192995.
68. E. Davidson, U.S. Patent Application, 20040167037, 2004.
69. D. A. Wilson and D. K. Crump, U.S. Patent 5569443, 1996.
70. S. Piché and F. Larachi, "Dynamics of pH on the Oxidation of HS^- with Iron(III) Chelates in Anoxic Conditions," *Chemical Engineering Science* 61(23) (2006): 7673–7683.
71. D. C. Olson, U.S. Patent 4443423, 1984.
72. D. McManus and A. E. Martell, "The Development, Chemistry and Application of a Chelated Iron, Hydrogen Sulphide Removal Process," *Recent Advances in Oilfield Chemistry* (1994): 207.
73. J. J. Weers and C. E. Thomasson, U.S. Patent 5223127, 1993.
74. B. C. Collins, P. A. Mestetsky, and N. J. Savaiano, U.S. Patent 5807476, 1998.
75. J. W. Bozzelli, G. D. Shier, R. L. Pearce, and C. W. Martin, U.S. Patent 4112049, 1978.
76. E. T. Kool and C. E. Uebele, U.S. Patent 4569766, 1986.
77. M. K. Pakulski, P. Logan, and R. Matherly, U.S. Patent Application 20050238556.
78. T. J. O'Brien and J. J. Weers, International Patent Application WO/2008/115704.
79. G. Westlund and D. Weller, International Patent Application WO/2009/035570.

16 Oxygen Scavengers

16.1 INTRODUCTION

Dissolved oxygen in water can cause destructive oxygen corrosion to metal pipes and process equipment. The corrosion by-products in turn cause formation damage by plugging. Thus, oxygen needs to be removed from oilfield waters.[1-3] In addition, inhibition of corrosion in CO_2/O_2 systems is more difficult that in CO_2 alone, hence the need for good oxygen removal.[3]

The most common uses of oxygen scavenger in oil and gas production are for seawater injection facilities, hydrotesting (the process of using water under pressure to test the integrity of pipelines and vessels), and acid stimulation.[4-6] For water injection systems, primary oxygen removal is normally either by use of a vacuum tower or a gas-stripping tower. These towers reduce the level of oxygen in the water from ~9 ppm to < 50 ppb. A patented technology using nitrogen gas to strip oxygen down to 5–15 ppb has been used on a number of seawater injection facilities since the early nineties.[7] This MINOX process uses no added chemicals and has the advantage of being lightweight.

Oxygen removal to < 10 ppb (preferably less than ~5 ppb to avoid significant oxygen corrosion) can be achieved by the addition of an oxygen scavenger downstream of the deaeration vessel.

16.2 CLASSES OF OXYGEN SCAVENGERS

There are many classes of oxygen scavengers that have been used, including:

- bisulfite, metabisulfite, and sulfite salts
- dithionite salts
- hydrazines including 1-aminopyrrolidine
- guanidines
- semicarbazide and carbohydrazides
- hydroxylamines
- oximes
- activated aldehydes
- polyhydroxyl compounds
- hydrogen with activated noble metal catalysts
- an enzyme that catalyzes the reaction between a substrate material and oxygen

However, current sales of oil industry oxygen scavengers for water injection and hydrotesting are dominated totally by sulfite, bisulfite, and metabisulfite salts, sometimes with added catalysts. The other classes of oxygen scavengers, which are largely organic nitrogen compounds, have mainly been applied to drilling fluids or boiler waters. These will briefly be discussed first followed by the bisulfite, metabisulfite, and sulfite salts.

16.2.1 Dithionite Salts

Dithionite salts such as sodium dithionite ($Na_2S_2O_4$) have been proposed for use in drilling and completion operations. Sulfur is in the +3 oxidation state in these salts and is oxidized through +4 (sulfite) to +6 (sulfate) on reaction with oxygen.[8–9]

16.2.2 Hydrazine and Guanidine Salts

Hydrazines (RNH_2NH_2) are basic and will raise the pH of the water. If the water contains calcium/magnesium ions and bicarbonate ions, there is the possibility of new or increased calcium/magnesium carbonate scale formation. Hydrazine (NH_2NH_2) is also a suspected carcinogen and requires special handling precautions. Hydrazine reacts fairly slowly with oxygen, but the reaction rate is accelerated by transition metal ion catalysts such as copper(II) and manganese(II) ions and also by increasing the temperature. The reason that transition metals are used rather than other metals is that they have two or more available oxidation states and can coordinate a dioxygen molecule fairly strongly via overlap of the metal d-orbitals with dioxygen p-orbitals. Unlike the bisulfites, hydrazine does not decompose at high temperatures, although this is rarely encountered. Hydrazine can be used as an oxygen scavenger in acid stimulation as hydrazinium salts. Other hydrazines such as phenylhydrazine can be used but are more expensive than bisulfite and metabisulfite salts. 1-Aminopyrrolidine is a cyclic 1,1-dialkylhydrazine, which has been shown to have excellent oxygen-scavenging properties in boiler applications, again in the presence of a transition metal ion catalyst.[10–11] Hydrazines have been used as oxygen scavengers to control external corrosion of oil-string casing, giving corrosion protection for 12–18 months after a well is completed.[12]

Guanidine salts ($H_2NC(=NH)NH_3^+X^-$) such as guanidine acetate have been proposed as oxygen scavengers primarily in seawater injection.[13] Semicarbazide ($H_2NNHCONH_2$) and carbohydrazides ($RCONHNH_2$) have also been used, especially in boiler water treatment, as they do not add to the dissolved inorganic solids as do the sulfur-based oxygen scavengers. A catalyst such as hydroquinone or cobalt ions is needed.[14]

16.2.4 Hydroxylamines and Oximes

Hydroxylamines such as the volatile liquid diethylhydroxylamine (DEHA, Figure 16.1) or N,N-bis-(2-hydroxyethyl)hydroxylamine are currently used in closed-loop water treatment.[15] They minimize oxygen pitting and corrosion potential, are not temperature dependent, and do not add to the dissolved inorganic

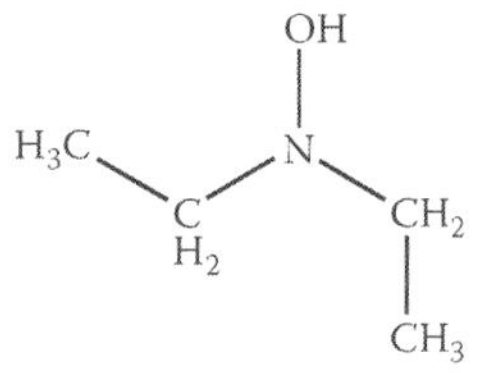

FIGURE 16.1 Diethylhydroxylamine.

solids. Transition metal catalysts, amines, tannin, *t*-butylcatechol, hydroquinone, or pyrogallol can be used as catalysts. DEHA has a slow oxygen reduction rate and can absorb only a small amount of oxygen per unit weight so that this compound is required to be added in a relatively large amount.[17]

Oximes (R=NOH and RR'NOH) such as methylethylketoxime and acetaldoxime have been proposed for use in acid gas-stripping facilities, as their addition reduces the formation of amine or glycol degradation products.[18]

16.2.5 ACTIVATED ALDEHYDES AND POLYHYDROXYL COMPOUNDS

Aldehydes such as formaldehyde are generally poor oxygen scavengers being slowly oxidized to carboxylic acids.[19] Salicylaldehyde and gallic acid are exceptions due to the activating nature of the hydroxyl groups on the aromatic ring.[20–21] Hydroquinone catalyzes their reaction with oxygen. Erythorbic acid can also be added to this category as it contains a tautomer of a ketone and four hydroxyl groups (Figure 16.2). This molecule is often used for oxygen scavenging in boiler waters. Magnesium ion or copper ion as catalysts, or in the presence of a pH control agent that maintains the pH above 7, improves the performance.[22] Other polyhydroxyl compounds such as gluconates and pyrogallol have also been used as oxygen scavengers. For example, salts of a keto-gluconic acid, or a salt of a stereoisomer of a keto-gluconic acid, have been proposed for use in boiler water systems or oilfield injection water or brine.[23]

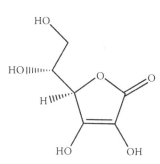

FIGURE 16.2 Erythorbic acid (also known as isoascorbic acid).

16.2.6 CATALYTIC HYDROGENATION

The reaction of oxygen with hydrogen activated by noble metal catalysts, such as finely divided palladium adsorbed on a resin, have been used to reduce oxygen concentrations in seawater to negligible levels in less than 60 s, lower than the oxygen corrosion-pitting potential.[24–25] Group 8 noble metals are extremely good adsorbers of hydrogen and oxygen, facilitating the reaction of these molecules to produce water.[26] This technology has been investigated on installations offshore Angola, Brazil and the Gulf of Mexico, and the North Sea.[24] The small "footprint" and low equipment weight and volume are potential advantages with this technology.

16.2.7 ENZYMES

An enzyme that catalyzes the reaction between a substrate material and oxygen has been claimed as a method to reduce oxygen levels and thereby corrosion.[27] The enzyme can be alcohol oxidase enzyme while the substrate material is a small alcohol. Crude oil or other hydrocarbon material is also added to the water.

16.2.8 BISULFITE, METABISULFITE, AND SULFITE SALTS

The most common oxygen scavengers in oilfield production are sulfite (M_2SO_3), bisulfite ($MHSO_3$), and metabisulfite ($M_2S_2O_5$) salts. The bisulfite ion, HSO_3^-, is also known as hydrogen sulfite, and it is salts of this ion that are principally used today. Sulfite salts have been used in the past but cannot be prepared in as high aqueous concentrations as the bisulfite salts. The most common bisulfite salts are ammonium bisulfite (NH_4HSO_3) and sodium bisulfite ($NaHSO_3$).[28] Ammonium bisulfite is more water-soluble than sodium bisulfite: solutions of up to 65 wt.% at 4°C (39°F) can be made. Being a salt of a weak acid and a strong base, sodium bisulfite forms a significantly basic aqueous solution whereas ammonium bisulfite is only very slightly basic at around pH 8. As a rule, to reduce the oxygen concentration in water from 9 ppm to 50 ppb requires addition of 60 ppm of 65 wt.% ammonium bisulfite.

The reaction of sulfite or bisulfite with oxygen is slow below approximately 130°C (266°F) without a catalyst, forming bisulfate ions as follows:

$$2HSO_3^- + O_2 \rightarrow 2HSO_4^-$$

The best catalysts to accelerate this process are transition metal ions.[29] The rate of oxygen scavenging for various transition metal ions has been quantified by equations.[30] Since seawater contains small amounts of these ions, it is not necessary to add further ions when using ammonium bisulfite for oxygen scavenging. However, when using sodium bisulfite, extra transition metal ions are added to the formulation. Cobalt(II) ions have been shown to be the best metal ion catalyst, and this ion is still used today in some areas, but it has been superceded by iron(III) salts in environmentally sensitive areas.[31] Manganese(II) ions have been shown to work synergistically at pH 5 with iron ions to improve the oxygen-scavenging rate of bisulfites.[29] There is evidence that the iron-catalyzed oxidation of bisulfite aqueous solution occurs by a free-radical chain mechanism via sulfate radicals, $SO_4\times$.[32] The natural or added transition metal ion catalysts can be deactivated, for example, by chelation.[33] Thus, if chelates such as polyphosphate corrosion inhibitors, polycarboxylate and polyphosphonate scale inhibitors, and polyaminocarboxylate scale dissolvers were present in the water to be injected, they would significantly reduce the performance of bisulfite oxygen scavengers by complexing with the transition metal ion catalysts.[28] Chlorine dioxide or its water-soluble salts, such as sodium or ammonium chlorite, also catalyze the oxygen-scavenging process by bisulfite salts.[34]

Despite their widespread use, bisulfite salt oxygen scavengers do have several drawbacks:

1. Nitrogen in ammonium bisulfite provides food for bacteria such as sulfate reducing bacteria (SRBs) in high-temperature environments, but proper use of biocide should stop this.
2. Aldehyde-based biocides such as glutaraldehyde reduce the scavenging efficiency of bisulfites.[35]

3. Hypochlorite and chlorine biocides react with bisulfites deactivating them. In practice, the total bisulfite demand for oxygen and any residual chlorine/hypochlorite is calculated.
4. Stannic ions, mannitol, ethanol, and organic acids also reduce the oxygen-scavenging ability of bisulfites.
5. The oxygen scavenging ability of sulfites and bisulfites is sensitive to the pH of the water. One study showed that in pH > 7, sulfite ions rapidly scavenge dissolved oxygen, and in pH < 6.0, the reaction is too slow for practical use.[36] In another study, ammonium bisulfite was shown to work well down to pH 6.5, although it performed better at an optimum pH of 7.5–9.[37] Raw seawater has a pH of 7.8. Sulfites and bisulfites will decompose in strongly acidic solutions so they cannot be used in normal acid stimulation packages.
6. The oxygen-scavenging rate is measurably slower in very cold seawater, close to 0°C.

In hydrotesting, it is essential to remove oxygen to prevent oxygen corrosion since the aqueous phase may be left in the system for years. Microbially induced corrosion may also take place. Addition of a biocide can eliminate microbially induced corrosion. However, not all biocides are compatible with sulfites and bisulfites. Second, biocides are of course toxic, and there may be toxicity regulations on the discharged water. An environment-friendly method of preventing corrosion in hydrotesting is to first add a bisulfite oxygen scavenger to the hydrotest water, then when the oxygen has been removed, add a limited amount of biocide, adjust the pH to about 9.5 with a base such as sodium hydroxide (which limits bacterial growth), and add a scale inhibitor.[38]

REFERENCES

1. M. Davies and P. J. B. Scott, *Oilfield Water Technology*, Houston, TX: National Association of Corrosion Engineers (NACE), 2006.
2. H. G. Byars and B. R. Gallop, "Injection Water + Oxygen = Corrosion and/or Well Plugging Solids," SPE 4253 (paper presented at the SPE Symposium on Handling of Oilfield Water, Los Angeles, CA, 4–5 December 1972).
3. R. L. Martin, "Corrosion Consequences of Oxygen Entry into Sweet Oilfield Fluids," SPE 71470 (paper presented at the SPE Annual Technical Conference and Exhibition, New Orleans, LA, 30 September–3 October 2001).
4. K. B. Flatval, S. Sathyamoorthy, C. Kuijvenhoven, and D. Ligthelm, "Building the Case for Raw Seawater Injection Scheme in Barton," SPE 88568 (paper presented at the SPE Asia Pacific Oil and Gas Conference and Exhibition, Perth, Australia, 18–20 October 2004).
5. S. Yntema, P. De Boer, R. A. Trompert, R. M. de Jonge, and B. J. Gellekom, "Oxygen-Free Acid Stimulation in an Underground Gas Storage Well Completed with Pre-Packed Screens," SPE 99846 (paper presented at SPE/IcoTA Coiled Tubing and Well Intervention Conference, Woodlands, TX, 4–5 April 2006).
6. S. L. Wellington, "Biopolymer Solution Viscosity Stabilization—Polymer Degradation and Antioxidant Use," SPE 9296, *SPE Journal* 14 (1974): 643.

7. N. Henriksen, European Patent Application EP0234771, 1986.
8. J. L. Watson and L. L. Carney, U.S. Patent 4059533, 1977.
9. C. J. Philips, European Patent 106666, 1984.
10. Y. Shimura, S. Taya, and T. Shiro, U.S. Patent Application 20030141483, 2003.
11. Y. Shimura, S. Taya, K. Uchida, and T. Sato, "The Performance of New Volatile Oxygen Scavenger and Its Field Application in Boiler Systems," Paper 00327 (paper presented at the Corrosion 2000, NACE International Conference).
12. F. W. Schremp, J. F. Chittum, and T. S. Arczynski, "Use of Oxygen Scavengers to Control External Corrosion of Oil-String Casing," SPE 1606, *Journal of Petroleum Technology* 13(7) (1961): 703.
13. F. Dawans, D. Binet, N. Kohler, and D. V. Quang, U.S. Patent 4454620, 1984.
14. M. Slovinsky, U.S. Patent 4269717, 1981.
15. B. Greaves, S. C. Poole, C. M. Hwa, and J. C.-J. Fan, U.S. Patent 5830383, 1998.
16. A. M. Rossi and P. R. Burgmayer, U.S. Patent 5256311, 1993.
17. Y. Shimura, J. Takahashi, U.S. Patent 7112284, 2006.
18. R. R. Veldman and D. Trahan, U.S. Patent 5686016, 1997.
19. E. J. Burcik and G. C. Thankur, "Reaction of Polyacrylamide with Commonly Used Additives," SPE 4164, *Journal of Petroleum Technology* 24 (1972): 1137–1139.
20. J. A. Muccitelli, U.S. Patent 4569783, 1986.
21. C. A. Soderquist, J. A. Kelly, and F. S. Mandel, U.S. Patent 4968438, 1990.
22. M. Slovinsky, Canadian Patent CA1186425, 1985.
23. H. L. Gewanter and R. D. May, U.S. Patent 5114618, 1992.
24. Article in *Offshore* 59(11) (1999).
25. E. Gobina, International Patent Application, WO/2001/085622.
26. L. A. Cantu and L. D. Harrison, "Field Evaluation of Catalytic Deoxygenation Process for Oxygen Scavenging in Oilfield Waters," SPE 14284, *SPE Production Engineering* 1988: 619–624.
27. Y. Wu, U.S. Patent 4501674, 1985.
28. N. Matsuka, Y. Nakagawa, M. Kurihara, and T. Tonomura, "Reaction Kinetics of Sodium Bisulfite and Dissolved Oxygen in Seawater and Their Applications to Seawater Reverse Osmosis,"*Desalination* 51(2) (1984): 163–171.
29. R. K. Ulrich, G. T. Rochelle, and R. E. Prada, "Enhanced Oxygen Absorption into Bisulphite Solutions Containing Transition Metal Ion Catalysts," *Chemical Engineering Science* 41(8) (1986): 2183–2191.
30. T. Chen and C. H. Barron, "Some Aspects of the Homogeneous Kinetics of Sulfite Oxidation," *Industrial & Engineering Chemistry Fundamentals* 11(4) (1972): 466.
31. J. Nakajima, M. Yamashita, and K. Kimura, U.S. Patent 6402984, 2002.
32. J. Ziajka, F. Beer, and P. Warneck, "Iron-Catalysed Oxidation of Bisulphite Aqueous Solution: Evidence for a Free Radical Chain Mechanism,"*Atmospheric Environment* 28 (1994): 2549–2552.
33. A. J. McMahon, A. Chalmers, and H. Macdonald, "Optimising Oilfield Oxygen Scavengers," Proceedings of the Chemistry in the Oil Industry VII, Royal Society of Chemistry, Manchester, UK, 2002, p. 163.
34. J. R. Stanford, J. H. Martin, and G. D. Chappell, U.S. Patents 3996135, 1976, and 4098716,l 1978.
35. T. G. Braga, "Effects of Commonly Used Oilfield Chemicals on the Rate of Oxygen Scavenging by Sulfite/Bisulfite," SPE 13556, *SPE Production Engineering*, 1987: 137.
36. E. S. Snavely, "Chemical Removal of Oxygen from Natural Waters," SPE 3262, *Journal of Petroleum Technology* 23 (1971): 443–446.
37. R. W. Mitchell, "The Forties Field Seawater Injection System," SPE 6677, *SPE Journal* (1978): 877.
38. R. Prasad, U.S. Patent 6815208, 2002.

17 Drag-Reducing Agents

17.1 INTRODUCTION

The flow of liquid in a conduit, such as a pipeline, results in frictional energy losses. As a result of this energy loss, the pressure of the liquid in the conduit decreases along the conduit in the direction of the flow. For a conduit of fixed diameter, this pressure drop increases with increasing flow rate. The effect of a drag reducer added to a liquid is to reduce the frictional resistant in turbulent flow (Reynold's number greater than about 2,100) compared with that of the pure liquid. Drag-reducing agents (DRAs) are sometimes known as friction reducers or flow improvers, although the latter term can be confused with wax inhibitors/pour-point depressants such as poly(meth) acrylic esters. DRAs interact with the turbulent flow processes and reduce frictional pressure losses such that the pressure drop for a given flow rate is less, or the flow rate for a given pressure drop is greater. In most petroleum pipelines, the liquid flows through the pipeline in a turbulent regime. Therefore, DRAs can perform very well in most pipelines. Because DRAs reduce frictional energy losses, increase in the flow capability of pipelines, hoses, and other conduits in which liquids flow can be achieved. DRAs can also decrease the cost of pumping fluids, the cost of equipment used to pump fluids, and provide for the use of a smaller pipe diameter for a given flow capacity. As crude oil is cooled to near its pour point, the effectiveness of DRAs may be reduced. The largest use of oil-soluble DRAs is for pipeline transportation of refined oils, not crude oils.

Ultrahigh molecular weight (UHMW) polymers are the most effective drag reducers but surfactants can also show good drag-reducing behavior, although usually at higher dose rates.[1] It is not unusual to see as high as 70–80% drag reduction with 20–30 ppm of added polymer. Generally, higher concentrations are needed in larger pipes. Fibers can also exhibit DRA properties.[2] The performance of DRAs in water and/or hydrocarbon fluids depends on many parameters such as fluid viscosity, pipe diameter, liquid and gas velocities, composition of the oil, pipe roughness, water cut, pipeline inclination, DRA concentration, type of DRA, shear degradation of DRA, and temperature, and even pH for aqueous DRAs.

One of the first field reports of using a DRA in the oil industry was in 1965 with the use of guar gum to reduce the cost of pumping aqueous fracturing fluids.[3] Since then, there have been a number of uses for DRAs in the oil and gas industry including fracturing, acid stimulation,[4] drilling fluids, water injection,[5] coiled tubing operations,[6] and oil transportation. This latter application is discussed in more detail later. DRAs are usually designed to be either oil-soluble (for application to oil lines) or water-soluble (for application to water lines). DRAs have also been studied for use in multiphase flow (oil and water, and sometimes gas, flowing together).[7] The external liquid phase (oil or water) will determine whether an oil-soluble or water-soluble DRA is most appropriate. Oil-soluble DRAs become less effective as the water cut increases.

Drag reducers for gas transportation have also been applied in the field. For example, injection of a film-forming amphiphilic corrosion inhibitor or simply a fatty acid amine has been claimed to reduce friction or drag on a gas in turbulent flow.[8] Field implementation of a corrosion inhibitor gas DRA has been reported.[9] The desired scenario is based on positively charged amine and amide functional groups providing strong binding to the metal surface and the long-chain hydrocarbon part serving as a compliant or lubricating surface to mitigate turbulence at the gas-phase boundary.

A common way to test the performance of a DRA in the laboratory is to use a flow loop and measure the pressure drop. Many of the references in this chapter, particularly patent references, describe equipment and test methods. Results from loop tests can be compared directly to field observations.[10] The use of a loop with submerged jet cell has also been described.[11] A fast screening multi-test apparatus has been described.[12] A simpler and smaller apparatus that has been much used to measure DRA performance is the rotating disk or screen extensional rheometer.[13–15] A specially adapted capillary viscosimeter has also been described.[16] If good field experience has already been acquired for a commercial DRA, laboratory tests may not be necessary before using it in other similar applications.

17.2 DRAG-REDUCING AGENT MECHANISMS

Despite considerable research in the areas of DRAs, there is no universally accepted model, which explains the mechanism by which polymers, or surfactants, reduce friction in turbulent flow. One early theory states that the stretching of randomly coiled polymers increases the effective viscosity.[17] By consequence, small eddies are damped, which leads to a thickening of the viscous sublayer and, thus, drag reduction. A more recent theory proposed that drag reduction is caused by elastic rather than viscous properties.[18] This conclusion was reached by observing drag reduction in experiments where polymers were active at the center of the pipe, where viscous forces do not play a role. Another group observed that the amount of drag reduction is limited by an empirical asymptote, called the "Virk asymptote,"[19] although others have found conflicting results.[20]

A later qualitative theory discusses turbulent flow in a pipeline as having three parts (Figure 17.1).[21] In the very center of the pipe is a turbulent core where one finds the eddy currents. It is the largest region and includes most of the fluid in the pipe. Nearest to the pipeline wall is the laminar sublayer. In this zone, the fluid moves laterally in sheets. Between the laminar layer and the turbulent core lies the buffer zone where turbulence is first formed. A portion of the laminar sublayer, called a "streak," occasionally will move to the buffer region. There, the streak begins to vortex and oscillate, moving faster as it gets closer to the turbulent core. Finally, the streak becomes unstable and breaks up as it throws fluid into the core of the flow. This ejection of fluid into the turbulent core is called a "turbulent burst." This bursting

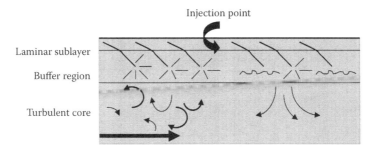

FIGURE 17.1 Injection of polymer DRA into turbulent flow suppressing energy bursts.

motion and growth of the bursts in the turbulence core results in wasted energy. Drag-reducing polymers interfere with the bursting process and reduce the turbulence in the core. The polymers absorb the energy in the streak, rather like a shock absorber, thereby, reducing subsequent turbulent bursts. As such, drag-reducing polymers are most active in the buffer zone. The overall effect may be to increase the thickness of the laminar sublayer, thereby reducing convective heat transfer.[22]

Clearly, molecular weight, aggregation, and chain flexibility (versus polymer rigidity) are all important factors affecting the performance of polymer DRAs. The theory regarding molecular weight is that the longer polymers will be best suited to break up turbulence bursts or eddies in the flow. The hydrodynamic volume (coil volume) of the polymer has been proposed as a better critical factor than molecular weight by some workers based on laboratory studies.[23] The polymer volume varies with the solvent and, if aqueous, sometimes with the pH and ionicity.

A remarkable aspect of the addition of polymers to multiphase flow is not only the drag reduction, which can be measured, but also the changes in the configurations of phases or flow patterns. For example, one study found that the injection of a concentrated solution of polyacrylamide (PAM) and sodium acrylate into an air-water flow in a horizontal pipe changed an annular pattern to a stratified pattern by destroying the disturbance waves in the liquid film. Drag reduction of 48% was measured for mean concentrations of 10–15 ppm.[24]

17.3 OIL-SOLUBLE DRAS

17.3.1 Background

The story of oil-soluble DRAs for oil transportation begins with their successful use on the Trans-Alaskan pipeline.[25–26] This first commercial drag reducer application began in July 1979. By 1980, flow through the TAPS line had increased to the 1.5 million bbl/d (9,940 m³/h) level. Approximately 200,000 bbl/d (1,300 m³/h) of this throughput was a direct result of injecting a drag-reducing additive. Since then, many projects have used DRAs to increase pipeline capacity or reduce the need for pumping stations.[27–28] For oil transportation, polymer DRAs are preferred over surfactants due to their performance being higher at a given concentration despite the higher unit cost.

17.3.2 Oil-Soluble Polymeric DRAs

In their extended configuration, polymers have a size, which is smaller than the smallest length scale of the turbulence. A well-known effect is the increase of the shear viscosity of a fluid due to polymers, which gives reason to suspect that polymers can affect turbulence on a microscale. However, UHMW polymers are active on both the microscale and macroscale of the turbulence.[29] Therefore, it is a key feature of DRAs that they are as long as possible, that is, have as high a molecular weight as possible.

Commercial DRA polymers for oil transportation that can be produced with UHMW are mostly based on Ziegler-Natta organometallic polymerization of alkenes (olefins).[30] Examples of cheap monomers that can be used include:

- isobutylene
- isoprene
- styrene
- hexene
- octene
- decene
- tetradecene

Only Ziegler-Natta polymerization is able to produce ultrahigh olefin polymer molecular weights of 10–30,000,000 Da (weight average), which are needed for good DRA performance. The term "ultrahigh molecular weight" corresponds to an inherent polymer viscosity of at least about 10 dl/g. Because of the extremely high molecular weight of DRA polymers, it is difficult to reliably and accurately measure the actual molecular weight, but inherent viscosity provides a useful approximation of molecular weight. Meth(acrylate) esters can also be polymerized to give UHMW polymers useful as DRAs.

17.3.2.1 Polyalkene (Polyolefin) DRAs

For many years, oil-soluble polymeric DRAs were based on small monomers such as isobutylene to form polyisobutylene (PIB; Figure 17.2).[31–33] Typical concentrations needed for drag-reducing behavior were about 10–30 ppm. More recently, researchers found that copolymers with larger monomers such as hexene, octene, decene, and tetradecene gave improved DRA performance.[34–35] This may be due to two factors. First, long linear polymers are liable to be degraded by the turbulence (shear) in the pipeline or at pumps. Increasing the molecular weight improves the drag-reducing performance but also increases the polymer degradation potential.[36] Long polyvinyl

FIGURE 17.2 A section of polyisobutylene (PIB).

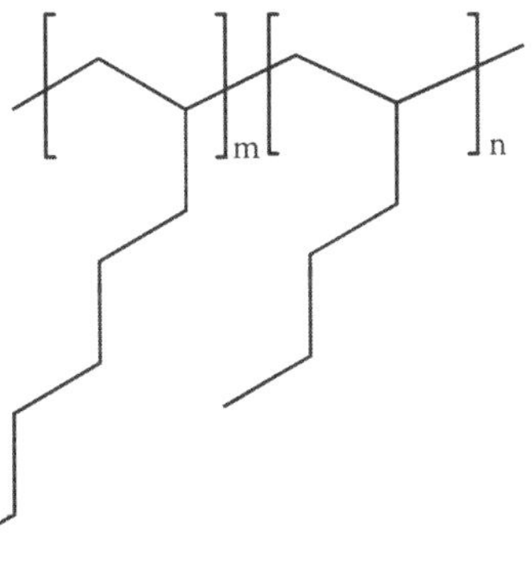

FIGURE 17.3 Copolymer of 1-hexene and 1-octene.

polymers with little or no side chains, such as polyethylene or polyisobutylene, have little shear degradation resistance but polymers with larger side chains, containing larger alkenes, have more resistance. Second, for a given polymer length, polymers with larger side chains have a larger polymer hydrodynamic volume (coil volume) and will have a better chance of breaking up turbulent flow. Copolymers of two alkenes rather than homopolymers of a single alkene appear to be preferred.[37] Copolymers appear to have less crystallinity than homopolymers because of the different monomers used. Lack of crystallinity is extremely advantageous in dissolution of the materials in the flowing hydrocarbon, with resultant increase in drag reduction.

Most larger alkenes manufactured are of the 1-isomer type (α-olefins). In fact, for many years only the α-olefin products made by Shell had been suitable for high-performance polymeric DRAs. Thus, copolymers of 1-hexene and 1-octene, 1-octene and 1-decene, or 1-decene and 1-tetradecene have been proposed as improved oil-soluble DRAs (Figure 17.3). As the monomer size increases, it becomes increasingly difficult to polymerize them to UHMWs due to steric congestion at the Ziegler-Natta metal catalyst center (usually titanium). However 1-alkene monomers with up to 16 carbon atoms can be used successfully, although 6–10 carbons seems to be the preferred range.

Another polyalkene polymer with a fairly large side chain is polystyrene (Figure 17.4). Experiments in benzene in pipes with polymers up to 7.1×10^6 Da in molecular weight gave good drag reduction. The polystyrene samples exhibited a high resistance to the loss of drag reduction via degradation in turbulent flow. The researchers concluded that drag reduction and degradation depend strongly on molecular weight distribution.[38] Polystyrenes with UHMWs (> 10^7 Daltons) would probably have performed better. UHMW copolymers of styrene and 1-alkenes have been claimed as DRAs,[39] as well as alkylstyrene copolymers such as *t*-butylstyrene-hexene-dodecene terpolymer.[40]

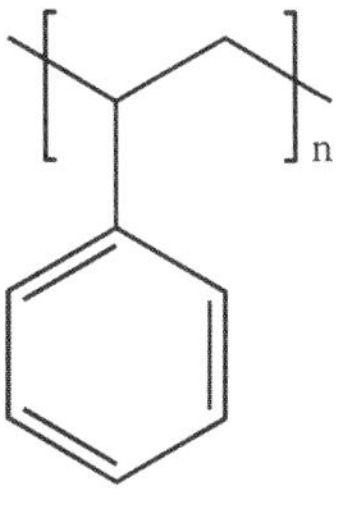

Another improvement on polyalkene DRAs has been to use one or more isomers of the 1-alkenes (α-olefins) in copolymers. For example, copolymers of 1-hexene

FIGURE 17.4 Polystyrene.

and 1-dodecene or copolymers of 1-octene and 1-tetradecene, whereby one or both monomers are isomerized before polymerization.[41] Using isomers of 1-alkenes was also shown to reduce the catalyst requirement by about 50%. Further, the dissolution of polymeric (DRAs) in flowing hydrocarbon fluids is improved by incorporating branching into the polymer DRAs.[42] The branches have an average chain length of at least four to eight carbon atoms. A branched polymer of the same molecular weight will have a smaller overall size because of its reduced radius of gyration (R_g), and thus dissolve more readily.

17.3.2.2 Poly(meth)acrylate Ester DRAs

The use of long side chains for improving the performance of DRA polymers is not limited to polyalkenes. For example, emulsion polymerization of alkyl methacrylates, alkyl acrylates, and alkyl styrenes give polyvinyl polymers with long side chains.[43] Alkyl (meth)acrylates can be used alone to make DRA polymers and usually contain pendant alkyl groups of at least six to eight carbon atoms. For example, poly(isodecylmethacrylate) was found to be superior to commercially available polyisobutylene as a DRA, especially in terms of shear stability (Figure 17.5).[44] Increasing the molecular weight from 10×10^6 to 26×10^6 Da increased the drag reduction for a given polymer concentration and pipe size.[45] This polymer was commercialized for use in fracturing fluids but not oil transportation.

Poly(dodecyl methacrylate)s with molecular weights greater than 10^7 Da have been shown to be good DRAs, again with good shear stability (Figure 17.6).[46] These alkylacrylate polymers are more expensive to manufacture on a large scale due to the price of the monomers. Another claimed example is behenyl acrylate copolymers (behenyl = C22).[47] Copolymers of alkenes and vinyl esters have also been reported such as a copolymer of 1-octene and methyl 10-undecenoate.[48] Hydrate-inhibited drag-reducing latexes can be prepared by polymerizing an alkyl(meth)acrylate such

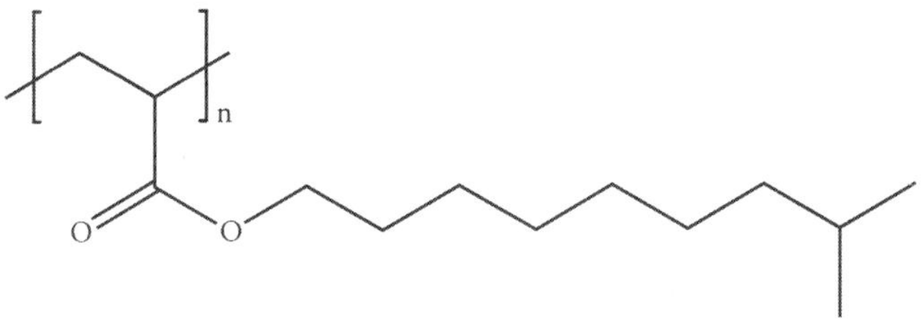

FIGURE 17.5 Poly(isodecylmethacrylate).

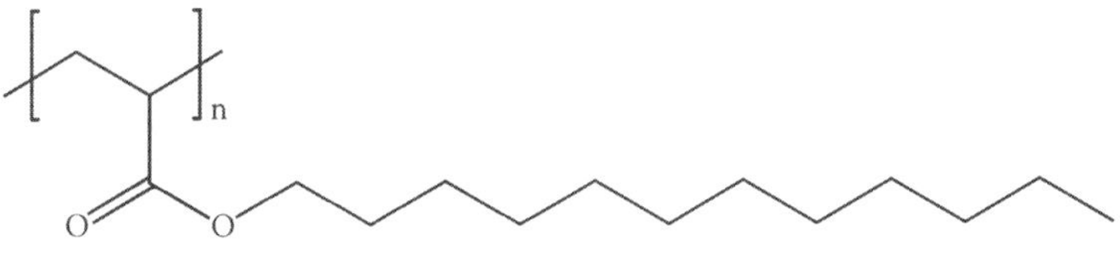

FIGURE 17.6 Poly(dodecylmethacrylate).

as 2-ethylhexyl methacrylate in an emulsion comprising water, THI (MEG), surfactant, initiator, and a buffer.[49]

Alkyl acrylate polymers have been claimed as superior DRAs (compared with polyolefins) for crude oils having a low API gravity and/or a high asphaltene content.[49] Examples are UHMW polymers of 2-ethylhexyl methacrylate, optionally copolymerized with other alkyl acrylates such as n-butyl acrylate.

17.3.2.3 Other Oil-Soluble DRA Polymers

A study on oil-soluble polymer DRAs with a few percent of polar-associating groups showed that these polymers performed better as DRAs than the homopolymer if there were interpolymer interactions but worse if there were intrapolymer interactions.[50] In a related article, the same team found that a mixture of a cationic and an anionic polymer, both made oil-soluble by copolymerization with alkenes, gave excellent drag reduction as well as shear degradation resistance. The molecular weights of the polymers do not need to be in the UHMW range, making them easier to handle and inject. An example is a mixture of a styrene/vinyl pyridine copolymer (cationic) and a neutralized sulfonated copolymer (anionic).[51] Others have reported better drag-reducing properties from hydrogen bonding–associating polymer systems.[52]

Polymer mixtures, which exhibit acid-base interactions have also been reported as DRAs, for example, a mixture of a copolymer of an α-olefin such as 1-octene and 10-undecenoic and a copolymer of styrene and vinyl pyridine. Such acid-base interacting polymers can provide improved drag reduction via polymeric networks rather than by high molecular weight. Consequently, such networks are less sensitive to flow degradation.[53] Later studies of hydrogen-bonding polymer complexes have been reported.[54]

17.3.2.4 Overcoming Handling, Pumping, and Injection Difficulties with UHMW DRA Polymers

Making an UHMW DRA polymer is only part of the manufacturing problem. The main difficulty lies in getting a finished product that has low viscosity, is free flowing, nonagglomerating, has high concentration, and allows smooth and easy injection of the material into a pipeline. Due to the extreme molecular weights, solutions of DRA polymers are extremely viscous. Hence, they have often been diluted down to below 10% active material in a hydrocarbon solvent to reduce the viscosity to a manageable level. Consequently, there are many patents detailing ways of improving the viscosity, handling, and dissolution into pipeline fluids of DRA polymers.[55] Many variations on emulsions, dispersions, and suspensions in water, oxygenated solvents, and hydrocarbon liquids have been proposed.[56-57] One preferred way is to manufacture the DRA polymer as solid particles using grinding methods and disperse them in a nonsolvent liquid or emulsion. This enables a higher concentration (25–30%) DRA product to be made than if the DRA was dissolved in a solvent. It also reportedly gives improved performance.[37,58-59] DRA polymers can also be microencapsulated to make them easy to handle and give them timed-release properties.[60] The use of solid particles of polymeric DRA overcomes the problem of shear degradation at the injection valve. Studies have shown that it is here that degradation is greatest, due to high extensional straining of polymer molecules, rather than in turbulent pipe flow.[61] Better latex suspension DRAs have been claimed that avoid transportation and handling of hazardous solvents and avoid settling and heat problems on site.[62]

Another improvement designed to overcome early shear degradation of the DRA polymer is to use a mixture of two DRA products.[63] One product is designed to dissolve faster than the other. The second product does not dissolve straightaway but only further into the pipeline. Hence, when the first product has lost its performance due to shear degradation, the second product is available to keep up the overall drag-reducing performance. One DRA product can be a precipitation slurry, which dissolves fairly quickly; the second product can be a ground polymer slurry, which dissolves more slowly.

A patented improvement on the use of solid particles as DRAs is to use a bimodal or multimodal particle size distribution.[64] Drag reducers having larger particle sizes dissolve more slowly than drag reducers having smaller particle sizes. By using at least bimodal particle size distributions, drag reduction can be distributed more uniformly over the length of the pipeline where smaller sized particles dissolve sooner or earlier in the pipeline and larger sized particles dissolve later or further along the pipeline. Another method to improve the handling and injection of a UHMW polymeric DRA is to grind the polymer with a wax crystal modifier and suspend it in a suspending fluid.[65] Freeze-protected, concentrated suspensions of low-viscosity polyolefin DRA particles that remain stable for long periods can be made by using a fatty acid–suspending medium, such as soya bean oil, and a modifier comprising of an oxygenated, polar organic compound, such as ethanol.[65]

A novel method of making an organometallic oil-soluble polymer DRA has been reported. Tri-n-butylstannyl fluoride exhibited 75% drag reduction in a capillary rheometer at 25,000 Reynolds number in hexane at 0.1% concentration, a similar result to polyisobutylene. Tri-n-butylstannyl chloride showed no drag reduction under these same conditions. Moreover, there was no loss of drag reduction of the fluoride product at high shear. The effectiveness of tri-n-butylstannyl fluoride is explained by the formation of a linear polymer, –Sn-F–Sn-F–, in which pentacoordinate tin is linked through fluorine bridges.[66]

17.3.2.5 Oil-Soluble Polymeric DRAs in Multiphase Flow

Polymeric DRAs have been studied in a two-phase flow.[67–68] UHMW polyolefin DRAs for gas condensate two-phase flow have been tested and drag reductions up to 65% observed. These polymers are typically used for oil flows.[69] Oil-soluble polymeric DRAs have been tested in multiphase flow using oil and CO_2 gas.[70] In sharp contrast with expectations, the drag reduction was recovered mainly from the accelerational component indicating that the DRA worked not only in the buffer zone but also in the mixing zone in the slug body. The accelerational drag reduction reached values as high as 88% out of total drag reduction.[71]

17.3.3 OIL-SOLUBLE SURFACTANT DRAs

Surfactant DRAs have been more extensively explored for water or multiphase flow applications. However, there are a few reports of nonpolymeric oil-soluble DRA studies.

In single-phase hexane flow tests, an alkyl phosphate ester performed better than three UHMW polymers (20–100 ppm) at concentrations > 200 ppm and friction velocities < 0.3 ft/s.[72] For these conditions, the ester yielded drag reduction levels up to 85% and exhibited negligible shear degradation. For concentrations less than 100 ppm and friction velocities above 0.3 ft/s, a polymer product was the superior drag reducer in the absence of degradation. However, in a two-phase flow with hexane and natural gas, the polymer products were superior and at a lower concentration.

Aluminum carboxylate DRAs have been described.[73] These additives are not subject to permanent shear degradation and do not cause undesirable changes in the emulsion or fluid quality of the fluid being treated, or undesirable foam. In addition, they are claimed to be easy to inject. Examples are aluminum dioctoate, aluminum distearate, and various mixtures. Another variation on this theme is to blend an aluminum monocarboxylate with at least one carboxylic acid on site to produce an aluminum dicarboxylate DRA. This avoids handling, transportation, and injection difficulties with very viscous solutions.[74] Some of the classes of surfactants described for drag reduction aqueous solutions (see below) can be hydrophobic enough for use as DRAs in hydrocarbon liquids or multiphase transportation.

17.4 WATER-SOLUBLE DRAs

Water-soluble DRAs can be divided into two categories:

- High molecular weight linear polymers
- Surfactants

Both of these categories are discussed below as well as the relationship between drag reduction and corrosion inhibition.

17.4.1 WATER-SOLUBLE POLYMER DRAs

Many classes of water-soluble polymer exhibit drag-reducing properties, including cationic, anionic, or nonionic polymers.[75-76] Water-soluble polymers that have been deployed as DRAs include:

- PAM and partially hydrolyzed polyacrylamide (PHPA)
- Other copolymers of acrylamide and acrylamide derivatives and acrylates
- Polyethyleneoxide (PEO)
- Polyvinyl alcohols
- Polysaccharides and derivatives, such as:
  - Guar gum
  - Hydroxypropylguar
  - Xanthan
  - Carboxymethylcellulose (CMC)
- Hydroxyethylcellulose (HEC)

Of these categories, UHMW acrylamide polymers and copolymers are the most-used DRAs in water-injection projects in the oil industry. The other class of water-soluble synthetic polymeric DRAs that has been well researched is the UHMW PEOs.

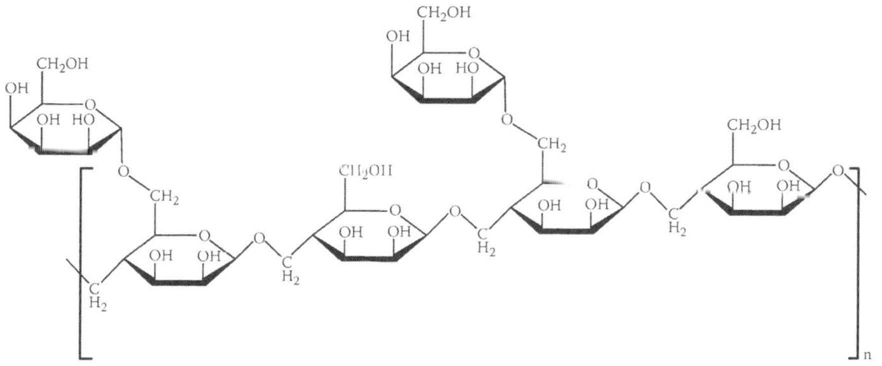

FIGURE 17.7 Guar gum.

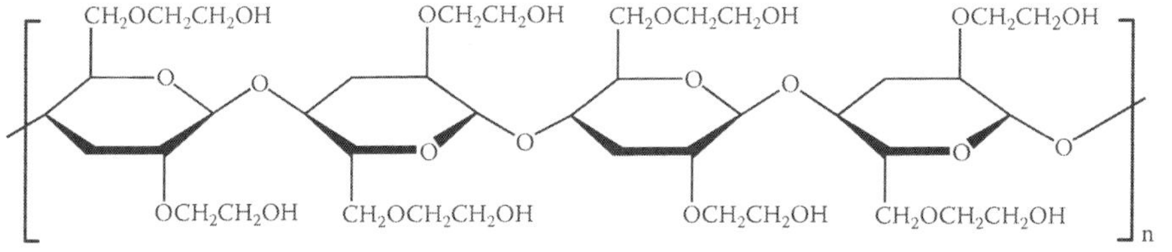

FIGURE 17.8 Hydroxyethylcellulose (HEC).

17.4.1.1 Polysaccharides and Derivatives

In general, synthetic water-soluble polymers perform better than the more biodegradable polysaccharides, probably due to the higher molecular weights obtainable and less shear problems.[77–78] Synergism has been observed between solutions of polysaccharides and PAMs.[79] Polysaccharides and derivatives are used as DRAs in the oil industry, for example, in fracturing operations. They are more susceptible to shear and biodegradation than the vinyl polymers. Guar gum and its derivatives are often considered the best of the natural polysaccharide DRAs (Figure 17.7). HEC (Figure 17.8) and CMC are also typical viscosifying, drag-reducing polysaccharide derivatives used in well operations. Polymer linearity is the key structural feature in polysaccharides for them to function as DRAs. This would explain why highly branched polysaccharides such as dextran and gum arabic are poor DRAs. Guar gum and other polysaccharides can be made more resistant to shear and biodegradation by grafting acrylamide on to them.[80–81]

17.4.1.2 Polyethyleneoxide Drag-Reducing Agents

PEO is made by metal-catalyzed polymerization of ethylene oxide to give very high molecular weights (up to 8,000,000 Da; Figure 17.9).[82–83] Base-catalyzed polymerization of ethylene oxide gives poly(ethyleneglycol)s with lower molecular weights, < 100,000 Da. PEO is commercially available. Its use as a DRA appears to be outside the oil industry, since it is prone to shear degradation when injected or under turbulent flow.[84] PEO is also used as a thickener.

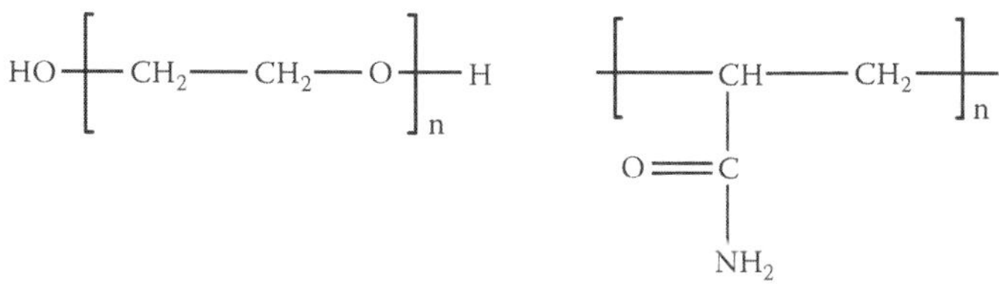

Drag-Reducing Agents

$$HO-[CH_2-CH_2-O]_n-H \qquad [-CH-CH_2-]_n$$
$$\qquad\qquad\qquad\qquad\qquad\qquad\qquad\qquad |$$
$$\qquad\qquad\qquad\qquad\qquad\qquad\qquad\qquad O{=}C$$
$$\qquad\qquad\qquad\qquad\qquad\qquad\qquad\qquad |$$
$$\qquad\qquad\qquad\qquad\qquad\qquad\qquad\qquad NH_2$$

FIGURE 17.9 Polyethyleneoxide (left) and polyacrylamide (right).

17.4.1.3 Acrylamide-Based DRAs

PAM, PHPA, and related acrylamide or acrylate polymers and copolymers have more practical use as DRAs than PEO, as they have a side chain and are less susceptible to shear degradation (Figure 17.9).[85] Thus, PAM and related derivatives are generally the preferred polymeric water-soluble DRAs in the oil industry for water injection. A typical dose is 20–30 ppm in the water phase. Molecular weights of acrylamide copolymers can be as high as 20,000,000 Da. PHPA, which is mostly a copolymer of acrylamide and acrylate monomers, has been extensively used in water-injection polymer-flooding projects to increase the injected volumes. Up to 70% drag reduction was obtained in single-phase flow laboratory experiments with a PHPA; lower values (50–60%) are normally obtained in the field.[86] In multiphase flow, drag reduction is usually lower. PHPA is made by partially hydrolyzing PAM. Alternatively, polymers with similar DRA performance can be made by copolymerizing acrylamide with smaller amounts of acrylate monomer.[87]

Several acrylamide copolymers, mostly with larger monomers showed improved performance over PAM, for example, acrylamide copolymerized with diacetoneacrylamide, sodium acrylamido-2-methylpropane sulfonate or sodium acrylamido-3-methylbutenoate (Figure 17.10). Polymer molecular weights of up to 28,000,000

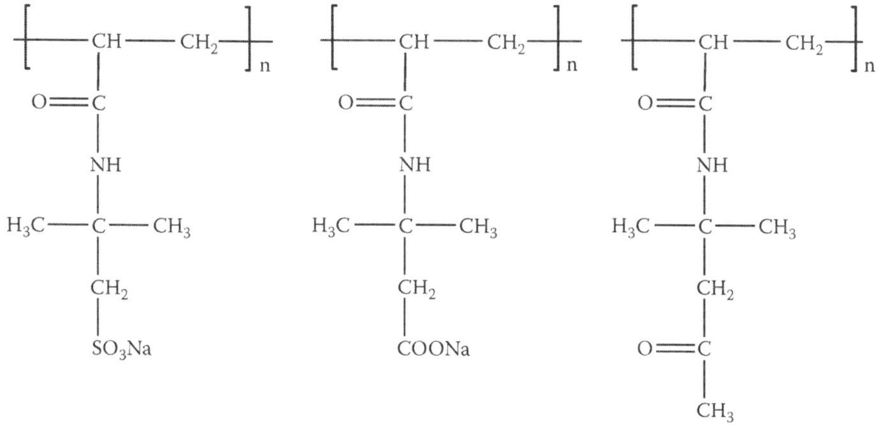

FIGURE 17.10 Sodium acrylamido-2-methylpropane sulfonate, sodium AMPS (left), sodium acrylamido-3-methylbutenoate (middle), and diacetoneacrylamide (right).

Da were obtained. These polymers were capable of intermolecular associations.[88] In addition, hydrophobically modified PAMs with associating groups in solution performed better than polyampholytes, which collapsed due to intramolecular associations.[89–90] Copolymers of acrylamide and alkyl poly(etheroxy)acrylates have also been claimed as improved acrylamide-based DRAs with large side chains.[91]

One drawback of PHPA or other polymers containing acrylate monomer is their reduced compatibility with high calcium brines. Various other ionic groups such as sulfonates (e.g., AMPS, Figure 17.10) that are more compatible can be incorporated into acrylamide copolymers. Another claimed example is acrylamide: (N-3-sulfopropyl)-N-methacryol-oxyethyl-N,N-dimethylammonium betaine copolymer, where the betaine groups ensure good calcium compatibility.[92] PAMs have also been shown to have reduced performance in the presence of ferric ions.[93]

Most polymer DRAs show reduced performance as low molecular additives (i.e., acids, bases, or salts) are dissolved in the aqueous solution. These additives screen the charges that are fixed along the chain backbone, which results in a decrease in the dimensions of the polymer molecule. The drag reduction diminishes as long as the chain continues to shrink. However, zwitterionic terpolymers of acrylamide/metal styrene sulfonate/methacrylamidopropyltrimethyl ammonium chloride show increasing viscosity and DRA performance as the ionicity of the aqueous increases.[94]

Another study showed that addition of urea (a hydrogen-bond breaker) to the water phase greatly reduced the drag-reducing effect of acrylamide copolymers, whereas an increase in ionic strength by addition of sodium chloride had the opposite effect.[95] Aggregation of polymer chains in solution is not necessary for DRA effect, but it will give a larger volume to break up any turbulence. Aggregation has been shown to occur in dilute PAM and PEO solutions. High-chain flexibility appears to give good drag-reducing effect. Thus, polymethacrylamide should perform worse than PAM at the same chain length.

Although early theories suggested that polymer branching would be detrimental to the drag-reducing effect, some studies have shown otherwise. For example, highly branched high-molecular-weight PAM showed enhanced performance and shear stability.[96] It is probably the higher hydrodynamic volume compared with linear PAM that is critical in this case. This is why linear polymers with large side chains are better DRAs than polymers with little or no side chains. However, it is difficult to get UHMW PAMs with a high percentage of large side chains. This is possible with polyolefins via Ziegler-Natta–catalyzed polymerization of olefins, but these organometallic titanium or zirconium catalysts are not robust in the presence of such a polar monomer as acrylamide.

As with oil-soluble UHMW polymeric DRAs, there are difficulties in handling water-soluble polymers of such high molecular weights because of the high viscosity in polar solvents. The normal way around this problem for acrylamide-based DRAs is to use a water-in-oil emulsion of the polymer. This greatly reduces the viscosity of the product. A novel alternative to avoiding highly viscous solutions is to use a mixture of an anionic acrylamide copolymer and a cationic acrylamide copolymer of relatively low molecular weight. These will associate and form a complex when injected into the aqueous medium, causing much higher DRA performance than using the copolymers separately.[97]

Simple monomeric surfactants affect the performance of DRAs such as PEO or PAM. At low surfactant concentration, the performance of PEO was worse, but at high concentration, the performance improved relative to pure PEO. This may be due to micellar structures at high concentration contributing to the drag-reducing effect.[98]

Water-soluble polymeric DRAs have also been proposed for use in multiphase flow although there do not appear to be any field applications reported yet.[43,99–100] For example, polymeric nanoemulsions having a hydrocarbon external phase and a PAM-based DRA aqueous internal phase have been claimed to reduce drag and friction in multiphase pipelines.[101] The presence of water-soluble polymer DRA in oil-water flow laboratory pipe experiments extended the region of stratified flow and delayed transition to slug flow.[102] It has also been claimed that PAMs that contain anionicity in the polymer backbone exhibit substantially lower emulsion creating tendency as compared with their cationically or neutrally modified congeners.[103] It should also be mentioned that reducing the drag of a turbulent flowing multiphase fluid susceptible to viscous emulsion problems can be carried out by injecting a demulsifier at the wellhead.[104]

17.4.2 WATER-SOLUBLE SURFACTANT DRAs

Surfactants have been known for a long time to be capable of reducing drag in turbulent flowing liquids.[105] Surfactant additives have dual effects on frictional drag. First, they introduce viscoelastic shear stress, which increases frictional drag. Second, they dampen the turbulent vortical structures, decrease the turbulent shear stress, and then decrease the frictional drag. Since the second effect is greater than the first one, drag reduction occurs.[106]

Surfactant DRAs have been proposed as alternatives to polymer DRAs for water injection. Surfactant DRAs are only active at higher concentrations than UHMW polymeric DRAs. For example, 200 ppm of a good surfactant DRA may be needed for the same drag performance as 20 ppm of a UHMW polymeric DRA. Up to 80% drag reduction has been observed with surfactant DRAs. The reason for the high concentration with surfactant DRAs is that the surfactants, or blends or surfactants, need to be above a critical micelle concentration so they can associate into large-enough micelles.[107] For some surfactants, these micelles can take on a rod-like nature. It is only these associated surfactant rodlike micelles and not individual surfactant molecules or spheroidal micelles that are capable of reducing eddies and bursts in turbulent flow. Some water-soluble surfactants, which show no drag-reducing properties at ambient temperature, may perform as DRAs at elevated temperature just below their cloud point. This may be due to aggregation of the surfactants but not total collapse of the surfactant-water interactions.[108]

Since the late nineties, there has been renewed interest by the oil industry in surfactant DRAs, especially for use in seawater injection but also for water circulation in bundle pipelines. This may be for several reasons. First, although a higher concentration of surfactant DRA is needed compared with a UHMW polymeric DRA, the cost of UHMW polymeric DRAs is quite high. Hence, a cheaper surfactant can compete on a cost-performance basis. The second reason is that surfactant DRAs are not susceptible to permanent shear degradation as are the UHMW polymeric DRAs and can therefore find applications where there is turbulent flow over long periods. If a thread-like micelle structure is damaged, it can repair itself further along the pipe.[109] For seawater injection, this means surfactants DRAs can be injected before the pumps. Third, some surfactant DRAs are biodegradable and may be more easily accepted by some environmental authorities.

Many surfactant classes are capable of showing drag-reducing effects. Due to the higher concentrations needed for good DRA performance, the surfactants need to be made from fairly cheap materials. Many studies have been carried out on simple cationic surfactants such as cetyltrimethylammonium chloride but anionic surfactants such as sodium dodecyl sulfate (SDS) have also been shown to exhibit drag reduction.[110] For example, 400 ppm SDS in distilled water was shown to reduce the pressure drop by 25–40%.[111]

For cationic surfactants the nature and size of the corresponding anion can significantly affect the micellar shape and drag-reducing performance. In one study with a cationic surfactant, the anion was varied among three isomeric counterions, 2-, 3-, or 4-chlorobenzoate.[112] Each isomer showed different types of rheological and drag-reduction behavior and different micellar structures. The 4-Cl system showed good drag reduction and a rod-like micellar network, while the 2-Cl system showed no drag reduction, low apparent extensional viscosity, and only spherical micelles. The 3-Cl system gave more complicated behavior.

Cationic surfactants with large anions appear to be beneficial for forming rod-like micelles. For example, cetyltrimethylammonium salicylate and cetylpyridinium salicylate are particularly preferred drag reducers although the anion could also be thiosalicylate, sulfonate, or hydroxynaphthenate (Figure 17.11).[113–115] Erucyl methyl bis(2-hydroxyethyl) ammonium chloride does not require a large counterion to form viscoelastic micelles.

Further examples of cheap surfactants that can be used as DRAs include fatty acids, alkoxylated derivatives of fatty acids, organic, and inorganic salts of fatty acids and alkoxylated derivatives thereof (Figure 17.12).[116] The more hydrophobic surfactants in this class can be used for drag reduction in hydrocarbon liquids. These DRAs, and the cationic surfactants previously mentioned, can also provide the additional function of corrosion inhibition, which can reduce the overall cost of production chemical consumption. Other cheap surfactant DRAs are maleated fatty acids and the esters thereof and the organic, inorganic, or amine salts thereof.[117] Particularly, preferred salts are imidazoline salts, which can provide corrosion protection.

Cooperative drag-reducing effects between polymers and surfactants have been reported.[118] For example, results with PEO mixed with a homologous series of

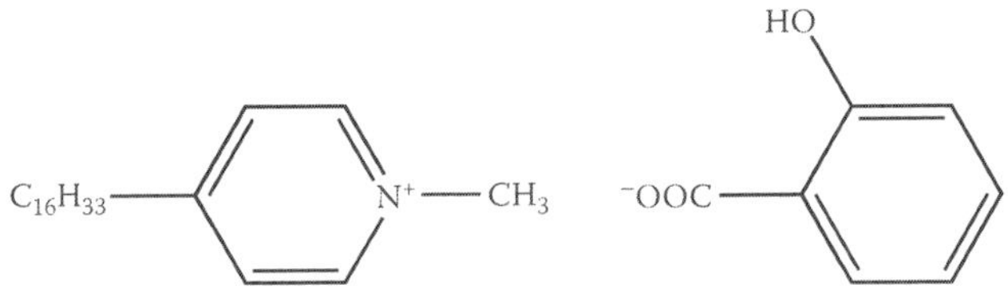

FIGURE 17.11 Cetylpyridinium salicylate.

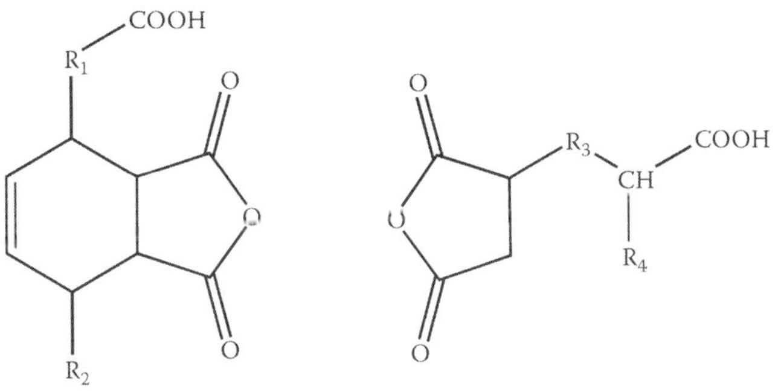

FIGURE 17.12 Maleated fatty acid surfactant DRAs. Monoesters of these products are also preferred as are imidazoline salts.

carboxylate soaps suggest a cooperative micelle formed between soap and polymer. Furthermore, the enhanced drag reduction obtained is consistent with the model of surfactant molecules hydrophobically bonded to the polymer chain in which repulsion between adjacent polar surfactant groups promotes expansion of the polymer coil.

One company has done considerable work on alkanolamide and zwitterionic surfactant water-soluble DRAs and blends with other surfactants. These surfactants are capable of forming long cylindrical micelles, ideal for drag reduction. Some of the first surfactant DRAs they developed were based on alkoxylated alkanolamides.[119] However, they performed well only within a limited temperature range and in very low salinity soft water, not seawater. Blends of the alkoxylated alkanolamides with alkoxylated alcohols or an ionic surfactant, sulfonated, amphoteric, or zwitterionic surfactant, have been claimed as improvements.[120–122] More recent work has concentrated on zwitterionic surfactants in double and triple blends. For example, a mixture of a zwitterionic surfactant in combination with an ether sulfate or ether carboxylate surfactant has been claimed (Figure 17.13). Specific examples are N-behenyl betaine mixed with sodium dodecyl ether sulfate.[123] This mixture also has a low sensitivity to hard water. However, the amount of the surfactant necessary to obtain an essential reduction of the drag has been shown to be above 500 ppm. In addition, the formation of micelles and, therewith, the reduction of drag was expected to be negatively affected by the presence of large amounts of electrolytes. Thus, this type of DRA was regarded as suitable to be used in injection waters, but not when the injection water is based on seawater. A further improvement was a triple blend comprising two

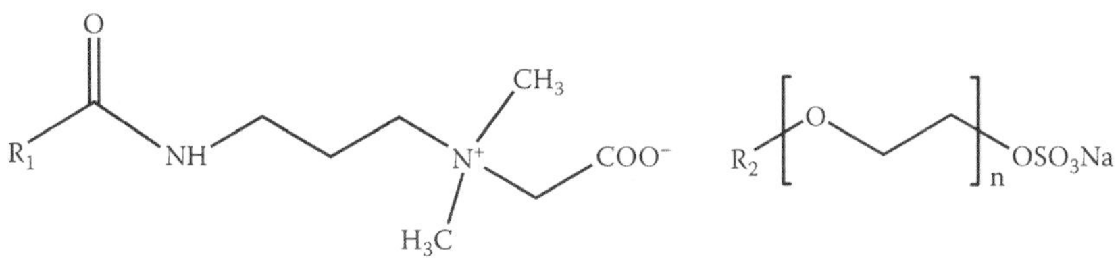

FIGURE 17.13 Examples of preferred zwitterionic (left) and alkyl ether sulfate surfactants for surfactant DRA blends.

zwitterionic surfactants, each comprising an acyl group and an anionic surfactant where the hydrophilic group is a sulfate, a sulfonate, or an ether sulfate.[124] This blend gave a very good drag-reducing effect at a concentration of 50–400 ppm, preferably 60–300 ppm, at large temperature intervals within the range of 2–70°C (36–158°F), even in water with an electrolyte content of as a high as 7 wt.%.

The zwitterionic N-alkylbetaine/anionic surfactant DRA blends have already found one field application in the North Sea, to boost the heating capacity of a 10-km-long pipe bundle heated with an aqueous glycol solution.[125–126] Tests with a polymer DRA failed due to degradation of the polymer in the pumps during circulation of the heating fluid. For the goal of using surfactant DRAs in seawater injection, laboratory tests of a combination of a zwitterionic and an anionic surfactant in synthetic seawater have been reported to give a drag reduction between 75 and 80% with 200 ppm of the surfactant blend at an average velocity of 1.9 m/s and between 50% and 55% at 2.9 m/s.[127] Due to the self-healing properties of the drag-reducing structures formed by surfactants, these may be added before the pump section—contrary to polymers, which are permanently destroyed by high shear forces. The surfactants are also biodegradable.

17.4.3 DRAG REDUCTION AND CORROSION INHIBITION

Application of a DRA can reduce flow-induced localized corrosion.[128] The mechanism can be twofold. First, if the DRA is a surfactant, it can additionally perform as a film-forming corrosion inhibitor. For example, cetyltrimethylammonium salicylate and cetylpyridinium salicylate have been shown to work in both ways.[129] Many film-forming corrosion inhibitors will show some drag-reducing properties above a critical micelle concentration.[130] Second, a DRA will reduce turbulence near the walls of the pipe. This can slow down erosion corrosion by itself or help prevent a film-forming corrosion inhibitor from being removed from the pipe wall.[11,131] One manufacturer of water-soluble polymer DRAs claims a corrosion reduction of 40% by using their polymer.

REFERENCES

1. B. A. Jubran, Y. H. Zurigat, M. S. Al-Shukri, and H. H. Al-Busaidi, *Polymer-Plastics Technology and Engineering* 45(4) (2006): 553.
2. A. Gyr and H. W. Bewersdorff, *Drag Reduction of Turbulent Flows by Additives (Fluid Mechanics and Its Applications)*, Dordrecht, The Netherlands: Kluwer Academic Publishers, 1995.
3. G. T. Pruitt, C. M. Simmons, G. H. Neil, and H. R. Crawford, "A Method to Minimize the Cost of Pumping Fluids Containing Friction Reducing Additives," SPE 997, *Journal of Petroleum Technology* 1965: 641.
4. R. A. Woodroof and R. W. Anderson, "Synthetic Polymer Friction Reducers Can Cause Formation Damage," SPE 6812 (paper presented at the 52nd Annual Fall Technical Conference and Exhibition, Denver, CO, 9–12 October 1977).
5. H. A. Al-Anazi, M. G. Al-Faifi, F. Tulbah, and J. Gillespie, "Evaluation of Drag Reducing Agent (DRA) for Seawater Injection System: Lab and Field Cases," SPE 100844 (paper presented at the SPE Asia Pacific Oil & Gas Conference and Exhibition, Adelaide, Australia, 11–13 September 2006).

6. M. Ke, Q. Qu, R. F. Stevens, N. Bracksieck, C. Price, and D. Copeland, "Evaluation of Friction Reducers for High-Density Brines and Their Application in Coiled Tubing at High Temperature," SPE 103037 (paper presented at the SPE Annual Technical Conference and Exhibition, San Antonio, TX, 24–27 September 2006).
7. R. L. J. Fernandes, B. M. Jutte, and M. G. Rodriguez, "Drag Reduction in Horizontal Annular Two-Phase Flow," *International Journal of Multiphase Flow* 30 (2004): 1051.
8. Y.-H. Li, U.S. Patent 5020561, 1991.
9. Y.-H. Li, G. R. Chesnut, R. D. Richmond, G. L. Beer, and V. P. Caldarera, "Laboratory Tests and Field Implementation of Gas-Drag-Reduction Chemicals," SPE 37256, *SPE Production & Facilities* 13 (1998): 53.
10. A. A. Hamouda, "Drag Reduction-Performance in Laboratory Compared to Pipelines," SPE 80258 (paper presented at the SPE International Symposium on Oilfield Chemistry, Houston, TX, 5–7 February 2003).
11. G. Schmitt, C. Bosch, H. Bauer, and M. Mueller, "Modelling the Drag Reducing Effect of CO_2 Corrosion Inhibitors," Paper 00002 (paper presented at the NACE CORROSION Conference, 2000).
12. K. Slater and M. Zamora, U.S. Patent Application 20080289435.
13. (a) M. E. Cowan, R. D. Hester, and C. L. McCormick, "Water-Soluble Polymers: LXXXII. Shear Degradation Effects on Drag Reduction Behavior or Dilute Polymer Solutions," *Journal of Applied Polymer Science* 82 (2001): 1211. (b) M. E. Cowan, C. Garner, R. D. Hester, and C. L. McCormick, "Water-Soluble Polymers: LXXXII. Correlation of Drag Experimentally Determined Drag Reduction Efficiency and Extensional Viscosity of High Molecular Weight Polymers in Dilute Aqueous Solution," *Journal of Applied Polymer Science* 82 (2001): 1222.
14. K. Lee, C. A. Kim, S. T. Lim, D. H. Kwon, H. J. Choi, and M. S. Jhon, "Mechanical Degradation of Polyisobutylene in Kersosene," *Colloid and Polymer Science* 280 (2002): 779.
15. C. A. Kim, D. S. Jo, H. J. Choi, C. B. Kim, and M. S. Jhon, "A High-Precision Disk Apparatus for Drag Reduction Characterization," *Polymer Testing* 20 (2000): 43.
16. M. S. Figueiredo, L. C. Almeida, F. G. Costa, M. D. Clarisse, L. Lopes, R. Leal, A. L. Martins, and E. F. Lucas, "Development of a Method to Evaluate the Performance of Aqueous Polymer Solutions as Drag Reduction Agents in Bench Scale," *Macromolecular Symposia* 245–246 (2006): 260.
17. J. L. Lumley, "Drag Reduction by Additives," *Annual Review of Fluid Mechanics* 1 (1969): 367.
18. P. G. De Gennes, *Introduction to Polymer Dynamics*, Cambridge: Cambridge University Press, 1990.
19. P. S. Virk, H. S. Mickley, and K. A. Smith, "The Ultimate Asymptote and Mean Flow Structure in Tom's Phenomenon," *ASME, Journal of Applied Mechanics* 37 (1970): 480.
20. P. Peyser and R. C. Little, "The Drag Reduction of Dilute Polymer Solutions as a Function of Solvent Power, Viscosity, and Temperature," *Journal of Applied Polymer Science* 15 (1971): 2623.
21. B. A. Jubran, Y. H. Zurigat, and M. F. A. Goosen, "Drag Reducing Agents in Multiphase Flow Pipelines: Recent Trends and Future Needs," *Petroleum Science & Technology* 23 (2005): 1403.
22. A. A. Hamouda and F. S. Evensen, "Possible Mechanism of the Drag Reduction Phenomenon in Light of the Associated Heat Transfer Reduction," SPE 93405 (paper presented at the SPE International Symposium on Oilfield Chemistry, Houston, TX, 2–4 February 2005).
23. A. P. Matjukhatov, B. P. Mironov, and I. A. Animisov, *The Influence of Polymer Additives on Velocity and Temperature Fields*, ed. B. Gampert, Berlin: Springer-Verlag, 1985, 107.

24. A. Al-Sarkhi and T. J. Hanratty, "Effect of Drag-Reducing Polymers on Pseudo-Slugs Interfacial Drag and Transition to Slug Flow," *International Journal of Multiphase Flow* 28 (2002): 1911–1927.
25. E. D. Burger, W. R. Munk, and H. A. Wahl, "Flow Increase in the Trans Alaska Pipeline Through Use of a Polymeric Drag-Reducing Additive," SPE 9419, *Journal of Petroleum Technology* 34 (1982): 377.
26. H. A. Wahl, W. R. Beatty, J. G. Dopper, and G. R. Hass, "Drag Reducer Increases Oil Pipeline Flow Rates," SPE 10446 (paper presented at the Offshore South East Asia 82 Conference, Singapore, 9–12 February 1982).
27. B. K. Berge and O. Solsvik, "Increased Pipeline Throughput Using Drag Reducing Additives (DRA): Field Experiences," SPE 36835 (paper presented at the SPE European Petroleum Conference, Milan, Italy, 22–24 October 1996).
28. J. U. Ibrahim and L. A. Braimoh, "Drag Reducing Agent Test Result for ChevronTexaco, Eastern Operations, Nigeria," SPE 98819 (paper presented at the 29th Annual SPE International Technical Conference and Exhibition, Abuja, Nigeria, 1–3 August 2003).
29. P. K. Ptasinski, B. J. Boersma, F. T. M. Nieuwstadt, M. A. Hulsen, B. H. A. A. Van den Brule, and J. C. R. Hunt, "Turbulent Flow of Polymer Solutions Near Maximum Drag Reduction; Experiments, Simulations and Mechanisms," *Journal of Fluid Mechanics* 490 (2003): 251.
30. M. P. Mack, U.S. Patent 4493903, 1985.
31. J. A. Lescarboura, J. D. Culter, and H. A. Wahl, "Drag Reduction with a Polymeric Additive in Crude Oil Pipelines," SPE 3087 (paper presented at the SPE 45th Annual Fall Meeting, Houston, TX, 4–7 October 1970).
32. H. J. Choi and M. S. Jhon, "Polymer-Induced Turbulent Drag Reduction," *Industrial & Engineering Chemistry Research* 35(9) (1996): 2993.
33. K. Lee, C. A. Kim, S. T. Lim, D. H. Kwon, H. J. Choi, and M. S. Jhon, "Mechanical Degradation of Polyisobutylene under Turbulent Flow," *Colloid & Polymer Science* 280 (2002): 779.
34. R. L. Johnston and S. N. Milligan, U.S. Patent 6596832, 2003.
35. E. Karhu, M. Karhu, L. Rockas, Leif, and H. Harjuhahto, U.S. Patent Application, 20020173569, 2002.
36. W. Brostow, "Drag Reduction and Mechanical Degradation in Polymer Solutions in Flow," *Polymer* 24 (1983): 631.
37. K. W. Smith, L. V. Haynes, and D. F. Massouda, U.S. Patent 5449732, 1995.
38. D. L. Hunston, "Effects of Molecular Weight Distribution in Drag Reduction and Shear Degradation," *Journal of Polymer Science: Polymer Chemistry Edition* 14 (1976): 713.
39. B. Liu, Bing, X. Bao, Y. Gao, C. Li, and G. Li, U.S. Patent Application 20070004837.
40. S. N. Milligan and K. W. Smith, U.S. Patent 6576732, 2003.
41. G. B. Eaton, M. J. Monahan, A. K. Ebert, R. J. Tipton, and E. Baralt, U.S. Patent 6730750, 2004.
42. J. R. Harris, U.S. Patent Application 20060281832, 2006.
43. S. Malik, S. N. Shintre, and R. A. Mashelkar, U.S. Patent 5080121, 1992.
44. D. E. Farley, "Drag Reduction in Non-Aqueous Solutions: Structure-Property Correlations for Poly(isodecylmethacrylate)," SPE 5308 (paper presented at the SPE International Symposium on Oilfield Chemistry, Dallas, TX, 16–17 January, 1975).
45. M. D. Holtmyer and J. Chatterji, *Polymer Engineering & Science* 20 (1980): 473.
46. Y. Ma, X. Zheng, F. Shi, Y. Li, and S. Sun, "Synthesis of Poly(Dodecyl Methacrylate)s and Their Drag-Reducing Properties," *Journal of Applied Polymer Science* 88 (2003): 1622.
47. W. Ritter and C. P. Herold, International Patent Application WO9002766, 1990.
48. D. N. Schulz, K. Kitano, T. J. Burkhardt, and A. W. Langer, U.S. Patent 4518757, 1985.

49. (a) K. W. Smith, W. R. Dreher, and T. I. Burden, International Patent Application, WO/2008/014190; (b) S. N. Milligan, R. L. Johnston, T. L. Burden, W. R. Dreher, K. W. Smith, and W. F. Harris, International Patent Application WO/2008/079642.
50. R. M. Kowalik, I. Duvdevani, D. G. Pfeiffer, R. D. Lundberg, K. Kitang, and D. N. Schulz, "Enhanced Drag Reduction via Interpolymer Associations," *Journal of Non-Newtonian Fluid Mechanics* 24 (1987): 1.
51. R. M. Kowalik, I. Duvdevani, D. G. Pfeiffer, and R. D. Lundberg, U.S. Patent 4508128, 1985.
52. S. Malik and R. A. Mashelkar, "Hydrogen Bonding Mediated Shear Stable Clusters as Drag Reducers," *Chemical Engineering Science* 50 (1995): 105.
53. R. M. Kowalik, I. Duvdevani, K. Kitano, and D. N. Schulz, U.S. Patent 4625745, 1986.
54. S. Malik and R. A. Mashelkar, *Chemical Engineering Science* 50 (1995): 195.
55. R. L. Johnston and L. G. Fry, U.S. Patent 5376697, 1994.
56. K. Fairchild, R. Tipton, J. F. Motier, and N. Kommareddi, U.S. Patent 5733953 1998.
57. G. B. Eaton and M. J. Monahan, U.S. Patent 5869570, 1999
58. (a) K. M. Labude, K. W. Smith, and T. L. Burden, U.S. Patent 6399676, 2002. (b) T. Mathew, N. S. Kommareddi, U.S. Patent Application 20080287568.
59. J. R. Harris, J. F. Motier, L.-C. Chou, and T. J. Martin, U.S. Patent 7119132, 2006.
60. N. S. Kommareddi, R. Dinius, N. Vasishtha, and D. E. Barlow, U.S. Patent 6841593, 2005.
61. T. Moussa and C. Tiu, "Factors Affecting Polymer Degradation in Turbulent Pipe Flow," *Chemical Engineering Science* 49 (1994): 1681.
62. W. F. Harris, F. William, K. W. Smith, S. N. Milligan, R. L. Johnston, and V. S. Anderson, International Patent Application WO/2006/081010.
63. J. F. Motier, J. R. Harris, L. C. Chou, and N. S. Kommareddi, U.S. Patent Application 20070021531.
64. J. R. Harris, L. C. Chou, G. G. Ramsay, J. F. Motier, N. S. Kommareddi, and T. Mathew; U.S. Patent Application 20060293196.
65. (a) K. M. Labude, K. W. Smith, and R. L. Johnston, U.S. Patent Application 20030187123. (b) B. A. Bucher, M. J. Monahan, and S. B. Erikson, International Patent Application WO/2008/073293.
66. A. P. Evans, "A New Drag-Reducing Polymer with Improved Shear Stability for Nonaqueous Systems," *Journal of Applied Polymer Science* 18 (1974): 1919.
67. D. Mowla and A. Naderi, "Experimental Study of Drag Reduction by a Polymeric Additive in Slug Two-Phase Flow of Crude Oil and Air in Horizontal Pipes," *Chemical Engineering Science* 61 (2006): 1549.
68. D. Mowla, M. Moshfeghian, and M. S. Hatamipour, "A Simple Model for Prediction of Pressure Drop in Horizontal Two-Phase Flow," *Iranian Journal of Science and Technology* 15 (1991): 177.
69. R. L. Fernandes, "Multiphase Drag Reduction. Part I: Proof-of-Concept Experiments," Internal Shell Report EP 2003-5028, Shell Rijswijk, 2003.
70. C. Kang and W. P. Jepson, "Effect of Drag-Reducing Agents in Multiphase, Oil/Gas Horizontal Flow," SPE 58976 (paper presented at the SPE International Petroleum Conference and Exhibition, Villahermosa, Mexico, 1–3 February 2000).
71. M. Daas, C. Kang, and W. P. Jepson, "Quantitative Analysis of Drag Reduction in Horizontal Slug Flow," SPE 62944 (paper presented at the SPE Annual Technical Conference and Exhibition, Dallas, TX, 1–4 October 2000).
72. N. D. Sylvester, R. H. Dowling, H. Paz-y-Mino, and J. Brill, "Drag Reduction in Two-Phase Gas-Liquid Flow," Prepared for the Materials Committee Pipeline Research Committee of Pipeline Research Council International, Inc. 1977.
73. V. Jovancicevic, S. Campbell, S. Ramachandran, P. Hammonds, and S. J. Weghorn, U.S. Patent Application 20040142825.

74. P. Hammonds, V. Jovancicevic, C. M. Means, C. Mitch, and D. Green, U.S. Patent Application 20040216780.
75. K. Oh-Kil and C. Ling Siu, *Drag Reducing Polymers: The Polymeric Materials Encyclopedia*, Boca Raton, FL: CRC Press Inc., 1996.
76. S. E. Morgan and C. L. McCormick, "Water-Soluble Copolymers: XXXII. Macromolecular Drag Reduction. A Review of Predictive Theories and the Effects of Polymer Structure," *Progress in Polymer Science* 15 (1990): 507.
77. J. W. Hoyt, "Drag Reduction in Polysaccharide Solutions," *Trends Biotechnology* 3 (1985): 17.
78. W. Interthal and H. Wilski, "Drag Reduction Experiments with Very Large Pipes," *Colloid & Polymer Science* 263 (1985): 217.
79. J. P. Malhotra, P. N. Chaturvedi, and R. P. Singh, "Drag Reduction by Polymer-Polymer Mistures," *Journal of Applied Polymer Science* 36 (1988): 837.
80. S. R. Deshmukh and R. P. Singh, "Drag Reduction Characteristics of Graft Copolymers of Xanthangum and Polyacrylamide," *Journal of Applied Polymer Science* 32 (1986): 6163.
81. R.P Singh, G. P. Karmakar, S. K. Rath, N. C. Karmakar, S. R. Pandey, T. Tripathy, J. Panda, K. Kanan, S. K. Jain, and N. T. Lan, "Biodegradable Drag Reducing Agents and Flocculants Based on Polysaccharides: Materials and Applications," *Polymer Engineering & Science* 40 (2000): 46.
82. R. C. Little and M. Wiegard, "Drag Reduction and Structural Turbulence in Flowing Polyox Soltions," *Journal of Applied Polymer Science* 14(2) (1969): 409.
83. S. A. Nosier, Y. A. Alhamed, A. A. Bakry, and I. S. Mansour, 2007, "Forced Convection Solid-Liquid Mass Transfer at a Surface of Tube Bundles under Single Phase Flow," *Chemical and Biochemical Engineering Quarterly* 21(3) (2007): 213.
84. S. U. S. Choi, Y. I. Cho, and K. E. Kasza, "Degradation Effects of Dilute Polymer Solutions on Turbulent Friction and Heat Transfer Behavior," *Journal of Non-Newtonian Fluid Mechanics* 41 (1992): 289.
85. D. H. Fisher and F. Rodriguez, "Degradation of Drag-Reducing Polymers," *Journal of Applied Polymer Science* 15 (1971): 2975.
86. P. K. Ptasinski, F. T. M. Nieuwstadt, B. H. A. A. van den Brule, and M. A. Hulsen, "Experiments in Turbulent Pipe Flow with Polymer Additives at Maximum Drag Reduction," *Flow Turbulence and Combustion* 66 (2001): 159.
87. B. L. Knight, J. S. Rhudy, and W. B. Gogarty, U.S. Patent 4236545, 1980.
88. C. L. McCormick, R. D. Hester, S. E. Morgan, and A. M. Safieddine, "Water-Soluble Copolymers: 30. Effects of Molecular Structure on Drag Reduction Efficiency," *Macromolecules* 23(8) (1990): 2124.
89. C. L. McCormick, R. D. Hester, S. E. Morgan, and A. M. Safieddine, "Water-Soluble Copolymers: 31. Effects of Molecular Parameters, Solvation, and Polymer Associations on Drag Reduction Performance," *Macromolecules* 23(8) (1990): 2139.
90. P. S. Mumick, R. D. Hester, and C. L. McCormick, "Water-Soluble Copolymers: 55. N-Isopropylacrylamide Copolymers in Drag Reduction: Effect of Molecular Structure, Hydration, and Flow Geometry on Drag Reduction Performance," *Polymer Engineering & Science* 34 (1994): 1429.
91. D. N. Schulz, R. M. Kowalik, J. Bock, and J. J. Maurer, U.S. Patent 4546784, 1985.
92. D. N. Schulz, D. G. Peiffer, R. M. Kowalik, and J. J. Kaladas, U.S. Patent 4560710, 1985.
93. J. W. Hoyt, "Effect of Ferric Ions on Drag Reduction Effectiveness of Polyacrylamide," *Polymer Engineering & Science* 20 (1980): 493.
94. D. G. Peiffer, R. D. Lundberg, R. M. Kowalik, and S. R. Turner, U.S. Patent 4460758, 1984.

95. C. L. McCormick, B. H. Hutchinson, and S. E. Morgan, "Water-Soluble Copolymers: 16. Studies of the Behavior of Acrylamide/N-(1,1-Dimethyl-3-Oxobutyl)Acrylamide Copolymers in Aqueous Salt Solution," *Makromolekulare Chemie* 188 (1987): 357.
96. O. K. Kim, R. C. Little, R. L. Patterson, and R. Y. Ting, "Polymer Structure," *Nature* 250 (1974): 408.
97. R. D. Lundberg, D. G. Peiffer, I. Duvdevani, and R. M. Kowalik, U.S. Patent 4489180, 1984.
98. R. L. Patterson and R. C. Little, *Journal of Colloid Interface Science* 53 (1975): 110.
99. R. N. Grabois and Y. N. Lee, U.S. Patent 5027843, 1991.
100. C. Kang and W. P. Jepson, "Multiphase Flow Conditioning Using Drag-Reducing Agents," SPE 56569 (paper presented at the SPE Annual Technical Conference and Exhibition, Houston, TX, October 1999).
101. J. Yang, Jiang, and S. J. Weghorn, U.S. Patent Application 20050209368.
102. T. Al-Wahaibi, M. Smith, and P. Angeli, *Journal of Petroleum Science and Engineering* 57 (2007): 334.
103. V. Jovancicevic, S. J. Weghorn, and P. R. Hart, U.S. Patent Application 20050049327.
104. H. JianZhong, "Reducing the Drag Force of the Multiphase Flow in Gathering Lines by Injecting Demulsifiers at the Wellhead," SPE 10576 (paper presented at the International Meeting on Petroleum Engineering).
105. A. V. Shenoy, *Colloids and Polymer Science* 262 (1984): 319.
106. B. Yu, F. Li, and Y. Kawaguchi, *International Journal of Heat & Flow* 25 (2004): 961.
107. J. Myska, P. Stepanek, and J. L. Zakin, *Colloids and Polymer Science* 275 (1997): 254.
108. A. V. Shenoy, *Rheology Acta* 15 (1976): 658.
109. J. Myska and J. L. Zakin, *Industrial & Engineering Chemistry Research* 36(12) (1997): 5483.
110. H. Inaba and N. Haruki, *Heat Transfer—Japanese Research* 27 (1998): 1.
111. R. J. Wilkiens and D. K. Thomas, "Influence of Gravity and Lift on Particle Velocity Statistics and Transfer Rates in Turbulent Vertical Channel Flow," *International Journal of Multiphase Flow* 33 (2007): 134.
112. B. Lu, X. Li, L. E. Scriven, H. T. Davis, Y. Talmon, and J. L. Zakin, *Langmuir* 14 (1998): 8.
113. H. W. Bewersdoff and D. Ohlendorf, *Colloids and Polymer Science* 266 (1988): 941.
114. B. A. Maria Oude Alink and V. Jovancicevic, U.S. Patent Application 20040206937.
115. L. Chaal, C. Deslouis, A. Pailleret, and B. Saidani, "On the Mitigation of Erosion-Corrosion of Copper by a Drag-Reducing Cationic Surfactant in Turbulent Flow Conditions Using a Rotating Cage," *Electrochimica Acta* 52 (2007): 7786.
116. V. Jovancicevic and K. A. Bartrip, U.S. Patent 6774094, 2004.
117. V. Jovancicevic and Y. S. Ahn, U.S. Patent 7137401, 2006.
118. R. L. Patterson and R. C. Little, *Journal of Colloids & Interface Science* 53 (1975): 110.
119. M. Hellsten and I. Harwigsson, U.S. Patent 5339855, 1994.
120. M. Hellsten and I. Harwigsson, U.S. Patent 5979479, 1999.
121. M.Hellsten, and I. Harwigsson, U.S. Patent 5911236 1999.
122. M. Hellsten and I. Harwigsson, U.S. Patent, 5902784, 1999.
123. M. Hellsten and H. Oskarsson, U.S. Patent Application 20040077734.
124. M. Hellsten and H. Oskarsson, International Patent Application WO/2004/007630.
125. E. Sletfjerding, A. Gladsø, Statoil, S. Elsborg, and H. Oskarsson, "Boosting the Heating Capacity of Oil-Production Bundles Using Drag-Reducing Surfactants," SPE 80238 (paper presented at the International Symposium on Oilfield Chemistry, Houston, TX, 5–7 February 2003).

126. M. Hellsten, *Journal of Surfactants and Detergents* 4 (2002): 65.
127. H. Oskarsson, I. Uneback, and M. Hellsten, "Surfactants as Flow Improvers in Water Injection," SPE 93116 (paper presented at the SPE International Symposium on Oilfield Chemistry, The Woodlands, TX, 2–4 February 2005).
128. G. H. Sedahmed, M. S. E. Abdo, M. A. Amer, and G. Abd El-Latif, *Intenational Communications in Heat and Transfer* 26(4) (1999): 531.
129. S. E. Campbell and V. Jovancicevic, "Performance Improvements from Chemical Drag Reducers," SPE 65021 (paper presented at the SPE International Symposium on Oilfield Chemistry, Houston, TX, 13–16 February 2001).
130. G. Schmitt, "Drag Reduction by Corrosion Inhibitors: A Neglected Option for Mitigation of Flow Induced Localized Corrosion," *Materials and Corrosion* 52 (2001): 329.
131. Hoerstmeier, G. Schmitt, and M. Bakalli, "Contribution of Drag Reduction to the Performance of Corrosion Inhibitors in One and Two Phase Flow," Paper 07615 (paper presented at the NACE CORROSION Conference, 2007).

APPENDIX 1

OSPAR Environmental Regulations for Oilfield Chemicals

The required ecotoxicological tests on all components of oilfield chemicals proposed for use in the North Sea offshore region are laid out in the OSPAR guidelines for the Northeast Atlantic implemented in 2001 under a harmonized mandatory control scheme.[1-3] Three categories of ecotoxicological tests on oilfield chemicals are required by OSPAR:

- acute toxicity[4]
- bioaccumulation
- biodegradation in seawater

It is the environmental properties of each individual production chemical (and not the finished product) in a proposed formulation that has to be determined.

The full OSPAR marine acute toxicity data set comprises:

(a) *Skeletonema costatum* (marine algae; ISO 10253)
(b) *Acartia tonsa* (marine copepod; ISP 14669 with recommendations given by OSPARCOM).
(c) *Corophium volutator* (marine amphipod; Paris Commission Guidelines 1995, Part A of the OSPAR Protocols on Methods for the Testing of Chemicals Used in the Offshore Industry). This test, called a "sediment reworker test," is only needed if the chemical has certain properties such being very bioaccumulative or surface-active or if known to adsorb to particles or be deposited in the sediment.
(d) *Scophthalmus maximus* (marine fish larvae; Paris Commission Guidelines 1995, Part B of the OSPAR Protocols on Methods for the Testing of Chemicals Used in the Offshore Industry: Protocol for a Fish Acute Toxicity Test). This test is not needed if the chemical is very toxic to *S. costatum* and/or *A. tonsa*.

S. costatum is the most sensitive species to toxicity. Therefore, toxicity tests of a new production chemical are generally first carried out on this species to gauge the level of toxicity. Surfactants, particularly cationic ones, are often some of the most toxic production chemicals.[5] Thus, besides biocides, film-forming surfactant corrosion inhibitors, such as quaternary ammonium compounds and imidazolines (which become cationic due to protonation in acidic-produced water) have often been the most toxic production chemicals. However, many service companies and chemical suppliers now offer a range of less toxic products, sometimes with a trade-off with performance.

The length of the test period varies for the four toxicity tests. For *S. costatum*, the result is given as EC50, i.e., the concentration of the test substance that gives 50% growth inhibition. For the other marine organisms, the result is given as LC50, i.e., the concentration of the test substance that gives 50% mortality (immobilization). Toxicity data is usually assessed in five categories: < 1 mg/l, > 1–10 mg/l, > 10–100 mg/l, > 100–1,000 mg/l, and > 1,000 mg/l, i.e., it is the dosage level that determines the toxicity of a chemical.[6] Full toxicity testing can be quite expensive for formulations containing several components such as corrosion inhibitors or demulsifiers.

The bioaccumulation potential is usually determined by high-performance liquid chromatography (Organization for Economic Cooperation and Development, OECD117) for water-soluble chemicals and the shake-flask method (OECD107) for more lipophilic chemicals.[7–9] The bioaccumulation is recorded as $\log P_{ow}$ (partition *n*-octanol/water). The $\log P_{ow}$ value is also used in the United Kingdom in the chemical hazard assessment and risk management (CHARM) model to estimate how a substance partitions between oil and water with the aim of predicting the environmental concentration (PEC). If the calculated or experimentally determined log-P_{ow} is ≥3, bioaccumulation is assumed unless experimental bioconcentration factor (BCF) tests indicate the opposite. A new method of measuring partition coefficients for nonionic surfactants has been published.[10]

Those substances with a $\log P_{ow}$ of > 4.5–5.0 are considered to be potentially highly bioaccumulative. Oil-soluble polymers will be expected to have high $\log P_{ow}$ values, but above a certain size, they are considered unlikely to pass lipophilic cell membranes and cause accumulative damage. This size is about a molecular weight of 700 Da.

Testing for biodegradability is an elemental part of environmental risk assessment of new chemicals, and the OECD guidelines for testing of organic chemicals is widely accepted as the European consensus methodology.[11] The adopted biodegradation strategy consists of a level 0 screening test for ready biodegradability in aerobic aqueous environments, followed by inherent biodegradability testing (level 1), and, finally, simulation testing (level 2). Positive screening (level 0) tests will make further testing (at the higher level) unnecessary. Level 0 testing for marine biodegradability is presented in the OECD 306 method, either according to the shake-flask dissolved organic carbon (DOC) die-away method or the closed-bottle BOD (biological oxygen demand) test method (OECD, 1992; the freshwater test protocol OECD 301 may, under certain circumstances, be accepted by the Center for Environment, Fisheries, & Aquaculture Science (CEFAS) in the United Kingdom and in the Netherlands. The protocol used should be declared as the values obtained in freshwater are generally higher than in seawater). Closed-bottle testing by means of respirometry is based on the stoichiometric relationship between the mineralized test compound and the respired oxygen. Biodegradation is calculated as the ratio between the theoretical oxygen demand (ThOD) and the measured oxygen consumed during degradation (BOD). If the oxygen consumed equals or exceeds 60% of the ThOD, the compound is (readily) biodegradable. One problem with this method is that the fraction of the test chemical that is assimilated into a new biomass is not taken into account. Thus, compensation must be included for test compounds that provide a high cellular yield (Y_{xs}) when calculating the degree of biodegradation. A more correct data analysis will be based on the ratio between the measured BOD and the calculated theoretical biological oxygen demand (ThBOD) of the test compound, given as $1 - Y_{xs}$.[12] For glucose, one of the recommended compounds used for positive controls, the ThBOD, is about 33% of the ThOD, leading to the incorrect conclusion that based on the OECD 306 method, glucose will never be readily biodegradable in seawater. Glucose is still used (successfully) as the positive control compound in laboratory testing; however, a high degree of biodegradation is only determined by including the postgrowth endogenous respiration phase in the total BOD estimation.

The OECD306 marine biodegradation test is usually carried out over 28 days. The chemical is considered persistent if the biodegradation is < 20% in 28 days and readily biodegradable if the biodegradation is > 60–70%. The OECD306 protocol states, "Owing to the relatively high test concentrations as compared with most natural systems, and consequently an unfavourable ratio between the concentrations of test substance and other carbon sources, the method is to be regarded as a preliminary test which can be used to indicate whether or not a substance is easily biodegradable. Accordingly, a low result does not necessarily mean that the test substance is not biodegradable in marine environments, but indicates that more work will be necessary in order for this to be established." Thus, chemical suppliers are encouraged to submit additional data to prove whether a substance is biodegradable. For substances shown to be < 60% or < 70% biodegradable (depending on the test end point) in an OECD 306 test, this could be an extended OECD 306 test such as over 60 days. In the United Kingdom, an OECD 301 test is acceptable.[13] If the OECD 306 test shows the substance to be < 20% biodegradable, an appropriate inherent test is recommended. Marine BODIS, a test that is specifically designed for poorly soluble substances, may be run.[14] Current regulations stipulate that biodegradation testing must be carried out on each component in a formulation separately, including any organic solvents, and not on the whole formulation.

The OECD306 protocol sets a standard temperature for carrying out the biodegradation test since biodegradation rates increase with increasing temperature. However, no standard is given for the concentration and type of organisms present in the seawater sample. This can vary significantly from location to location and depth at which the seawater is taken and will affect the rate of biodegradation. The concentration of the test substance will also affect the rate of biodegradation. The OECD306 closed-bottle method suggests using 2–10 mg/l of the test compound, but higher concentrations can be used especially if one opts to have a headspace of air in the bottle and measure the pressure drop. The shake-flask method (the alternative OECD306 method) suggests using 5–40 mg/l DOC. This is equivalent to (depending on the compounds carbon content) 10–80 mg/l of the test compound. Generally, the biodegradation is faster at lower concentrations.[7] Water-soluble substances, particularly those that are nonpolymeric, generally degrade faster than insoluble substances, as the latter tend to adsorb strongly to solid phases.[15] Further, if the substance is very toxic or partially degrades to toxic compounds, it may kill the very organisms that carry out the biodegradation.

In the next two sections, the offshore environmental regulations in the United Kingdom (which are very similar to those of the Netherlands) and Norway are described in some detail to illustrate the differences in how they would assess production chemicals. The regulations in the North Sea countries are under constant revision, so the reader should check with the relevant national pollution authorities for any updates.

A.1 UNITED KINGDOM AND THE NETHERLANDS NORTH SEA ECOTOX REGULATIONS

The Offshore Chemical Notification Scheme (OCNS) is administered by CEFAS, an agency of the U.K. government's Department for Environment, Food, and Rural Affairs (DEFRA).[16] Since 2007, CEFAS (United Kingdom) have been administrating the Netherlands OCNS schemes. The OCNS conducts hazard assessments on chemical products that are used offshore. The CHARM model calculates the ratio of predicted effect concentration against no effect concentration (PEC/NEC) and is expressed as a hazard quotient (HQ), which is then used to rank the product. Data used in the CHARM assessment include percentage of component in product, expected product dose rate, toxicity, biodegradation, and bioaccumulation. The HQ is converted to a color banding, which is then published on the Definitive Ranked Lists of Approved Products. There are six color bands: gold, silver, white, blue, orange, and purple (in increasing order of hazardousness). Products not applicable to the CHARM model (i.e., inorganic substances, hydraulic fluids, or chemicals used only in pipelines) are assigned an OCNS grouping A–E, with A being the greatest potential environmental hazard and E being the least.

Substitution is an important component of the OSPAR Harmonized Mandatory Control Scheme, and the United Kingdom is obliged to implement a strategy to replace chemicals that have been identified as candidates for substitution, or contain components that have been identified as candidates for substitution.[17] An offshore chemical should be substituted if it:

- is listed in Annex 2 of the OSPAR Strategy with regard to hazardous substances;
- is considered by the authority, to which the application has been made, to be of equivalent concern for the marine environment as substances covered by the previous subparagraph;
- is inorganic and has a LC50 or EC50 < 1 mg/l;
- has a biodegradation < 20% during 28 days or meets two of the following three criteria:

- biodegradation in 28 days < 70% (OECD 301A, 301E) or < 60% (OECD 301B, 301C, 301F, 306);
- either bioaccumulation $\log P_{ow} > 3$ and molecular weight of the substance is < 600 or BCF > 100
- toxicity LC50/EC50 < 10 mg/l

A.2 NORWEGIAN OFFSHORE ECOTOX REGULATIONS

In Norway, chemicals are ranked in one of four color categories, green, yellow, red, and black in decreasing order of environmental acceptability. Yellow chemicals are further subdivided into yellow 1, yellow 2, and yellow 3, the first being the most environmentally acceptable.[18] "Green" chemicals, which are on OSPAR's PLONOR list, are allowed to be used offshore Norway, and "black" chemicals are not allowed. There are several categories of "black" chemicals, including endocrine disrupters[5,19-20] and carcinogens, and the following:

- chemicals with < 20% biodegradation and $\log P_{ow} \geq 5$
- chemicals with < 20% biodegradation and toxicity EC50 or LC50 ≤ 10 mg/l

"Red" chemicals are classified as hazardous to the environment, and if in use they should be prioritized for replacement. Besides chemicals on OSPAR's tainting list, "red" chemicals include:

- inorganic chemicals with toxicity EC50 or LC50 ≤ 1 mg/l;
- organic chemicals with < 20% biodegradation.
- organic chemicals or mixtures which meet two of the following three criteria
  - < 60% biodegradation in 28 days
  - bioaccumulation, log Pow ≥ 3
  - acute toxicity EC50 or LC50 ≤ 10mg/l"

Like "green" chemicals, "yellow" chemicals are generally allowed to be used offshore Norway without the need for special approval. "Yellow" chemicals are chemicals that do not come under the other color categories or must meet the following requirements:

- biodegradation > 20%;
- $\log P_{ow} < 3$;
- toxicity EC50 or LC50 > 10 mg/l.

If the molecular weight of a chemical exceeds 700 Da, then it is not considered to have the potential to bioaccumulate (it is too large to cross cell membranes) and need not be tested for this. For chemicals with a biodegradation between 20 and 60%, there is additionally a focus on the products of biodegradation in case they are persistent and toxic.[21] The Norwegian authorities' requirements involve a literature evaluation of the hazards associated with the biodegradation products of the test chemical. The percentage of biodegradation of the chemical should also continue to increase after 28 days.

REFERENCES

1. OSPAR Guidelines for Completing the Harmonized Offshore Chemical Notification Format (2005-13) (http://www.ospar.org).
2. S. Glover and I. Still, "HMCS (harmonized mandatory control scheme) and the issue of substitution" (paper presented at the Ninth Annual International Petroleum Environmental Conference, 22–25 October 2002).
3. M. Thatcher and G. Payne, "Impact of the OSPAR Decision on the Harmonized Mandatory Control System on the Offshore Chemical Supply Industry," Proceedings of the Chemistry in the Oil Industry VII Symposium, Royal Society of Chemistry, Manchester, UK, 13–14 November 2001.
4. G. M. Rand, *Fundamentals of Aquatic Toxicology: Effects, Environmental Fate and Risk Assessment*, 2nd ed., Philadelphia, Pa.: Taylor & Francis, 1995.
5. G.-G. Ying, Fate, "Behavior and Effects of Surfactants and Their Degradation Products in the Environment," *Environment International* 32 (2006): 417.
6. M. A. Ottoboni, *The Dose Makes the Poison*, New York: Wiley-Blackwell, 1996.
7. Organisation for Economic Cooperation and Development, *OECD Guideline for the Testing of Chemicals 117—Partition Coefficient (n-Octanol/Water), High Performance Liquid Chromatography (HPLC) Method*, 1989.
8. Organisation for Economic Cooperation and Development, *OECD Guideline for the Testing of Chemicals 107—Partition Coefficient (n-Octanol/Water), Shake Flask Method*, 1995.
9. A. J. Millais, R. J. Rycroft, M. A. Tolhurst, and D. A. Sheahan, "Bioconcentration: Comparison of Methods for Assessing Potential Hazards of Offshore Chemicals," Proceedings of the Chemistry in the Oil Industry X Symposium RSC/EOSCA, Manchester, UK, 5–7 November 2007.
10. A. Karcher, H. Wiggins, I. Robb, and J. M. Wilson, "A Method for Measuring n-Octanol/Water Partition Coefficients for Non-Ionic Surfactants," Proceedings of the Chemistry in the Oil Industry X Symposium RSC/EOSCA, Manchester, UK, 5–7 November 2007.
11. N. Nyholm, "The European System of Standardized Legal Tests for Assessing the Biodegradability of Chemicals," *Environmental Toxicology and Chemistry* 10 (1991): 1237.
12. L. Del Villano, R. Kommedal, and M. A. Kelland, "Class of Kinetic Hydrate Inhibitors with Good Biodegradability," *Energy Fuels* 22 (2008): 3143.
13. M. M. Jordan, N. Feasey, C. Johnston, D. Marlow, and M. Elrick, "Biodegradable Scale Inhibitors: Laboratory and Field Evaluation of 'Green' Carbonate and Sulfate Scale Inhibitors with Deployment Histories in the North Sea," Proceedings of the Chemistry in the Oil Industry X Symposium RSC/EOSCA, Manchester, UK, 5–7 November 2007.
14. ECETOC Technical Report No. 20 (1986) Annex III of OECD 1992 301 and ISO Guidance Document ISO 10634.
15. R. S. Boethling, E. Sommer, and D. DiFiore, "Designing Small Molecules for Biodegradability," *Chemical Reviews* 107 (2007): 2207.
16. http://www.cefas.co.uk/offshore-chemical-notification-scheme-(ocns)/hazard-assessment.aspx.

17. D. Sheahan, J. Girling, P. Neall, R. Rycroft, S. Thompson, M. Tolhurst, L. Weiss, M. Kirby, and A. Millais, "Evaluating and Forecasting Trends in Chemical Use and Impacts by the UK Offshore Oil and Gas Industry; Management and Reduction of Use of Those Substances Considered of Greatest Environmental Concern, CEFAS," Proceedings of the Chemistry in the Oil Industry X: Oilfield Chemistry, Royal Society of Chemistry, Manchester, UK, 5–7 November 2007.
18. (a) "Supplementary Guidance for the Completing of Harmonized Offshore Chemical Notification Format (HOCNF) 2000 for Norwegian Sector," *Harmonized Offshore Chemical Notification Format OSPAR Recommendation 2000/5*. (b) http://www.sft.no/regelverk___36586.aspx.
19. J. Beyer, A. Skadsheim, M. A. Kelland, K. Alfsnes, S. Sanni, "Ecotoxicology of Oilfield Chemicals: The Relevance of Evaluating Low-Dose and Long-Term Impact on Fish and Invertebrates in Marine Recipients," SPE 65039 (paper presented at the SPE International Symposium on Oilfield Chemistry, 13–16 February 2001, Houston, TX).
20. J. M. Getliff, and S. G. James, "The Replacement of Alkylphenol Ethoxylates to Improve the Environmental Acceptability of Drilling Fluid Additives," SPE 35982 (paper presented at the International Conference on Health, Safety & Environment, New Orleans, LA, June 1996).
21. http://www.sft.no/arbeidsomr/petroleum/skim/skim_biodegradationproducts.pdf.

Index

A
α-alkonylphenones, 157
α,β-unsaturated aldehydes, 154
Acartia tonsa, 405
acetic acid
 acids for carbonate formation, 150–151
 carbonate scale removal, 86
 corrosion inhibitors containing sulfur, 158
 emulsified acids, 175
 gelled or viscous acids, 173
 heat-generating chemicals, 250
 lead scale removal, 91
 naphthenate deposition control, 190
 oxidizing biocides, 340
 temperature-sensitive acid-generating chemicals and enzymes, 173–174
 thermchemical packages, 267–268
 weak organic acids, 172
acetophenone, 157–158
acetylenic alcohols, 154, 156, 158
acetylenic sulfides, 156
acidic head groups, nonpolymeric surfactants, 120–122
acidizing
 anionic polymers, 330
 matrix, 3
 polymer RPM, 36
 scale inhibitor squeeze treatments, 80
 types of scale, 53
acid stimulation
 additives, 153–162
 alcohols, 162
 antisludging agent, 161
 asphaltene, 161
 axial placement, treatments, 162–171
 buffered acids, 172
 calcium sulfate scale inhibitors, 162
 carbonate formations, 150–151
 chemical compatibility issues, 7
 clay stabilizer, 161
 conjugate ion pairs, 162
 corrosion, 3, 153–159
 damage potential, 152–153
 demulsifiers, 162
 drag reducers, 162
 emulsified acids, 174–175
 ester quaternary surfactants, 161
 fines fixing agent, 161

 foam diverters, 166–167
 foamed acids, 173
 foaming agents, 162
 fracturing, 149–150
 fundamentals, 149, 153
 gelled acids, 173
 hydrogen sulfide scavengers, 162
 iron control agents, 159–160, 161
 matrix acidizing, 150
 nitrogen-based corrosion inhibitors, 154–155
 oil-wetting surfactants, 172
 oxygen-containing corrosion inhibitors, 155–158
 polymer gel diverters, 164–166
 radial placement, treatments, 171–175
 reducing problems by, 3
 sandstone formations, 151–152
 solid particle diverters, 163
 sulfur-containing corrosion inhibitors, 158–159
 surfactants, 162
 temperature-sensitive acid-generating chemicals and enzymes, 173–174
 types and uses, 150–152
 unsaturated linkages, corrosion inhibitors, 155–158
 viscoelastic surfactants, 167–171
 viscous acids, 173
 water-wetting agents, 161
 weak organic acids, 172
 weak sandstone-acidizing fluorinated agents, 172
acrolein
 aldehydes, 342, 367–368
 anthraquinone as control biocide, 352
 cross-linked polymer gels, 31
 nitrate and nitrite treatment, 353
 nitrogen heterocyclics, 213
 nonoxidizing organic biocides, 342–343
 sulfide scale remover, 89–90
acrylamide
 acrylamide-based DRAs, 393–394
 alkylamines, diamines, and triamines, 349
 amides, 211
 anionic polymers, 330
 cationic polymers, 323–325
 cross-linked polymer gels, 30–31
 cross-linked polymer RPMs, 37

413

414 Index

diallyldimethylammonium chloride polymers, 323
emulsified gels as DPRs, 33
environment-friendly cationic polymeric flocculants, 327
hydrophobically modified synthetic polymers, 36, 37
metal ion cross-linking, 26
(N-3-sulfopropyl)-N-methacrlyol-oxyethyl-N,N-dimethylammonium betaine copolymer, 394
organic cross-linking, 28–29
polycarboxylates, 69
poly(di)alkyl(meth)acrylamide KHIs, 238
polymer gel diverters, 164–165
polymer gel water shut-off treatments, 30
polymer injection, 25
polysaccharides and derivatives, 392
in situ monomer polymerization, 31
types of polymer RPM, 34–35
water control using microparticles, 39
water-soluble polymer DRAs, 391
acrylamides, 26–27, 34, 323–325, 393–395
acrylamidopropyltrimethylammonium chloride, 35
acrylates, 39, 323–325, 391, 394
acrylic acid
acrylamide or acrylate-based cationic polymers, 324
alkylphenol-aldehyde resin alkoxylates, 300
amidoamines and imidazolines, 210
anionic polymers, 330
comb polymers, 274
corosion inhibitors for acidizing, 153
low-dosage naphthenate inhibitors, 191
nonpolymeric surfactant ADs with acidic head groups, 122
polyalkoxylates of polyols or glycidyl ethers, 301
polycarboxylates, 67, 69
polymer RPM, 34
polyphosphonates, 66
polysulfonates, 72
quaternary ammonium and iminium salts and zwitterionics, 206
quaternary phosphonium compounds, 344
reducing toxicity, 12
sulfide scale inhibition, 74
vinyl polymers, 303–304
acryloxyethyltrimethyl ammonium chloride (AETAC), 325
1-acryloyl-4-methyl piperazine, 325
2-(acryloyloxy)ethyltrimethylammonium chloride, 327
acute toxicity, 9, 90, 207, 327–328, 343, 405, 409
additives, acid stimulation

alcohols, 162
antisludging agent, 161
asphaltene, 161
calcium sulfate scale inhibitors, 162
clay stabilizer, 161
conjugate ion pairs, 162
corrosion inhibitors, 153–159
demulsifiers, 162
drag reducers, 162
ester quaternary surfactants, 161
fines fixing agent, 161
foaming agents, 162
fundamentals, 153
hydrogen sulfide scavengers, 162
iron control agents, 159–160, 161
nitrogen-based, 154–155
oxygen-containing, 155–158
sulfur-containing type, 158–159
surfactants, 162
unsaturated linkages, 155–158
water-wetting agents, 161
AFP, *see* Antifreeze proteins (AFPs)
alcohols, 162
aldehydes
acidizing treatments, 162
activated, and polyhydroxyl compounds, 379
alkylphenol-aldehyde resin alkoxylates, 298–300
alkylphenol-aldehyde resin oligomers, 129–131
biocides, 342–343
biodegradable polycarboxylates, 71
bisulfite, metabisulfite, and sulfite salts, 380
cationic polymers, 326
classes of oxygen scavengers, 377
corrosion inhibitors containing sulfur, 158
corrosion inhibitors for acidizing, 154
cross-linked polymer RPMs, 38
dithiocarbamates, 349
gels using natural polymers, 27
hydrate-philic pipeline AAs, 247
increasing biodegradability, 13
iron control agents, 160
nitrogen heterocyclics, 213
nonoxidizing biocides, 341–343
nonregenerative scavengers, 365, 367–371
oligomeric (resinous) and polymeric AIs, 128
organic cross-linking, 28
oxygen-containing corrosion inhibitors, 156–158
polymer gel diverters, 165
polymers, 278
polyvinyl alcohol or polyvinylamine gels, 30
quaternary phosphonium compounds, 344
reaction products of, 369–371
sulfide scale removal, 90
sulfur compounds, 215

Index

aldehydes, activated, 379
aliphatic ether bonds, 12
aliphatic hydrocarbons, 8
alkanesulfonic acids, 119
alkylamidoamine oxide surfactants, 167
alkylamines
    acrylamide or acrylate-based cationic polymers, 324
    amide and imide nonpolymeric surfactant ADs, 123
    amine salts of (poly)carboxylic acids, 205
    maleic copolymers, 276–277
    nitrogen-based corrosion inhibitors, 155
    nonoxidizing biocides, 349
    nonoxidizing organic biocides, 341
    nonpolymeric surfactant ADs with acidic head groups, 120
    oil-miscible scale inhibitors, 81
    oxidizing biocides, 340
    oxygen-containing corrosion inhibitors, 158
    polyhydroxy and ethoxylated amines and amides, 213
    quaternary ammonium and iminium salts and zwitterionics, 207
    reaction products, 370–371
    scale inhibitor squeeze treatments, 78
    sulfide scale removal, 90
alkylarylsulfonic acid, 118, 126, 174
alkylaryl surfactants, 12
alkylated phenols, 8
alkyldimethylbenzylammonium surfactants, 345
alkyl fumarate, 277
alkyl group, 12
alkylguanidinium salts, 345
alkylimidazoline, 126–127, 208–209
alkyl maleimide, 277
alkyl maleimide, α-olefin copolymers, 277
alkyl(meth)acrylamide, 325
alkyl(meth)acrylate, 325, 388
alkylphenol-aldehyde Resin Alkoxylates, 296, 298–300, 302
alkylphenol-aldehyde resin alkoxylates, 298–300
alkylphenol-aldehyde resin oligomers, 129–131
alkylphenol formaldehyde resins, 130, 275
alkylphenols, 9, 117, 125–127, 204, 298–299
alkylphenylcarboxylic acids, 122
alkylphenylethoxylates, 121, 123
alkyl poly(etheroxy)acrylates, 394
alkylpolyglucosides, 217
alkylpyrrolidones, 117, 123
alkyl sarcosinates, 32
alkyl taurate anionic surfactants, 168
aluminum chloride, 24, 172
aluminum citrate, 26
aluminum dicarboxylate, 391
amide nonpolymeric surfactants, 123–124
amides, 211–212

amidines, 214, 372
amidoamine oxide surfactant, 167–168, 171
amidoamines, 207–211
amidoimidazolines, 208
amidomethionine derivatives, 215
amidophosphonates, 66
amine oxide
    acidizing treatments, 162
    hydrate-philic pipeline AAs, 246
    nonpolymeric phosphonates and aminophosphonates, 63
    products, 372
    quaternary ammonium and iminium salts, 206
    viny lactam KHI polymers, 234
    viscoelastic surfactant gels, 32
    viscoelastic surfactants, 168
    zwitterionics, 169
amines, 369–372
amine salts, (poly)carboxylic acids, 205
aminoacid alkylphosphonic acids, 64
aminobenzoic acids, 28
aminobenzothiazoles, 159
aminoethylpiperazine (AEP), 329, 370
aminophosphinate polymers, 67
aminophosphonates, 59–61, 64, 73–74, 85, 87
    bis-hexamethylene triamine-penta(methylene phosphonic) acid, 55
    carbonate and sulfate scale inhibition, 59–61
    1,2-diaminoethanetetrakis(methylenephosphonic acid) (EDTMP), 63
    diethylenetriaminepentakis(methylenephosphonic acid) (DTPMP), 63
    dihexamethylenetriaminepentakis(methylene phosphonic acid) (DHTPMP), 63
    ethanolamine-N,N-bis (methylene phosphonates) (EBMP), 63
    nonpolymeric phosphonates and aminophosphonates, 64
    performance testing of scale inhibitors, 85
    scale control, 62–65
    sulfate scale removal, 87
    sulfide scale inhibition, 73–74
3-aminopropyltriethoxysilane, 23, 186
1-aminopyrrolidine, 377–378
ammonium bifluoride, 151, 172
ammonium bisulfite, 380–381
ammonium nitrite, 250, 267–269
amphoteric polymers/terpolymers, 34–35
AMPS
    acrylamide-based DRAs, 394
    cross-linked polymer RPMs, 38
    kinetic hydrate inhibitors, 239
    polymer gel water shut-off treatments, 30
    polymer RPM, 34–35
    polysulfonates, 71–73

sulfide scale inhibition, 74
types of polymer RPM, 34–35
anionic acrylamide, 35
anionic polymers, 330
anionic surfactants
    acidizing treatments, 162
    foam diverters, 166
    hydrate-philic pipeline AAs, 245
    reducing toxicity, 12
    viscoelastic surfactant gels, 32
    viscoelastic surfactants, 167–170
    water-soluble surfactants DRAs, 396, 398
anthraquinone, 351–352
anti-agglomerants, KHIs
    emulsion pipeline class, 243
    fundamentals, 242–243
    gas wells, 248
    hydrate-philic pipeline class, 243–247
    natural surfactants, 247–248
    nonplugging oils, 247–248
antifoams, 313–316
antifreeze proteins (AFPs), 240
antimicrobials, 335, 348
antimony chloride, 154–156, 158, 160
anti-sludging agents, 161
α-olefin copolymers, 277
α-olefin products
    1-decene, 387
    1-hexene, 387
    maleic anhydride polymers, 276
    1-octene, 387–389
    sulfonate surfactants, 40
    1-tetradecene, 387–388
aragonite, 53
ARN acids, 189
arsenic, 15–16, 154
arylamine groups, 12
asphaltene
    acidic head groups, nonpolymeric surfactant ADs, 120–122
    additives, acid stimulation, 161
    alkylphenol-aldehyde resin oligomers, 129–131
    alkylphenols and related ADs, 125–126
    amide nonpolymeric surfactant ADs, 123–124
    biodegradability, inhibitors, 14
    chemical compatibility issues, 7, 8
    copolymers, 135
    dispersants, 62, 114–127, 278
    dissolvers, 137–140
    downstream issues, 7
    fundamentals, 111–114
    imide nonpolymeric surfactant ADs, 123–124
    inhibitors, 114–116
    ion-pair surfactant ADs, 126–127
    lignosulfonate polymers, 135
    low molecular weight, nonpolymeric dispersants, 116–127
    low-polarity nonpolymeric aromatic amphiphiles, 117–118
    miscellaneous nonpolymeric ADs, 127
    oligomeric (resinous) and polymeric AIs, 128–137
    polyamide/imide AIs, 131–135
    polyester, 131–135
    polymeric inhibitors, 135, 137
    removal and removal performance, 3, 6
    semisynthetic type, 135
    sulfonic acid-based nonpolymeric surfactant ADs, 118–120
    summary, 137
    tetrapyrrolitic patterns, 137
asphaltic sludge, 137, 152
axial placement, acid stimulation
    foam diverters, 166–167
    fundamentals, 162–163
    polymer gel diverters, 164–166
    solid particle diverters, 163
    viscoelastic surfactants, 167–171
azide ions, 351
azide salts, 353–354
2,2'-azo(bisamidinopropane)dihydrochloride, 37
azo groups, biodegradability, 12

B

bacteria, chemical control, 337–338
bactericides, 335
barite
    phosphino polymers and polyphosphinates, 67
    polyphosphonates, 65
    polysulfonates, 72
    scale inhibition, carbonates and sulfates, 59
    squeeze treatments, 78
    sulfate scale removal, 87–89
    sulfate scales, 56
    types of scale, 53
benzene, 8
betaine surfactants, 167–169
biguanides, 346
bioaccumulation
    amidoamines and imidazolines, 210
    chemical sand control, 186
    designing greener chemicals, 11
    environment-friendly cationic polymeric flocculants, 327
    environment-friendly film-forming corrosion inhibitors, 202–203
    fundamentals, 11
    nonpolymeric phosphonates and aminophosphonates, 64

Index

North Sea region, 11
OSPAR Commission, 9, 405–406, 408–409
biocides
  aldehydes, 342–343
  alkylamines, 349
  anthraquinone, 351–352
  azide salts, 353–354
  bacteria control, chemical, 337–338
  biostats, 351–354
  cationic polymers, 345–346
  classes, 338–351
  2-(decylthio)ethanamine, 350
  diamines, 349
  dithiocarbamates, 349
  fundamentals, 335–337, 351
  hydrogen sulfide reduction, 4
  isothiazolones/isothiazolinones, 347–348
  metronidazole, 347
  molybdate ions, 354
  nitrate treatment, 352–353
  nitrite treatment, 352–353
  nonoxidizing type, 341–351
  organic biocides, 346–347
  organic thiocynanates, 348–349
  oxazolidines, 350
  oxidizing type, 339–341
  permanganate ions, 354
  phenolics, 349
  polymer RPM, 35
  quaternary ammonium compounds, 345
  quaternary phosphonium compounds, 343–345
  selenate, 354
  sodium azide, 353–354
  specific surfactant classes, 351
  thiones, 347–348
  transition metal oxyanions, 354
  triamines, 349
  triazine derivatives, 350
  tungstate ions, 354
  vanadate ions, 354
  water injection systems, 5
biodegradability
  acids for carbonate formations, 151
  amides, 212
  demulsifiers, 304–306
  environment-friendly cationic polymeric flocculants, 327
  fundamentals, 12–15
  greener chemicals, designing, 12–15
  hydrate-philic pipeline AAs, 245
  hyperbranched polyesteramide KHIs, 235
  increasing, 12–15
  isothiazolones (or isothiazolinones) and thiones, 348
  kinetic hydrate inhibitor classes, 239
  nonpolymeric phosphonates and aminophosphonates, 64
  organic cross-linking, 29
  OSPAR Commission, 405
  phosphate esters, 62
  phosphino polymers and polyphosphinates, 67
  polyaminoacids and other polymeric water-soluble corrosion inhibitors, 216
  polycarboxylates, 67, 69–71
  quaternary ammonium and iminium salts and zwitterionics, 207
  scale control, 14
  sulfur compounds, 215
  testing, 405
  vinyl lactam KHI polymers, 233
  wax solvents, 267
biodegradable polycarboxylates, 70–71
biodegradation
  biodegradable polycarboxylates, 71
  chemical sand control, 186
  choice factors, 6
  emulsifiers, 304
  increasing biodegradability, 12–15
  OSPAR Commission, 9–11, 405–410
  phosphino polymers and polyphosphinates, 67
  polycarboxylates, 67
  polysaccharides and derivatives, 392
  squeeze treatments, 79
  stability issues, 6
  sulfate scale removal, 88–89
  vinyl lactam KHI polymers, 233
biofilms
  aldehydes, 342–343
  bacteria control, 337–338
  biocides, 335–336
  corrosion control, 197
  isothiazolones (or isothiazolinones) and thiones, 348
  nitrate and nitrite treatment, 353
  nonchemical solution, 2
  oxidizing biocides, 341
  quaternary ammonium and iminium salts and zwitterionics, 206
  quaternary phosphonium compounds, 344
biofouling, 1, 335–338, 348, 353
biogradability, 304–306
biopolymers, 25–27, 35
biostats, 4, 337–338, 348, 351, 353–354
  anthraquinone, 351–352
  azide salts, 353–354
  fundamentals, 351
  molybdate ions, 354
  nitrate treatment, 352–353
  nitrite treatment, 352–353
  permanganate ions, 354

selenate, 354
sodium azide, 353–354
transition metal oxyanions, 354
tungstate ions, 354
vanadate ions, 354
biosurfactants, 12, 247, 265
1,7-bis(3-aminopropyl)ethylene diamine, 64
4,4′-bis(chloromethyl)diphenyl ether, 306
bismuth salts, 155
bis-oxazolidines, 369
bis-phenols, 300
bis-(triethoxy silylpropyl)amine, 23, 186
bisulfite salts, 380–381
branched polyvinyl imidazoline acid salts, 323, 327
breaker
    acrylamide-based DRAs, 394
    alkylphenol-aldehyde resin alkoxylates, 299
    demulsifiers, 293
    flocculants, 319
    hydrophilic polymers as RPMs, 34
    impurities specifications, 3
    polymer gel diverters, 164–165
    squeeze treatment, 83
    viscoelastic surfactants, 170–171
bromides, organic, 341, 346, 349
bromine chloride, 339–340
1-bromo-1-(bromomethyl)-1,3-propanedicarbonitrile, 346–347
2-bromo-2-nitropropanediol (BNPD), 346
bubble point, 113, 115, 261–262
buffered acids, 172
buffered acid systems, 152
bundle pipeline system, 226
butyl glycol, 161, 234
butyl p-hydroxybenzoate, 299

C

calcite, 53, 85, 89
calcium carbonate scale, 54–55
calcium fluoride, 53, 152
calcium sulfate, 162
caprolactam, 27, 233–236, 244, 247
carbamates, water-in-oil demulsifiers, 302
carbohydrazides, 166, 377–378
carbonates
    aminophosphonates, 62–65
    biodegradable polycarboxylates, 70–71
    formations, acid stimulation, 149–151
    fundamentals, 59–61
    nonpolymeric phosphonates, 62–65
    phosphate esters, 61–62
    phosphino polymers, 66–67
    polycarboxylates, 67, 69–71
    polyphosphates, 61
    polyphosphinates, 66–67
    polyphosphonates, 65–66
    polysulfonates, 71–73
    scale control, 59–73
    scale removal, 85–87
carbon chain branching, 12
carbon disulfide, 266, 270
carboxylates, 371–372
carboxymethylcellulose (CMC), 71, 391
carboxymethylimino-3-hydroxybutane diacid (CIMM), 87
carboxymethylinulin, 71
catalytic hydrogenation, 379
cathodic protection, 197–198
cationic acrylamide polymers, 34–35
cationic monomers
    acrylamide or acrylate-based cationic polymers, 324–325
    cationic polymers, 326
    environment-friendly cationic polymeric flocculants, 327–328
    polycarboxylates, 70
    polysulfonates, 72
    squeeze treatments, 77, 79
    water control using microparticles, 39
cationic polymers
    acrylamide/acrylate-based type, 324–325
    alkylphenol-aldhyde resin oligomers, 129
    anionic polymers, 330
    classes of water-in-oil demulsifiers, 296
    diallyldimethylammonium chloride polymers, 323–324
    dithiocarbamates, 328–329
    environment-friendly, 327–328
    flocculants, 321–329
    nonoxidizing biocides, 345–346
    nonoxidizing organic biocides, 341
    other types, 325–327
    polyalkoxylate block copolymers and ester derivatives, 298
    polyamine polyalkoxylates and related cationic polymers, 301
    reducing toxicity, 12
    squeeze treatments, 77, 79
    types of polymer RPM, 36
    water-in-oil demulsifiers, 301
cationic polysaccharides and chitosan, 323
cationic starches, 34
cationic surfactants, 12
CDF, see Critical dilution factor (CDF)
celestite, 53
cement squeezing, 21
centrifugation, 2, 293
cetylpyridinium salicylate, 396, 398
cetyltrimethylammonium bromide (CTAB), 167
cetyltrimethylammonium chloride, 396
cetyltrimethylammonium salicylate, 396, 398
chelants, 27, 87

Index

chelates, 55, 87, 371–372
chemical compatibility issues, 7
chemical control, sand, 185–187
chemical hazard assessment and risk management (CHARM) model, 406, 408
chemical prevention, gas hydrate control
  anti-agglomerants, 242–248
  compatibility, 236
  emulsion pipeline class, 243
  fundamentals, 227
  gas wells, 248
  hydrate-philic pipeline class, 243–247
  hyperbranched polyesteramides, 235–236
  kinetic hydrate inhibitors, 231–242
  natural surfactants, 247–248
  nonplugging oils, 247–248
  operational issues, 230–231
  other classes, 238–240
  performance testing, 240–242
  poly(di)alkyl(meth)acrylamide, 237–238
  pyroglutamate, 236–237
  thermodynamic hydrate inhibitors, 227–231
  vinyl lactam polymers, 233–235
chemical prevention, wax
  comb polymers, 274–277
  dispersants, 279–281
  ethylene polymers and copolymers, 273–274
  inhibitors, 270–272
  maleic copolymers, 276–277
  (meth)acrylate ester polymers, 274–276
  miscellaneous polymers, 278–279
  pour-point depressants, 270–272
  test methods, 269–270
chemical removal, wax, 266–269
chemical scale removal, 85–91
chitosan, 29, 323, 328
chloramines, 339–340
chlorine, 12, 91, 336, 338–341, 380–381
chlorine dioxide, 91, 339–340, 380
chloroisocyanurates, 339, 341
choice factors, 5–8
chromium(III) (acetate, propionate), 26, 27, 165
cinnamaldehyde, 154, 156–159
cis-1,2-cyclohexanedicarboxylic anhydride, 235
classes, biostats
  aldehydes, 342–343
  alkylamines, 349
  cationic polymers, 345–346
  2-(decylthio)ethanamine, 350
  diamines, 349
  dithiocarbamates, 349
  fundamentals, 338–339, 341–342
  isothiazolones/isothiazolinones, 347–348
  metronidazole, 347
  nonoxidizing type, 341–351
  organic biocides, 346–347
  organic thiocynanates, 348–349
  oxazolidines, 350
  oxidizing type, 339–341
  phenolics, 349
  quaternary ammonium compounds, 345
  quaternary phosphonium compounds, 343–345
  specific surfactant classes, 351
  thiones, 347–348
  triamines, 349
  triazine derivatives, 350
classes, film-forming inhibitors
  amides, 211–212
  amidoamines, 207–211
  amine salts, (poly)carboxylic acids, 205
  ethoxylated amines/amides, 212–213
  fundamentals, 203–204
  imidazolines, 207–211
  nitrogen heterocyclics, 213
  phosphate esters, 204–205
  polyaminoacids, 216–217
  polyhydroxy, 212–213
  polymeric water-soluble inhibitors, 216–217
  quaternary ammonium iminium salts, 205–207
  sulfur compounds, 213–216
  zwitterionics, 206–207
classes, oxygen scavengers
  aldehydes, activated, 379
  bisulfite salts, 380–381
  catalytic hydrogenation, 379
  dithionite salts, 378
  enzymes, 379
  fundamentals, 377–378
  guanidine salts, 378
  hydrazine salts, 378
  hydroxylamines, 378–379
  metabisulfite salts, 380–381
  oximes, 378–379
  polyhydroxyl compounds, 379
  sulfite salts, 380–381
classes, water-in-oil demulsifiers, 296–297
classification of chemicals, 2
clathrate, 225, 232
clay acid, 152, 172
clay stabilizer, 161
coacervation, 36
coalescence, 2, 292, 294, 302, 321
coatings, 2, 58, 197–198
cocodiamine, 345, 349
cold flow, 226–227, 264–265
comb polymers, 36, 271–278, 281
compatibility
  acrylamide-based DRAs, 394
  asphaltene dissolvers, 137
  biodegradable polycarboxylates, 70–71
  choice factors, 5
  halite scale inhibition, 75

kinetic hydrate inhibitors, 236
metal ion cross-linking, 27
nonpolymeric phosphonates and
    aminophosphonates, 64
operational issues, 5, 6–7
performance testing of scale inhibitors, 84
phosphino polymers and polyphosphinates,
    66
polycarboxylates, 69–71
potential formation damage from acidizing,
    138
squeeze treatments, 80
testing corrosion inhibitors, 202
thermodynamic hydrate inhibitors, 228
types of polymer RPM, 35
condensed tannins, 323, 326
conjugate ion pairs, 162
continuous injection
    anthraquinone as control biocide, 352
    deploying scale inhibitors, 76
    environment-friendly film-forming corrosion
        inhibitors, 202
    film-forming corrosion inhibitors, 200
    oil-miscible scale inhibitors, 81
    oligomeric (resinous) and polymeric AIs, 128
    performance testing of scale inhibitors, 84
Corophium volutator, 405
corrosion
    acids, 3
    during acid stimulation, 3
    amides, 211–212
    amidoamines, 207–211
    amine salts, (poly)carboxylic acids, 205
    classes, 203–217
    cracking, 197, 213, 363
    drag-reducing agents, 398
    environmental trends, 202–203
    ethoxylated amines/amides, 212–213
    film-forming inhibitors, 200–217
    fundamentals, 4, 195–197, 200, 203–204
    imidazolines, 207–211
    inhibitor additives, acid stimulation, 153–159
    inhibitors, 198–201, 209
    methods, 197–198
    nitrogen heterocyclics, 213
    phosphate esters, 204–205
    polyaminoacids, 216–217
    polyhydroxy, 212–213
    polymeric water-soluble inhibitors, 216–217
    quaternary ammonium iminium salts,
        205–207
    removal performance, 6
    sulfur compounds, 213–216
    testing, 201–202
    work process, 200–201
    zwitterionics, 206–207
cost, choice factors, 5, 6

cracking corrosion, 197, 213, 363
CRA (corrosion resistant alloy), 4, 197
Critical Dilution Factor (CDF), 11
cross-linked polymer gels, permanent shut-off
    acrylamides, 26–27
    biopolymers, 26–27
    fundamentals, 24–25
    improvements, 31
    natural polymers, 27
    organic cross-linking, 27–29
    polymer gel water shut-off treatments, 30–31
    polymer injection, 25–31
    polyvinyl alcohol gels, 30
    polyvinylamine gels, 30
    in situ monomer polymerization, 31–32
cross-linked polymers
    chemical sand control, 185
    demulsifiers with improved biodegradability,
        306
    polycarboxylates, 67
    relative permeability modifier, 38–39
    squeeze treatments, 79
    water control using microparticles, 39
cross-linker
    acrylamide or acrylate-based cationic
        polymers, 325
    alkylphenol-aldehyde resin alkoxylates, 300
    carboxylate-containing acrylamides and
        biopolymers, 26–27
    cross-linked polymer RPMs, 37
    demulsifiers with improved biodegradability,
        305
    emulsified gels as DPRs, 33
    gels using natural polymers, 27
    hydrophobically modified synthetic polymers
        as RPMs, 37
    organic cross-linking, 28
    organic polymer gels for permanent shut-off,
        24
    polymer gel diverters, 165
    polymer gel water shut-off treatments, 30
    polymer injection, 25
    in situ monomer polymerization, 31
    squeeze treatments, 80
cyclohexanone, 140
cystamine, 214, 216

D

DADMAC, *see* Diallyl dimethyl ammonium
    chloride (DADMAC)
damage potential, acid stimulation, 152–153
deaerator, 336, 340
2-(decylthio)ethanamine, 341, 350
deep water, 53, 226, 263, 265, 267
defoamers, 4, 304, 306, 313–316
demulsification methods, 293

Index

demulsifiers
  additives, acid stimulation, 162
  alkylphenol-aldehyde resin alkoxylates, 298–300
  biogradability, 304–306
  carbamates, 302
  cationic polymers, 301
  classes, 296–297
  demulsification methods, 293
  dual-purpose type, 306–307
  ester derivatives, 298
  fundamentals, 291–293
  glycidyl ether polyalkoxylates, 300–301
  hyperbranched polymers, 302–303
  polyalkoxylate block copolymers, 298
  polyalkoxylate derivatives, 302
  polyamine polyalkoxylates, 301
  polyol polyalkoxylates, 300–301
  polysilicones, 304
  polyurethanes, 302
  selection parameters, 295–296
  test methods, 295–296
  theory and practice, 293–294
  vinyl polymers, 303–304
  water-in-oil type, 293–307
dendrimers, 304, 327
denitrifying bacteria (hNRB), 335
Denmark, 9
deoilers, 3, 293
deploying scale inhibitors
  continuous injection, 76
  emulsified type, 81–82
  fundamentals, 75–76
  nonaqueous type, 80–83
  oil-miscible type, 81
  placement, squeeze treatment, 83–84
  solid scale inhibitors, 80–83
  squeeze treatments, 76–84
  water-free in organic solvent bends, 81
  well treatments, combination, 79–80
deployment techniques, wax control, 281–282
deposition, 3, 261–263
desalting, 291–292, 295
desulfation, 57–58
diacetoneacrylamide, 393
1,3-dialkylthioureas, 158
diallyldimethylammoniumchloride (DADMAC), 34, 323–325
diallyldimethylammonium chloride polymers, 72, 323–324
diamines
  alkylphenol-aldehyde resin alkoxylates, 300
  amine salts of (poly)carboxylic acids, 205
  cationic polymers, 346
  nonoxidizing biocides, 349
  nonoxidizing organic biocides, 341

nonpolymeric phosphonates and aminophosphonates, 64
polyamine polyalkoxylates and related cationic polymers, 301
polyhydroxy and ethoxylated amines/amides, 212
1,2-diaminoethane, 208, 326
2,2-dibromo-3-nitrilopropionamide (DBNPA), 346
2-(1,2-dicarboxyethylimino) butanediacid (IDS), 87
2-(1,2-Dicarboxyethylimino)-3-hydroxybutane diacid (HIDS), 87
1,2-dichloroethane, 326
dicyclohexylamine nitrite, 200
diepoxy- or diglycidyl compounds, 300–301, 305
diethanolamides, 121, 123, 243
diethanolamine (DEA), 364
diethylene glycol (DEG), 228
diethylenetriamine
  amide and imide nonpolymeric surfactant ADs, 123–124
  amides, 212
  amidoamines and imidazolines, 208, 211
  cationic polymers, 326
  environment-friendly cationic polymeric flocculants, 327
  polyamine polyalkoxylates and related cationic polymers, 301
  reaction products, 371
  resins and elastomers, 22
  sulfate scale remover, 88
  sulfur compounds, 215
diethylenetriamine pentaacetic acid (DTPA), 88
diethylenetriphosphinic acid, 66
diethylhydroxylamine, DEHA, 378
diglycidyl ethers, 300, 305
diglycolamine (DGA), 364
diisopropanolamine, 135, 235, 279
dilational elasticity, 293
dimethylaminoethylacrylate benzyl chloride quaternary salt, 325
dimethylaminoethyl(meth)acrylate, 325
dimethylaminoethylmethacrylate (DMAEMA), 34, 233
dimethylaminopropylacrylamide, 349
dimethylaminopropylamine, 370
dimethylaminopropylmethacrylamide, 349
dimethyl diallyl ammonium chloride, 34
2,2-dimethylolpropionic acid, 304
1,4 dimethylpiperazinone, 372
diphosphate esters, 191
dipropylene glycol methyl ether, 161
disodium ethylene-1,2-bisdithiocarbamate, 349
dispersants, 114–127, 279–281
disproportionate permeability reducer, 33
disproportionate permeability reducer (DPR), 21

dissolvers, 137–140
dithiocarbamates (DTCs)
   biocides, 349
   flocculants, 321, 328–329
   iron control agents, 160
   mercury and arsenic production, 15–16
   nonoxidizing organic biocides, 341
   sulfur compounds, 214
3,3′-dithiodipropionic acid (DTDPA), 213
dithionite, 377–378
dithionite salts, 377–378
diutan, 164
diverters, 83, 162–164, 170, 173
divinylbenzene, 39, 275
DMAEMA, 34, 36, 40, 233
dodecylamine polyglycol ether, 306
dodecylbenzenesulfonic (DDBSA), 117, 190
downstream issues, 7
drag-reducing agents (DRAs)
   acrylamide-based, 393–395
   additives, acid stimulation, 162
   background, 385
   chemical compatibility issues, 7
   corrosion control, 198
   corrosion inhibition, 398
   fundamentals, 4, 383–384
   handling issues, 389–390
   injection issues, 389–390
   mechanisms, 384–385
   multiphase flow, 390
   oil-soluble type, 385–391
   other types, 389
   polyalkene type, 386–388
   polyethyleneoxide, 392
   polymeric type, 386–390
   polymers, 391–395
   poly(meth)acrylate ester type, 388–389
   polyolefin type, 386–388
   polysaccharides and derivatives, 392
   pumping issues, 389–390
   surfactants, 390–391, 395–398
   topside injection issues, 7
   UHMW polymers, 389–390
   water injection systems, 5
   water-soluble type, 391–398
drag reduction, 197–198, 383–385, 387–391, 393–398
dual-purpose water-in-oil demulsifiers, 306–307
dynamic tension, 293

E

ecotoxicological regulations, 8–11, 405–410
EDTA, ethylenediaminetetraacetic acid, 172
eicosene copolymers, 277
elastomers, 7, 22, 138, 202

electrochlorinator, 336
electrostatic coalescence, 292
embrittlement, 195, 200
emulsified acids
   antisludging agent, 161
   asphaltene dissolvers, 139
   axial placement of acid treatments, 163
   oil-wetting surfactants, 172
   radial placement, acid stimulation, 174–175
   radial placement of acidizing treatments, 171
emulsified gels, 22, 33
emulsion pad, 189–190
emulsion pipeline class, 243
emulsions
   compatibility issues, 7
   deploying scale inhibitors, 81–82
   resolving, 3
   unlocking, 294
end-capped polymers, 66–67, 191
endocrine disrupters, 9, 117, 125, 299, 409
environmental issues
   cationic polymers, 327–328
   choice factors, 5
   OSPAR Commission, 405–410
   regulations, 8–11
   trends, film-forming inhibitors, 202–203
enzymes
   asphaltene dissolvers, 137
   biodegradability, 13
   hydrogen sulfide scavengers, 372
   oxygen scavenger classes, 379
   polymer gel diverters, 165
   radial placement of acidizing treatments, 171
   temperature-sensitive acid-generating, 173–174
   wax control, 265
epichlorohydrin
   cationic polymers, 326, 346
   dithiocarbamates, 329
   hydrate-philic pipeline AAs, 246
   hyperbranched polymers, 302
   polyamine polyalkoxylates, 301
   polyphosphonates, 65
   resins and elastomers, 22
   sand control, 185
epoxidized fatty esters, 304
erosion, 185, 196–198, 201–202, 398
erosion corrosion, 196–198, 201–202, 398
erucylamidopropyl betaine, 67
erucyl methyl bis(2-hydroxyethyl) ammonium chloride (EMHAC), 168–169, 396
erythorbic acid, 160, 344, 379
ester derivatives, 207, 298
ester quaternary surfactants, 157, 161
esters of polyhydric alcohols, 117, 127
ethane-1,2-bisphosphinic acid, 66
1,2-ethanediol diacetate, 173

Index　423

ethanol, 210, 228, 364, 381, 390
ethoxylated amines/amides, 212–213
ethoxylates, 12, 14, 117
ethyl acetate, 173
ethylene, 273
ethylenediaminetetraacetic acid (EDTA), 172
ethyleneglycol-hemiformals, 343
ethylene glycol (MEG), 198, 228
ethylene glycol methacrylate phosphate, 62
ethylene polymers and copolymers, 273–274
ethylhexylacrylate ester, 324
2-ethylhexyl methacrylate, 36, 389
ethyl salicylate, 299
etidronic acid, 63
eutrophication, 61, 64

F

facilities design, 2
fatty amines
　amide and imide nonpolymeric surfactant ADs, 123
　amidoamines and imidazolines, 208
　foam diverters, 166
　ion-pair surfactant ADs, 126
　phosphate esters, 205
　polyhydroxy and ethoxylated amines/amides, 212
　viscoelastic surfactants, 168
field operator, responsibilities, 5, 9
film-forming corrosion inhibitors, 11–12
film-forming inhibitors
　amides, 211–212
　amidoamines, 207–211
　amine salts, (poly)carboxylic acids, 205
　classes, 203–217
　environmental trends, 202–203
　ethoxylated amines/amides, 212–213
　fundamentals, 200, 203–204
　imidazolines, 207–211
　nitrogen heterocyclics, 213
　phosphate esters, 204–205
　polyaminoacids, 216–217
　polyhydroxy, 212–213
　polymeric water-soluble inhibitors, 216–217
　quaternary ammonium iminium salts, 205–207
　sulfur compounds, 213–216
　testing, 201–202
　work process, 200–201
　zwitterionics, 206–207
fines
　acidizing treatments, 161
　biodegradable polycarboxylates, 70
　corrosion control, 198
　demulsifiers, 291
　fixing agent, 161

nonchemical techniques, 2
potential formation damage from acidizing, 153
sand control, 185
sandstone formation, 151
sulfate scale removal, 87
water and gas control, 21
weak sandstone-acidizing fluorinated agents, 172
floc, 4, 319–320, 325, 329–330
flocculants
　acrylamide/acrylate-based type, 324–325
　anionic polymers, 330
　cationic polymers, 323–329
　diallyldimethylammonium chloride polymers, 323–324
　dithiocarbamates, 328–329
　environment-friendly, 327–328
　fundamentals, 319–320, 321–322
　other types, 325–327
　performance testing, 322–323
　resolving impurities, 3–4
　theory, 320
flocculation
　acrylamide or acrylate-based cationic polymers, 325
　alkylphenol-aldehyde resin oligomers, 131
　amide and imide nonpolymeric surfactant ADs, 124
　asphaltene control, 113, 137
　asphaltene dispersants and inhibitors, 114
　cationic polymers, 327
　diallyldimethylammonium chloride polymers, 324
　flocculants, 321
　mercury and arsenic production, 16
　production chemistry, 2
　theory, 320
　unlocking emulsions, 294
　water-in-oil demulsifiers, 294
flotation, 319–320, 323, 325
flow assurance, 2–3
flow improvers, 281
fluorosilicones, 314–315
fluorosurfactants, 40
foam
　compatibility issues, 7
　control, 313–316
　diverters, 166–167
　resolving impurities, 4
foamed acids, 166–167, 171, 173
foam-forming surfactants, 166
foaming agents, 162
foaming surfactants, 41
formaldehyde, polymer RPM, 35
formation damage, 138, 153

formic acid
  carbonate formations, 150–151
  carbonate scale removal, 86
  corrosion inhibitors for acidizing, 154
  iron control agents, 160
  nitrogen-based corrosion inhibitors, 155
  sulfide scale removal, 89–90
  sulfur compounds, 215
  temperature-sensitive acid-generating chemicals and enzymes, 174
  weak organic acids, 172
fouling, 1–3, 7, 189–190, 230, 292
fracturing, carbonate formations, 149–150
furfuryl alcohol, 22, 28, 185

G

gallic acid, 28, 326, 379
galvanic corrosion, 196–197
gas control, *see* Water and gas control
gas cusping, 40
gas hydrate anti-agglomerants, 7
gas hydrate control, 227
  anti-agglomerants, 242–248
  chemical prevention, 227–248
  compatibility, 236
  emulsion pipeline class, 243
  fundamentals, 225–227, 231–232, 242–243
  gas wells, 248
  heat-generating chemicals, 249–250
  hydrate-philic pipeline class, 243–247
  hyperbranched polyesteramides, 235–236
  kinetic hydrate inhibitors, 231–242
  natural surfactants, 247–248
  nonplugging oils, 247–248
  operational issues, 230–231
  other classes, 238–240
  performance testing, 240–242
  plug removal, 248–250
  poly(di)alkyl(meth)acrylamide, 237–238
  pyroglutamate, 236–237
  thermodynamic hydrate inhibitors, 227–231, 249
  vinyl lactam polymers, 233–235
gas lift injection, 76
gas reservoirs, shallow, 8
gas shut-off, 40–41
gas wells, 248
gelled acid diverters, 164
gelled acids, 173
gelling, wax control, 263–264
gemini quaternary surfactants, 154
general corrosion, 196–197, 205, 207, 213
glucomannan, 327, 330
glutamic acid N,N-diacetic acid, 87
glutaraldehyde

  aldehydes, 342–343, 367
  bisulfite, metabisulfite, and sulfite salts, 380
  2-(decylthio)ethanamine and its hydrochloride, 350
  gels using natural polymers, 27
  isothiazolones (or isothiazolinones) and thiones, 348
  nitrate and nitrite treatment, 352
  nonoxidizing organic biocides, 342
  organic thiocyanates, 349
  polyvinyl alcohol or polyvinylamine gels, 30
  quaternary phosphonium compounds, 345
glutaric acid, 28
glycidyl ether polyalkoxylates, 300–301
glyoxal, 28, 30, 38, 156, 300, 367
glyoxylic acid, 71, 90, 156, 299
GoM, *see* Gulf of Mexico (GoM)
graft copolymer, 135, 233
graft polymers, 15, 128, 327
gravel packs, 2, 82
gravity separation, 292–293
greener chemicals, designing, 11–15
guanidines, 377
guanidine salts, 378
guar, 71, 84, 164
guar gum, 383, 391–392
Gulf of Mexico (GoM), 10–11, 115, 197, 269, 379

H

halite, 53, 57, 74–75, 85
halogens, 12
Hammerschmidt equation, 229
Harmonised Mandatory Control Scheme (HMCS), 9
hazards, choice factors, 6
health, handling and storage, 5, 6
heat-generating chemicals, 249–250
heterocyclics, 13, 204, 213
hexahydrophthalic anhydride, 235
hexamethylenetetramine (HMTA), 27–28, 367
hexanal, 28
HLB (hydrophilic-lipophilic balance), 161
hold time, 231, 240, 242
hot-oiling, 266–268
HPHT, 56–57, 74, 207
huff-puff stimulation, 24
hydantoins, 339, 341
hydrate inhibitors, 7, 76, 198, *see also* Kinetic hydrate inhibitors (KHIs); Low-dosage hydrate inhibitors (LDHIs); Thermodynamic hydrate inhibitors (THIs)
hydrate-philic pipeline class, 243–247
hydraulic fracturing, 36, 165, 170
hydrazines, 377–378
hydrazine salts, 378
hydrochloric acid

Index

acids for carbonate formations, 150–151
acids for sandstone formations, 151
carbonate scale removal, 85
foam diverters, 166
heat-generating chemicals, 250
inorganic gels, 23
naphthenate deposition control, 190
oxidizing biocides, 339
oxygen-containing corrosion inhibitors, 157
sulfide scale removal, 89–90
hydrocyclones, 293, 319–320
hydrofluoric acid (HF), 151
hydrogen peroxide, 91, 269, 339, 366
hydrogen sulfide (H_2S)
    additives, acid stimulation, 162
    aldehydes, 367–371
    amines, reaction products, 369–372
    chemical compatibility issues, 7
    corrosion, 4
    fundamentals, 4, 363–365, 373
    issues, 1
    metal carboxylates and chelates, 371–372
    nonregenerative scavengers, 365–372
    oxidizing chemicals, 366–367
    reaction products, 369–371
    solid scavengers, 366
    triazines, reaction products, 369–371
hydrophilic polymers, 33–39, 203
hydrophobic tail
    alkylphenol-aldehyde resin alkoxylates, 298
    bioaccumulation, 11
    environment-friendly film-forming corrosion inhibitors, 203–204
    film-forming corrosion inhibitors, 200–201
    hydrate-philic pipeline AAs, 244–245
    nitrogen-based corrosion inhibitors, 154
    polyhydroxy and ethoxylated amines/amides, 212
    quaternary ammonium and iminium salts and zwitterionics, 207, 209
    reducing toxicity, 12
    sulfur compounds, 214
hydroquinone, 28, 299, 378–379
hydrostatic loading, 21
hydrotesting, 5, 377, 381
hydroxyacids, 135
hydroxyaminocarboxylic acids, 151
2-hydroxy-3-chloropropyl trimethylammonium chloride, 326
1-hydroxyethane-1,1-diphosphonic acid (HEDP), 63
hydroxyethylcellulose (HEC), 164, 391
hydroxyethylethylenediamine triacetic acid (HEDTA), 86
hydroxyethyliminodiacetic acid (HEIDA), 86
hydroxyethyl(meth)acrylate, 303
hydroxylamines, 377–379
hydroxyl polyesters, 306
hydroxyl radicals, 339–340, 372
hydroxymethylpyridines, 132
3-hydroxy-1-phenyl-1-propanone, 158
hydroxypropyl acrylate monomers, 69
hydroxypropylguar, 391
hyperbranched polyesteramides, 134, 232, 235–236, 244, 279, 302
hyperbranched polymers, 296, 302–304, 327
hypobromite, 339–340
hypochlorite, 336, 338–341, 347
hypophosphite, 66, 155
hypophosphorous acid, 66, 160

I

imbibition, 25
imidazoles, 12, 34, 128, 135, 347, 366
imidazolines, 207–211
imide nonpolymeric surfactants, 123–124
imides, 131–135
3,3-iminobis(N,N-dimethylpropylamine), 235
impurities specifications, 3–4
incompatibility, *see* Compatibility
induction time, 242
inhibitors
    asphaltene, 14, 114–116, 135, 137
    calcium sulfate scale, 162
    corrosion, 153–159, 201
    corrosion control during production, 198–200
    film-forming, 11, 12, 200–217
    kinetic hydrate, 13–15, 231–242
    low-dosage hydrate, 3, 6
    naphthenate control, 191–192
    nitrogen-based corrosion, 154–155
    nonaqueous scale, 80–83
    oxygen-containing corrosion, 155–158
    passivating corrosion, 199
    performance testing, 84–85
    polymeric, 135, 137
    solid scale, 80–83
    squeeze treatment, placement, 83–84
    sulfur-containing corrosion, 158–159
    thermodynamic hydrate, 249
    threshold hydrate, 227
    unsaturated linkages, corrosion, 155–158
    vapor phase corrosion, 199
    water-free scale, organic solvent bends, 81
    wax, 3, 6, 8, 14
    wax prevention, chemical, 270–272
injection, *see also* Water injection
    deploying scale inhibitors, 76
    fundamentals, 4
    oil-soluble drag-reducing agents, 389–390
injection (subsea), 7, 270, *see also* Seawater injection
injection (topside), 7

inorganic gels, 22–24, 40
inorganic scales, 3, 53, 57, 59, 191–192, 227
in situ monomer polymerization, 31–32
insulation, 1, 226, 264
iodide, 154–156, 159–160
ionic liquids, 88, 117, 127, 240
ion pair amphiphile, 245
ion-pair surfactants, 126–127
iron control agents, 159–160, 161
isopropylbenzoate, 139
isothiazolinones, 341, 347–348
isothiazolones, 341, 347–348, 351
isothiazolones (isothiazolinones), 341

K

kaolinite, 40, 77, 79
kinetic hydrate inhibitors (KHIs), *see also* Hydrate inhibitors
 biodegradability, 13–14, 14, 15
 compatibility, 236
 fundamentals, 231–232
 hydrate plugging prevention, 227
 hyperbranched polyesteramides, 235–236
 performance issues, 8
 performance testing, 240–242
 poly(di)alkyl(meth)acrylamide, 237–238
 pyroglutamate, 236–237
 testing corrosion inhibitors, 202
 vinyl lactam polymers, 233–235

L

lauryl methacrylate-hydroxyethylmethacrylate copolymer, 130
LDHIs, *see* Low-dosage hydrate inhibitors (LDHIs)
lead
 alkylphenols and related ADs, 125
 chemical scale removal, 85, 91
 sulfide scale inhibition, 73
 sulfide scales, 57
 types of scale, 53–54
lignin, 15
lignosulfonate, 27, 128, 135
lignosulfonate polymers, 135
limonene, 267
linear alkyl group, 12
localized corrosion
 corrosion control during production, 196–197
 drag reduction, 398
 film-forming corrosion inhibitors, 201, 204
 methods of control, 198
 sulfur compounds, 213
 testing corrosion inhibitors, 202
logPow, 11, 406, 409
long-term shutdowns, 7

low-dosage hydrate inhibitors (LDHIs), 3, 6, 206, 227, *see also* Hydrate inhibitors
low-dosage inhibitors, 191–192
low leak-off control acids (LCA), 165
low molecular weight, nonpolymeric asphaltene dispersants
 acidic head groups, nonpolymeric surfactants, 120–122
 alkylphenols, 125–126
 amide nonpolymeric surfactants, 123–124
 fundamentals, 116–117
 imide nonpolymeric surfactants, 123–124
 ion-pair surfactants, 126–127
 low-polarity nonpolymeric aromatic amphiphiles, 117–118
 miscellaneous nonpolymeric type, 127
 sulfonic acid-based nonpolymeric surfactants, 118–120
low-polarity nonpolymeric aromatic amphiphiles, 117–118

M

magnesium oxychloride, 24
magnets, 58–59, 264–265
maleated fatty acid surfactant, 397
maleic acid
 biodegradability, 15
 maleic copolymers, 277
 phosphino polymers and polyphosphinates, 165
 polyalkoxylate block copolymers and ester derivatives, 298
 polycarboxylates, 69
 polyphosphonates, 65
 polysulfonates, 72
 sulfide scale removal, 90
maleic anhydride
 alkylphenol-aldehyde resin alkoxylates, 300
 alkylphenol-aldehyde resin oligomers, 130–131
 amide and imide nonpolymeric surfactant ADs, 124
 amine salts of (poly)carboxylic acids, 205
 biodegradable polycarboxylates, 70
 comb polymers, 274
 maleic copolymers, 276–277
 oligomeric (resinous) and polymeric AIs, 128
 polyalkoxylate block copolymers and ester derivatives, 298
 polyester and polyamide/imide AIs, 131–134
 polyphosphonates, 66
 vinyl polymers, 303
maleic copolymers, 131, 239, 276–277
maltenes, 111, 113, 247
Mannich acrylamide polymers, 324
Mannich base, 154

Index

material choice, 2
material safety data sheet (MSDS), 6
matrix acidizing
    acid stimulation, 149, 150
    axial placement of acid treatments, 163
    carbonate formations, 150–151
    emulsified acids, 174
    foam diverters, 166
    polymer gel diverters, 164
    potential formation damage, 153
    purpose, 3
    radial placement of acid treatments, 171
    sandstone formations, 151–152
    viscoelastic surfactants, 167
mechanical isolation, 21
mechanical pigging, *see* Pigging
Menidia beryllina (silverside minnow), 10–11
mercaptoalcohol (MA), 213
mercaptobenzothiazoles, 214
2-mercaptoethyl sulfide, 214
mercury, 15–16
metabisulfite, 22, 377–378, 380–381
metal carboxylates and chelates, 371–372
methacrylamide, 69, 233, 237–238, 324
methacrylamidopropyltrimethylammonium chloride, 34–35
methacrylamidopropyltrimethyl ammonium chloride, 394
(meth)acrylate copolymers, 131
(meth)acrylate ester polymers, 274–276
methacrylatetrimethoxysilane, 39
meth(acrylic) acid, 72, 277
methacryloxyethyltrimethylammonium chloride, 70
methacryloylpyrrolidine, 238
methanol
    aldehydes, 368
    chemicals in acidizing treatments, 162
    chemical sourcing, 5
    compatibility, 7
    downstream issues, 7
    gas-well AAs, 248
    hydrogen sulfide scavengers, 364–365
    kinetic hydrate inhibitors, 231
    nonpolymeric ADs, 127
    oil-miscible scale inhibitors, 81
    pollution, 7
    reaction products, 370
    regulations, 8
    thermodynamic hydrate inhibitors, 228–230, 249
    types of scale, 53
    viscoelastic surfactants, 170
3-methoxypropylamine (MOPA), 369
methyl acetate, 173
methylacrylamidopropyltrimethylammmonium chloride, 325

methylamine, 73, 89, 369
methylenebisacrylamide, 323
methylene bis(thiocyanate), 343, 348, 350
methyl formate, 173
methylorthosilicate, 33
1-methylpiperazine, 235
5-methyl salicylic acid, 171
metronidazole, 207, 341, 347
MIC (minimum inhibitor concentration), 230
microbial treatment, 264–265, 352
microemulsions, 23, 80–81, 140, 175, 324
microgels, 38
microparticles, water and gas control, 39–40
milling, 2, 88
MINOX process, 377
mixed scales, control, 57
modified acrylamidopropylsulfonic acid (AMPS) polymers, 239
modified synthetic polymers, 36–37
molybdate, 155, 199, 217, 351, 353–354
monobutyl glycol ether (MBGE), 90, 161, 190
monoethanolamine (MEA), 134, 326, 364
monoethylene glycol (MEG), 198, 228
morpholine, 205, 216, 372
multiphase flow
    acrylamide-based DRAs, 393, 395
    drag-reduction, 383, 385
    film-forming corrosion inhibitors, 200
    fouling problems, 3
    oil-soluble drag-reducing agents, 390
    oil-soluble polymeric DRAs, 390
    oil-soluble surfactant DRAs, 390–391
multiphase flow-modeling, 249
multiphase fluids, 226, 231, 263, 293
multiphase pipelines, 225, 248, 395
multiphase production, 234, 368
multiphase system, 248
multiphase transportation
    kinetic hydrate inhibitors, 231
    natural surfactants and nonplugging oils, 248
    thermodynamic hydrate inhibitors, 227, 230
    wax control strategies, 264
multiphase wellstreams, 202
mutual solvent, 8, 77–79, 90, 153
Mysidopsis bahia (mysid shrimp), 10–11

N

nanofiltration, 57, 337
naphthenate
    demulsifiers, 292
    flow assurance, 2
    fundamentals, 189–190
    hydrate-philic pipeline AAs, 244
    low-dosage inhibitors, 191–192
    nonchemical techniques, 2
    reducing problems, 3

scale control, 53
thermodynamic hydrate inhibitors, 230
using acids, 190
naphthylmethyl quinolium chloride, 154
National Pollution Discharge Elimination System (NPDES), 10
naturally occurring radioactive material (NORM scale), 56, 89
natural polymers
  biodegradability, 14
  cross-linked polymer gels, 27
  metal ion cross-linking, 26
  polymer gel diverters, 164
  types of polymer PRM, 34–35
natural surfactants, 3, 247–248, 291
N-(3-chloro-2-hydroxypropyl) trimethyl ammonium chloride, 328
N-decanoyl-L-aspartic acid, 212
NEP, see N-ethyl pyrrolidone (NEP)
N-erucyl-N,N-bis(2-hydroxyethyl)-N-methylammonium chloride, 37
Netherlands
  biodegradability, 14
  OSPAR Commission, 9, 408–409
  REACH regulations, 10
N-ethyl pyrrolidone (NEP), 13, 124, 139
N-hydroxyethylsuccinimide, 134
nitrate treatment, 337, 352–353
nitrilotriacetamide, 75
nitrilotriacetic acid (NTAA), 73, 75, 160
nitrite and nitrate ions, 351
nitrite treatment, 352–353
Nitrogen Generating System (SGN), 268
nitrogen heterocyclics, 213
nitro groups, 347
nitroso groups, 12
N-methyldiethanolamine (MDEA), 364
N-methyl-N-vinyl acetamide, 233
N-methylolacrylamide, 32
N-methyl pyrrolidone, 118, 228, 365
N,N-bis(3-aminopropyl)dodecylamine, 349
N,N'-bis(3-aminopropyl)ethylene diamine, 64
N,N-dimethyl acrylamide, 164
N,N-dimethyl-N-ethyl-N-propylammonium tribromide, 345
n-nonyl acrylate, 37
N-oleyl-diamino-1,3-propane, 124
N-oleyl-1,3-propylenediamine, 209
nonaqueous scale inhibitors, 80–83
nonchemical scale control, 57–59
nonchemical techniques, 5
non-ionic surfactants, 12, 81, 161, 166, 173
nonoxidizing biocides
  aldehydes, 342–343
  alkylamines, 349
  cationic polymers, 345–346
  2-(decylthio)ethanamine, 350

diamines, 349
dithiocarbamates, 349
fundamentals, 341–342
isothiazolones/isothiazolinones, 347–348
metronidazole, 347
organic biocides, 346–347
organic thiocynanates, 348–349
oxazolidines, 350
phenolics, 349
quaternary ammonium compounds, 345
quaternary phosphonium compounds, 343–345
specific surfactant classes, 351
thiones, 347–348
triamines, 349
triazine derivatives, 350
nonplugging oils, 247–248
nonpolymeric phosphonates, 62–65
nonregenerative scavengers
  aldehydes, 367–371
  amines, reaction products, 369–372
  fundamentals, 365, 373
  metal carboxylates and chelates, 371–372
  oxidizing chemicals, 366–367
  reaction products, 369–371
  solid scavengers, 366
  triazines, reaction products, 369–371
nonyl phenol, 118–119, 298
North-East Atlantic area, 9
North Sea and region, 11
  biodegradable polycarboxylates, 70
  catalytic hydrogenation, 379
  classes of water-in-oil demulsifiers, 298
  corrosion control, 197
  cross-linked polymer RPMs, 39
  demulsifiers with improved biodegradability, 304
  flocculants, 319–320
  gas hydrate control, 226
  gas hydrate plug removal, 249
  impurities, 4
  inorganic gels, 24
  metal ion cross-linking, 26
  naphthenate deposition control using acids, 190
  natural surfactants and nonplugging oils, 248
  nitrate and nitrite treatment, 353
  nitrogen-based corrosion inhibitors, 155
  nonpolymeric phosphonates and aminophosphonates, 64
  organic cross-linking, 29
  OSPAR Commission, 9, 11, 405, 408–409
  oxidizing chemicals, 366–367
  PAH regulations, 8
  polyaminoacids and other polymeric water-soluble corrosion inhibitors, 217
  polycarboxylates, 67, 70

Index

solid scavengers, 366
squeeze treatment, 84
sulfate scale removal, 87
sulfate scales, 56
sulfide scales, 56
water production, 21
water-soluble surfactant DRAs, 398
wax solvents, 267
Norway, 9, 14, 29, 409–410
N-phosphonomethylated amino-2-hydroxypropylene polymers, 65
N-stearyl methyl-1-diamino-1,3-propane, 124–125
N-vinylimidazole, 135
N-vinyllactams, 135
N-vinyl pyrrolidone, 27, 34, 39, 275

O

octadecene-maleic anhydride copolymer, 130
oil discharge level, 8
oil-in-water demulsifier, 319, 325
oil-miscible deploying scale inhibitors, 81
oil solubility, 8, 244, 372
oil-soluble drag-reducing agents
  background, 385
  handling issues, 389–390
  injection issues, 389–390
  multiphase flow, 390
  other types, 389
  polyalkene type, 386–388
  polymeric type, 386–390
  poly(meth)acrylate ester type, 388–389
  polyolefin type, 386–388
  pumping issues, 389–390
  surfactants, 390–391
  UHMW polymers, 389–390
oil-wetting surfactants, 171–172
oleylamidopropyl betaine, 67
oleyl amine, 122
oligomer
  cationic polymers, 346
  environment-friendly cationic polymeric flocculants, 327
  nonpolymeric phosphonates and aminophosphonates, 62
  oxygen-containing corrosion inhibitors, 156
  phosphino polymers and polyphosphinates, 66–67
  polyaminoacids and other polymeric water-soluble corrosion inhibitors, 216
  reaction products, 371
  silicones and fluorosilicones, 315
  vinyl lactam KHI polymers, 233
oligomeric (resinous) and polymeric asphaltene inhibitors

alkylphenol-aldehyde resin oligomers, 129–131
copolymers, 135
fundamentals, 128
lignosulfonate polymers, 135
polyamide/imide AIs, 131–135
polyester, 131–135
polymeric inhibitors, 135, 137
semisynthetic type, 135
tetrapyrrolitic patterns, 137
operational issues, THIs, 230–231
organic biocides, 346–347
organic bromides, 341, 346, 349
organic cross-linking, 27–29
organic thiocyanates, 341, 348–349
organosilane, 23, 87, 161, 186–187
orthoester-based polymers, 305
ortho-phthalaldehyde, 342–343
OSPAR Commission
  designing greener chemicals, 11
  environmental and ecotoxicological regulations, 9–10, 405–410
  fundamentals, 9–10, 405–408
  Netherlands, 408–409
  North Sea, 408–409
  Norway, 409–410
  oxidizing chemicals, 366–367
  REACH regulations, 10
  United Kingdom, 408–409
oxadiazole, 158
oxazolidines, 342, 350, 369
oxidizing biocides, 339–341
oxidizing chemicals, 366–367
oximes, 377–379
oxygen additives, acid stimulation, 155–158
oxygen scavengers
  aldehydes, activated, 379
  bisulfite salts, 380–381
  catalytic hydrogenation, 379
  classes, 377–381
  dithionite salts, 378
  enzymes, 379
  fundamentals, 377–378
  guanidine salts, 378
  hydrazine salts, 378
  hydroxylamines, 378–379
  metabisulfite salts, 380–381
  oximes, 378–379
  polyhydroxyl compounds, 379
  sulfite salts, 380–381
  water injection systems, 5

P

PAH (polyaromatic hydrocarbons), 8
PAM, *see* Polyacrylamide (PAM)
paraffin, *see* Wax

partially hydrolyzed polyacrylamide (PHPA), 25, 34–35, 166, 330, 391
passivating corrosion inhibitors, 199
pentaerythritol, 13, 300
peracetic acid, 91, 339
performance and performance testing
    choice factors, 5–8
    flocculants, 322–323
    hydrate-philic pipeline class, 247
    kinetic hydrate inhibitors, 240–242
    scale control, 6
    testing scale inhibitors, 84–85
permanganate ions, 354
pH, 7
phenoformaldehyde, 27
phenolics, 22, 341, 349
phenyl esters, 28, 126
phosphate esters
    amidoamines and imidazolines, 210
    biodegradability, 13
    carbonates and sulfates, 60
    film-forming corrosion inhibitors, 203–205
    low-dosage naphthenate inhibitors, 191
    quaternary ammonium and iminium salts and zwitterionics, 206
    scale control, 61–62
    squeeze treatments, 80
    sulfur compounds, 214
    wax dispersants, 280
phosphinopolycarboxylic acid (PPCA), 55
phosphino polymers, 60, 66–67
2-phosphonobutane-1,2,4-tricarboxylic acid (PBTCA), 62, 86, 89
phosphonocarboxylic acid esters, 62, 121
phosphonosuccinic acid, 62, 121
phosphoric acid, 61, 121, 190, 297, 300
phthalic acid, 167
phthalic anhydride, 124
pigging, 2–3, 197–198, 264, 271, 274
pipeline heating, 264
piperazinone, 372
pitting and crevice corrosion, 196–197
planktonic organisms, 337–338
PLONOR, 9, 409
plug removal, 248–250
p-methoxyphenol, 299
p-N,N-dibutylaminophenol, 299
polar crude fractions, 281
polonium, 54
polyacrylamide (PAM)
    anionic polymers, 330
    biodegradability, 14, 385
    chemical sand control, 185
    corrosion inhibitors, 217
    drag reducers, 162, 385
    foam diverters, 166
    gelled or viscous acids, 173
    hyperbranched polyesteramide KHIs, 235
    polymer gel diverters, 164, 165
    polymer injection, 25
    polymer RPM, 34–35
polyacrylates, 14, 66–67, 166
polyalkanolamines, 279, 323, 325
polyalkene, 387, 388
polyalkene (polyolefin) DRAs, 386–387
polyalkoxylate block copolymers, 296, 298
polyalkoxylate derivatives, 302
polyalkyleneimines, 323
polyalkylene polyamines, 215, 326
polyalkylenesuccinimide copolymers, 132
polyalkylmethacrylate esters, 275
polyalkyloxazolines (or polyacylethyleneimines), 239
polyallylamides, 239
polyallylammonium chloride, 323, 327
polyamides, 131–135
polyamidoamines, 302, 327
polyamine
    alkylphenol-aldehyde resin oligomers, 130
    amide and imide nonpolymeric surfactant ADs, 123–124
    amides, 212
    amidoamines and imidazolines, 208, 211
    amine-based products, 372
    cationic polymers, 326–327
    chemicals in acidizing treatments, 161
    demulsifiers with improved biodegradability, 304
    diallyldimethylammonium chloride polymers, 324
    dithiocarbamates, 329
    film-forming corrosion inhibitors, 204
    gas-well AAs, 248
    nitrogen heterocyclics, 213
    organic cross-linking, 28
    phosphate esters, 205
    polyamine polyalkoxylates and related cationic polymers, 301
    polyester and polyamide/imide AIs, 135
    polymer gel diverters, 164
    polyphosphonates, 65
    sulfur compounds, 215
    vinyl lactam KHI polymers, 234
    water-in-oil demulsifier, 296
polyamine polyalkoxylates, 296, 301
polyaminoacids, 70, 187, 203, 214, 216–217
polyaminocarboxylic acids, 149, 152
polyaromatic hydrocarbons, see PAH (polyaromatic hydrocarbons)
polyaspartamides, 239
polyaspartates, 15, 70, 187, 216–217
poly(behenylacrylate), 275
polycarboxylates, 59, 64, 67, 69–71, 380

Index

polycarboxylic acids, 61, 66–67, 69, 133, 304–305
polycardanol, 129
polycyclic residues, 12
polyDADMAC, 79
polydialkyldiallyl polymers, 40
poly(di)alkyl(meth)acrylamide, 237–238
polydiallyldimethylammonium chloride (polyDADMAC), 14, 60, 79, 323
polydimethyldiallylammonium chloride, 161, 187
poly(dimethylsiloxane) (PDMS), 314
poly(dodecyl methacrylate)s, 388
polyelectrolyte, 171, 319
polyepoxysuccinic acid, 71
polyester pyroglutamate polymer, 232
polyesters
   biodegradability, 15
   chemical sand control, 185
   demulsifiers with improved biodegradability, 304, 306
   hyperbranched polymers, 302
   kinetic hydrate inhibitors, 232
   (meth)acrylate ester polymers, 275
   oligomeric (resinous) and polymeric AIs, 128
   polyamide/imide AIs, 131–135
   pyroglutamate KHI polymers, 236–237
   resins and elastomers, 22
   solid particle diverters, 163
polyetheramines, 235, 248
polyethoxylates, 13, 119, 304
polyethyleneamines, 211, 301
polyethyleneglycol, 233, 327, 365
poly(ethylene glycol), 306
polyethyleneimine (PEI), 25, 135, 235, 275, 279, 302
polyethyleneoxide (PEO), 31, 235, 391–392
polyethylene oxide-propylene oxide, 235
poly-☒-glutamic acid, 330
polyglycerol, 304
polyglycolic acid, 174
polyglycol phosphonates, 65
polyglycols, 266, 302, 314–316
polyglycosides, 166, 305
polyglyoxylic acid, 71
polyhydroxy, 23, 204, 212–213
poly-3-hydroxybutyrate (PHB), 15
polyhydroxyl compounds, 377, 379
poly(12-hydroxy stearic acid), 135
polyisobutylene-alkyl maleimide polymers, 278
polyisobutylene (PIB), 123, 133–134, 275, 386–388, 390
polyisobutylene succinic anhydride, 123
polyisobutylene succinimide, 123
poly(isodecylmethacrylate), 388
polyisopropylmethacrylamide, 232
polylactide acid (PLA), 15, 174
polymaleates, 67, 192

poly(maleic anhydride/ethyl vinyl ether), 277
polymaleimides, 239
polymer gel, 30–31, 164–166, *see also* Cross-linked polymer gels
polymeric oil-soluble drag-reducing agents, 386–390
poly(meth)acrylate ester, 388–389
polymethylenepolyaminedipropionamides, 211
poly(octadecyl acrylate), 272
polyolefin, 386–388
polyolefin/maleic anhydride copolymers, 130, 132
polyol polyalkoxylates, 300–301
poly[oxyethylene(dimethyliminio)ethylene(dimethyliminio)ethylene dichloride], 346
polyoxymethylene polymers, 343
polyphosphates, 60, 61, 67, 199, 214, 380
polyphosphinates, 60, 66–67
polyphosphonates, 60, 65–66, 380
polypropoxylates, 13, 315
poly(ricinoleic acid), 135
polysaccharides
   anionic polymers, 330
   biodegradability, 14–15
   biodegradable polycarboxylates, 71
   cationic polymers, 323
   cross-linked polymer RPMs, 38
   derivatives, 392
   environment-friendly cationic polymeric flocculants, 327–328
   foam diverters, 166
   oxidizing biocides, 339
   polymer gel diverters, 164
   water-soluble drag-reducing agents, 392
   water-soluble polymer DRAs, 391
polysilicones, 296, 304
polysuccinimide, 70, 217
polysulfonates, 60, 71–73
polytartaric acid, 71
polyurethane alkoxylates, 302
polyurethanes, 185, 296, 302
polyvinyl alcohol gels, 30
polyvinyl alcohols, 25, 30–31, 321, 391
polyvinylalkanamides, 239
polyvinylamine, 25, 30, 326
polyvinylamine gels, 30
polyvinylammmonium chloride, 323
polyvinylcaprolactam (PVCap), 13, 14, 233
polyvinyloxazolines, 239
polyvinylpyrrolidone (PVP), 13, 31, 231
"pose little or no risk," *see* PLONOR
potassium dimethyl dithiocarbamate, 214
potassium formate, 174, 228–229, 249
potassium hexacyanoferrate (HCF), 75
potassium permanganate, 91

pour point depressants (PPDs), 264, 270–272, 281–282, 306, 383
precipitation squeeze treatment, 78
price, choice factors, 5, 6
production chemistry
　arsenic production, 15–16
　bioaccumulation, 11
　biodegradability, 12–15
　choice factors, 5–8
　classification of chemicals, 2
　corrosion, 4
　deposition issues, 3
　drag-reducing agents, 4
　environmental and ecotoxicological regulations, 8–11
　facilities design, 2
　flow assurance, 2–3
　fouling problems, 2–3
　greener chemicals, designing, 11–15
　H_2S savenger chemicals, 4
　impurities specifications, 3–4
　injection of chemcials, 4
　issues, 1, 5
　material choice, 2
　mercury production, 15–16
　nonchemicals techniques used, 5
　North Sea region, 11
　OSPAR Commission, 9–10
　REACH regulations, 10
　resolution of issues, 1–2
　scale control, 5
　service companies, 5
　squeeze treatments, 4–5
　toxicity reduction, 12
　United States, 10–11
1,2,3-propanetriol diacetate, 174
propargyl alcohol, 90, 156, 159
proppants, 40, 80, 82, 149
pumping issues, 389–390
PVT models, 229
PWRI (produced water reinjection), 57, 76, 335
pyridines, 138, 207, 327
pyridinium, 12, 154–155, 158, 207
pyroglutamate, 232, 236–237
pyrophosphates, 200

Q

Quantitative Structure Activity Relationship (QSAR) analysis, 201
quaternary ammonium compounds, 345
quaternary ammonium iminium salts, 205–207
quaternary ammonium surfactants, 12
quaternary gas hydrate anti-agglomerants, 7
quaternary phosphonium compounds, 191, 341, 343–345

quaternary surfactant anti-agglomerants, 10, 13–14, 245
quinolines, 327
quinolinium, 12

R

radial placement, acid stimulation
　buffered acids, 172
　emulsified acids, 174–175
　foamed acids, 173
　fundamentals, 171
　gelled acids, 173
　oil-wetting surfactants, 172
　temperature-sensitive acid-generating chemicals and enzymes, 173–174
　viscous acids, 173
　weak organic acids, 172
　weak sandstone-acidizing fluorinated agents, 172
REACH regulations, 10, 245
reaction products, 369–371
relative permeability modifier (RPM)
　axial placement of acid treatments, 163
　cross-linked polymers, 38–39
　fundamentals, 21, 32–34
　hydrophilic polymers, 33–39
　modified synthetic polymers, 36–37
　squeeze treatments, 80
　types, 34–36
resolution of issues, 1–2
resorcinol, 28, 118, 299
resorcinol octadecyl ether, 299
retarded acids, 150–151, 171–175
reverse emulsion breakers, 3, 293
Reynold's number, 383
rheology, 167, 294
risks, choice factors, 6
rock adsorption, 35
RPM, *see* Relative permeability modifier (RPM)
RRF, 34
RSN (Relative Solubility Number), 295

S

safety, handling and storage, 5
salicyl alcohol, 28
salicylaldehyde, 379
salicylic acid, 28, 126, 171
salt tolerance, 35
sand consolidation, 3, 22–24, 36, 185–187
sand control, 185–187
sandstone-acidizing fluorinated agents, 172
sandstone formations, 151–152
sarcosinate surfactants, 122
(S)-aspartic acid-N-monoacetic acid, 87

scale control
 aminophosphonates, 62–65
 biodegradability, 14
 biodegradable polycarboxylates, 70–71
 calcium carbonate scale, 54–55
 carbonate scale removal, 85–87
 chemical compatibility issues, 7
 chemical scale removal, 85–91
 continuous injection, 76
 deploying scale inhibitors, 75–84
 emulsified type, 81–82
 fundamentals, 5, 53
 halite scale inhibition, 74–75
 inhibition of scale, group II carbonates and sulfates, 59–73
 lead scale removal, 91
 mixed scales, 57
 nonaqueous type, 80–83
 nonchemical scale control, 57–59
 nonpolymeric phosphonates, 62–65
 oil-miscible type, 81
 performance, 6, 84–85
 phosphate esters, 61–62
 phosphino polymers, 66–67
 placement, squeeze treatment, 83–84
 polycarboxylates, 67, 69–71
 polyphosphates, 61
 polyphosphinates, 66–67
 polyphosphonates, 65–66
 polysulfonates, 71–73
 sodium chloride (halite) scale, 57
 solid scale inhibitors, 80–83
 squeeze treatments, 76–84
 sulfate scale, 55–56, 87–89
 sulfide scale, 56–57
 sulfide scale removal, 89–91
 types, 53–57
 water-free in organic solvent bends, 81
 well treatments, combination, 79–80
scleroglucan, 27, 34–35, 164
Scophthalmus maximus, 405
screens, 82, 185
seawater injection
 aldehydes, 357
 biocides, 336
 corrosion control during production, 196
 hydrazine and guanidine salts, 378
 nitrate and nitrite treatment, 352
 nonchemical scale control, 57
 oxidizing biocides, 340
 oxygen scavengers, 377
 sulfate scales, 56
 water-soluble surfactant DRAs, 395–396, 398
selection, water-in-oil demulsifiers, 295–296
selenate, 351, 354
self cross-linking water-soluble polymer, 31
self-diversion, 83
semicarbazide, 377
semisynthetic oligomeric (resinous) and polymeric AIs, 135
separator
 asphaltene control, 114
 biocides, 336
 corrosion control, 198
 defoamers and antifoams, 313
 demulsifiers, 292
 demulsifier selection, 295
 film-forming corrosion inhibitors, 7
 foam, 4, 313
 impurities, 3
 injection location, 7
 naphthenate deposition control, 189–190
 nonchemical techniques, 1–2
 polygycols, 315
 silicones and fluorosilicones, 315
 sulfide scale inhibition, 74
 water-in-oil demulsifiers, 293
sequestering agents, 54, 75, 160
service companies, 5
shallow gas reservoirs, 8
shear degradation, 30–31, 383, 387, 389–393, 395
shear sensitivity, 35, 169
shear viscosity, 293, 386
shutdowns, 7, 190, 226, 261, 264, 313, 319
silicate gel, 23, 33
silicones, 314–315
silverside minnow *(Menidia beryllina)*, 10–11
Skeletonema costatum, 405, 406
sludging, 7, 86, 157, 159, 161–162, 172, *see also* Asphaltene
soaps, 189–190, 397
sodium acrylamido-3-methylbutenoate, 393
sodium aluminate, 278
sodium azide, 353–354
sodium bisulfite, 22, 380
sodium chloride (halite) scale, 57
sodium dichloroisocyanuric acid (SDCC), 341
sodium dodecylbenzene sulfonate (SDBS), 169
sodium persulfate, 165, 269
sodium salicylate, 167–168
solid particle diverters, 163
solid scale inhibitors, 80–83
solid scavengers, 366
solids wetting, 294
solvents, wax removal, 266–267
sorbitan ester, 32
squeeze modeling programs, 77
squeeze treatments, 4–5, 6, 76–84
stability, choice factors, 5, 6
starch, 30, 34, 239, 327
stearyl acrylate, 275
strategies, wax control, 264–266

stress cracking, 153, 195, 214, 363
styrene
    (meth)acrylate ester polymers, 275
    oil-soluble DRA polymers, 389
    oil-soluble polymeric DRAs, 386
    polyalkene (polyolefin) DRAs, 387
    poly(meth)acrylate ester DRAs, 388
    polysulfonates, 71
    vinyl polymers, 303–304
    vinyl pyridine copolymer, 233, 389
    water control using microparticles, 39
subcoolings
    anti-agglomerants, 242–243
    chemical wax prevention, 269
    gas hydrate control, 227
    hydrate-philic pipeline AAs, 245, 247
    hyperbranched polyesteramide KHIs, 235
    kinetic hydrate inhibitors, 231–232
    performance testing of KHIs, 240, 242
    pyroglutamate KHI polymers, 236
    thermodynamic hydrate inhibitors, 229
    vinyl lactam KHI polymers, 234
succinoglycan, 83, 164
sugars, 23, 235
sulfate-reducing bacteria (SRB), 56, 197, 345, 363
sulfates
    aminophosphonates, 62–65
    biodegradable polycarboxylates, 70–71
    chemical scale removal, 87–89
    fundamentals, 59–61
    nonpolymeric phosphonates, 62–65
    phosphate esters, 61–62
    phosphino polymers, 66–67
    polycarboxylates, 67, 69–71
    polyphosphates, 61
    polyphosphinates, 66–67
    polyphosphonates, 65–66
    polysulfonates, 71–73
    scale control, 55–56, 59–73
sulfide, 56–57, 73–74, 89–91
sulfite ions, 165, 381
sulfites
    bisulfite, metabisulfite, and sulfite salts, 380–381
    dithionite salts, 378
    nitrate and nitrite treatment, 352
    oxygen scavengers, 377–378
    quaternary phosphonium compounds, 344
    resins and elastomers, 22
sulfite salts, 380–381
sulfonated alkylnaphthalenes, 119–120
sulfonic acid-based nonpolymeric surfactants, 118–120
sulfosuccinates (surfactants), 168
sulfur, 158–159, 213–216
supersaturation, 56–57, 59, 75, 84–85
Surfactant Alternating Gas technique (SAG), 40

surfactants
    additives, acid stimulation, 162
    nonoxidizing biocides, 351
    oil-soluble drag-reducing agents, 390–391
    reducing toxicity, 12
    water-soluble drag-reducing agents, 395–398
syneresis, 27, 30

T

t-butyl acrylate ester, 28, 33
t-butylstyrene, 387
temperature-sensitive acid-generating chemicals and enzymes, 173–174
teraphthalaldehyde, 28
terpene, 140, 267
tertiary amines, 12
testing
    corrosion inhibitor, 201, 209
    film-forming inhibitors, 201–202
    gas hydrate control, chemical prevention, 240–242
    water-in-oil demulsifiers, 295–296
    wax, chemical prevention, 269–270
tetrabutylammonium bromide (TBAB), 234
tetrafluoroboric acid, 152
tetrahydropyrimidines, 209
tetrakishydroxymethylphosphonium (THPS), 89
tetralin, 137–138
tetramethylammonium salts, 31
tetrapropylenes, 12
tetrapyrrolitic patterns, 137
thermochemical packages, 267–269
thermochemical sulfate reduction (TSR), 363
thermodynamic hydrate inhibitors (THIs), 3, 7, 53, 227, 230–231, *see also* Hydrate inhibitors
thiadiazole, 158
thiazolidines, 214
thioacetamide, 165
thiocarboxylic acids, 210, 214
thioethanol, 156
thioglycolic acid, 90, 155, 160, 210
thiones, 341, 347–348
thiophosphates, 200, 214
thiophosphorus compounds, 214
thiosulfate, 89, 155, 204, 213, 372
thiosulfate salts, 155
THIs, *see* Thermodynamic hydrate inhibitors (THIs)
threshold hydrate inhibitors, 227
toluene
    Ads and AIs summary, 137
    asphaltene control, 111
    asphaltene dissolvers, 138–140
    biocides, 336
    chemical wax prevention, 270

Index 435

demulsifier selection, 295
flocculants, 320
hot-oiling and related techniques, 266
low-polarity nonpolymeric aromatic amphiphiles, 118
sulfonic acid–based nonpolymeric surfactant ADs, 118
wax inhibitors and pour-point depressants, 271
wax solvents, 267
toxicity (acute), 9, 90, 207, 327–328, 343, 405, 409
Trans-Alaskan pipeline, 385
transition metal oxyanions, 354
triamines, 329, 341, 349
triazines
aldehydes, 368–369
chemical compatibility issues, 7
nonoxidizing biocides, 350
nonregenerative scavengers, 365
oxidizing chemicals, 366
reaction products, 369–371
tributylamine oxide, 234
tributyl tetradecyl phosphonium chloride (TTPC), 345
triethanolamine, 62, 124, 216, 240, 277, 325, 365
triethanolamine phosphate monoester, 62
triethyleneglycol (TEG), 64, 304
triethylenetetraamine hexaacetic acid (TTHA), 88
trimethylamine, 88, 200
trimethylolpropane, 300
tri-*n*-butylstannyl fluoride, 390
tris(hydroxymethyl)phosphines (THP), 73
trisodium ethylenediaminodisuccinate, 86
trisodium hydroxyethylethylenediamine triacetate, 169
trithiones, 214
tungstate, 199, 351, 354
tungstate ions, 351, 354
turbulent flow
acrylamide-based DRAs, 395
drag-reducing agents, 383–384
polyalkene (polyolefin) DRAs, 387
polyethyleneoxide DRAs, 392
polymer gel water shut-off treatments, 31
testing corrosion inhibitors, 202
water-soluble surfactant DRAs, 395–396
twin-tail quaternary surfactant anti-agglomerants, 13–14, 245

U

ultrahigh-molecular weight (UHMW) polymers, 7, 383, 386–395
umbilical tubes, 7
10-undecenoic, 389

underdeposit corrosion, 196
United Kingdom, 9, 406–409
United States
2-(decylthio)ethanamine and its hydrochloride, 350
regulations, 10–11
sulfate scale remover, 88
thermochemical packages, 267
wax solvents, 266
unsaturated linkages, 155–158
urea
acrylamide-based DRAs, 394
aldehydes, 368
cationic polymers, 326
chemical sand control, 186
emulsified scale inhibitors, 81
inorganic gels, 23–24
oxygen-containing corrosion inhibitors, 156
quaternary phosphonium compounds, 344
squeeze treatments, 78–79

V

vanadate ions, 199, 354
vapor phase corrosion inhibitors (VpCIs), 199
vapor pressure depressant, VPD, 76
VDA, 169–170
VES, *see* viscoelastic surfactants (VES)
vinyl alkanoate, 131, 239
1-vinyl-4-alkyl-2-pyrrolidone polymers, 134
vinyl formamide, 34–35
vinylidene-1,1-diphosphonic acid (VDPA), 65
vinyl lactam polymers, 233–235
vinyl phosphonic acid (VPA), 38, 65–66
vinyl polymers
environment-friendly cationic polymeric flocculants, 327
hydrophobically modified synthetic polymers as RPMs, 37
polysaccharides and derivatives, 392
types of polymer RPM, 34–35
water-in-oil demulsifiers, 296, 303–304
vinyl pyridine, 233, 275, 389
4-vinylpyridine, 135
vinyl pyrrolidone, 36, 128, 233–234, 277
vinyl sulfonate, 34, 72, 85
vinyl sulfonic acid, 66, 71
vinyltrimethoxysilane, 39, 323
viscoelasticity, 37, 167, 294
viscoelastic surfactant gels, 32
viscoelastic surfactants (VES)
axial placement, acid stimulation, 167–171
axial placement of acid treatments, 163
carbonate scale remover, 86
squeeze treatment, 83
water and gas control, 22
viscous acids, 171, 173

viscous fingering, 150
volume, maximizing, 36

W

water and gas control
  acrylamides, 26–27
  biopolymers, 26–27
  cross-linked polymer gels, 24–32
  cross-linked polymers, 38–39
  disproportionate permeability reducer, 32–39
  elastomers, 22
  emulsified gels, 33
  fundamentals, 21–22
  gas shut-off, 40–41
  hydrophilic polymers, 33–39
  improvements, 31
  inorganic gels, 22–24
  modified synthetic polymers, 36–37
  natural polymers, 27
  organic cross-linking, 27–29
  polymer gel water shut-off treatments, 30–31
  polymer injection, 25–31
  polyvinyl alcohol gels, 30
  polyvinylamine gels, 30
  relative permeability modifier, 32–39
  resins, 22
  in situ monomer polymerization, 31–32
  types, 34–36
  viscoelastic surfactant gels, 32
  water control using microparticles, 39–40
water-based drag reducers, 5, *see also* Drag reducing agents (DRAs)
water clarifiers, 3, 293, 329
water-free scale inhibitors, organic solvent bends, 81
water injection
  acrylamide-based DRAs, 393
  anthraquinone as control biocide, 352
  bacteria control, 338
  biocides, 336
  classes of biocides, 339
  corrosion control, 198
  drag-reducing agents, 383
  nitrate and nitrite treatment, 353
  nonchemical scale control, 57
  oxidizing biocides, 339
  oxygen scavengers, 377
  production chemical service companies, 5
  water-soluble polymer DRAs, 391
  water-soluble surfactant DRAs, 395
water-in-oil demulsifiers
  alkylphenol-aldehyde resin alkoxylates, 298–300
  biogradability, 304–306
  carbamates, 302

  cationic polymers, 301
  classes, 296–297
  dual-purpose type, 306–307
  ester derivatives, 298
  glycidyl ether polyalkoxylates, 300–301
  hyperbranched polymers, 302–303
  polyalkoxylate block copolymers, 298
  polyalkoxylate derivatives, 302
  polyamine polyalkoxylates, 301
  polyol polyalkoxylates, 300–301
  polysilicones, 304
  polyurethanes, 302
  selection parameters, 295–296
  test methods, 295–296
  theory and practice, 293–294
  vinyl polymers, 303–304
water solubility, 203, 205, 213–214, 327, 329, 347
water-soluble drag-reducing agents
  acrylamide-based, 393–395
  polyethyleneoxide, 392
  polymers, 391–395
  polysaccharides and derivatives, 392
  surfactants, 395–398
water-soluble inhibitors, polymeric, 216–217
water-wetting agents, 161
wax
  chemical prevention, 269–282
  chemical removal, 266–269
  comb polymers, 274–277
  deployment techniques, 281–282
  deposition, 261–263
  dispersants, 279–281
  diverters, 83
  ethylene polymers and copolymers, 273–274
  flow improvers, 281
  fundamentals, 261–262
  gelling, 263–264
  hot-oiling, 266
  inhibitors, 3, 6, 8, 14, 270–272
  maleic copolymers, 276–277
  (meth)acrylate ester polymers, 274–276
  miscellaneous polymers, 278–279
  polar crude fractions, 281
  pour-point depressants, 270–272
  removal, 3
  solvents, 266–267
  strategies, 264–266
  test methods, 269–270
  thermochemical packages, 267–269
  viscosity, 263–264
wax appearance temperature (WAT), 230, 261
weak organic acids, 171, 172
weak sandstone-acidizing fluorinated agents, 172
well treatments, 79–80
wormholes, 150–151, 165

Index

X

xanthan
- anionic polymers, 330
- chemical scale remover, 85
- environment-friendly cationic polymeric flocculants, 327
- gelled or viscous acids, 173
- gels using natural polymers, 27
- polymer gel diverters, 164–165
- squeeze treatment, 83–84
- types of polymer RPM, 34–35
- water-soluble polymer DRAs, 391

xylene
- ADs and AIs summary, 137
- alkylphenol-aldehyde resin alkoxylates, 299
- asphaltene dissolvers, 138, 140
- chemicals in acidizing treatments, 161
- emulsified acids, 174
- flocculants, 320
- hot-oiling and related techniques, 266
- low-polarity nonpolymeric aromatic amphiphiles, 118
- (meth)acrylate ester polymers, 275–276
- wax inhibitors and pour-point depressants, 271
- wax solvents, 267

Z

Ziegler-Natta polymerization, 386
zirconium (IV), 27, 164–165, 394
zwitterionics, 204, 206–207